THE AUTHORS

PROFESSOR W.G. HALE, B.Sc., Ph.D., D.Sc., F.I.Biol., is Dean of the Faculty of Science and Head of the Department of Biology at Liverpool Polytechnic. He is author of *Waders*, *Eric Hosking's Waders*, co-author of *Basic Biology* (with J.P. Margham) and co-editor of *Wildfowl and Waders in Winter*. He has published widely on the taxonomy and ecology of Collembola and wading birds, and is currently writing *Estuaries*, a volume in the Collins New Naturalist series. He has a wide experience of examining at first degree, M.Phil., Ph.D. and D.Sc. levels in both polytechnics and universities.

Dr J.P. MARGHAM, B.Sc., Ph.D., Dip.Gen., M.I. Biol., is Principal Lecturer and Course Leader for the B.Sc. Honours Applied Biology Degree at Liverpool Polytechnic. He is co-author of *Basic Biology* and editor (with T.M. Jeves) of a series of booklets on A-level Biology options. He is Chief Examiner for the A-level Genetics option for the Northern Universities Joint Matriculation Board, an Assistant Examiner for Social and Environmental Biology and an experienced examiner for core A-level Biology. His research interests centre on the genetics of resistance to insecticides.

COLLINS
REFERENCE

DICTIONARY OF
BIOLOGY

W.G. HALE AND J.P. MARGHAM

COLLINS
London and Glasgow

First published 1988

© W.G. Hale, J.P. Margham

Diagrams by Karen Glover

British Library Cataloguing in Publication Data

Hale, W.G.
  Collins dictionary of biology.
  1. Biology—Dictionaries
  I. Title II. Margham, J.P.
  574′.03′21   QH 302.5

ISBN 0 00 434351 4

Printed in Great Britain by Collins, Glasgow

For
Professor Arthur J. Cain
who told us both what a species is!

CONTENTS

PREFACE

This dictionary has been produced particularly with A-level (or equivalent) and undergraduate students in mind, but will also prove useful to anyone with an interest in biology. It provides a comprehensive coverage of terms used in both general and specialized literature, and reference is made to biographical details of some of the biologists who have contributed to important discoveries or concepts.

Statistical terms and tests are briefly explained where they apply to biological situations but there is little overlap with topics in the physical sciences. Readers are recommended to consult other volumes of this series should they fail to find particular entries in this dictionary. Where appropriate, cross-references are made in the text by reproducing key words in SMALL CAPITAL LETTERS.

Acknowledgements

We should like to thank particularly Mrs Janie Thomas for reading through and commenting on the manuscript, especially from an A-level point of view. In addition, many of our colleagues have helped with particular entries and we should like to extend our thanks to Drs Peter Wheeler, Terry Marks, John Carter, John Haram, Ian Hodkinson, Terry Jeves, Nick Lepp, Venetia Saunders, George Sharples, Philip Smith, Graham Triggs, Tony Whalley and Mike Thomas.

Edwin Moore of Collins helped a great deal in preparing the manuscript for publication. Jim Carney (also of Collins) helped prepare the diagrams, which were drawn by Karen Glover.

Our thanks are also due to Mrs Karen Bernard for typing the manuscript and to Mrs Susan Abraham and Mrs Valerie Guthrie for operating the word processor.

A

Å or **A,** *abbrev. for* ANGSTROM.

a-, *prefix.* denoting not, without, away.

A (amino acid) site, *n.* the location at which specific TRANSFER RNA molecules become attached to RIBOSOMES during the TRANSLATION phase of protein synthesis.

ab-, *prefix.* denoting from, out of, away, apart.

A-band, *n.* the dark band of the muscle sarcomere which corresponds to the thick myosin filaments. See MUSCLE CONTRACTION.

abaxial, *adj.* (of a leaf surface) facing away from the stem of the plant.

abdomen, *n.* that part of the body of vertebrates containing the viscera, i.e. the kidneys, liver, stomach and intestines. In mammals it is separated from the thorax, which contains the heart and lungs, by the diaphragm. In arthropods it is that part of the body directly behind the thorax; in many other invertebrates the abdomen is divided into segments that have a superficial similarity.

abducens nerve, *n.* the cranial nerve of vertebrates that supplies the external rectus muscle of the eye (see EYE MUSCLE); it is mainly MOTOR in function.

abductor or **levator,** *n.* any muscle that moves a limb away from the body. An example of an abductor is the abductor pollicis, which moves the thumb outward. Compare ADDUCTOR.

abiogenesis, see SPONTANEOUS GENERATION.

abiotic factor, *n.* any contribution to the environment that is of a non-living nature, e.g. climate.

ABO blood group, *n.* a classification of blood based on natural variation in human blood types, identified and named by Karl Landsteiner (1868–1943) in 1901. There are four groups: A, B, AB and O, each classified by a particular combination of ANTIGEN(S) on the red blood cells (see also H–SUBSTANCE) and naturally occurring ANTIBODIES in the BLOOD PLASMA. The relative frequency of the four ABO groups in the GENETIC POLYMORPHISM has been investigated in most human populations and differs widely between races (see Fig. 1 on page 2).

Antigens and antibodies of the same type cause AGGLUTINATION when mixed, resulting in difficulties in blood transfusion (see UNIVERSAL DONORS and UNIVERSAL RECIPIENTS). Although possessing no A or B antigens, Group O individuals have an H-antigen (see H–SUBSTANCE) which is a precursor to the A and B types. H, A and B antigens are found

1

Blood group	Antigen (RBC)	Antibody (plasma)	Typical % frequency in:		
			UK	China	Australian Aborigines
A	A	Anti-B	41	31	57
B	B	Anti-A	9	28	0
AB	AB	none	4	7	0
O	none	Anti-A + Anti-B	46	34	43

Fig. 1. **ABO blood group.** Main features of the ABO blood types.

Genotype	Phenotype
A/A, A/O	Group A
B/B, B/O	Group B
A/B	Group AB
O/O	Group O

Fig. 2. **ABO blood group.** Inheritance of the ABO groupings.

also in human body secretions such as saliva and semen, often a useful fact in forensic tests. See SECRETOR CONDITION.

Inheritance of grouping is controlled by a single autosomal gene (see AUTOSOME) on chromosome 9 with three major ALLELES, A, B and O (sometimes written as I^A, I^B and I^O). See Fig. 2. Four types of the A group are now known, making six multiple alleles (see MULTIPLE ALLELISM) at this LOCUS.

abomasum, *n.* the true digestive part of the RUMINANT STOMACH, secreting enzymes for the digestion of food prior to passing it on to the small intestine. The abomasum is HOMOLOGOUS with the monogastric stomach of non-ruminants.

aboral, *adj.* away from or opposite the mouth in those groups of animals that have no clear-cut DORSAL or VENTRAL surfaces.

abortion, *n.* the spontaneous or induced expulsion of a foetus before it becomes viable outside the uterus or womb.

abscess, *n.* a collection of pus surrounded by an inflamed area in any tissue or organ of an animal.

abscisic acid, *n.* a plant hormone. It is called *abscisin* when found in ageing leaves and it is partly responsible for leaf ABSCISSION. It is also called DORMIN when in buds and seeds, and causes induction of dormancy. See Fig. 3.

abscission, *n.* the process by which plant organs are shed. The process occurs in the stalks of unfertilized flowers, ripe fruits and deciduous leaves in autumn, or in diseased leaves at any time. It is due to the formation of an *abscission layer* of thin-walled cells in the stalk base which

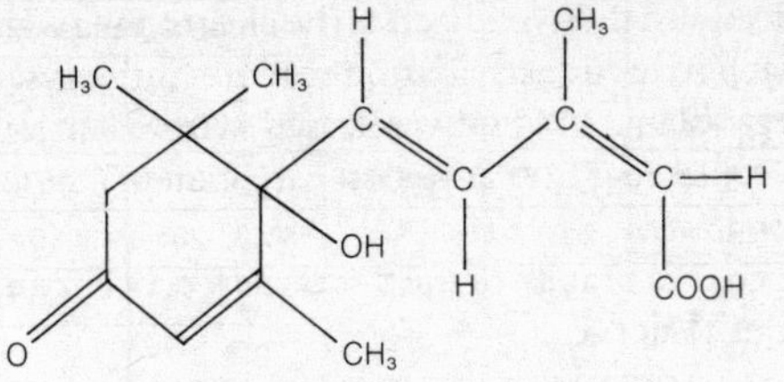

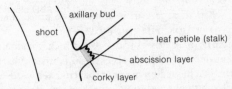

Fig. 3. **Abscisic acid.** Molecular structure.

shoot

axillary bud

leaf petiole (stalk)

abscission layer

corky layer

Fig. 4. **Abscission.** Abscission layer in a leaf stalk.

rupture under strain, e.g. wind. A layer of cork forms beneath the abscission layer to seal the plant surface. Abscission is controlled by plant hormones present: a low concentration of AUXIN, high amounts of ETHYLENE, and (in some plants) a high concentration of ABSCISIC ACID all stimulate production of the abscission layer. See Fig. 4.

absolute refractory period, *n.* the brief period during the discharge of a nerve impulse when the neuron cannot fire again.

absolute zero, *n.* the lowest possible temperature for all substances, at which their molecules possess no heat energy. $-273.15°C$ is usually accepted as absolute zero.

absorption, *n.* the process by which energy or matter passively enters a system, e.g. the take up of nutrient material from the gut system of animals into the blood stream, or the process by which chlorophyll absorbs light for the process of photosynthesis. Compare ADSORPTION.

absorption spectrum, *n.* that part of the light spectrum which is absorbed by a pigment. For example, chlorophyll (see Fig. 9) absorbs red and blue light and thus appears green.

abyssal, *adj.* (of organisms) inhabiting deep water below 1000 m.

acanth- or **acantho-,** *prefix.* denoting spiny.

Acanthodii, *n.* an order of placoderm fossil fish, whose members bear spines along the anterior margin of the fins.

Acarina, *n.* an order of the ARACHNIDA, containing ticks and mites.

acceptor molecule, *n.* a molecule that has a high affinity for electrons, usually passing them on to another acceptor molecule in a series (called an ELECTRON TRANSPORT SYSTEM). As each acceptor receives an electron it becomes reduced and then oxidized as the electron is given up (see REDOX POTENTIAL). Each reduction–oxidation reaction is catalysed by a different enzyme, energy being gradually released with each electron transfer. Acceptor molecules (e.g. CYTOCHROME) are vital in AEROBIC RESPIRATION and PHOTOSYNTHESIS.

accessory nerve, *n.* a branch of the VAGUS nerve, occurring in TETRAPODS as the 11th cranial nerve.

acclimation, see ACCLIMATIZATION.

acclimatization or **acclimation,** *n.* the process by which a range of adjustments occur in the body, e.g. in TEMPERATURE CONTROL and RESPIRATION, when an organism is subjected to unusual environmental stresses. See ADAPTATION, PHYSIOLOGICAL.

accommodation, *n.* **1.** the process by which the focus of the eye changes so that distant and near objects can be made sharp. In fish and amphibians this is achieved by moving the lens backwards and forwards; in birds, reptiles, humans and some other mammals the shape of the lens is changed by the CILIARY MUSCLES. When the ciliary muscles contract, the tension lessens on the ligaments and the lens thickens as it rounds off through its own elasticity, enabling the eye to focus on near objects. When the ciliary muscles relax, the ligaments tighten, and the lens becomes stretched and thin and normally focuses on infinity. See Fig. 5.

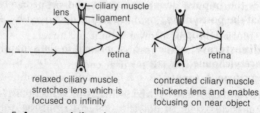

relaxed ciliary muscle stretches lens which is focused on infinity

contracted ciliary muscle thickens lens and enables focusing on near object

Fig. 5. **Accommodation.** Accommodation in the vertebrate eye.

2. the process by which the sensitivity of excitable membranes to DEPOLARIZATION during a NERVE IMPULSE depends upon the rate at which the current increases. As a result, greater depolarizations are required to produce an ACTION POTENTIAL when the current increases slowly than when there is a sudden increase in current.

accrescent, *adj.* (of plant structures such as the CALYX) becoming larger after flowering.

accumulator, *n.* any plant, such as a nitrogen fixer, which enriches the habitat with nutrients.

acellular, *adj.* (of an organism) having a body which is not composed of cells. Acellular organisms may have a complex structure differentiated into specialized areas and ORGANELLES. Such organisms are also described as *unicellular*, although this adjective suggests a simple structure, which is not the case.

acentric chromosome, *n.* a chromosome formed by the joining of two broken pieces of chromosomes that lack CENTROMERES; the acentric chromosome is lost at the next division. Compare METACENTRIC CHROMOSOME.

acephalous, *adj.* without a head.

Acetabularia, *n.* a genus of marine algae. Individuals may reach a size of 30 mm and are ACELLULAR with a body consisting of three distinct areas: head, stalk and a base containing the single NUCLEUS. Experiments by the German biologist Joachim Hammerling (b.1901) using grafts between two species demonstrated that the nucleus alone is responsible for body morphology and produces a chemical substance which can control development. Such results can now be interpreted as indications of the role of DNA and MESSENGER RNA.

acetabulum, *n.* one of two cup-shaped sockets in the PELVIC GIRDLE articulating with the ball of the FEMUR.

acetic acid or **ethanoic acid,** *n.* a clear colourless liquid with a pungent odour of vinegar. Formula: CH_3COOH.

acetylcholine (ACh), *n.* a TRANSMITTER SUBSTANCE that is secreted at the ends of CHOLINERGIC nerve fibres on the arrival of a NERVE IMPULSE. ACh then passes the impulse across a SYNAPSE and immediately ACh has depolarized the postsynaptic membrane it is destroyed by the enzyme CHOLINESTERASE. Compare ADRENERGIC.

acetylcholinesterase, *n.* the enzyme present in the synaptic cleft that destroys acetylcholine (see END PLATE).

acetylcoenzyme A (acetyl-CoA), *n.* an organic compound important in the process of OXIDATION of energy-rich compounds to CO_2 and water. In the presence of oxygen the three-carbon pyruvic acid produced in GLYCOLYSIS is broken down in the MITOCHONDRIA to produce CO_2, two hydrogen atoms (reducing NAD to $NADH_2$), and an active form of two-carbon acetic acid bonded to *coenzyme A*, called acetylcoenzyme A. Metabolism of fats and proteins also produces acetyl-CoA, either via pyruvic acid or directly. Thus acetyl-CoA is common to the metabolism of all major types of food, forming a step leading to the KREBS CYCLE. See Fig. 6 on page 6.

ACh, see ACETYLCHOLINE.

achene, *n.* a small, dry, INDEHISCENT, one-seeded fruit with a thin PERICARP seen, for example, on the surface of strawberry 'fruits'. See Fig. 7 on page 6.

achlamydeous, *adj.* (of a flower) lacking both petals and sepals.

achondroplasia, *n.* the commonest form of human dwarfism. It results

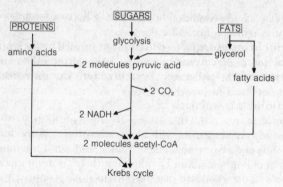

Fig. 6. **Acetylcoenzyme A.** Production of acetyl-CoA from food.

Fig. 7. **Achene.** The achenes of a strawberry.

from abnormal cartilage development, producing numerous bone defects: stunted trunk, shortened and deformed limbs, and bulging skull. The condition is controlled by an autosomal dominant gene (see DOMINANCE, GENETIC) and thus affected children born to normal parents must result from a mutation in one or other parent. The majority (around 80%) of achondroplastics die in early childhood but those surviving are fertile, have the normal range of intelligence and can live more or less normal lives. The typical circus dwarf is achondroplastic.

acid, *n.* any chemical substance that acts as a proton donor. Acids dissolve in water with the formation of hydrogen ions which may be replaced by metals to form salts.

acid–base balance, *n.* the maintenance of a constant internal environment in the body through BUFFER systems maintaining a balance between acids and bases.

acid dyes, *n.* those dyes that contain an acidic organic component which stains materials such as cytoplasm and collagen when combined with a metal.

acidophil or **acidophile** or **acidophilic** or **acidophilous,** *adj.* **1.** (of a cell such as a leucocyte) having tissues that are easily stainable with acid dyes. **2.** (of organisms, particularly microorganisms) preferring an acidic environment.

acidosis, *n.* a body condition in which there is excessive acidity in body fluids, normally regulated by the kidney.

acid rain, *n.* any rain that contains pollutants such as dissolved sulphur compounds, and as a result adversely affects plant and animal life on precipitation. The pollutants are released into the atmosphere by the burning of fossil fuels such as coal or oil.

acinus, *n.* the terminal sac of an alveolar gland. The gland has a multicellular structure with sac-shaped secreting units.

acoelomate, *adj.* lacking a COELOM, the true body cavity. This occurs in some invertebrate groups such as coelenterates, platyhelminths, nemertines and nematodes.

acorn worm, *n.* a hemichordate, one of the three groups of invertebrate chordates collectively referred to as PROTOCHORDATES.

acoustic, *adj.* pertaining to hearing.

acoustico-lateralis system, see LATERAL-LINE SYSTEM.

acquired characters, *n.* characteristics of an organism which, according to Lamarck (see LAMARCKISM) become well developed through use or through environmental influences during an organism's lifetime, and which are inherited by its progeny. However, no clear experimental evidence has yet been found to support the theory. Current opinion is that the environment does play a large part in evolution, though only through NATURAL SELECTION.

Acrania, *n.* an outdated taxonomic term often used as a synonym of PROTOCHORDATE, but in some classifications limited to synonymy with CEPHALOCHORDATE.

Acrasiales, *n.* a grouping in some classifications within the Myxomyceta which includes those slime moulds that are cellular or communal and which do not coalesce.

acrocentric, *adj.* (of a chromosome) having the CENTROMERE near one end, making one chromosome arm much longer than the other.

acrocephaly, *n.* the possession of a dome-shaped head.

acromegaly, *n.* a chronic disease characterized by enlargement of the head, hands and feet. It is caused by over-secretion of growth hormones.

acropetal, *adj.* **1.** (of plant structures, e.g. leaves and flowers) produced one after another from the base of the stem to the apex. **2.** (of substances, e.g. water) moving from the base of a plant to the apex.

acrosome, *n.* a membrane-bound sac derived from the GOLGI APPARATUS, located at the tip of the sperm. The acrosome plays an important role in fertilization. It contains enzymes that, when the sac ruptures upon contacting the egg, aid in penetration of the egg membrane so enabling transfer of the sperm nucleus to the egg cell. See Fig. 8 on page 8.

A-C soil, see RENDZINA.

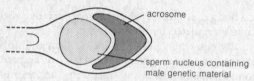

Fig. 8. **Acrosome.** Head of a typical mammalian sperm.

ACTH, see ADRENOCORTICOTROPIC HORMONE.

actin, *n*. a contractile protein found in the muscles of all animals from protozoa to vertebrates and in the MICROFILAMENTS of all cells. The energy for contraction is derived from ATP.

actinomorphic, *adj*. (of regular flowers) divisible into two or more planes to give identical halves. In animals such a structure is referred to as radially symmetrical (see RADIAL SYMMETRY).

Actinomycetes, *n*. a group of PROCARYOTIC organisms in the form of Gram-positive bacteria (see GRAM'S STAIN) which develop branching HYPHAE (0.5–1.0 μm in diameter). The hyphae in turn form a MYCELIUM. Reproduction is either by total FRAGMENTATION of the hyphae and/or by production of SPORES in specialized areas of the mycelium. Most species are SAPROPHYTES, AEROBES, MESOPHILIC, and grow optimally at neutral pH.

actinomycin D, *n*. a substance that inhibits the transcription of RNA from DNA. When isolated from soil bacteria and used pharmaceutically it acts as an ANTIBIOTIC.

Actinopterygii, *n*. a class of bony fishes characterized by having fins with rays. They are the most successful group of fishes and are subdivided into two small orders, the Chondrostei (including sturgeons and paddlefishes) and the Holostei (including the garpikes and bowfin), and a third large order called the Teleostei, which are the dominant fishes of the world today and the most numerous of all vertebrates.

Actinosphaerium, *n*. a PROTOZOAN, known as the sun-animalcule, which is related to *Amoeba*.

Actinozoa, *n*. a less usual term for the ANTHOZOA.

action potential, *n*. the electrical potential present between the inside and outside of a nerve or muscle fibre when stimulated. In the resting state (RESTING POTENTIAL) the muscle or the nerve fibre is electrically negative inside and positive outside. With the passage of the impulse the charges are reversed and the wave of potential change which passes down the fibre is the most easily observable aspect of an impulse. The impulse (DEPOLARIZATION) is short-lived and lasts for about 1 millisecond, after which time the resting potential is restored. See NERVE IMPULSE.

action spectrum, *n*. a range of wavelengths of light within which a physiological process can take place. For example, the range of wavelength of light in which the chlorophyll in green plants is able to synthesize carbohydrates by photosynthesis corresponds closely to the ABSORPTION SPECTRUM, showing that most of the light absorbed by chlorophyll is actually used in photosynthesis. See Fig. 9.

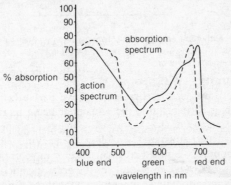

Fig. 9. **Action spectrum.** The absorption and action spectra of chlorophyll.

activated sludge, *n.* any material used in sewage treatment, consisting mainly of bacteria and protozoans, that purifies the sewage and increases by multiplication during the process. Part of the new sludge is used in the continuing treatment of the sewage.

activation energy, *n.* the energy required to initiate a reaction. Chemical bonds holding molecules together are difficult to break, requiring extra 'activation' energy to push the bonded atoms apart. This extra energy makes the bonds less stable so that the molecule releases not only the activation energy but also the energy unlocked when the chemical bonds break, forming an EXERGONIC REACTION.

Activation energy can be applied externally as heat, but this is inappropriate for living organisms. Instead, they rely on biological catalysts (ENZYMES) which decrease the activation energy needed for the reaction to take place. See Fig. 10 (p. 10). See also ENDERGONIC REACTION.

activator, *n.* any drug that increases the activity level of the person being treated.

active absorption/uptake, *n.* the uptake of substances through the active metabolism of a plant, often against a concentration gradient, as opposed to *passive uptake* along a concentration gradient.

active centre, *n.* **1.** the part of an ENZYME molecule that interacts with and binds the substrate, forming an enzyme substrate. **2.** the part of an antibody molecule that interacts with and binds the antigen, forming an antibody/antigen complex.

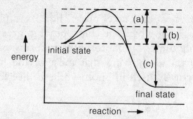

Fig. 10. **Activation energy.** (a) Activation energy required without enzymes. (b) Activation energy required with enzymes. (c) Energy from exergonic reaction.

active immunity, see ANTIBODY.

active ingredient, *n.* the chemically active part of a manufactured product such as an HERBICIDE, the remainder being inert.

active site, *n.* an area of ENZYME surface which has a shape complementary to a particular SUBSTRATE, enabling the enzyme and substrate to become temporarily bonded to form an enzyme–substrate complex. Such a *lock-and-key mechanism* explains the great specificity of enzymes for substrates and also why changes in enzyme three–dimensional shape (by pH, temperature) cause alterations to enzyme activity. See Fig. 11.

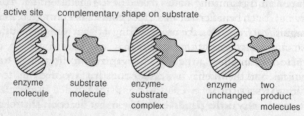

Fig. 11. **Active site.** Lock-and-key mechanism of enzyme activity.

active state, *n.* the condition in which muscle exists immediately before and during contraction which makes it nonextensible. It is caused by the attachment of MYOSIN bridges to ACTIN filaments.

active transport, *n.* movement of a substance from a region of low concentration to another of higher concentration, i.e. against the CONCENTRATION GRADIENT. Such transport typically occurs in cell membranes, which are thought to contain carriers which move molecules from one side of the membrane to the other. Since these processes involve movement up a free-energy gradient, they require the expenditure of energy from the breakdown of ATP and are therefore sensitive to factors affecting metabolism (temperature, oxygen, pH, etc.). Compare DIFFUSION. See Fig. 12.

activity, *n.* the ability of a substance to react with another.

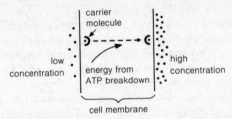

Fig. 12. **Active transport.** Active transport across a membrane.

actomyosin, *n*. a protein formed by the interaction of ACTIN and MYOSIN. It is closely concerned with the process of MUSCLE contraction.

acuity, *n*. sharpness of vision; the resolving power of the eye.

acuminate, *adj*. (of leaves) tapering to a point, narrowing sharply towards the tip and then less sharply near the point.

acute, *adj*. **1.** (of plant structures such as leaves) sharply pointed. **2.** (of a disease) coming quickly to a crisis. **3.** (of a radiation dose) applied at a high level in a short space of time. Compare CHRONIC.

ad-, *prefix*. denoting to, next to.

adaptation, genetic, *n*. any characteristic that improves the chances of an organism transmitting GENES to the next generation (i.e. producing offspring). Such beneficial changes are genetically controlled and can be distinguished from alterations occurring within one generation (see ADAPTATION, PHYSIOLOGICAL) which may not lead to genetic change. Adaptations can affect any level of organization, from cells to whole organisms, and their behaviour. Adaptions are favoured by the process of NATURAL SELECTION.

adaptation, physiological, *n*. change in the reaction that organisms show as a result of prolonged exposure to conditions that normally produce a different or more extensive reaction. For example, mountain sickness (headache, nausea, fatigue) in man passes off as he becomes acclimatized to high altitudes because of adjustments that occur in the respiratory and circulatory systems. See ACCLIMATIZATION.

adaptive enzyme, *n*. an enzyme that is produced by an organism, e.g. bacteria, only in the presence of its normal substrate or a closely similar substance. Such enzymes are not produced in circumstances to which the organism is not adapted. See OPERON MODEL.

adaptive radiation, *n*. the evolution from an ancestral form of a wide range of related species occupying and exploiting many different types of available habitat. For example, mammals, evolving from the Tertiary period, have occupied many habitats and now have flying, running, swimming and burrowing forms, often involving changes to the PENTADACTYL LIMB structure.

adaxial, *adj.* (of a leaf) facing the stem.

Addison's disease, *n.* a disease caused by a deficiency of adrenocortico-steroid hormones (e.g. CORTISONE, ALDOSTERONE) produced by cells of the ADRENAL GLAND cortex. Named after Thomas Addison (1793–1860), the English physician who first described it. The major symptoms of the disease are lowered blood pressure, lowered blood-sugar levels, reduced kidney function, loss of weight, extreme muscular weakness, and a brownish pigmentation of the skin and mucous membranes.

additive genes, *n.* those in which more than one gene controls a character, with each allele making a definite and measurable contribution towards the character; there is no DOMINANCE between alleles of one gene and no EPISTASIS between different loci (see LOCUS). Many pigment systems are controlled by additive genes, producing a wide spread of variability.

adductor, *n.* **1.** also called a **depressor.** a muscle that pulls a structure or limb towards the main part of the body. An example is the adductor mandibulae, which closes the jaws in amphibians. **2.** a muscle that pulls two structures together, e.g. the two valves of a shell.

adenine, *n.* one of four types of nitrogenous bases found in DNA, having the double-ring structure of a class known as PURINES (see Fig. 13).

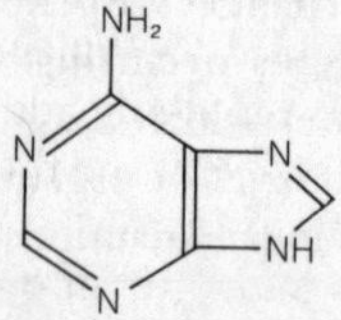

NH_2

Fig. 13. **Adenine.** Molecular structure.

Adenine forms part of a DNA unit called a NUCLEOTIDE and always forms COMPLEMENTARY BASE PAIRING with a DNA PYRIMIDINE base called THYMINE (see Fig. 14). When pairing with RNA during TRANSCRIPTION, adenine is complementary to URACIL. Adenine also occurs in RNA molecules, ATP and AMP.

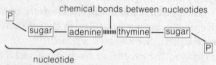

Fig. 14. **Adenine.** Complementary pairing. P = phosphate group.

adenohypophysis, *n.* that part of the PITUITARY GLAND derived embryologically from the HYPOPHYSIS.

adenosine, *n.* a nitrogen-containing compound consisting of an ADENINE base attached to a ribose sugar. Adenosine forms part of NUCLEOTIDES making up NUCLEIC ACIDS and ATP.

adenosine diphosphate, see ADP.

adenosine monophosphate, see AMP.

adenosine triphosphate, see ATP.

adenovirus, *n.* a type of virus responsible for several acute infections of the respiratory system. The name derives from the adenoid tissues, where the virus is often found.

adenyl cyclase, *n.* the enzyme that catalyses the formation of cyclic AMP from ATP by the removal of pyrophosphate.

ADH (antidiuretic hormone), *n.* a hormone secreted by neurosecretory cells of the HYPOTHALAMUS and released by the posterior lobe of the PITUITARY GLAND. ADH stimulates the resorption of water through the distal–convoluted tubule of the KIDNEY nephron in mammals and thus limits the water content and the overall volume of urine.

adipose tissue, *n.* a fatty CONNECTIVE TISSUE, the matrix of which contains large, closely packed, fat-filled cells. Adipose tissue is important in energy storage, occurring round the liver and kidneys, and where it occurs in the dermis of SKIN it insulates the body from heat loss.

adjacent disjunction, see TRANSLOCATION HETEROZYGOTE.

adjuvant, *n.* a substance added to enhance a physical or chemical property, e.g. adjuvants are commonly added to ANTIGENS, improving the IMMUNE RESPONSE in the recipient and thus increasing the production of ANTIBODIES.

adnate, *adj.* (of a structure) joined to an organ of a different kind.

adoral, *adj.* (of an organism) the side on which the mouth is situated.

ADP (adenosine diphosphate), *n.* a molecule consisting of an ADENOSINE unit onto which are attached two phosphate groups, which are joined by a high–energy bond. ADP can be converted to ATP by the addition of inorganic phosphate and about 34 kJ energy. See Fig. 15. See ATP for molecular structure.

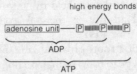

Fig. 15. **ADP.** General structure of ADP and ATP. P = phosphate group.

adrenal cortical hormones, *n.* hormones secreted by the cortex of the ADRENAL GLANDS. They are generally of three main types: (a) *mineralocorticoids*, e.g. ALDOSTERONE and deoxycorticosterone, which are concerned with salt and water balance; (b) GLUCOCORTICOIDS, e.g. CORTISONE and hydrocortisone, which aid the formation of carbohydrates from fat and protein; and (c) *sex hormones*, particularly ANDROGENS in both male and female mammals.

adrenal gland, *n*. an endocrine organ consisting of a medulla (central part) secreting ADRENALINE and NORADRENALINE and a cortex (outer zone) secreting ADRENAL CORTICAL HORMONES. The two parts are closely associated in mammals, but are sometimes separated into distinct organs in other vertebrates, e.g. fish. The activity of the medulla is controlled by the sympathetic nervous system, and that of the cortex by ADRENOCORTICOTROPIC HORMONE secreted by the pituitary gland. In mammals there is a pair of adrenal glands situated anteriorly to the kidneys; other vertebrates have more than two adrenals.

adrenaline or **epinephrine,** *n*. a hormone secreted by the medulla (central part) of the ADRENAL GLAND. It prepares the body for emergency action (FIGHT-OR-FLIGHT REACTION); it increases the cardiac frequency, constricts the vessels supplying the skin and gut, increases the blood pressure, increases blood sugar, dilates the blood vessels of the muscles, heart and brain, widens the pupils, and causes hair erection. It is usually secreted with NORADRENALINE, whose effects are similar. Both hormones are also secreted by the ADRENERGIC nerve endings of the sympathetic nervous system.

adrenergic, *adj*. (of nerve endings) secreting ADRENALINE and NON-ADRENALINE on the arrival of a NERVE IMPULSE. These substances then stimulate the effector nerve fibres in the SYMPATHETIC NERVOUS SYSTEM of many vertebrates in much the same way as ACETYLCHOLINE acts as a transmitter substance in CHOLINERGIC nerve fibres.

adrenocorticotropic or **adrenocorticotrophic hormone (ACTH)** or **corticotrophin,** *n*. a small, proteinaceous hormone secreted by the anterior lobe of the PITUITARY. It controls the secretion of other hormones of the cortex of the ADRENAL GLAND.

adsorption, *n*. the taking up of gas or liquid by a surface or interface. In physical adsorption, molecules are held by VAN DER WAAL'S FORCES of attraction; in chemical adsorption there is exchange or sharing of electrons. Compare ABSORPTION.

adventitious, *adj*. **1.** (of a root) growing laterally from a stem rather than the main root, e.g. prop roots of maize, clinging roots of climbing vines. **2.** (of a bud) not developing in a leaf axil, as in *Begonia* where such buds can be produced from leaf wounds.

Aepyornis, *n*. a very large, flightless, recently extinct bird from Madagascar. It was larger than an ostrich and the egg content was in excess of 9 litres.

aerenchyma, *n.* cork-like tissue with large air-filled cavities between cells, present in the stems and roots of certain water plants to enable adequate gaseous exchange even below water.

aerial respiration, *n.* a process of gas exchange occurring in terrestrial and some aquatic organisms, by which oxygen is absorbed from the air and carbon dioxide released. Respiratory surfaces are usually internal (e.g. the MESOPHYLL of leaves, the TRACHEAE of insects, the LUNG BOOKS of spiders and scorpions, and the LUNGS of land vertebrates), although some respiratory organs are external, e.g. the SKIN of amphibia. All surfaces rely on a water layer for gas exchange and may also depend on a blood system for transport of gases to and from distant parts of the body.

aerobe, *n.* any organism (typically a microorganism) that can survive only in the presence of oxygen required for AEROBIC RESPIRATION.

aerobic respiration, *n.* a type of CELLULAR RESPIRATION that requires oxygen. GLUCOSE is broken down to release energy in a series of steps which can be grouped into three main stages:

Stage 1: glucose is converted to PYRUVIC ACID in a process called GLYCOLYSIS which takes place in the cell CYTOPLASM. A stable glucose molecule is first energized by the addition of a phosphate group from two ATP molecules (PHOSPHORYLATION) and then broken down to two molecules of three-carbon phosphoglyceraldehyde (PGAL). Each PGAL molecule is oxidized by removal of two hydrogen atoms which are picked up by an NAD molecule. Since oxygen is present, NADH can undergo a MITOCHONDRIAL SHUNT to enter an ELECTRON TRANSPORT SYSTEM (ETS); see Fig. 16. Four molecules of ATP are then synthesized in

Stage	CO_2 molecules	H atoms	ATP molecules via substrate-level phosphorylation	ATP molecules via ETS		Total ATP	Water molecules via ETS
				FADH	NADH		
1	0	4	4 (2)	0	6	10 (8)	2
2	2	4	0	0	6	6	2
3	4	16	2	4	18	24	8
Total	6	24	6 (4)	34		40 (38)	12

Fig. 16. **Aerobic respiration.** The products of one molecule of glucose undergoing aerobic respiration; net figures in brackets.

SUBSTRATE–LEVEL PHOSPHORYLATION over several steps, giving a net gain of two ATP molecules per molecule of glucose. Glycolysis is completed with the production of two molecules of three-carbon pyruvic acid (*pyruvate*).

Stage 2: oxidation and DECARBOXYLATION of pyruvic acid in the MITOCHONDRIA to form two molecules of two-carbon ACETYLCOENZYME A (acetyl-CoA). CO_2 is released in this process, together with two

AEROBIC RESPIRATION

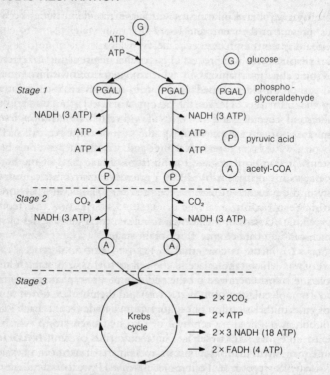

Fig. 17. **Aerobic respiration.** Summary of products in aerobic respiration.

hydrogen atoms per pyruvic acid molecule, which are picked up by NAD and passed down an ETS located on the inner membranes of mitochondrial cristae. Three molecules of ATP are produced per NADH, with oxygen acting as the final acceptor of hydrogen, producing water.

Stage 3: entry of acetyl–CoA into the KREBS CYCLE (TCA cycle). Each molecule of acetyl–CoA can turn the cycle once. As each glucose molecule is broken down to two acetyl–CoA molecules, the cycle will turn twice per glucose molecule, yielding 2×2 molecules of CO_2 and 2×8 atoms of hydrogen. Six pairs of hydrogen atoms are picked up by NAD to produce 18 (6×3) molecules of ATP via the ETS. The remaining two pairs of hydrogen atoms are accepted by FAD molecules which move to the ETS to produce 4 (2×2) molecules of ATP. One molecule of ATP is produced directly by substrate–level phosphorylation for each turn of the cycle. For each molecule of glucose undergoing aerobic respiration the products are as shown in Fig. 16. The three stages are summarized in Fig. 17.

Note that the net production of ATP per molecule of glucose is 38 molecules since two were required at the start of glycolysis. Of these 38 ATP molecules only two (about 5%) are synthesized without oxygen (i.e. anaerobically); the other 36 are the product of aerobic respiration.

Fats and proteins can also undergo aerobic respiration, entering the reactions at various stages; see ACETYLCOENZYME A for details.

aerotaxis, *n.* the movement of an organism with reference to the direction of air or oxygen.

aerotropism, *n.* plant growth response affected by the presence of air. For example, negative aerotropism is growth away from air, as in the growth of the POLLEN TUBE from the stigma to the ovary in flowering plants.

aestivation, *n.* **1.** (in animals) a state of dormancy during the summer or dry season. Compare HIBERNATION. **2.** (in plants) the arrangement of the various parts of the bud of a flower.

aetiology or **etiology,** *n.* the study of causes (usually of a disease).

afferent neurone, *n.* a nerve fibre that conducts impulses towards the central nervous system from receptor cells. Compare EFFERENT NEURONE.

affinity, *n.* the relationship of one organism to another in terms of its evolution.

African sleeping sickness, *n.* a fatal infection of the nervous and lymphatic systems that is endemic in certain parts of Africa and is caused by a flagellate protozoan called *Trypanosoma,* particularly *T. brucei gambiense* (West Africa) and *T. brucei rhodesiense* (East Africa). The vector of the flagellate is the tsetse fly *Glossina,* which also feeds on cattle, the latter acting as a reservoir for the parasite. African sleeping sickness is not to be confused with ENCEPHELITIS which is caused by a virus.

afterbirth, *n.* the remains of the placenta and blood that are voided from the uterus of EUTHERIAN mammals after the young is born.

after ripening, *n.* a period through which apparently fully-formed seeds (and the spores of some fungi) need to pass before they can germinate, being a form of DORMANCY that may involve the production of growth hormones or the destruction of growth inhibitors to produce suitable conditions for further development to commence.

agamospermy, *n.* reproduction that does not involve FERTILIZATION or MEIOSIS but produces an embryo by asexual means other than VEGETATIVE REPRODUCTION.

agar, *n.* the carbohydrate product of some seaweeds which, on forming a gel with water and allowed to solidify, is used as a culture medium for microorganisms.

agave, *n.* a semi-woody perennial native of the American continent. Pulque and aquamiel are produced from it as fermented beverages, and mescal and tequila may in turn be distilled from these.

agglutin, *n.* a type of ANTIBODY causing AGGLUTINATION.

agglutination, *n.* a clumping together of cells, usually as a result of reaction between specific ANTIGENS and ANTIBODIES in blood and lymph, forming a natural defence against foreign materials, including bacterial cells. Transfusion of blood between persons of different ABO BLOOD GROUPS is also subject to the risk of agglutination (see UNIVERSAL DONOR and UNIVERSAL RECIPIENT).

agglutinogen, *n.* a surface antigen that induces the formation of agglutins in cells (including bacteria) and binds them to produce an AGGLUTINATION reaction.

aggregate fruit, *n.* a fruit consisting of a collection of simple fruits, derived from a flower with several free CARPELS. An example is the blackberry.

aggregation, *n.* a clumping or clustering of individuals of a population in a nonrandom distribution. Such clumping can occur (a) as a response to local habitat differences, e.g. some organisms preferring, say, damp to dry areas; (b) as a response to daily or seasonal weather changes; (c) by reproductive processes, e.g. individuals hatching from the same egg batch in some animals, or poor seed dispersal in plants; (d) as a result of social attraction, e.g. colonial nesting in birds.

aggressins, *n.* enzymes that are produced by parasitic bacteria. The aggressins dissolve tissues and enable the bacteria to enter the host tissues.

aggression, *n.* a type of behaviour that includes both threats and actual attacks on other animals, though often limited to threat display. See also AGONISTIC BEHAVIOUR.

Agnatha, *n.* aquatic, jawless, fish-like vertebrates characterized by fewer than two pairs of limbs (fins). The group is often given the status of a subphylum to separate it from other vertebrates with jaws which are included in the subphylum Gnathostomata. The Agnatha includes lampreys, hagfish, and the fossil OSTRACODERMI.

agonistic behaviour, *n.* a broad grouping of behaviour patterns that not only includes all aspects of AGGRESSION, including threat and actual attack, but also the consequent aspects of appeasement and flight.

agouti, *n.* a fur coloration in which there are alternating light and dark bands of colour on the individual hairs, giving a speckled brownish appearance. Such a fur coloration is found in mammals such as rabbits, rats and mice.

agranulocyte, *n.* a type of LEUCOCYTE (white blood cell) with non-granular cytoplasm and a large spherical nucleus. They are produced either in the LYMPHATIC SYSTEM or the bone marrow. Agranulocytes form about 30% of all leucocytes and are of two types, LYMPHOCYTES and MONOCYTES.

agriculture, *n.* the cultivation of the soil for any aspect of farming or horticulture.

agronomy, *n.* the study of the cultivation of field crops, with particular emphasis on improving their productivity and qualitative features.

A horizon, *n.* the uppermost zone of soil, containing leaf litter at the top, raw humus in the middle, and a dark-coloured humus layer at the bottom. See also B HORIZON.

AIDS, *n. acronym for* Acquired Immune Deficiency Syndrome, a serious human disease caused by a virus that can destroy the body's natural defence system (see IMMUNE RESPONSE) and cause death through the body becoming unable to fight any infection. The level of response to infection varies between different people; not everyone who carries the virus develops AIDS but all infected individuals can pass it on, by two main methods: (a) through sexual intercourse, and (b) by transfer of blood cells, as when drug users share needles. At present there is no known vaccine against the AIDS virus, although several are being developed.

air bladder or **gas bladder** or **swim bladder,** *n.* a gas-containing sac present in the roof of the abdominal cavity of bony fish that enables them to maintain position off the bottom with little or no movement.

airborne pathogen, *n.* any disease-causing organism which can be transmitted through the air.

air pollution, *n.* the presence of contaminants in the form of dust, fumes, gases or other chemicals in the atmosphere in quantities which adversely affect living organisms. See ACID RAIN.

air sac, *n.* a thin-walled extension to the lungs present in the thorax, abdomen and bones of birds. Air sacs also occur as diverticula (outgrowths) of the TRACHEA in insects, where, as in birds, they are important in respiration. See Fig. 18.

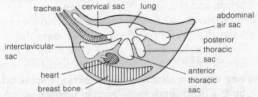

Fig. 18. **Air sac.** The air-sac system of a bird (left-side).

air space, *n.* any part of the interior of an organism occupied by air.

alanine, *n.* one of 20 AMINO ACIDS common in proteins. It has a NON-

POLAR structure and is relatively insoluble in water. The ISOELECTRIC POINT of alanine is 6.0. See Fig. 19.

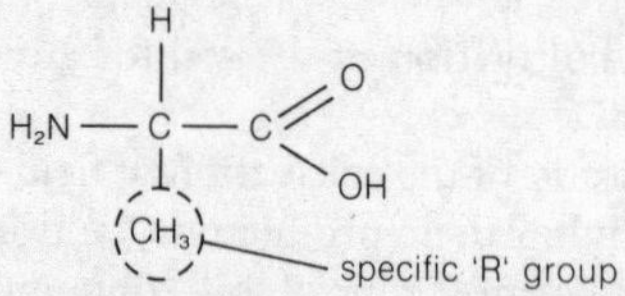

Fig. 19. **Alanine.** Molecular structure.

alary muscles, *n.* a series of small muscles found in the pericardial wall of insects. Their contraction causes blood to flow into the pericardium from the perivisceral cavity and then into the heart.

alar, *adj.* pertaining to the wings.

albinism, *n.* an inherited condition found in many organisms, whose chief feature is a lack of MELANIN pigment in structures that are normally coloured. In humans the syndrome of albinism comprises:

(a) pinkish skin in which the pigment–containing cells (MELANOPHORES) are present but contain little or no melanin. The pinkish coloration comes from the underlying blood vessels.

(b) eyes with a deep red pupil and pink iris, both due to a lack of normal pigmentation and heavy vascularization.

(c) *photophobia* (fear of light) brought on by excess light entering the eye and reflecting onto the retina. Albinos usually wear dark glasses.

(d) pale yellow hair.

Albinism is an INBORN ERROR OF METABOLISM produced by a blockage in tyrosine metabolism due to the absence of a functional tyrosinase enzyme. The condition is controlled by a single RECESSIVE GENE on an AUTOSOME. Approximately 1 in 20,000 of the population are albinos. See Fig. 20.

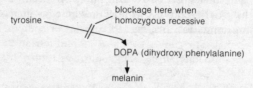

Fig. 20. **Albinism.** Blockage in tyrosine metabolism.

albumen or **egg white** *n.* a solution of protein in water which is secreted by the oviduct of birds and some reptiles. Albumen surrounds the embryo and yolk which it cushions within the shell of the egg. It is eventually absorbed by the embryo as food.

alcohol, *n.* an organic molecule in which a hydrogen atom of a hydrocarbon is replaced by a hydroxyl group (OH). See Fig. 21.

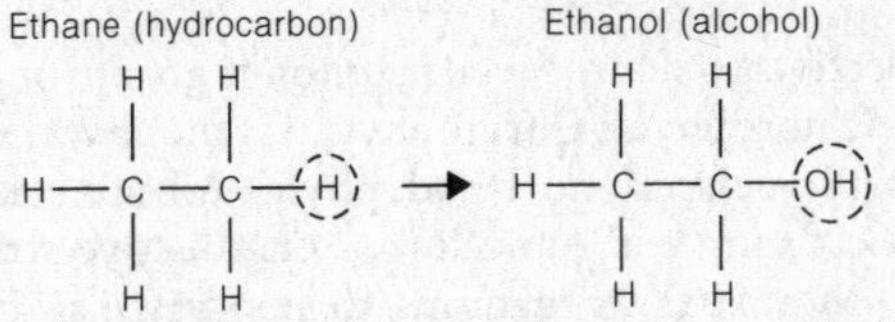

Ethane (hydrocarbon)　　　　　　Ethanol (alcohol)

Fig. 21. **Alcohol.** Molecular structure of an alcohol.

alcoholic fermentation, *n*. a type of CELLULAR RESPIRATION found in plants and some unicells which does not require oxygen, resulting in the production of ethanol (an alcohol) from GLUCOSE and the release of small amounts of energy. The main details are described under ANAEROBIC RESPIRATION but, briefly, fermentation involves GLYCOLYSIS to produce PYRUVIC ACID and then (in the absence of oxygen) the breakdown of pyruvic acid to ethanol via ethanal with the release of CO_2. See Figs. 22 and 34.

$$C_6H_{12}O_6 \longrightarrow 2CH_3CH_2OH + 2CO_2 + 168kJ \text{ energy}$$

Glucose　　　　　　Ethanol

Fig. 22. **Alcoholic fermentation.** The production of ethanol from glucose.

Alcyonaria, *n*. a division of the Anthozoa containing the soft corals, e.g. sea pens, sea pansies, deadmen's fingers and precious coral.

aldehyde group, *n*. any organic compound containing the group CHO, a carbonyl group fixed to one hydrogen and one oxygen atom.

aldose, *n*. a sugar having an aldehyde group attached to the first carbon atom, e.g. D-glyceraldehyde.

aldosterone, *n*. a hormone of the cortex of the ADRENAL GLAND. It is responsible for the relative concentration of sodium and potassium ions in the body. It promotes the reabsorption of sodium ions from the ascending limb of the LOOP OF HENLE in the kidney, with the elimination of potassium ions, and increases the uptake of sodium ions by the alimentary canal. The concentration of sodium ions in the blood thus rises and potassium ions fall, enabling ionic regulation of body fluids (see SODIUM PUMP).

aleurone layer, *n*. a layer of cells just below the TESTA of some seeds (e.g. barley). Aleurone releases large quantities of hydrolytic enzymes (AMYLASES, PROTEASES and NUCLEASES) for digestion of the ENDOSPERM. Thus food material is made available in soluble form for embryo growth. Enzyme release is brought on by the plant hormone GIBBERELLIN, which is released by the embryo when the seed is soaked in water prior to germination.

aleuroplast, *n.* a colourless PLASTID that stores protein, found in many seeds.

algae, *n.* a collective term for several taxonomic groups of plants, namely Charophyta, Chlorophyta, Chrysophyta, Cyanophyta, Euglenophyta, Phaeophyta, Pyrrophyta and Rhodophyta. All are relatively simple photosynthetic forms with unicellular reproductive structures. They range from UNICELLULAR organisms to non-vascular filamentous or thalloid plants. Algae occur in both marine and fresh water, while others are terrestrial, living in damp situations on walls, trees, etc.

algal bloom, *n.* an extensive growth of algae in a water body, usually as a result of the phosphate content of fertilizers and detergents.

alien, *n.* an organism, usually a plant, that is not native to the environment in which it occurs, and that is thought to have been introduced by man.

alimentary canal or **gut** or **enteric canal** or **enteron** or **gastrointestinal tract,** *n.* the tubular passage that extends from MOUTH to ANUS. It has several distinct functions:

(a) ingestion of food via the mouth, utilizing teeth and tongue;

(b) DIGESTION of food beginning in the mouth and continuing in the STOMACH and SMALL INTESTINE;

(c) ABSORPTION of nutrients in soluble form, occurring mainly in the small intestine, with water being taken in via the large INTESTINE;

(d) elimination of faecal material via the anus, the faecal material containing a mixture of undigested food and excreted material (e.g. BILE SALTS).

See DIGESTIVE SYSTEM.

aliquot, *n.* a measured portion of the whole; a sample of known size taken from a preparation.

alisphenoid, *n.* the bone forming in the middle region of the cranium of mammals.

alkali, *n.* a soluble BASE or a solution of a base.

alkaline, *adj.* having the properties of or containing an ALKALI.

alkaline tide, *n.* a period when the alkalinity of the body and urinary system is affected by excessive secretion of digestive hydrochloric acid.

alkalinity, *n.* **1.** the quality or state of being alkaline. **2.** the amount of alkali or base in a solution, often expressed in terms of pH.

alkaloids, *n.* basic organic compounds containing nitrogen that are found in some families of plants (e.g. Papaveraceae). Alkaloids have poisonous and medicinal properties. Examples include nicotine, quinine, morphine, and cocaine.

alkalosis, *n.* the state in which there is excessive body alkalinity.

alkaptonuria, see GARROD, A.E.

allantoic chorion, *n.* the fusion of the ALLANTOIS with the mesoderm of the CHORION which in eutherian mammals develops into the PLACENTA. See AMNION (Fig. 30) for diagram.

allantoin, *n.* the heterocyclic end product of purine catabolism in some reptiles, and mammals other than primates.

allantois, *n.* a membranous growth from the central side of the hind gut of the developing vertebrate embryo that extends outside the embryo proper and is covered by a layer of connective tissue containing many blood vessels. In birds and reptiles the allantois acts as a respiratory surface with the allantoic cavity (see Fig. 30) being used for the storage of excretory materials. In placental mammals the allantoic blood vessels carry blood to and from the PLACENTA for respiration, nutrition and excretion. In all groups most of the allantois is detached at birth.

allele, *n.* a particular form of GENE. Alleles usually occur in pairs, one on each HOMOLOGOUS CHROMOSOME in a DIPLOID cell nucleus. When both alleles are the same the individual is described as being a HOMOZYGOTE; when each allele is different the individual is a HETEROZYGOTE. The number of allelic forms of a gene can be many (MULTIPLE ALLELISM), each form having a slightly different sequence of DNA bases but with the same overall structure. Each diploid can, however, carry only two alleles at one time. See also DOMINANCE (1).

allele frequency or **gene frequency,** *n.* the proportion of a particular ALLELE of a gene in a population, relative to other alleles of the same gene. For example, if a gene has two alleles, A and a, and the frequency of A is 0.6, then the frequency of a will be $1.0 - 0.6 = 0.4$. The allele frequency can be calculated from the GENOTYPE FREQUENCY; see Fig. 23.

genotypes:	AA	Aa	aa
frequency:	0.25	0.40	0.35

frequency of A allele = 0.25 + 0.40/2 = 0.45
frequency of a allele = 0.35 + 0.40/2 = 0.55
                          ‾‾‾‾
                          1.00

Fig. 23. **Allele frequency.** Calculation of allele frequency from genotype frequency.

allelopathic substance, *n.* a secretion or excretion of an organism that has an inhibitory effect on others.

Allen's rule, *n.* a rule stating that in warm-blooded animals (HOMOIOTHERMS) there tends to be a reduction in size of protruberant parts of the body (such as legs, bill, tail) in cooler climates, for example, bill length within some species of birds, such as the redshank. This size reduction is an extension of BERGMANN'S RULE and is also a mechanism to reduce heat loss. The rule was devised by J.A. Allen in 1877.

allergen, *n.* an antigen that produces an allergic response.

allergy, *n*. the overreaction of the IMMUNE RESPONSE of the body to minute traces of foreign substances (antigens). The reaction is usually visible in the form of rashes, itching, breathing difficulties, etc. Many of these symptoms can be attributed to specific antigens, e.g. in hay fever, ANTIBODIES react against pollen (antigen) and cause local damage with the release of HISTAMINE. *Antihistamine* drugs are one method of counteracting the effects of histamine.

allochronic species, *n*. species which do not occur together in terms of time, i.e. species which occur in different geological times.

allochthonous, *adj*. (of peat) not made *in situ*, e.g. peat which has been deposited from plants grown elsewhere. Compare AUTOCHTHONOUS.

allogamy, see CROSS-FERTILIZATION.

allogenic, *adj*. (of vegetational SUCCESSION) affected by outside factors.

allograft, see HOMOGRAFT.

allometric growth, *n*. **1.** unequal growth rate in different portions of the body of an organism that gives rise to the final shape. **2.** growth of a particular structure at a constantly greater rate than the whole.

allopatric, *adj*. (of a population) being geographically separate (or nearly so) from another population of the same species. Compare SYMPATRIC.

allopatric speciation, *n*. the genetic differentiation of populations which are geographically separate to the point where they become separate SPECIES.

allopolyploid, *n*. a type of POLYPLOID in which the number of chromosomes is doubled in a hybrid between two species. The resulting *allotetraploid* or *amphidiploid* is fertile since HOMOLOGOUS CHROMOSOMES of both types can now pair at MEIOSIS to produce viable GAMETES (see Fig. 24). Such a process has been important in the production of new species, particularly in plants. For example, the successful marshland grass *Spartina anglica* is a fertile allotetraploid developed from an infertile hybrid (*S. townsendii*) produced originally by a cross between *S. alterniflora* and *S. maritima*. Compare AUTOPOLYPLOID.

all-or-none-law, *n*. a law stating that certain tissues respond in a similar way to stimuli no matter how strong the stimuli are, i.e. they either react by giving a response (all), or do not react and give no response (none). Nerve fibres normally act in this way.

allosteric enzyme, *n*. a type of enzyme which has two alternative forms: one active with a functional BINDING SITE, the other inactive where the binding site has been altered in shape so as to become non-reactive. Such enzymes have a QUATERNARY PROTEIN STRUCTURE which becomes changed by the addition of a 'modulator' molecule producing the inactive shape, such a process resulting in NONCOMPETITIVE INHIBITION. See Fig. 25.

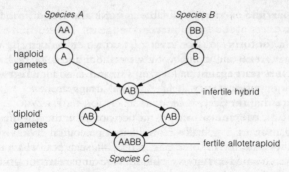

Fig. 24. **Allopolyploid.** The creation of a fertile allotetraploid.

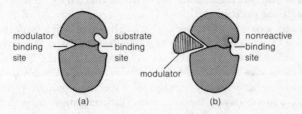

Fig. 25. **Allosteric enzyme.** Active (a) and inactive (b) forms.

allosteric inhibitors, *n*. substances which prevent an enzyme from changing into an active form by combining, not with the ACTIVE SITE, but with some other part of the enzyme.

allotetraploid, *n*. a form of polyploidy; see ALLOPOLYPLOID.

allotype, *n*. one of a range of ANTIBODIES containing different sequences of amino acids but all having an area with an identical amino acid sequence and thought to be produced by different alleles of the same gene (see MULTIPLE ALLELISM).

allozyme, *n*. a form of protein coded by a particular ALLELE at a single gene LOCUS that is detectable by ELECTROPHORESIS.

alluvial soil, *n*. a young soil derived from river, estuarine or marine deposits, and having a high fertility.

alpha helix, *n*. a twisted polypeptide chain which forms a helical structure in many proteins, with 3.6 amino-acid residues per turn of the helix. Successive turns of the helix are linked by weak hydrogen bonds and the structure is much more stable than an untwisted polypeptide chain. Long-chain alpha-helix construction is characteristic of structural 'fibrous' proteins, found in hair, claws, fingernails, feathers, wool and horn. Proteins that are intracellular are usually of the 'globular' type with short segments of alpha helixes.

alpha particle, *n.* a type of subatomic particle found in the atomic nucleus.

alpha taxonomy, *n.* that level of TAXONOMY concerned with the characterization and naming of a species. See also BETA TAXONOMY.

alternate arrangement, *n.* (in plants) structures arranged in two rows, but not opposite; it may include a spiral arrangement.

alternate disjunction, *n.* see TRANSLOCATION HETEROZYGOTE.

alternation of generations, *n.* the occurrence within the life cycle of an organism of a sexually reproducing generation followed by an asexual generation. In some COELENTERATES, such as *Obelia*, the sexual organs are carried on MEDUSAS, and subsequent fertilization gives rise to an asexually producing HYDROID stage; here both generations are DIPLOID. See Fig. 26a.

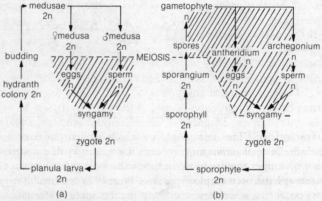

Fig. 26. **Alternation of generations.** (a) Life cycle of *Obelia* (Coelenterata). (b) Life cycle of fern (Pteridophyta). n = haploid number of chromosomes.

Amongst the plants, ferns have a diploid SPOROPHYTE generation (the plant itself). This undergoes MEIOSIS to form spores that give rise to a HAPLOID GAMETOPHYTE generation known as a PROTHALLUS, which reproduces sexually. Male and female GAMETES fuse to form a ZYGOTE, which then develops into a new fern plant (sporophyte). See Fig. 26b. Some definitions of the term require a haploid generation alternating with a diploid generation, so excluding all animal examples.

altruistic behaviour, *n.* behaviour which involves an unselfish regard for others. In organisms other than man altruism probably only exists where the benefits to the individual performing the behaviour are greater than those to the individual to whom the particular behaviour is directed, i.e. *phenotypic altruism.* For example, parental care of the young is clearly of genetic benefit to the individual parent, since it ensures that its genes are transmitted. See also KIN SELECTION.

alveolus, *n.* (*pl.* alveoli) **1.** a small air-filled sac which occurs in large numbers in the LUNGS of vertebrates and whose function is to increase surface area for gaseous exchange (see AERIAL RESPIRATION). **2.** groups of secretory cells found at the internal ends of ducts of mammary and other glands. **3.** an inflated area of the follicle which surrounds the depressions from which cilia (see CILIUM) arise in some ciliates, e.g. *Paramecium*. **4.** a tooth cavity in the jaw bone.

amber codon, *n.* a TERMINATION CODON of messenger RNA (UAG), the first stop signal discovered in the GENETIC CODE.

ambergris, *n.* a substance, used in the manufacture of perfume, found in the alimentary canal of sperm whales in the form of a grey waxy secretion.

ambly-, *prefix.* denoting dull.

ambient, *adj.* surrounding, prevailing. The term is used most often in connection with temperature levels during experiments.

ambulacrum, *n.* one of the five radial bands on which tube feet occur on the outside surfaces of ECHINODERMS.

Ames test, *n.* a technique devised in the USA by Bruce Ames and an associate that is designed to screen environmental chemicals for mutagenicity. The test uses histidine-requiring mutant strains of *Salmonella typhimurium* and measures the frequency of BACK MUTATIONS that no longer require histidine supplements in their food supply. The Ames test has been employed widely since 1975 as a check for potential CARCINOGENS, since these chemicals usually act as MUTAGENS.

Ametabola, see APTERYGOTA.

amine, *n.* an organic base formed by replacing one or more of the hydrogen atoms of ammonia by organic groups.

amino acid, *n.* a building block of protein, containing a carboxyl group (COOH) and an amino group (NH_2), both attached to the same carbon atom. Over 80 amino acids are known to occur naturally, with 20 found commonly in proteins (see Fig. 27 on p. 28), each with a different side chain, called an 'R' group (see Fig. 28). Each of these common amino acids is described under its own heading.

Many amino acids can be synthesized in the body from other amino acids by a process called TRANSAMINATION, although most organisms have a number of ESSENTIAL AMINO ACIDS that must be taken in with the diet. Each amino acid is coded by at least one triplet of DNA bases (see GENETIC CODE), and the string of amino acids making up a protein is joined by PEPTIDE BONDS to form a POLYPEPTIDE CHAIN. Amino acids are soluble in water but vary considerably in their solubility. When in

AMINO-ACID SEQUENCE

Ala = Alanine	Leu = Leucine
Arg = Arginine	Lys = Lysine
Asn = Asparagine	Met = Methionine
Asp = Aspartic acid	Phe = Phenylalanine
Cys = Cysteine	Pro = Proline
Gln = Glutamine	Ser = Serine
Glu = Glutamic acid	Thr = Threonine
Gly = Glycine	Trp = Tryptophan
His = Histidine	Tyr = Tyrosine
Ileu = Isoleucine	Val = Valine

Fig. 27. **Amino Acid.** The 20 amino acids commonly found in protein.

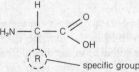

Fig. 28. **Amino Acid.** Structure of a generalized amino acid.

solution they are ionized (see ZWITTERION) and generally are electrically neutral with a pH known as the ISOELECTRIC POINT. They are amphoteric, i.e. acting as acids or bases if the pH is shifted.

amino–acid sequence, *n.* the order in which AMINO ACIDS are placed along a protein molecule. Secondary and tertiary protein structure is highly dependent upon amino–acid sequence in the POLYPEPTIDE CHAIN, which affects the bonding together of the molecule. The 'blueprint' for this sequence is contained in DNA, whose bases contain the amino acid code; three bases provide the code for each amino acid. By the processes of TRANSCRIPTION and TRANSLATION the DNA–base sequence is mirrored in the amino–acid sequence, there being COLINEARITY between DNA and protein. See Fig. 160 (FRAMESHIFT) for diagram.

aminoacyl-tRNA, *n.* the molecule produced when an AMINO ACID is activated into its aminoacyl form and attached to its specific TRANSFER RNA molecule, the whole process being catalysed by a specific aminoacyl-tRNA synthetase enzyme.

aminopeptidase, *n.* an enzyme that catalyses the sequential hydrolysis of amino acids in a polypeptide chain from the N-terminal.

amino sugar, *n.* a MONOSACCHARIDE in which an amino group has replaced an hydroxyl group on one or more occasions.

amitosis, *n.* nuclear division in a cell that occurs without the formation of a SPINDLE or the appearance of chromosomes, and from which daughter nuclei with identical chromosome sets are probably not produced. Amitosis occurs for example in the protozoan *Paramecium* when the MACRONUCLEUS divides.

ammocoete, *n.* the larva of a lamprey. It feeds by ciliary action similar to AMPHIOXUS.

ammonia, *n.* a colourless gas, which is the main form in which nitrogen is utilized in living cells. Formula: NH_3.

ammonite, *n.* a name given to a large group of fossil cephalopod molluscs of late Palaeozoic and Mesozoic time that possessed a spiral shell very similar to that of *Nautilus*.

amniocentesis, *n.* a technique for the diagnosis of congenital abnormalities before birth. In this procedure a sample of 10–15 cm³ of amniotic fluid surrounding the foetus is extracted through the abdominal wall using a surgical syringe, usually at about the 16th week of pregnancy. The sloughed-off embryonic cells contained in the fluid can be cultured to enable KARYOTYPE analysis, which shows the number and condition of the chromosomes. The fluid can also be analysed for the presence of chemicals which indicate biochemical and other defects (e.g. spina bifida). See Fig. 29.

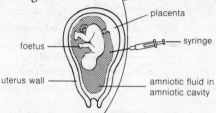

Fig. 29. **Amniocentesis.** Sampling amniotic fluid by means of a surgical syringe.

amnion, *n.* an embryonic, fluid-filled sac which occurs in reptiles, birds and mammals (AMNIOTES). Formed from ECTODERM and MESODERM and containing a coelomic space (see COELOM), the amnion grows around the embryo and eventually roofs over and completely encloses it. It provides a fluid-filled space necessary for the development of embryos of land animals, and it also acts as a protective cushion. See Fig. 30.

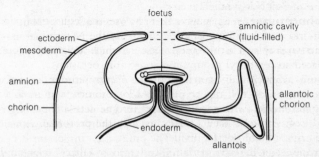

Fig. 30. **Amnion.** A vertebrate embryo lying within the amnion.

amniote, *n*. any land vertebrate which possesses an AMNION, CHORION and ALLANTOIS, i.e. reptiles, birds and mammals.

amniotic cavity, *n*. the space within the AMNION.

Amoeba, *n*. a genus of unicellular PROTOZOANS in the class Rhizopoda. Amoebae are characterized by their changing shape brought about by the projection of PSEUDOPODIA which have a locomotory function. *Amoeba* is often wrongly quoted as an example of a primitive organism, low on the evolutionary scale. However, as in many PROTISTA, functions that are carried out by organ systems in so-called 'higher' forms are here carried out within a single cell. Therefore they cannot be considered 'primitive'; they are highly evolved over millions of years, albeit on a different scale from higher organisms. See Fig. 31.

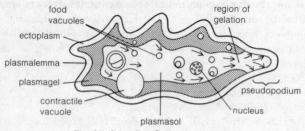

Fig. 31. **Amoeba.** General structure.

amoebic dysentry, *n*. a form of intestinal infection marked by severe diarrhoea and sometimes bleeding. It is caused by the parasitic amoeba *Entamoeba histolytica*. The disease is transmitted by cysts present in the faeces that are swallowed by a new host.

amoebocyte, *n*. a type of cell found in the blood and other body fluids of invertebrates that is capable of AMOEBOID MOVEMENT within the fluid, and which often acts as a PHAGOCYTE.

amoeboid, *adj*. moving and feeding in the manner of *Amoeba*, by the action of PSEUDOPODIA.

amoeboid movement, *n*. movement by means of PSEUDOPODIA, being a projection of the surface of a cell caused by alternate formation of SOL and GEL protoplasm in the region of the temporarily formed 'false foot'. See AMOEBA.

amorph allele, *n*. an ALLELE of a gene that appears to code for no detectable product. It can be produced by a DELETION mutation.

AMP (adenosine monophosphate), *n*. a molecule consisting of an ADENOSINE unit onto which is attached a phosphate group with a low-level energy bond (see Fig. 32). The molecule is important because it functions as the basic structure for ADP and ATP, both of which contain HIGH-ENERGY BONDS (34 kJ/mole) that serve as a short-term energy store

Fig. 32. **AMP.** The low-level energy bond of AMP.

for the cell. See ATP (Fig. 55) for molecular structure.

amphetamine, *n.* a drug that stimulates the CENTRAL NERVOUS SYSTEM and inhibits sleep. Its structure is: 1-phenyl-2-aminopropane.

amphi-, *prefix.* denoting both kinds, around.

amphibian, *n.* a member of the vertebrate class Amphibia, containing frogs, toads, newts, salamanders and the burrowing caecilians. Descended from fish-like animals, they colonized the land in the late DEVONIAN PERIOD. They are thought to be the immediate ancestors of the reptiles, though present-day amphibians are little like their Devonian ancestors. Most amphibians, although normally terrestrial animals, return to water during the breeding season. Fertilization is external (the eggs lacking a shell and embryonic membranes), and the eggs are usually laid and develop in water.

amphicribral bundle, *n.* a VASCULAR BUNDLE in which the PHLOEM surrounds the XYLEM, as in some ferns.

amphidiploid, see ALLOPOLYPLOID.

amphimixis, *n.* true sexual reproduction, as opposed to APOMIXIS.

Amphineura, *n.* a primitive group of molluscs thought to be very like the ancestral form. Present-day forms are similar to Ordovician fossils.

Amphioxus, *n.* the disused scientific, though better-known, name for *Branchiostoma* (the lancet), a PROTOCHORDATE found in many parts of the world. Its embryonic development has been extensively studied and is used in texts as an example of a link between invertebrates and vertebrates. It feeds by ciliary action, trapping food particles in mucus.

amphiploid, *n.* a POLYPLOID that is generally fertile, formed by the doubling of chromosomes in a species hybrid, e.g. *Primula kewensis* (2n = 36) is an amphiploid of *P. floribunda* (2n = 18).

Amphipoda, *n.* an order of crustaceans that includes shrimps and sand hoppers.

amphivasal bundle, *n.* a VASCULAR BUNDLE in which the XYLEM surrounds the PHLOEM, as in some monocotyledons.

amphistylic, *adj.* a type of jaw suspension in which the upper jaw is attached by two points to the neurocranium, as in some cartilaginous fish.

amphoteric, *adj.* (of chemicals) capable of acting either as bases or acids. Amino acids and some metal oxides (such as zinc) are amphoteric.

amplexicaul, *adj.* (of plants) clasping the stem. For example, the bracts of the henbit, *Lamium amplexicaule*, are amplexicaul.

amplification, genetic, *n.* the synthesis of many copies of DNA from one master region, a mechanism that ensures the rapid production of large quantities of protein when required.

ampulla, *n.* any small vesicle or sac-like offshoot, particularly the dilation at the end of the semicircular canal of the EAR, which houses sensory epithelium. Other examples include the internal expansion of the echinoderm tube-foot, and the pit housing the medusoid stage in the calcareous skeleton of *Hydrocorallina*.

amylase, *n.* a digestive enzyme which enables the splitting of starch into smaller subunits of MALTOSE, usually in alkaline conditions (see Fig. 33).

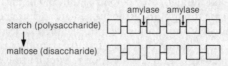

Fig. 33. **Amylase.** The splitting of starch to form maltose.

Amylases are found in the saliva of most mammals (and were previously called *ptyalin*), but their functioning is limited to the brief time that food remains in the mouth. Once food is swallowed, the acid stomach conditions prevent further amylase activity. The principal location of starch digestion is in the duodenum, where *pancreatic amylase* is poured into the gut lumen from the pancreas as part of the pancreatic juice; pancreatic amylase has a greater digestive efficiency than salivary amylase. Amylases are common in plants, particularly in association with the starch stores of seeds and underground perennating organs such as rhizomes, tubers and taproots.

amylopectin, *n.* a carbohydrate polymer, of high molecular weight, composed of branched chains of GLUCOSE units.

amyloplast, *n.* a type of cell inclusion found in many plant tissues, particularly storage organs such as the potato tuber. Amyloplasts contain starch enclosed in a UNIT MEMBRANE, the whole structure being a type of LEUCOPLAST. Besides serving as a starch store, amyloplasts are thought by some scientists to function also as a gravity-seeking device, helping the roots to push through the soil in the correct direction (see GEOTROPISM).

amylose, *n.* a form of starch composed of unbranched chains of GLUCOSE units.

an- or **a-,** *prefix.* denoting not, without.

ana- or **an-,** *prefix.* denoting **1.** up, upwards. **2.** again. **3.** back, backwards.

anabolism or **synthesis,** *n.* a type of metabolism in which complex chemicals are synthesized from simpler building blocks, a process which is endergonic (i.e. requires energy). The classic example of an anabolic process is PHOTOSYNTHESIS in which solar energy is incorporated into complex compounds such as glucose and its derivatives. Animals also carry out many anabolic reactions, building up complex molecules such as proteins from simpler subunits (amino acids) obtained from heterotrophic nutrition (see HETEROTROPH).

anadromous, *adj.* (of fishes such as the salmon) migrating up rivers to spawn in the shallow waters near the source. Compare CATADROMOUS.

anaemia, *n.* a deficiency in the number of red blood cells, their volume, or the haemoglobin content.

anaerobe, *n.* an organism able to metabolize in the absence of free oxygen, obtaining energy from the breakdown of glucose in ANAEROBIC RESPIRATION. Some anaerobes are *obligate*, i.e. they cannot survive in oxygen, e.g. bacteria causing food poisoning (see BOTULISM). Others (the majority) can live either in the presence or absence of oxygen and are called *facultative*. When oxygen is present, respiration in these types is of the aerobic type involving the KREBS CYCLE to release maximum energy; when oxygen is absent they rely solely on energy released in anaerobic respiration.

anaerobic respiration, *n.* a type of cell respiration that takes place in ANAEROBES, and in which energy is released from glucose and other foods without the presence of oxygen. The reactions fall into two stages:

Stage 1: GLYCOLYSIS, in which glucose is converted to two molecules of pyruvic acid (pyruvate) in the general cell cytoplasm. The same reactions occur as in AEROBIC RESPIRATION, but in anaerobes the absence of oxygen prevents the two molecules of reduced NAD produced from being oxidized via the ELECTRON TRANSPORT SYSTEM (ETS) in the MITOCHONDRIA. Instead, ATP is produced from ADP by SUBSTRATE-LEVEL PHOSPHORYLATION. Thus the net output of ATP in anaerobic respiration is 2 molecules (4 minus 2 used in the initial phosphorylation).

Stage 2: once pyruvate has been produced, two alternative pathways can occur. In plants and many microorganisms the pyruvate is broken down to ethanol via ethanal (acetaldehyde) in a process called ALCOHOLIC FERMENTATION, which requires hydrogen from NADH (see Fig. 34 on page 34). In animals the pyruvate is changed via a single step into LACTIC ACID, a process called lactic-acid fermentation, which again requires hydrogen from NADH.

The role of NADH in both fermentations should be noted. Since the amount of NAD present in the cell is limited, glycolysis would quickly come to a halt if anaerobic respiration stopped at pyruvate. By going on to ethanol or lactic acid, NAD is freed in the fermentations to return to glycolysis and thus allow glucose CATABOLISM to continue. The yield of ATP in anaerobic respiration is poor because, firstly, the ETS cannot be

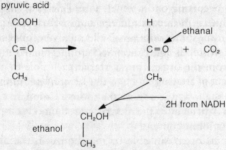

Fig. 34. **Anaerobic respiration.** Breakdown of pyruvic acid to ethanol.

used without oxygen, and, secondly, the end products still contain large amounts of energy. As a result, the free energy released and stored is only a fraction of the amount in the complete oxidation of glucose, as is shown in Fig. 35.

	Anaerobic respiration		Aerobic respiration
	alcoholic fermentation	lactic acid fermentation	
Free energy available from reactions (kJ/mole)	168	198	2874
Number of ATP molecules synthesized	2	2	38
Total energy stored in ATP (kJ)	$2 \times 34 = 68$	$2 \times 34 = 68$	$38 \times 34 = 1292$
Efficiency of storage	40%	34%	45%
$\dfrac{\text{Energy stored}}{\text{Total in glucose}} \times 100$	2.4%	2.4%	45%

Fig. 35. **Anaerobic respiration.** Energy release in anaerobic and aerobic respiration.

anagenesis, *n.* progressive EVOLUTION.

anal, *adj.* pertaining to the anus.

analgesic, *n.* a substance reducing pain without causing unconsciousness.

analogous structure, *n.* a structure that has the same function as another but different origins, e.g. the eye of man and the eye of the octopus. Analogy between structures does not imply evolutionary relationships but may imply CONVERGENCE.

anamniote, *n.* a vertebrate whose embryo lacks an AMNION.

anandrous, *adj.* (of flowers) lacking stamens.

anaphase, *n.* a stage of nuclear division in eucaryotic cells (see

EUCORYOTE), occurring once in MITOSIS and twice in MEIOSIS. The main process involved is the separation of chromosomal material to give two groups of chromosomes which will eventually become new cell nuclei. This important step is controlled by SPINDLE MICROTUBULES (or fibres) which run from the organizing centre at each pole to every chromosome, the point of attachment being the kinetochore of the CENTROMERE (see METAPHASE). Various theories for chromosomal movement have been put forward, including (a) active repulsion of chromosomes, (b) the idea that when sliding past each other the microtubules may act as tiny muscles (the 'sliding filament' theory), and (c) a suggestion that the microtubules are disassembled at the poles, so 'reeling in' the attached chromosomes.

anaphylaxis, *n.* a hypersensitive reaction that can occur after a second exposure to an ANTIGEN. Such anaphylactic 'shock' responses are varied, ranging from reddening and itching of the skin through to respiratory failure and death.

anapsid, *n.* a member of the Anapsida, a group of reptiles in which the skull lacks a temporal FOSSA. The Anapsida include *Seymouria*, tortoises and turtles.

anastomose, *adj.* joining to form loops, interconnecting, as in capillary blood vessels.

anatomy, *n.* the science of the structure of living organisms.

anatropous, *adj.* (of flowering plants) having the OVULE inverted so that the MICROPYLE points towards the PLACENTA.

ancylostomiasis or **ankylostomiasis,** *n.* a human disease in which the lining of the small intestine becomes heavily infested with adults of the hookworm *Ancylostoma*, causing a lethargic, anaemic state.

andro-, *prefix.* denoting male.

androdioecious, *adj.* having male and hermaphrodite flowers on separate plants. Compare ANDROMONOECIOUS.

androecium, *n.* the male parts of an ANGIOSPERM flower composed of two or more stamens which, since they are concerned with reproduction, are called *essential* organs.

androgen, *n.* one of several types of male hormone that stimulate the development and maintenance of the male's SECONDARY SEXUAL CHARACTERISTICS. Naturally occurring androgens (e.g. TESTOSTERONE) are STEROIDS which are produced mainly in the testis (see INTERSTITIAL CELLS), but also to a small extent in the OVARY and ADRENAL CORTEX.

andromonoecious, *adj.* having male and hermaphrodite flowers on the same plant. Compare ANDRODIOECIOUS.

anemophily, *n.* the transfer of pollen from male to female plant organs by means of the wind. The process usually involves cross-pollination between different plants, rather than self-pollination (see POLLINATION). Wind pollination is very wasteful of male gemetes and is fairly rare in ANGIOSPERMS (most having evolved other, more efficient methods in partnership with insects), but common in GYMNOSPERMS. Well-known anemophilous plants include the grasses (a group containing cereals such as maize, wheat and barley), stinging nettles, docks, plantains and many types of pine.

Various features of anemophily can be recognized in angiosperms: flowers small and inconspicuous, often green with reduced petals; flowers with no nectar or perfume; stamens often long and pendulous (e.g. hazel catkins) or projecting up above the herbaceous plant (e.g. plantain, docks) to ensure efficient pollen dispersal; flowers frequently appear early in spring before leaves can interfere with pollination; stigmata are often branched and feathery to catch the airborne pollen. Compare ENTOMOPHILY.

aneuploidy, *n.* a condition where more or less than a complete set of chromosomes is found in each cell of an individual. Compare EUPLOIDY. Typically aneuploids have one extra or one missing chromosome. For example, in DOWN'S SYNDROME affected individuals have three number-21 chromosomes rather than the normal two, a condition known as TRISOMY. See CHROMOSOMAL MUTATION.

aneurism or **aneurysm,** *n.* dilation of an artery wall, a sac formed by abnormal dilation of the weakened wall of a blood vessel.

angiosperm, *n.* any member of the class Angiospermae, in which the seeds are enclosed in an ovary. The class contains the most advanced vascular plants. Each member of the class is either a MONOCOTYLEDON (e.g. grasses, tulips) or a DICOTYLEDON (e.g. apple, primrose). See PLANT KINGDOM for a more complete classification.

angiotensin, *n.* a substance, produced by the action of RENIN on a GLOBULIN protein molecule, that is found in blood plasma and which then stimulates the cortex of the ADRENAL GLAND to release ALDOSTERONE, causing a general constriction of smooth muscle.

angstrom or **angstrom unit,** *n.* a measurement of length, equal to one ten thousandth of a MICRON. Symbol: Å or A. Nowadays, objects of such size are measured in nanometres (10^{-9}m), where $1Å = 0.1$ nm.

angular bone, *n.* in bony fish, reptiles and birds, a membrane bone of the lower jaw; in mammals it has become separated to form the TYMPANIC BONE.

angustiseptate, *adj.* (of a fruit) having the septum crossing the diameter at its narrowest point.

animal kingdom, *n.* a category of living organisms comprising all animals. The classification of animals into major groups is now generally agreed by most biologists, unlike the situation in the PLANT KINGDOM where there are several alternative systems of classification. However, the Protozoa remain problematical since in some modern classifications these organisms are placed with unicellular algae in the Protista, which is given the status of a kingdom. The system below places the Protozoa in the animal kingdom.

Subkingdom Protozoa – single-celled animals
Subkingdom Parazoa – sessile, aquatic, porous animals
   Phylum Porifera – sponges
Subkingdom Metazoa –
   Phylum Coelenterata – two-layered, sac-like animals possessing a MESOGLOEA and stinging cells
      Class Hydrozoa – hydra, colonial hydroids, siphonophores
      Class Scyphozoa – jellyfish
      Class Anthozoa – corals
   Phylum Platyhelminthes – flattened, worm-like animals
      Class Turbellaria – flatworms
      Class Trematoda – flukes
      Class Cestoda – tapeworms
   Phylum Nematoda – roundworms
   Phylum Nemertea – bootlace worms
   Phylum Annelida – segmented worms
      Class Polychaeta – paddle worms
      Class Oligochaeta – earthworms
      Class Hirudinea – leeches
   Phylum Mollusca – soft-bodied animals often possessing a shell
      Class Gasteropoda – snails
      Class Lamellibranchiata – bivalves
      Class Cephalopoda – squids and octopus
   Phylum Arthropoda – jointed-limbed animals with an exoskeleton
      Class Crustacea – water fleas, barnacles, crabs, etc.
      Class Myriapoda – centipedes, millipedes
      Class Arachnida – mites, ticks, scorpions, spiders
      Class Insecta – insects
   Phylum Echinodermata – spiny animals with a penta-radiate symmetry
      Class Asteroidea – starfish
      Class Ophiuroidea – brittle stars
      Class Echinoidea – sea urchins
      Class Holothuroidea – sea cucumbers
      Class Crinoidia – feather stars and sea lilies

Phylum Chordata – animals possessing a notochord at some stage during their development
Subphylum Protochordata – see squirts, *Amphioxus*
Subphylum Vertebrata
  Class Agnatha – jawless fishes, e.g. lamprey
  Class Chondrichythyes – sharks
  Class Choanichthyes – coelacanth, etc.
  Class Actinopterygii – bony fish
  Class Amphibia – salamanders, frogs, toads
  Class Reptilia – reptiles
  Class Aves – birds
  Class Mammalia – mammals

animal pole, *n.* the point on the surface of an animal ovum nearest to the nucleus and normally at the opposite side of the egg to the aggregation of yolk droplets at the VEGETAL POLE. See Fig. 36.

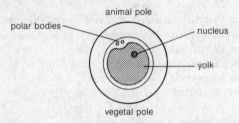

Fig. 36. **Animal pole.** The animal pole in a generalized ovum.

anion, *n.* a negatively charged ion that is attracted to the ANODE during electrolysis.

aniso-, *prefix.* denoting unlike or unequal.

anisogamy, *n.* the state in which the GAMETES are different from each other, i.e. male and female. Usually the former is smaller and more active than the latter. Compare ISOGAMETE.

Anisoptera, *n.* a suborder of the order Odonata containing the larger dragonflies in which the fore and hind wings are of different shapes and the nymphs possess a rectal gill.

ankylostomiasis, see ANCYLOSTOMIASIS.

annealing, see MOLECULAR HYBRIDIZATION.

annelid, *n.* a member of the animal phylum Annelida, a group containing the segmented worms, e.g. earthworms, polychaete worms and leeches. Annelids are characterized by the presence of METAMERIC SEGMENTATION, a COELOM, well-developed blood and nervous systems, and NEPHRIDIA.

annual plant, *n.* any plant that germinates from seed, grows to maturity and produces new seed all within one year or growing season. Since the life-cycle duration is so short, annuals are usually herbaceous rather than woody. Examples include groundsel, shepherd's purse. See BIENNIAL, PERENNIAL.

annual rhythm, *n.* any activity or process that an organism experiences on a yearly basis, e.g. courtship, migration, shedding of leaves.

annual rings, *n.* a series of concentric circles found in the heartwood of trees, indicating the approximate age of the tree. Each ring is formed by the contrast in texture between spring wood and autumn wood. Spring wood in ANGIOSPERMS contains a high proportion of large, thin-walled XYLEM vessels and few fibres for maximum water transport when the leaves are newly formed. Autumn wood contains fewer, smaller, thicker-walled vessels with many fibres for greater mechanical support. GYMNOSPERMS show similar contrasting structures in spring wood and autumn wood, although their water-conducting tissues contain TRACHEIDS, not vessels. Microscopic examination of annual rings (a science called DENDROCHRONOLOGY) has revealed many interesting facts about past climates and also allows wood structures (e.g. building timbers) to be accurately dated.

annular, *adj.* (of plants) ring-shaped, as in annular thickening of XYLEM vessels.

annular thickening, *n.* ring-shaped thickening in the internal wall of XYLEM cells or TRACHEIDS, providing mechanical strength but also allowing longitudinal stretching.

Annulata or **Annelida,** *n.* the ringed or segmented worms. See ANNELID.

annulus, *n.* **1.** the ring of cells in the moss or fern capsule which splits to allow the liberation of spores. **2.** a ring of tissue surrounding the stalk in the fruiting bodies of BASIDIOMYCETE fungi.

anode, *n.* a positively charged electrode to which negatively charged ions move. Compare CATHODE.

anom- or **anomo-,** *prefix.* denoting irregular.

Anopleura, see LOUSE.

anoxia, *n.* lack of oxygen in tissues.

ant- or **anti-,** *prefix.* denoting opposed to, against.

ant, *n.* any insect of the family Formicidae of the order Hymenoptera, characterized by a narrow abdominal construction. All ants are colonial and each colony may contain several castes (specialized individuals). The queen is the only egg-laying individual, fertilization taking place during a nuptial flight involving several colonies. The queen then sheds her wings and starts a colony in which workers are wingless. Social activities include the tending and milking of APHIDS. See Fig. 37 on page 40.

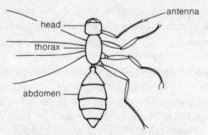

Fig. 37. **Ant.** Winged female ant.

antagonistic muscle, *n*. any muscle which acts in opposition to another, e.g. a contractor to a relaxor or vice versa. See Fig. 38.

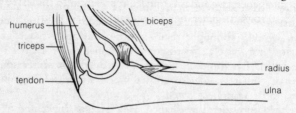

Fig. 38. **Antagonistic muscle.** Antagonistic muscles in the arm, biceps and triceps.

antagonism, *n*. **1.** the inhibiting or nullifying action of one substance or organism on another, e.g. the antibiotic effect of penicillin, or the exhaustion of a food supply by one organism at the expense of another. **2.** the normal opposition between certain muscles (see ANTAGONISTIC MUSCLE).

ante-, *prefix*. denoting before in time or position; in front of.

anteater, *n*. any of a group of ant-eating mammals that includes the spiny anteater, scaly anteater, Cape anteater (aardvark), and marsupial anteater.

antenna, *n*. (*pl*. antennae) usually one of a pair of many-jointed, whip-like structures present on the head of many arthropods, particularly insects (first appendage on head) and crustaceans (second appendage). Antennae have a sensory function, though in some crustaceans they are used for attachment or swimming.

antennal gland, *n*. an excretory organ that occurs at the base of the antennae in some crustaceans.

antennule, *n*. the foremost of two pairs of antennae that occur in some crustaceans; most antennae in other arthropods are homologous with the antennules of crustaceans.

anterior, *n*. **1.** (in animals) that part of the animal which procedes first in forwards movement, usually the head. **2.** (in plants) that part of the INFLORESCENCE which is furthest away from the main stem. Compare POSTERIOR.

anterior root, *n.* the VENTRAL ROOT of a nerve.

antesepalous, *adj.* (of plant parts) positioned opposite the insertion of the SEPALS.

antihelminthic, *adj.* (of a drug) removing parasitic worms when administered to the host.

anther, *n.* that part of the STAMEN of a flower that produces the male GAMETES and that is located at the end of a flexible stalk (the filament). Each anther usually has four 'pollen sacs' in which POLLEN GRAINS (HAPLOID) develop by a process of MEIOSIS. The pollen-grain nucleus divides into generative and tube nuclei, representing a truncated version of the GAMETOPHYTE generation found more clearly in lower plants. See Fig. 39. See also EMBRYO SAC.

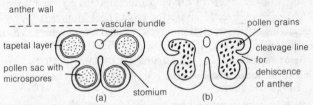

Fig. 39. **Anther.** Transverse sections of (a) young and (b) mature anthers.

antheridium, *n.* (*pl.* antheridia) the male sex organ of bryophytes, pteridophytes, algae and fungi. See Fig. 40.

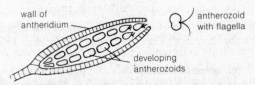

Fig. 40. **Antheridium.** Generalized structure, longitudinal section.

antherozoid or **spermatozoid,** *n.* a mobile male GAMETE having flagella (see FLAGELLUM).

anthesis, *n.* **1.** the opening of a flower bud. **2.** the period from flowering to fruit setting.

antho-, *prefix.* denoting bud.

Anthocerotae, *n.* a class of the division BRYOPHYTA containing the hornworts.

anthocyanins, *n.* a group of water-soluble pigments that are responsible for most reds, purples and blues in plants, particularly in flowers but also in leaves, stems and fruits.

anthozoan or **actinozoan,** *n.* any of the sessile COELENTERATES of the class Anthozoa, including sea anemones, sea pens and corals, that have no MEDUSOID stage (SEE MEDUSA) and possess a body cavity that is more complex than in other coelenterates.

anthrax, *n.* a fever of the spleen in cattle and sheep caused by toxins released from the bacterium *Bacillus anthracis*. The disease can spread to humans when infected animal products such as wool and bristles are handled, giving rise to malignant skin lesions and pustules.

anthropo-, *prefix.* denoting man or human.

Anthropoidea, *n.* a suborder of the primates that includes monkeys, anthropoid apes and humans. The group is subdivided into Old-World forms (Catarrhini) – for example, baboon, chimpanzee, man – and New-World forms (Platyrrhini) – for example, marmoset, howler monkey. Old-World forms have parallel nostrils directed downwards and two premolar teeth, and often have cheek pouches; New-World forms have nostrils diverging anteriorly, three premolar teeth, and no cheek pouches.

anthropomorphism, *n.* the attribution of human characteristics to animals other than man.

anti-, see ANT-.

antiauxin, *n.* any chemical substance that prevents the action of AUXINS.

antibiotic, *n.* any substance produced by a microorganism that even in low concentrations can inhibit or kill other microorganisms. For example, PENICILLIN produced by the fungus *Penicillium notatum* prevents the reproduction of many bacteria by preventing cell-wall synthesis. Antibiotics are frequently the products of 'secondary' metabolism in that, while not of major importance, their formation presumably offers a selective advantage to the organism. The amount of antibiotic produced per gram of producer can be greatly enhanced by optimal culturing conditions and strong selection pressure over many generations. Unfortunately, most antibiotics are not lethal to viruses. Furthermore, continued use of an antibiotic against a generally susceptible strain of bacteria will favour survival of the few resistant members of the bacterial population, resulting eventually in an antibiotic-resistant strain.

antibody, *n.* a type of protein called an IMMUNOGLOBULIN that reacts with a specific ANTIGEN, and serves as part of a vital body-defence mechanism. Various reactions can occur between antigen and antibody. If the antigen is a TOXIN (as in snake venom, or as produced by bacteria causing, for example, botulism and tetanus, neutralizing antibodies are called *antitoxins*. If the antigen is adhering to the surface of a cell, antibodies called *agglutinins* cause clumping or AGGLUTINATION of cells, while another antibody type (*lysins*) cause disintegration or LYSIS of the cell in conjunction with COMPLEMENT. Other antibodies (*opsonins*) facilitate uptake of antigens by PHAGOCYTES in the blood, while *precipitins* cause soluble antigens to precipitate.

Antibodies are produced in the lymphoid tissues of the body, e.g. LYMPH NODES, by a type of LYMPHOCYTE called B-CELLS. Most antibodies are produced during exposure to an antigen, such a response being termed *active immunity*. Specific antibodies to rare or synthetic antigens can be manufactured as easily as those to common antigens. A few antibodies are produced even without the apparent presence of the appropriate antigen. Such 'natural' antibodies include several involved in blood grouping, e.g. A and B antibodies in the ABO BLOOD GROUP. Young mammals have limited capacity to produce antibodies in the first few weeks of life, but can obtain some *passive immunity* by receiving maternal antibodies via the mother's milk. This fact has been used to encourage human mothers to breast-feed their infants rather than bottle-feed. Most antibodies circulate in the blood and other body fluids, but most if not all body secretions also contain antibodies, mainly of the IgA type.

anticlinal, *adj*. (of plant cell division) along a plane at right angles to the outer surface.

anticoagulant, *n*. a substance that hinders AGGLUTINATION or clotting of blood cells. BLOOD CLOTTING is a vital reaction to damage in the circulatory system, one of the steps in the process being dependent on the presence of calcium ions. However, the tendency of blood to clot can be overcome by suitable chemicals that remove calcium, such chemicals being used as anticoagulants when storage of blood is required. Many blood-sucking parasites such as mosquitoes secrete anticoagulants into the host's bloodstream, so enabling easy transport of blood into the parasite's digestive system.

anticodon, see TRANSFER RNA.

antidiuretic hormone, see ADH.

antigen, *n*. a substance (usually a protein or carbohydrate) that, when introduced into the body, induces an IMMUNE RESPONSE which includes the production of specific ANTIBODIES. Antigens can be toxins (as in snake venom) or molecules on cell surfaces (e.g. A/B antigens on red blood cells).

antigiberellin, *n*. any substance that causes the growth of short thick stems, i.e. that has the opposite effect to GIBBERELLINS. For example, maleic hydrazide is used to retard grass growth and thus reduce the frequency of cutting.

antihistamine, see ALLERGY.

antilymphocytic serum, *n.* a serum produced by injecting LYMPHOCYTES into a horse and collecting the antibodies produced. The serum is capable of destroying the original lymphocytes and allowing grafting of tissues to the animal so treated from another.

antimycin, *n.* a poison that blocks the flow of electrons from cytochrome b to cytochrome c in the ELECTRON TRANSPORT SYSTEM (see Fig. 141).

antipodal cells, *n.* the three haploid cells that remain at the non-micropylar end of the EMBRYO SAC subsequent to division of the cells. After the formation of the egg cell they play little or no further part in seed development.

antisepsis, see ANTISEPTIC.

antiseptic, *n.* literally, a substance that counteracts putrification; more normally, a substance that prevents infection in a wound. *Antisepsis* is carried out by disinfection or sterilization using non-toxic, non-injurious substances, and has the effect of killing or inactivating microorganisms which cause infection.

antiserum, see SERUM.

antisporulant, *n.* a substance that reduces or prevents spore production by a fungus while enabling vegetative growth to continue.

antitoxin, *n.* a type of ANTIBODY that neutralizes TOXINS.

anucleate, *adj.* without a nucleus or nuclei.

anuran, *n.* any amphibian of the order Anura, such as the batrachians (including frogs and toads), which are characterized by large hind legs and the lack of a tail.

anus, *n.* the posterior opening of the ALIMENTARY CANAL, often surrounded by a ring of muscle called the anal SPHINCTER, which is under nervous control (see AUTONOMIC NERVOUS SYSTEM) and which regulates the release of gut contents to the outside (DEFECATION).

anvil, *n.* **1.** the INCUS. **2.** any stone used by a song thrush to break open land snails such as *Cepaea nemoralis*. Collecting shells at such a site can show which colour forms of the snails are selected by the searching thrushes.

aorta, *n.* the 'great artery' of the mammalian BLOOD CIRCULATORY SYSTEM. It carries oxygenated blood from the left ventricle of the heart around the AORTIC ARCH and along the *dorsal aorta* which runs the length of the trunk, giving rise to several branches to individual body organs. The ventral aorta is the main artery in fish and lower chordates which carries blood from the heart to the gills.

aortic arch, *n*. one of a series (up to six) of paired arteries that join the ventral and dorsal aorta in vertebrates or their embryos. In fish the arches form the efferent and afferent branchial vessels, and in higher vertebrates the carotid, systemic and pulmonary arches.

aortic body, *n*. an area of nervous tissue in the wall of the AORTA, close to the heart, which is sensitive to small changes in blood CO_2 levels. Excess partial pressure of CO_2 triggers nerve impulses from the aortic body to cardiac/respiratory centres in the hindbrain via afferent nerve fibres (see BREATHING). A drop in CO_2 pressure causes inactivation of the aortic body (an example of a negative FEEDBACK MECHANISM).

ap- or **aph-** or **apo-,** *prefix*. denoting away from, apart.

ape, *n*. any tailless primate of the family Simiidae, including gorillas, chimpanzees, orang-utans and gibbons.

apetalous, *adj*. (of plants) without petals.

aph-, see AP-.

aphan-, *prefix*. denoting obscure.

Aphaniptera or **Siphonaptera,** *n*. the insect order containing the fleas, whose members are laterally compressed, blood–sucking forms which lack wings but have well developed hind legs modified for jumping.

Aphetohyoidea or **Placodermi,** *n*. a group of fossil fish possessing a primitive jaw suspension (ligaments only) and a functional pair of first gill slits.

aphid, *n*. any member of the family Aphididae, order Hemiptera. Commonly called *green fly*. Aphids feed on plant juices by using piercing and sucking mouthparts and are of considerable economic importance because they can act as vectors of plant viruses. See Fig. 41.

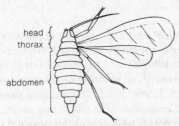

Fig. 41. **Aphid.** Winged female.

aphotic, *adj*. characterized by or growing in the absence of light.

aphotic zone, *n*. the zone of an ocean or sea into which light does not penetrate and in which, therefore, photosynthesis cannot take place. It is usually below about 100 m. See also SEA ZONATION.

aphyllous, *adj*. without leaves.

apical, *adj*. of, at, or being the apex.

apical dominance, *n.* a condition in plants where the stem apex prevents the development of side shoots from lateral buds near the apex. The dominance is controlled by the presence of high concentrations of plant hormones (AUXINS) at the apex, produced by the *apical bud*. Further down the stem the auxin concentration is reduced and strong lateral branches are formed. Removal of the apical bud causes lateral branches to develop (e.g. after pruning rose bushes).

apical growth, *n.* growth resulting from an APICAL MERISTEM.

apical meristem, *n.* a region of dividing tissue at the growing tips of roots and shoots that is responsible for increase in length of the plant body. MERISTEM cells are small, with thin walls and CYTOPLASM containing few vacuoles. They undergo active MITOSIS when growth is occurring (depending on the season) and the new cells look similar to the parental ones but become more vacuolated and elongated. Eventually, further away from the meristem, the cells undergo the process of CELL DIFFERENTIATION, taking on the appearance of mature plant cells.

Meristematic tissues are delicate and in need of protection. In shoots the apices are covered by tiny leaves forming apical buds at the tips of the branches and axillary buds at the nodes where the leaves join the stem. Root structures are less complex, each root-tip meristem being protected by a shield of loose cells called a ROOT CAP. Compare CAMBIUM.

apical region, *n.* a tip of a root or shoot in a plant where growth in length takes place by activity of the APICAL MERISTEM.

aplanospore, *n.* an asexual, nonmotile spore found in some algae and some fungi.

apnoea, *n.* the absence or suspension of breathing, as occurs in diving mammals.

apo-, see AP-.

apocarpous, *adj.* (of an ovary of a flowering plant) having the CARPELS free from each other. Compare SYNCARPOUS.

Apoda or **Gymnophiona,** *n.* an order of worm-like, burrowing amphibians (caecilians) that lack limb girdles and limbs. They possess small functionless eyes and are found in SE Asia, India, Africa and Central America.

apodeme, *n.* an inward projection of the exoskeleton of arthropods that serves as a site for muscle attachment.

apodous, *adj.* lacking legs.

apogamy, *n.* a form of asexual reproduction found in ferns where the GAMETOPHYTE, which is diploid, gives rise directly to the SPOROPHYTE.

apomict, *n.* any plant which is produced by APOMIXIS.

apomixis, *n.* the development of an embryo in plants without FERTILIZATION or MEIOSIS. This method of pseudosexual reproduction is regularly used in plants such as garlic and the citruses. Similar processes occur in animals, where the term used is PARTHENOGENESIS. See also APOSPORY, APOGAMY and AGAMOSPERMY.

apomorph, *n.* a derived character.

apophysis, *n.* a PROCESS of vertebrate bones to which muscles, tendons or ligaments are attached.

apoplast, *n.* **1.** a PLASTID which lacks CHROMATOPHORES. The adjective *apoplastic* is applied to individual protozoans that lack colour in a group which is generally coloured. *Apoplasty* occurs when cell division is so fast as to outpace plastid division, producing individuals which are formed without plastids. **2.** those areas of the plant that are outside the SYMPLAST, comprising the parts outside the PLASMALEMMA, such as cell walls and the dead tissues of XYLEM.

aporepressor, *n.* a regulatory protein that inhibits the activity of specific genes when in the presence of a COREPRESSOR.

aposematic coloration, see WARNING COLORATION.

aposematic selection, *n.* a type of FREQUENCY–DEPENDENT SELECTION operating on a species that has several morphological forms, at least one of which is a mimic of a different 'model' species that is distasteful. See MIMICRY.

apospory, *n.* a situation occurring in some plants where MEIOSIS is omitted and an ordinary diploid SPOROPHYTE cell gives rise to the spore which produces a diploid GAMETOPHYTE.

apostatic selection, *n.* a type of FREQUENCY–DEPENDENT SELECTION occurring in a species with several morphological forms, and classically used in relation to prey species that are visually POLYMORPHIC. The amount of predation of each type is directly related to the relative MORPH frequencies.

apothecium, *n.* the fruiting body of some ascomycete fungi (cup fungi) and lichens, containing asci (see ASCUS). Often lightly coloured, they can reach up to 40 cm in diameter, but are usually only a few millimetres across.

appeasement display, *n.* behaviour designed to pacify or make peace following a display of aggressive behaviour. For example, a defeated dog will turn on its back and present its underbelly to the victor, who then ceases combat.

appendage, *n.* any projection from the body of an animal, e.g. legs, mouthparts, antennae.

appendicularia or **tadpole larva,** *n.* the larva of tunicates (ASCIDIACEA). It possesses a tail, NOTOCHORD and tubular nerve cord, and is indicative of the evolutionary relationships of the adults.

appendicular skeleton, *n.* the part of the skeleton attached to the vertebral column, i.e. the limbs or fins.

appendix or **vermiform appendix,** *n.* a small fingerlike projection from the tip of the CAECUM in the mammalian gut. See Fig. 42. In herbivores the caecum and appendix are important, containing bacteria which facilitate the digestion of cellulose. In humans and other non-herbivores the caecum and appendix have no function. The appendix can become infected and must be removed surgically (*appendectomy*) if in this condition.

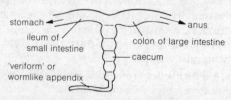

Fig. 42. **Appendix.** Location of appendix.

apposition, *n.* growth in cell-wall thickness brought about by the successive deposition of layers of material.

apposition image, *n.* the mosaic image formed in a compound eye (see EYE, COMPOUND).

appressed, *adj.* (of an organ) close to another but not united with it.

Apterygidae, *n.* the kiwis, primitive flightless birds of New Zealand. They are nocturnal and the feathers are modified to appear hair-like.

Apterygota or **Ametabola,** *n.* a subclass of primitive insects that have never evolved wings, and that undergo little or no METAMORPHOSIS. There is little change in form from INSTAR to instar. The subclass includes silverfish (Thysanura), springtails (Collembola), and Protura. Compare PTERYGOTA.

aptosochromatosis, *n.* the process of colour change without moult in birds (e.g. greenshank), in which fully formed feathers change colour without abrasion.

aquaculture, *n.* the manipulation of the reproduction, growth rates and mortality of aquatic organisms useful to man to enhance their yield; the aquatic equivalent of AGRICULTURE.

aquatic respiration, *n.* a process in which freshwater and marine organisms carry out gas exchange with the water that surrounds them. Fully saturated water contains only about 0.02% as much oxygen per unit volume as air. However, since the tissues are in direct contact with the gas-carrying medium flowing past, the system is very efficient and often a relatively small respiratory surface is required.

Many aquatic organisms (particularly the less advanced groups) have no special respiratory organs, relying on a large surface area for gas exchange (e.g. protozoa, algae, nematodes). Others have developed specialized structures which are either external (e.g. gills of the ragworm *Nereis*), or internal (e.g. gills of fish). GILLS are richly supplied with blood vessels for transport to and from the body tissues, and many organisms with gills expend large quantities of energy creating a regular flow of water over the respiratory surfaces.

aqueous habitat, *n.* any habitat in which water is the medium in which the organisms live.

aqueous humour, *n.* fluid that fills the space between the cornea and the lens of the vertebrate EYE (see Fig. 151).

arable farming, *n.* the cultivation of land to produce crops (as compared with the growing of livestock).

arachnid or **arachnoid,** *n.* any member of the class Arachnida in the phylum Arthropoda. The class contains scorpions, spiders, harvestmen, ticks, mites and king crabs. They lack antennae, usually possess four pairs of walking legs, and, excepting *Limulus* (king crab), are air-breathing.

arachnoid, *n.* **1.** the middle of the three membranes (see MENINGES) that cover the brain and spinal cord. **2.** see ARACHNID.

Araneida, *n.* an order of the class Arachnida, containing the spiders.

arboreal, *adj.* appertaining to trees.

arbovirus, *n.* any virus that is ARTHROPOD-borne, e.g. the yellow-fever virus carried by the mosquito *Aedes aegypti*.

arch- or **archaeo-** or **arche-** or **archi-,** *prefix.* denoting ancient.

Archaeopteryx, *n.* the earliest known fossil bird (genus *Archaeopteryx*) found in the Jurassic period about 140 million years BP. The bird is characterized by the presence of teeth in the jaws, claws on the wing and a tail containing vertebrae.

Archaeornithes, *n.* the order of extinct reptile-like birds that contains ARCHAEOPTERYX (the only known genus).

arche-, see ARCH-.

Archegoniatae, *n.* a nonsystematic grouping of primitive plants, e.g. BRYOPHYTA, in which the female reproductive organ is an ARCHEGONIUM.

archegonium, *n.* (*pl.* archegonia) the female sex organ of BRYOPHYTA, PTERIDOPHYTES and most GYMNOSPERMS. See Fig. 43 on page 50.

archenteron, *n.* that part of the developing embryo of animals (BLASTULA) formed by an invagination of MESODERM and ENDODERM cells into the blastocoel which ultimately forms the gut. The archenteron opens to the exterior via the blastopore. See Fig. 44 on page 50.

archesporium, *n.* the cell or group of cells from which SPORES are derived.

archetype, *n.* the hypothetical ancestral type from which other forms are

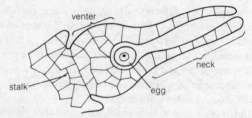

Fig. 43. **Archegonium.** Mature archegonium of *Marchantia* (Hepaticae).

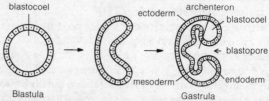

Fig. 44. **Archenteron.** Formation of the archenteron in an amphibian embryo.

thought to be derived; it usually lacks specialized characteristics.

archi-, see ARCH-.

Archiannelida, *n.* a class of annelid worms, possibly evolved from the same stock as the POLYCHAETE worms; all are marine and of simplified structure.

Archechlamydeae, *n.* a subclass of the ANGIOSPERMS where the PERIANTH is incomplete or the parts of the COROLLA are entirely separate.

arci-, *prefix.* denoting bow-shaped.

arcuate, *adj.* curved in the form of a bow to the extent of a quadrant of a circle or more.

areolar tissue, *n.* connective tissue consisting of a gelatinous matrix that includes large fibroblasts which synthesize COLLAGEN white fibres and yellow elastic fibres, ameoboid MAST CELLS which secrete the matrix, fat-filled cells, and phagocytic MACROPHAGES. Areolar tissue occurs beneath skin and between adjacent tissues and organs. See Fig. 45.

arginine, *n.* one of 20 AMINO ACIDS common in proteins. It has an extra basic group, and is alkaline in solution. The ISOELECTRIC POINT of arginine is 10.8. See Fig. 46.

arginine phosphate, see PHOSPHAGEN.

arista, *n.* **1.** a bristle at the base of the antenna in insects. In some types, e.g. syrphids, it is larger than the rest of the antenna. **2.** a bristle-like structure in the flowering GLUMES of grasses (an AWN).

aristate, *adj.* having an ARISTA (2).

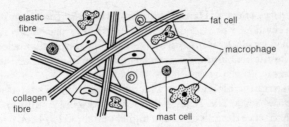

Fig. 45. **Areolar tissue.** Generalized structure.

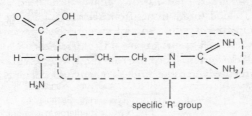

specific 'R' group

Fig. 46. **Arginine.** Molecular structure.

Aristotle's lantern, *n*. a five-sided, globular structure supporting the mouth and jaws of sea urchins.

arithmetic mean or **mean,** *n*. a number (symbol: $\bar{x}$) that is computed by calculating the sum of a set of numbers ($\sum x$) and dividing the sum by the number of terms (n). See Fig. 47.

$$
\begin{array}{r}
\bar{x}\ values \\
13 \\
9 \\
12 \\
\underline{10} \\
\sum \bar{x} = \underline{44}
\end{array}
$$

Fig. 47. **Arithmetic mean.** In this example, the mean of 4 values of $\bar{x}$ is 11.

Arnon, Daniel, (1910–) American biochemist. With his co-workers in the 1950s, he first discovered electron passage from one electron-acceptor to another in the ELECTRON TRANSPORT SYSTEM of PHOTOSYNTHESIS, particularly the role of ferredoxin.

arsenic, *n*. a chemical element in the form of a grey metal, more familiar in the extremely poisonous form of arsenious trioxide. It was at one time used in arsenical soap for the preservation of animal and bird skin in museums, so that particular care should be taken in handling old museum skins.

artefact, *n*. something that appears during preparation or examination of material which is not present in the natural state. Two scientists from the

University of Surrey, Harold Hillman and Peter Sartory, have suggested on the evidence provided by solid geometry, that some structures described by electron microscopy, e.g. Golgi apparatus, nuclear pores, endoplasmic reticulum, are artefacts of the preparation of material.

Artenkreis, *n.* a SUPERSPECIES.

arterial arch, *n.* blood vessels which link the lateral dorsal AORTA with the ventral aorta. Six arches are present in vertebrate embryos and 4–6 remain to serve the gills in fish. In higher vertebrates 3, 4 and 6 remain as the carotid, systemic and pulmonary arches.

arteriole, *n.* any of the narrow, thin-walled arteries that carry oxygenated blood to the tissues from the heart, forming part of the BLOOD CIRCULATORY SYSTEM. Arterioles progressively get smaller as they approach the target tissue, giving off side branches which become capillaries. Some arterioles connect up directly to VENULES via SHUNT VESSELS, particularly in tissues with a constant blood supply such as the skin. See Fig. 48.

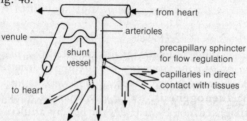

Fig. 48. **Arteriole.** Arterioles, capillaries and venules.

arteriosclerosis, *n.* a pathological condition or any of a group of diseases in which there is an increase in the thickness of the arterial walls, a reduction in elasticity of the vessel, and a constriction of diameter which affects the blood flow; the classic 'hardening of the arteries' of the elderly.

artery, *n.* any of the large vessels with thick, elasticated muscular walls that carry oxygenated blood to the tissues from the heart, forming part of the BLOOD CIRCULATORY SYSTEM. See Fig. 49. Arteries become less massive as the vessels approach their target organs, eventually becoming reduced to ARTERIOLES. See VEIN.

arthro-, *prefix.* denoting a joint.

arthropod, *n.* any member of the animal phylum Arthropoda, containing jointed-limbed organisms that possess a hard EXOSKELETON. It is the largest phylum in terms of the numbers of species and includes insects, crustaceans, spiders, centipedes and many fossil forms.

articul- or **articulo-,** *prefix.* denoting joint.

articular bone, *n.* a bone present in the lower jaw of birds and reptiles

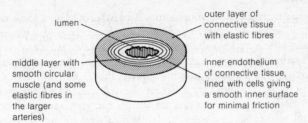

lumen

outer layer of
connective tissue
with elastic fibres

middle layer with
smooth circular
muscle (and some
elastic fibres in
the larger
arteries)

inner endothelium
of connective tissue,
lined with cells giving
a smooth inner surface
for minimal friction

Fig. 49. **Artery.** Transverse sections.

that forms the joint with the skull (*quadrate*). In mammals it forms the
MALLEUS.

articulate, *vb.* to connect by means of a joint.

articulation, see JOINT.

artificial classification, see CLASSIFICATION.

artificial insemination (AI), *n.* the introduction of spermatozoa into
the reproductive tract of a female animal without COPULATION in order
to achieve fertilization. The method is used routinely in animal breeding
to transfer genetic material over large distances and to achieve many
fertilizations in different females from one ejaculation of semen. In the
case of humans, AI is used sometimes in attempts to achieve fertilization
where normal methods have failed.

artificial parthenogenesis, *n.* the development of an egg into an
embryo stimulated not by fertilization but by artificial means, e.g.
cooling, treating with acid, mechanical damage.

artificial respiration, *n.* the maintenance of breathing by artificial
means. Examples include mouth-to-mouth resuscitation, physical
compression and distention of the thorax, the use of a respirator.

artificial selection, *n.* a SELECTION process in which man chooses
particular organisms from which to breed, based upon their PHENOTYPE,
so aiming to alter the average GENOTYPE and phenotype of the resulting
progeny (a form of DIRECTIONAL SELECTION). This process is carried out
by plant and animal breeders whose job is to enhance certain features of
the organisms with which they work, e.g. greater resistance to root rot
in tomatoes, or higher milk yield in cattle. Such selection depends upon
the presence of GENETIC VARIABILITY in the chosen population. See also
HERITABILITY.

artio-, *prefix.* denoting an even number.

Artiodactyla, *n.* the even-toed ungulates (i.e. animals possessing hoofs
with two or four toes), such as cattle, pigs, deer and camel. Compare
PERISSODACTYLA.

arum lily, *n.* the plant *Arum maculatum*. It is a remarkable example of

COEVOLUTION in that the shoot is modified to form a large, foul-smelling BRACT which is attractive to small flies. The bract (or *spathe*) is swollen at its base, which encloses the male and female flowers. Once attracted by the spathe, flies carrying pollen enter the swollen bract chamber searching for nectar, and so pollinate the female flower. CROSS-POLLINATION is ensured by differential maturation times for male and female flowers; when the flies leave they carry pollen from the male flowers which have ripened while the flies were in the chamber.

Ascaris, *n.* a genus of parasitic roundworm (phylum Nematoda) that occasionally inhabits the small intestine of man. *A. lumbricoides* can reach a length of 30 cm and when in large numbers can block the gut and kill the host.

ascending, *adj.* sloping or curving upwards, e.g. the ascending limb of the loop of Henle in kidney tubules.

ascidian, *n.* any PROTOCHORDATE of the class Ascidiacea (subphyllum Tunicata), such as the sea squirt, the adults of which are degenerate and sedentary. See also APPENDICULARIA.

asco-, *prefix.* appearing in the shape of a leather wine bottle.

ascocarp, *n.* the fruiting body of an ASCOMYCETE fungus. See ASCUS.

ascomycete, *n.* any fungus of class Asctomycetes (now included in the subdivisions Ascomycotina and Deuteromycotina) characterized by the formation of ascospores during sexual reproduction. The hyphae are septate and the asci (see ASCUS) are grouped in a visible fruiting body.

ascorbic acid or **vitamin C,** *n.* a water-soluble organic compound, found in citrus fruits, green vegetables and tomatoes. Formula: $C_6H_8O_6$. Its best-known role is in the formation of COLLAGEN, the chief component of CONNECTIVE TISSUE. Deficiency in ascorbic acid results in the disease called SCURVY which is characterized by changes in collagen leading to many further effects. Ascorbic acid acts as a VITAMIN in man, but rats and most other mammals can manufacture their own supply from D-glucose.

ascospore, *n.* a HAPLOID spore resulting from MEIOSIS in the ASCUS within the fruiting body of ASCOMYCETE fungi.

ascus or (formerly) **theca,** *n.* (*pl.* asci) a cell present in the fruiting body of ASCOMYCETE fungi in which the fusion of HAPLOID nuclei occurs during sexual reproduction. This is normally followed by MEIOSIS, giving rise to four haploid cells, after which MITOSIS produces eight ASCOSPORES. The precise arrangement of ascospores within the ascus enables the events at meiosis to be fully analysed (see TETRAD ANALYSIS). The asci are usually enclosed within an aggregation of hyphae termed an ASCOCARP, a number of different types being recognized, e.g. perithecium, cleistothecium, apothecium. See Fig. 50.

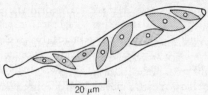

Fig. 50. **Ascus.** Containing ascospores of *Neurospora crossa*.

aseptic techniques, *n.* techniques performed in an environment which is as free of disease-causing microorganisms as possible, e.g. in a hospital operating theatre. To achieve this end the atmosphere is filtered, all instruments and outer clothing are sterilized (usually with high temperatures in an AUTOCLAVE) and the operators wear face masks and gloves. Such techniques should not be confused with the use of ANTISEPTICS.

asexual reproduction, *n.* a process by which organisms multiply without the formation and fusion of specialized sex cells (GAMETES). Each feature of asexual reproduction has its advantages and disadvantages. See Fig. 51.

Feature	Advantage	Disadvantage
1. Only involves one individual.	No need to find a partner.	No chance of mixing genes to produce a new combination.
2. Produces individuals which are identical with parent.	When the environment is constant a genetically favoured type will spread rapidly.	Genotype may not be able to adjust to a change in environment.
3. Progeny often produced in large numbers near parent.	Environment is suitable since parent survives.	Overcrowding may result. Widespread dispersal is difficult.

Fig. 51. **Asexual reproduction.** Advantages and disadvantages of asexual reproduction.

The genetically identical products of asexual reproduction are called CLONES. Asexual reproduction is a feature of lower animals and occurs in all groups of plants, including ANGIOSPERMS, usually in addition to sexual reproduction. Perhaps the fixed location of many plants makes this type of reproduction particularly suitable. There are several types of asexual reproduction. These are dealt with under their own headings but are summarized here:

(a) fission (see BINARY FISSION) where the entire organism splits into two (e.g. bacteria).

(b) BUDDING: new individuals produced as outgrowth of the parent (e.g. yeast, *Hydra*).

(c) FRAGMENTATION: mechanical separation of the plant body into segments (some algae).

(d) SPORULATION: producing asexual spores (e.g. fungi).

(e) VEGETATION PROPAGATION: parts of higher plants producing new whole organisms (e.g. potato tubers).

asexual spore or **isopore**, *n*. a spore which does not arise through sexual reproduction.

asparagine, *n*. one of 20 AMINO ACIDS common in proteins. It has a polar 'R' structure and is water soluble. See Fig 52. The isoletric point of asparagine is 5.4.

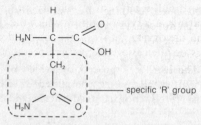

Fig. 52. **Asparagine.** Molecular structure.

aspartic acid or **aspartate**, *n*. one of 20 AMINO ACIDS common in proteins. It has an extra carboxyl group and is therefore acidic in solution. The isoelectric point of aspartic acid is 2.8 See Fig. 53.

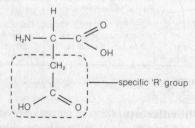

Fig. 53. **Aspartic acid.** Molecular structure.

aspect, *n*. the direction in which an object faces. For example, a north-facing slope has a northern aspect.

asperous, *adj*. rough to the touch.

asphalt lake, *n*. a pool formed largely of bitumen (almost entirely carbon and hydrogen, but some oxygen, nitrogen and sulphur) found mainly in southern USA and South America, e.g. La Brea tar pits in which trapped animals have been preserved since the Miocene period.

asphyxia, *n*. suffocation, lack of oxygen.

aspirin or **acetylsalicylic acid,** *n*. an analgesic that relieves pain without loss of consciousness.

assimilation, *n*. the intake by organisms of new materials from the outside and their incorporation into the internal structure of the organism.

association, *n*. a small natural grouping of plants which live together. The term originally denoted a whole habitat, e.g. a pine forest, but is now usually applied to much smaller groupings. Compare CONSOCIATION.

association centre, *n*. an area of nervous coordination in invertebrates, that distributes stimuli received from sensory receptors.

assortative mating, *n*. a type of cross in which the choice of partner is affected by the GENOTYPE, i.e. mating is nonrandom. For example, human matings are often assortative with respect to racial features since persons of one racial group may tend to have children by partners of the same group. This choice of mates means that genotypic frequencies predicted by the HARDY-WEINBERG LAW may not be found. See also RANDOM MATING, INTERBREED.

aster- or **astero-,** *prefix*. denoting a star.

aster, *n*. a group of blind-ending SPINDLE MICROTUBULES radiating out from the CENTRIOLES of dividing cells in lower plants and all animals. The aster function is uncertain, but does not seem to be concerned with spindle formation. See Fig. 54.

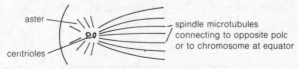

aster

spindle microtubules connecting to opposite pole or to chromosome at equator

centrioles

Fig. 54. **Aster.** Generalized form.

Asteroidea, *n*. a class of star-shaped ECHINODERMS which have a basic pentaradiate shape (a five-sided RADIAL SYMMETRY), e.g. starfish.

asynchronous fibrillar muscle, see FLIGHT.

atavism, *n*. the recurrence of a characteristic possessed by an ancestor after an absence for several generations.

athlete's foot or **tinea** or **ringworm,** *n*. a fungal disease caused by *Epidermophyton floccosum*. The fungus can cause irritation of other parts of the body apart from the feet and is parasitic or pathogenic on nails and skin in general. Athlete's foot is most common in adolescent males and infection is caused usually by walking barefoot on infected floors.

atlas, *n*. the first vertebra of TETRAPODS.

atmometer, *n*. an instrument for measuring the rate of evaporation of water into the air. This is simply carried out by measuring the level of water in a calibrated cylinder at frequent intervals.

atmosphere, *n.* the gaseous envelope surrounding a particular structure, or the gaseous content of a given structure or container.

atom, *n.* the smallest particle of matter possessing the properties of an element.

atomic weight, *n.* the weight of an atom of an element in relation to hydrogen, which is considered as one.

ATP (adenosine triphosphate), *n.* a molecule consisting of ADENINE and ribose sugar onto which are attached three phosphate groups, two being joined by high-energy bonds (see Fig. 55). HYDROLYSIS of these special bonds results in the release of energy. ATP molecules are formed in two main processes, both involving the addition of inorganic phosphate to ADP via a high energy bond:

$$ADP + P + 34 \text{ kJ energy} = ATP + H_2O$$

ATP can be formed during CELLULAR RESPIRATION, either in the general cytoplasm during GLYCOLYSIS or in the MITOCHONDRIA via the KREBS CYCLE and the ELECTRON TRANSPORT SYSTEM if oxygen is present. ATP is also formed during photosynthesis in the CHLOROPLASTS of green plants, again using an electron transport system. ATP molecules act therefore as short-term 'biological batteries', retaining energy until required for such processes as active transport, synthesis of new materials, nerve transmission, and muscle contraction. An active cell requires more than two million molecules of ATP per second to drive its biochemical machinery.

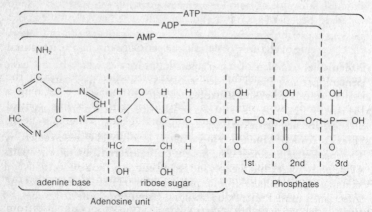

Fig. 55. **ATP.** The structure of ATP, ADP and AMP.

atrioventricular node (AVN), *n.* see HEART.

atrium, *n.* (*pl.* atria) a passage or chamber in the body, particularly one or more of the chambers in the heart receiving blood from the body. All vertebrates except fish have hearts with two atrial cavities. See Fig. 56. See also HEART and HEART, CARDIAC CYCLE.

Fig. 56. **Atrium.** Primitive heart.

atrophy, *n.* the reduction in size of an organ or tissue mass, often after disuse.

atropine, *n.* a poisonous chemical obtained from the deadly nightshade that prevents depolarization of the postsynaptic membrane and therefore prevents synaptic transmission, in a similar way to CURARE. It is used medicinally in preanaesthetic medication and to treat peptic ulcers, renal and biliary colic, etc.

attached-X chromosome, *n.* a single chromosome formed by the joining of two X-chromosomes at a common CENTROMERE. Such a chromosome acts as a single entity so that a carrier has a double X complement plus another sex chromosome, a form of NON-DISJUNCTION.

attachment, *n.* an enlargement of the base of an algal THALLUS by which the plant is anchored to the SUBSTRATE.

attenuate, *adj.* (of a leaf or plant structure) gradually tapering to a long slender point.

attenuation, *n.* the loss of virulence of a microbial pathogen so that although still alive, it is no longer pathogenic. In this state it is able to stimulate beneficial ANTIBODY production when used as a vaccine. Various procedures are used to attenuate the virulence of a pathogen, e.g. the ageing of cultures, or passing the pathogen through an unnatural host.

attenuator, *n.* a regulatory NUCLEOTIDE sequence occurring in the LEADER REGION of an operon (see OPERON MODEL) and containing a TRANSCRIPTION stop signal. The attenuator acts as a fine control mechanism for the activity of structural genes, sometimes being operative, sometimes not, depending on conditions in the cell.

auditory canal or **external auditory meatus,** *n.* the canal leading from the PINNA (outer ear) to the ear drum.

auditory capsule, *n.* the bony or cartilaginous capsule that encloses the middle and inner parts of the ear in vertebrates.

auditory nerve, *n.* the 8th cranial nerve. It carries sensory impulses from the inner ear.

auditory organ, *n.* the sense organ that detects sound. In vertebrates it also provides information about the animal's relationship to gravity and its acceleration.

Auerbach's plexus, *n*. a network of nerve fibres and ganglia in the gut musculature, bringing about PERISTALSIS when stimulated by pressure of food in the gut.

auricle, *n*. **1.** a former term for the ATRIUM of the heart. **2.** an alternative term for the PINNA (outer ear). **3.** (also called *auricula*) an ear-shaped part or appendage, such as that occurring at the base of some leaves.

Australian fauna, See MARSUPIAL.

Australopithecus, *n*. a genus of early Pleistocene primate that was hominid in some features but ape-like in others, such as the skull. Southern African in origin, *Australopithecus* was upright in posture.

autecology, *n*. the ecology of individual species as opposed to community ecology. Compare SYNECOLOGY.

authority, *n*. the author of a scientific name usually cited in most taxonomic works, e.g. *Tringa totanus* (Linnaeus).

auto-, *prefix*. pertaining to oneself, independently.

autocatalytic, *adj*. (of a substance) catalysing its own production. The more substance is made, the more catalyst is available for further production.

autochthonous, *adj*. (of peat) derived from plants that lived on the site of its formation.

autoclave, *n*. **1.** an apparatus in which objects or materials are sterilized using air-free saturated steam under pressure at temperatures in excess of 100°C. **2.** a sealed vessel in which chemical reactions can occur at high pressure.

autocoid, *n*. an alternative name for HORMONE.

autoecious, *adj*. (of parasites, esp. the rust fungus) completing the entire life cycle on a single species of host.

autogamy, *n*. **1.** the process by which the two parts of a divided cell nucleus reunite, as in some protozoans. **2.** self-fertilization in plants.

autogenic, *adj*. (of VEGETATIONAL SUCCESSION) affected by plant communities themselves altering their environment.

autograft, *n*. the grafting upon an organism of a part of itself.

autoimmunity, *n*. a situation where the immunological defences of the body become sensitized to parts of the same body, resulting in self-destruction in that area. In other words, 'self' ANTIGENS become mistaken for 'foreign' antigens, so setting up an IMMUNE RESPONSE. Such an occurrence is clearly dangerous and may be associated with ageing. An example is rheumatoid arthritis in man.

autolysis, *n*. the breakdown of tissues, usually after death, by their own enzymes.

autonomic, *n.* **1.** appertaining to that part of the nervous system controlling involuntary muscles (see AUTONOMIC NERVOUS SYSTEM). **2.** (in plants) movements arising from internal stimuli, e.g. protoplasmic streaming, spiral growth of stem apices.

autonomic nervous system, *n.* the part of the nervous system that controls the involuntary activities of the body. There are two main parts:

(a) The *sympathetic nervous system*, in which complexes of SYNAPSES form ganglia alongside the vertebrae, and the preganglionic fibres from the central nervous system are therefore short; the fibres are ADRENERGIC.

(b) The *parasympathetic nervous system*, in which the ganglia are embedded in the wall of the effector so that the preganglionic fibres are long and the postganglionic fibres short; these are CHOLINERGIC.

The sympathetic and parasympathetic systems innervate the same end organs, but the effects produced by the two systems generally oppose one another, for example:

Sympathetic system	*Parasympathetic system*
inhibits peristalsis	stimulates peristalsis
stimulates contraction in sphincters of bladder and anus	inhibits contraction in sphincters of bladder and anus
inhibits bladder contraction	stimulates bladder contraction
stimulates pacemaker, speeding up heart	inhibits pacemaker, slowing down heart
stimulates arterial constriction	inhibits arterial constriction, causing dilation
inhibits contraction of bronchioles, causing dilation	stimulates contraction of bronchioles
inhibits contraction of iris muscle, causing dilated pupil	stimulates contraction of iris muscle, causing reduced pupil

autopolyploid, *n.* a type of POLYPLOID in which there has been duplication of the number of each chromosome, all chromosomes coming from the same original species. For example, in Fig. 57 on page 62, A represents one complete set of chromosomes. Like an ALLOPOLYPLOID, an autopolyploid is a mechanism for creating new species, particularly in plants. Allopolyploids are more successful, perhaps because autopolyploid chromosomes have pairing difficulties at MEIOSIS.

autoradiograph, *n.* a photographic representation of radioactive areas

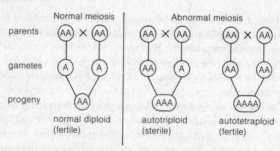

Fig. 57. **Autopolyploid.** Types of autopolyploid compared with normal diploid.

in a tissue. Tissues labelled with an ISOTOPE (e.g. ^{14}C) are placed on a photographic plate which is irradiated in the same way as photographic film is exposed to light. Radioactive areas can be seen in the photographs.

autosexing, *n.* a technique by which the sex of an offspring can be determined without examination of the genitalia. For example, the sex of young poultry chicks is difficult to determine anatomically. However, a 'Barring' gene (see BARR BODY) allows the colour of females (one X-chromosome) to be distinguished from that of males (two Xs).

autosomal gene, *n.* a gene found on an AUTOSOME.

autosomal inheritance, *n.* a pattern of transmission of ALLELES that are located on AUTOSOMES, i.e. not on sex chromosomes, so that the sex of the parent does not affect the result of a mating, RECIPROCAL CROSSES giving identical results.

autosome, *n.* a type of chromosome found in all cells *not* concerned with SEX DETERMINATION. Chromosomes are of two types: autosomes and SEX CHROMOSOMES. Autosomes carry the major part of genetic information in cells, including information on sexual characteristics. See SEX LINKAGE for a comparison of the inheritance of autosomal and sex-linked genes.

autostylic jaw suspension, *n.* the attachment of the upper jaw to the skull by a number of attachments called the *otic process*, the *basal process* and the *ascending process*. It occurs in lungfish and terrestrial vertebrates. Compare HYOSTYLIC JAW SUSPENSION.

autotomy, *n.* the deliberate casting of part of the body when attacked, as when a lizard casts it tail.

autotroph, *n.* an organism that can manufacture its own organic requirements from inorganic materials independent of other sources of organic substrates. Autotrophs are either phototrophic (see PHOTO-AUTOTROPH) or CHEMOAUTOTROPHIC, energy being derived either by photosynthesis where chlorophyll is present, or from inorganic oxidation where it is absent (e.g. hydrogen sulphide is oxidized by sulphur bacteria). Autotrophs are primary producers (see PRIMARY PRODUCTION). Compare HETEROTROPH.

autotrophic, *adj*. pertaining to an AUTOTROPH.

autozooid, *n*. an independent coelenterate polyp that is capable of feeding itself.

auxin, *n*. a type of plant hormone involved in the growth of cells and several other functions. The most important auxin is indolacetic acid (IAA; see Fig. 188), but many other substances have been classified as auxins, using a BIO-ASSAY method developed by FRITZ WENT.

The effects of auxins depend on their concentration in the plant. They are most concentrated at the shoot tip and least concentrated in the root, except for small amounts at the root tip. The major effects of auxin are summarized below:

(a) Encourages cell growth by elongation, producing a softening of the MIDDLE LAMELLAE of cell walls.

(b) Stimulates cell division in the PHLOEM of the VASCULAR BUNDLES, so encouraging new growth.

(c) Promotes positive PHOTOTROPISM in shoots by growth of tissues towards the light source.

(d) Promotes GEOTROPISM in all parts of the plant (positive in roots, negative in shoots, due to unequal distributions of the hormone).

(e) Induces APICAL DOMINANCE by suppressing lateral buds.

(f) Induces lateral root formation.

(g) Stimulates fruit development, enabling seedless fruits to be produced artificially.

(h) Suppresses ABSCISSION in leaves and fruit.

(i) Encourages the formation of wound tissues in injured or diseased plants.

auxotroph, *n*. a mutant strain of microorganism that requires growth factors in addition to those required by the WILD TYPE, i.e. will not grow on MINIMAL MEDIUM.

Avery, Oswald T, (1877–1955) American physician, who with Colin MacLeod and Maclyn McCarty in 1944, demonstrated that the 'transforming principle' suggested by GRIFFITH to explain bacterial transformation was DNA. Thus DNA alone could cause a heritable change in bacteria, strongly suggesting that DNA (rather than protein) is the basic genetic material of cells.

Aves, *n*. a class of vertebrates, the BIRDS.

avirulent, *n*. lacking virulence i.e. the ability of an organism to cause disease is absent.

awn, see ARISTA (2).

axenic, *adj.* descriptive of a pure culture where no other organism is present.

axial skeleton, *n.* the part of the SKELETON containing the SKULL and VERTEBRAL COLUMN.

axil, *n.* a location on a plant stem which is the angle between the leaf and the upper part of the stem. (See Fig. 58).

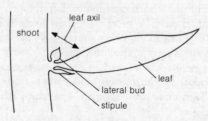

Fig. 58. **Axil.** A leaf axil.

axile, *adj.* (of plant structures) attached to the central axis.

axillary, *adj.* (of plant structures) arising in the angle of a leaf or bract.

axis cylinder, *n.* an AXON.

axolol, *n.* the larval stage of the aquatic American salamander *Amblystoma*. It is capable of breeding in the larval form so demonstrating a phenomenon known as NEOTONY.

axon, *n.* the process of a NERVE cell which conducts impulses from the nerve cell body.

axoneme, *n.* the complex of MICROTUBULES and associated tubes found in the shaft of a CILIUM or FLAGELLUM.

axoplasm, *n.* the cytoplasm which occurs within the AXON.

azo dye, *n.* any of a class of artificial dyes obtained from aromatic amines.

B

baboon, *n.* a PRIMATE of the genus *Papio* which is characterized by the presence of a dog-like muzzle, a short tail and callosities on the buttocks. There are five species in Africa and Asia.

Bacillariophyceae, *n.* the diatoms, a class within the division Chrysophyta. They are unicellular, occur in both marine and freshwater environments and their silicified cell walls are preserved in diatomaceous earth and peat. Diatoms are one of the most important PRIMARY PRODUCERS in the sea and are responsible for the formation of petroleum deposits.

bacillus, *n*. the general name for a rod-shaped BACTERIUM, but more specifically a generic name for a group of spore-producing forms, e.g. the hay bacillus *Bacillus subtilis*.

backbone, see VERTEBRAL COLUMN.

backcross, *n*. a genetical term describing a mating between a HYBRID organism and one of the original parental types. The results of a backcross depend upon the parental type. See Fig. 59. A backcross to the recessive parental is diagnostic (since it reveals the genotype of the hybrid) and is called a TESTCROSS.

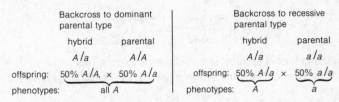

Fig. 59. **Backcross.** Results from different parental types.

back mutation, *n*. an inherited change (MUTATION) in a mutant gene that causes it to regain its WILD TYPE function. Compare FORWARD MUTATION.

bacteria, *n*. (*sing*. bacterium) unicellular or (more rarely) multicellular PROCARYOTE organisms. Some are AUTOTROPHIC, contain BACTERIOCHLOROPHYLLS and bacterioviridin, carrying out photosynthesis anaerobically. Bacteria have various shapes, occurring as cocci (spherical), bacilli (rod-shaped) and spirilla (helical) which range in size from 1 μm to about 500 μm, but usually 1–10 μm in diameter. They are present in soil, water, air and as free-living SYMBIONTS, PARASITES and PATHOGENS. While some bacteria are useful in NITROGEN and SULPHUR CYCLES, many cause diseases of plants, animals and man, e.g. ANTHRAX, TETANUS. Reproduction occurs asexually but genetic transfer can take place by CONJUGATION; other genetic changes may be brought about by MUTATION or the acquisition of a PLASMID (see Fig. 60 on page 66). Identification usually involves biochemical tests rather than visual study and classification often involves the use of NUMERICAL TAXONOMY.

bactericide, *n*. a substance causing death to BACTERIA, for example an ANTIBIOTIC.

bacteriochlorophyll, *n*. a form of CHLOROPHYLL possessed by photosynthetic bacteria (green and purple types). In addition to small amounts of bacteriochlorophyll, green bacteria mostly possess another type of chlorophyll called *bacterioviridin*.

bacteriology, *n*. the study of BACTERIA.

bacteriophage or **phage,** *n*. a VIRUS that attacks BACTERIA. Each

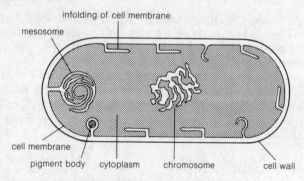

Fig. 60. **Bacteria.** A generalized bacterium.

bacteriophage is specific to one form of bacterium.

bacteriostasis, *n.* the inhibition of growth in bacteria without killing them.

bacteriostatic, *adj.* (of a substance) inhibiting the growth of bacteria but not killing them.

balance, *n.* the maintenance of stability and specific orientation by an organism in relation to the immediate environment. Organs of balance occur at the end of each SEMICIRCULAR CANAL in the inner ear where there is a swelling, an AMPULLA; this is a RECEPTOR which has a group of sensory cells, hairs from which are embedded in a gelatinous cap – the CUPULA. Since the semicircular canals are at right angles to each other the ampullae are sensitive to movement in any plane, as the canal fluid moves the cupula in a direction opposite to that of the movement of the head. Head position is given by receptors containing calcareous OTOLITHS in the UTRICLE and SACCULE, and these react to gravity in relation to the position of the head. Nerve fibres lead to the brain from ampullae, saccule and utricle.

balanced diet, *n.* an intake of the various types of food in such proportions as to promote good health.

balanced polymorphism or **stable polymorphism,** *n.* a GENETIC POLYMORPHISM in which the various morphs are maintained in a stable frequency over several generations due, possibly, to constant NATURAL SELECTION pressures.

balancer, see HALTERE.

Balanoglossus, *n.* a genus of worm-like, burrowing PROTOCHORDATE.

baleen or **whalebone,** *n.* horny plates which hang from the roof of the mouth in some whales (Mystacoceti), e.g. right whales and rorquals. The plates filter out krill (small shrimplike crustaceans) from the sea water as it is forced out of the mouth.

Banks, Barbara, biochemist who, in the early 1970s, argued that ATP when hydrolysed required rather than gave up energy to break covalent bonds. Most biologists consider that energy is provided when ATP is hydrolysed but it is not now thought to be stored simply in the terminal phosphate bond.

Banting, Sir Frederick Grant (1891–1941) Canadian physician and Nobel prizewinner. He is best-known for his discovery, with C.H. BEST, of the hormone INSULIN, in 1921, during his work on the internal secretion of the pancreas.

banyan tree, *n. Ficus benghalensis,* an Indian tree best known for the production of large aerial roots that are let down from its larger branches and, in effect, form secondary trunks which give extra support and allow the tree to spread.

barbiturate, *n.* any UREIDE such as phenobarbital, amytal, seconal, etc. Barbiturates have a depressant effect on the CNS, usually producing sleep.

barbs, *n.* lateral lamellae which project from the quill or calamus of the FEATHER and form the vane.

barbules, *n.* **1.** hook-like processes on the barbs of FEATHERS which interlock to form the continuous vane. **2.** teeth found in the capsules of some mosses.

bark, *n.* the outer, living part of a woody stem, consisting of three layers:

(a) an inner layer called secondary PHLOEM, containing the elements of primary phloem plus horizontal ray cells which function in transporting materials across the stem.

(b) a middle layer of cork CAMBIUM, a group of meristematic (dividing) cells originating in the PARENCHYMA cells of the outer stem cortex. As the cells divide, the outer ones develop into cork cells and the inner ones give rise to parenchyma-like tissue.

(c) CORK, an outer region of cells forming a waterproof and protective layer broken only by LENTICELS.

barnacle, *n.* the common name for any of the CRUSTACEAN class Cirripedia, most of which are SESSILE, grow on rocky substrates, and have a CALCAREOUS outer covering.

baroreceptor or **baroceptor,** *n.* a pressure RECEPTOR that responds to changes in blood pressure, found particularly in the carotid sinus and aortic arch.

Barr body, *n.* sex CHROMATIN particles found in the interphase nucleus of buccal epithelial cells in some female mammals and probably derived from an X-chromosome (see INACTIVE-X HYPOTHESIS). First described by Bertram Barr in 1969, Barr bodies can be used as a sex marker, always occurring in numbers one less than the total number of X-chromosomes. Thus normal human males and TURNER'S SYNDROME females have no Barr bodies, whilst KLINELFETER'S SYNDROME males have two. See Fig. 61.

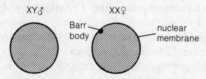

Fig. 61. **Barr body.** Sex chromatin in nuclei from human male and female buccal epithelial human cells.

basal body, *n.* a structure similar to (and homologous with) a CENTRIOLE connected to the axial filament of the flagellum and cilium, e.g. in sperms.

basal metabolic level, see BASAL METABOLIC RATE.

basal metabolic rate (BMR) or **basal metabolic level (BML),** *n.* the minimal rate of METABOLISM in a resting organism in an environment with a temperature the same as its own body heat, whilst not digesting or absorbing food. The rate is commonly expressed in terms of energy per unit surface area per unit time, usually as $kJ \ m^{-2} \ h^{-1}$. See STANDARD METABOLIC RATE, REGULATORY HEAT PRODUCTION.

base, *n.* a chemical substance which has a tendency to accept protons (H^+); the base dissolves water with the production of hydroxyl ions and reacts with acids to form salts.

base analogue, *n.* any chemical that has a similar structure (i.e., analogous) to one of the purine or pyrimidine bases in DNA or RNA. Such analogues can become incorporated into the nucleic acid and may act as a MUTAGEN. For example, 5-bromouracil (5Bu) is an analogue of thymine and can be incorporated into DNA in place of thymine. In its normal state 5Bu acts like thymine and pairs with adenine (see COMPLEMENTARY PAIRING) but the analogue sometimes undergoes a chemical change called a *tautomeric shift* and now pairs with guanine. At the next DNA replication an incorrect base will be incorporated into the DNA, causing a TRANSITION SUBSTITUTION mutation (adenine – guanine). See Fig. 62.

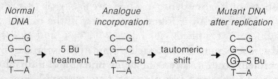

Fig. 62. **Base analogue.** The action of 5-bromouracil on DNA.

base deletion, *n.* removal of a NUCLEOTIDE base, a purine or a pyrimidine from the DNA structure, producing a DELETION MUTATION that can have serious consequences for the related protein. See FRAMESHIFT.

base insertion, *n*. addition of an extra NUCLEOTIDE base to a DNA molecule, producing an INSERTION MUTATION that can have serious consequences for the related protein. See FRAMESHIFT.

basement membrane, *n*. a membrane lying between an EPITHELIUM and its underlying CONNECTIVE TISSUE.

base pairing, *n*. hydrogen bonding between NUCLEOTIDE bases of DNA. See COMPLEMENTARY PAIRING.

base substitution, *n*. replacement of a NUCLEOTIDE base in a DNA molecule with another base, possibly producing a SUBSTITUTION MUTATION.

basic dyes, *n*. dyes with a basic organic grouping which stain when coloured with an acid; they are used mainly for nuclear staining where they combine with nucleic acids, e.g. fuchsin as used in the Feulgen stain for DNA.

basidiomycete, *n*. any fungus of the subdivision Basidiomycotina, a group containing over 14,000 species. They have a septate mycelium, and many forms such as mushrooms produce complex sporophores. Sexually derived spores (basidiospores) are produced externally on basidia, usually in groups of four, and this is characteristic of the group Basidiomycotina. Members are important as food (mushrooms), in plant diseases (rusts and smuts) and in dry rot of timber.

basidium, *n*. the (microscopic) structure on which the sexually produced basidiospores of BASIDIOMYCETE fungi are formed during sexual reproduction. See Fig. 63.

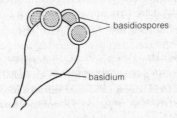

basidiospores

basidium

Fig. 63. **Basidium.** The basidium of *Scleroderma*.

basifixed, *adj*. (of anthers) joined by the base and not capable of independent movement.

basilar membrane, *n*. the tissue which bears the auditory hair cells in the COCHLEA of the EAR. See TECTORIAL MEMBRANE.

basipetal, *adj*. **1.** (of a stem) producing structures one after the other from the apex down to the base, so that the oldest are at the apex. **2.** (of substances, e.g. AUXINS) movement away from the apex to lower parts of the plant.

basipodite, *n*. the segment of the crustacean limb proximally attached to the coxopodite and distally supporting the exopodite and endopodite.

basophil leucocyte, *n*. a type of white blood cell (of the GRANULOCYTE group) which takes up the stain of a basic dye. Basophils form a tiny proportion of the adult human leucocytes (about 1%), being produced in the bone marrow and have a function which is probably associated with the IMMUNE RESPONSE, e.g. in chickenpox infection.

bastard wing, *n*. that part of the wing of a bird attached to the thumb of the forelimb.

bat, *n*. a flying mammal of the order CHIROPTERA, the only true flying vertebrate apart from the birds.

Batesian mimicry, see MIMICRY.

Bateson, William (1861–1926) British pioneer of genetics. He 'rediscovered' Mendelian inheritance in 1900, using, for example, poultry and sweet peas. See MENDELIAN GENETICS.

batrachian, *n*. see ANURAN.

B-carotene, *n*. a CAROTENE which is a precursor of vitamin A, producing two molecules of vitamin A per molecule of B-carotene.

B-cell, *n*. a LYMPHOCYTE formed in the bone marrow of mammals that, after stimulation with a particular ANTIGEN, divides to produce daughter cells that manufacture large amounts of ANTIBODIES which are passed into the bloodstream.

B-complex or **vitamin B complex,** *n*. a group of water-soluble VITAMINS which function mostly as COENZYMES in the metabolic reactions taking place in animal cells. The complex consists of: THIAMINE (B_1), RIBOFLAVIN (B_2), PANTOTHENIC ACID (B_5), PYRIDOXINE (B_6), BIOTIN, NICOTINIC ACID, FOLIC ACID and COBALAMIN (B_{12}).

bdelloid, *adj*. leech-like.

Beadle, George Wells (1903–) American geneticist. With Edward Tatum, he began the modern study of biochemical genetics using the red bread mould *Neurospora crassa*. They investigated nutritional mutants in the fungus (AUXOTROPHS), and coined the ONE GENE/ONE ENZYME HYPOTHESIS.

beak, *n*. **1.** also called *bill*. the jaws and associated horny covering in a bird or turtle. **2.** any pointed projection in plant fruits. **3.** a projecting jawbone in fish such as pike. **4.** the tip of the UMBO in bivalve molluscs. **5.** the jaws of CEPHALOPODS, e.g. of an octopus.

bee, *n*. any member of the Apoidea in the insect order HYMENOPTERA. They possess membranous wings, usually a hairy body and sucking or chewing mouthparts, e.g. *Apis melifica*, the honey bee.

beetle, *n.* any member of the insect order COLEOPTERA.

behaviour, *n.* **1.** the total activities of a living organism (usually an animal) ranging from simple movement to complex patterns involved with courtship, threat, camouflage, etc. **2.** the observable response of an organism to stimuli from the environment. See INSTINCT, LEARNING.

Benedict's Test *n.* a procedure used to detect the presence of REDUCING SUGARS in a solution. Benedict's Reagent consists of a single solution of copper sulphate, sodium citrate and sodium carbonate in water. The reagent is added to the test solution and heated. A red precipitate of Cu(I) oxide indicates the presence of a reducing sugar.

benign, *adj.* nonmalignant, as in a growth.

benthos, *n.* the organisms that live on the bed of the sea, or a lake, ranging from the high-water mark to the deepest trenches. Benthic organisms are subdivided into littoral (to 40 m deep), sublittoral (41–200 m), bathyal (200–4000 m), abyssal (4001–6000 m) and hadal (> 6000 m). See SEA ZONATION.

Bergmann's rule, *n.* a rule stating that the individuals from populations of warm-blooded species of animals (HOMOIOTHERMS) which occur in cooler climates tend to be larger on average than individuals of the same species in warmer climates. This is because the *surface area/volume ratio* in large animals is smaller, so that heat loss is consequently reduced. The rule is named after the German biologist W. Bergmann, and was formulated in 1847. See also ALLEN'S RULE.

beri-beri, *n.* a human disease caused by vitamin B_1 (THIAMINE) deficiency in which affected individuals suffer from wasting of muscles, paralysis, mental confusion and sometimes heart failure.

Bernard, Claude (1813–78) French physiologist. He was the father of modern experimental physiology, and the first person to appreciate the importance of the internal environment in the functioning of organisms.

berry, *n.* a type of succulent, fleshy FRUIT produced by some plants, in which seeds are embedded in the pulp. The fruit is formed from the swollen tissue of the PERICARP, e.g. tomato, grape, date, gooseberry, citrus fruits.

Best, Charles Herbert (1899–1978) Canadian physician. He was the codiscoverer of INSULIN in 1921, with Sir Frederick G. BANTING. He also discovered histaminase, carried out extensive studies on insulin and diabetes, and worked on heparin and thrombosis, and choline and liver damage.

beta taxonomy, *n.* the level of TAXONOMY concerned with arranging species into a natural CLASSIFICATION, which logically follows ALPHA TAXONOMY.

B horizon, *n.* the second main zone of a soil profile that accumulates leeched substances. See also A HORIZON.

bi-, *prefix*. denoting two.

bibliographical reference, *n*. a precise form of reference to an organism in taxonomic literature which must include the name of the author, the date of publication of the organism's name, together with the title of the book or journal and exact page references.

biceps, *n*. a group of muscles present in the upper forelimb of TETRAPODS which are flexors of the elbow joint. In the hind limb the biceps femoris is a flexor of the knee and an elevator of the femur. See ANTAGONISM.

bicollateral bundle, *n*. a VASCULAR BUNDLE of plants with two phloem groups, one internal and one external to the xylem, e.g. marrow, tomato.

bicuspid, 1. *n*. a type of tooth in which the crown is formed into two distinct points or cusps, found typically in premolar teeth. **2.** *adj*. (of a plant structure) possessing two points or cusps. **3.** *adj*. formerly called *mitral*. (of a valve) allowing a flow of blood from the left ATRIUM to the left VENTRICLE.

biennial, 1. *adj*. occurring every two years. **2.** *n*. a plant that lives for two seasons (or years), germinating from seed and growing to maturity in the first season with food storage in swollen roots for overwintering. In the second season flowers, fruit and seeds are produced before death occurs. Such plants are herbaceous rather than woody, e.g. Canterbury Bell (*Campanula*) and the carrot. See ANNUAL, PERENNIAL.

bifid, *adj*. divided into two, forked.

biflagellate, *adj*. having two flagella, as in *Chlamydomonas*.

bilateral cleavage or **radial cleavage,** *n*. the type of CLEAVAGE which gives rise to BILATERAL SYMMETRY. It occurs in echinoderms and CHORDATES. Compare SPIRAL CLEAVAGE.

bilateral symmetry, *n*. an animal body structure in which there is a head and a rear with the body organs arranged so that a section through the midline from dorsal (upper) to ventral (lower) surfaces would divide the organism into almost identical right and left halves. Most higher invertebrates (e.g. PLATYHELMINTHS, ANNELIDS, ARTHROPODS) and all vertebrates are bilaterally symmetrical. Compare RADIAL SYMMETRY.

bile, *n*. a thick, brown-green fluid secreted by the liver which is alkaline in its reactions, containing bile salts, bile pigments, CHOLESTEROL and inorganic salts. Bile is transferred from liver to the DUODENUM via the bile duct which in many mammals contains a reservoir called the gall bladder. The bile pigments (bilirubin and biliverdin) result from the breakdown of HAEMOGLOBIN in red blood cells, giving the bile its coloration which in turn affects the colour of the FAECES. The amount of cholesterol excreted in bile depends upon the blood fat level, the cholesterol in the bile normally being kept in solution by the bile salts. Reduction in the bile salt concentration can cause cholesterol to be deposited in the gall bladder, contributing to the formation of gallstones. Although bile contains no digestive enzymes, bile salts are also responsible for the EMULSIFICATION of fats in the duodenum, lowering the surface tension of the fatty film surrounding fatty food particles, so producing a larger surface area on which digestive enzymes (LIPASES) can work. Secretion of bile from the liver is stimulated by the hormone SECRETIN which is produced in the wall of the duodenum. See Fig. 64. See also CHOLECYSTOKININ–PANCREOZYMIN.

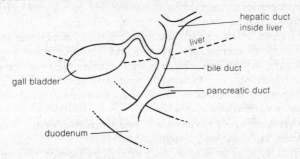

Fig. 64. **Bile.** The human gallbladder and bile duct.

bile duct, *n.* the duct through which bile passes from the liver or gall bladder to the duodenum.

bile salts, *n.* the sodium salts secreted in bile, sodium taurocholate and sodium glycocholate, which greatly lower surface tension and are important in emulsifying fats.

bilharzia or **schistosomiasis,** *n.* a human disease common in Egypt and other warm parts of the developing world, in which affected individuals suffer from body pains followed by severe dysentery and anaemia, leaving them weak and highly susceptible to other diseases. The disease is caused by the MIRACIDIUM larva of the blood fluke *Schistosoma*, which lives as an adult in the abdominal veins and has a water snail as a secondary host, transmission being effected via urine.

bilirubin, *n.* a BILE pigment.

biliverdin, *n.* a BILE pigment.

bill, see BEAK.

bimodal distribution, *n.* a statistical pattern in which the frequencies of values in a sample have two distinct peaks, even though parts of the distribution may overlap. For example, the sexual differences between men and women for such characters as height and weight produce a bimodal distribution. See Fig. 65.

BIMOLECULAR LEAFLET

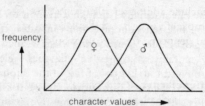

Fig. 65. **Bimodal distribution.** The peaks (or modes) represent the character values of height and weight for men and women.

bimolecular leaflet, *n.* a double layer of LIPID molecules that has many of the properties of the PLASMA MEMBRANE and on which Danielli and Davson based their theory of membrane structure.

binary fission, *n.* an asexual method of cell reproduction (see ASEXUAL REPRODUCTION) carried out by PROCARYOTES and some primitive EUCARYOTES, in which the chromosomal material is replicated and then the cytoplasm splits by CYTOKINESIS. Such processes differ from MITOSIS and MEIOSIS in that the chromosomal separation does not involve cellular MICROTUBULES forming a spindle.

binary system, *n.* any system where two alternatives occur, as in some taxonomic keys (see KEY, IDENTIFICATION).

binding site, *n.* an area of an ENZYME to which other molecules can attach. Such an area can be an ACTIVE SITE for particular substrates or for other molecules, as in COMPETITIVE INHIBITION. Alternatively, some enzymes possess two binding sites, one for substrate attachment the other for attachment of other molecules as in noncompetitive inhibition and allosteric inhibition (see ALLOSTERIC INHIBITORS).

binocular vision, *n.* a type of vision where the image of an object viewed falls on the retinas of both eyes simultaneously. Examples occur in many vertebrates particularly in primates and predators such as owls and cats. Such vision facilitates judgement of distance and in this the FOVEA, an area of acute vision, is of particular importance. Binocular vision results in a stereoscopic or 3-D effect, the slightly different positions of the two eyes being important in that they view the object from slightly different angles.

binomial expansion equation, *n.* an equation stating that $(p+q)^2 = 1.0$, in which p and q are variables. The equation is used in population genetics (see POPULATION sense 2). Expansion of the term produces

$$p^2 + 2pq + q^2 = 1.0,$$

the genotype frequencies expected in a population that has a gene with two alleles at frequencies of p and q respectively. The equation can be expanded to include additional alleles of a gene (see HARDY–WEINBERG LAW). For example, if a gene has three alleles with frequencies of p, q and r, the expected genotype frequencies would be:

$$(p+q+r)^2 = p^2 + 2pq + 2pr + q^2 + 2qr + r^2.$$

binomial nomenclature, *n.* the basis of the present scientific nomenclature of animals and plants, each of which is given a generic name followed by a specific name, in Greek, Latin or often Latinized English. The generic name invariably has an initial capital letter, and the specific name, even if it is the name of a person, an initial small letter, both names being in italics, or underlined. Thus the robin is named *Erithacus rubecula*. All scientific names used before the publication of the 10th edition of LINNAEUS's *Systema Naturae* (1758) are no longer applicable, and the names given since then have priority by date as a rule, the earliest name for an organism being given preference over others. Often the scientific name is followed by the name of the person allocating the name and the date, e.g. *Erithacus rubecula* (L.) 1766. L. is an abbreviation for Linnaeus, and the brackets indicate a change from the genus in which he originally placed it; where genera and species are redefined, change of generic name is allowable. The robin was originally named *Motacilla rubecula* L. 1766. *Motacilla* is now the genus including wagtails, a group not closely related to robins which were subsequently placed in the genus *Erithacus*.

bio– or before a vowel **bi–,** *prefix.* denoting life or living organisms.

bio–assay, *n.* a technique in which the presence of a chemical is quantified by using living organisms, rather than by carrying out chemical analysis. For example, the amount of a toxic substance (such as an insecticide) in a sample could be estimated by comparing mortality in organisms treated with the unknown sample, against an agreed standard figure of mortality in organisms treated with known quantities of the substance.

biochemical evolution, *n.* the evolutionary processes concerned with the formation of biomolecules and attributes of living cells, such as metabolic pathways.

biochemical mutant, *n.* a mutant which has a defect in a biochemical pathway, such as an AUXOTROPH. See also one gene/one enzyme bypothesis.

biochemical oxygen demand (BOD), *n.* an empirical standardized laboratory test designed to measure the oxygen requirements (or demands) of a given effluent. It is an approximate measure of biochemically degradable organic matter in a water sample. In its simplest form BOD is measured by taking a water sample of known volume, incubating it in the dark for a known period (usually five days), and measuring the oxygen levels at the start and end of the period. During the test the oxygen levels will have been reduced by bacteria in proportion to the amount of organic matter present.

biochemistry, *n.* the study of the chemistry of living organisms.

biocide, see SYSTEMIC BIOCIDE.

biodegradation, *n.* the breaking down of inorganic and organic substances by biological action, a process usually involving bacteria and fungi, which are known as SAPROBIONTS when the substrate is biological.

biodeterioration, *n.* the unwanted breakdown of materials such as foodstuffs, surface coatings, rubber, lubricants, by microorganisms, resulting in significant financial losses in many industries. Compare MICROBIAL DEGRADATION.

biogenesis, *n.* the principle that life is propagated only by existing life, rather than by SPONTANEOUS GENERATION. The principle's original formulation was that organisms are able to generate only other forms similar to themselves, but where ALTERNATION OF GENERATIONS occurs the statement is not entirely true; biogenesis has thus come to mean that a species gives rise only to similar forms and these forms can only be derived from parents which are similar.

biogeographical region, *n.* the regions of the earth's surface which contain distinctive groups of animals and plants that are normally prevented from moving out of those regions by various natural barriers. See Fig. 66.

biokinetic zone, *n.* the temperature range over which organisms can survive (5°–60°C).

biological classification, *n.* the arrangement of organisms into taxa (see TAXON) on the basis of their genetic relationships.

biological clock or **internal clock,** *n.* an internal mechanism (as yet poorly understood) by which many plants and animals keep a sense of time, making possible a rhythmic pattern of behaviour. Many organisms have such 'clocks' producing activity cycles of approximately 24 hours (CIRCADIAN RHYTHMS) which, however, can be affected by external influences that 'set' the clock (*entrainment*). An individual's clock can be re-entrained if placed in a new time zone, as happens with rapid, long-distance travel.

Biological clocks affect not only whole organism activities (e.g. sleeping) but also cellular patterns of activity (e.g. varying METABOLIC RATES). See also DIURNAL RHYTHM.

biological control, *n.* a method of pest control using natural predators

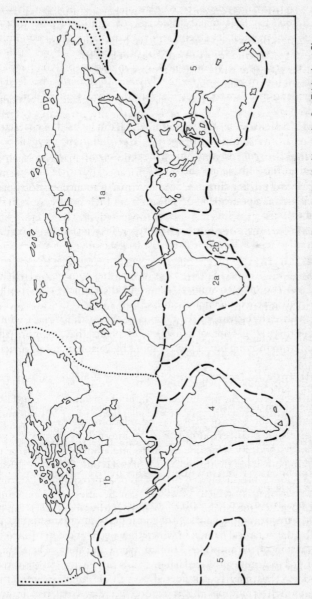

Fig. 66. **Biogeographical region.** The faunal regions of the world: 1a, Palaearctic; 1b, Nearctic; 2a, Ethiopian; 2b, Madagascan; 3, Oriental; 4, Neotropical; 5, Notogaea; 6, Wallacea.

(harmless to other organisms) to regulate the size of pest populations. The technique has been mainly used against arthropod pests causing economic or medical problems. For example, red spider mites (*Tetranychus urticae*), which are pests of glasshouse crops such as cucumbers, are routinely controlled by application of commercially available samples of predatory mites (*Phytoseilus persimilis*) that feed on the red spider mites, thus reducing their numbers below an economically harmful level.

biological indicator, *n.* any organism which can be used to characterize particular environmental properties, e.g. pollution, oxygenation. For example, *Sagitta elegans* characterizes oceanic waters, *S. setosa* characterizes the continental shelf.

biological magnification, *n.* the increasing concentration of a substance such as a pesticide in organisms from their predecessors in the FOOD CHAIN.

biological oxygen demand (BOD), see BIOCHEMICAL OXYGEN DEMAND.

biological rhythm, *n.* any regularly occurring sequence of events in living organisms, either internal (ENDOGENOUS), e.g. heartbeat, or external (EXOGENOUS), e.g. the seasons and tides. See CIRCADIAN RHYTHM.

biological speciation, *n.* the formation of new species, based on the REPRODUCTIVE ISOLATION of the constituent populations.

biological warfare, *n.* the use of living organisms, particularly microorganisms, or their products, to induce illness or death in a population.

biology, *n.* the study of living organisms.

bioluminescence, *n.* the production of light of various wavelengths by living organisms (often wrongly termed *phosphorescence*). The phenomenon is found in such divergent groups as bacteria, fungi, fireflies (COLEOPTERA) and various marine organisms. See LUMINESCENCE.

biomass, *n.* the total mass of organisms in a given ENVIRONMENT or at a certain TROPHIC LEVEL. It is measured as either live or dry weight per unit area, and the alternative used should be stated.

biomass pyramid, see PYRAMID OF NUMBERS.

biome, *n.* a major regional ecological community of organisms usually defined by the botanical habitat in which they occur and determined by the interaction of the substrate, climate, fauna and flora. The term is often limited to denote terrestrial habitats, e.g. tundra, coniferous forest, grassland. Oceans may be considered as a single biome (the marine biome), though sometimes this is subdivided, e.g. coral reef biome. There is no sharp distinction between adjacent biomes.

biometry or **biometrics,** *n*. the analysis of biological data by means of statistical or mathematical techniques.

bionomics, see ECOLOGY.

biophysics, *n*. the physics of biological processes and the application of methods used in physics to biology.

biopoiesis, *n*. the production of living from nonliving material.

biopsy, *n*. the surgical removal of small amounts of tissue for examination to aid a diagnosis.

biosphere, *n*. that part of the earth's surface and its immediate atmosphere which is occupied by living organisms.

biosynthesis, *n*. the process by which more complex molecules are formed from simpler ones by living organisms, e.g. PHOTOSYNTHESIS.

biosystematics, *n*. that part of the study of systematics concerned with variation within a species and its general evolution.

biota, *n*. the living organisms present in a specific region or area, ranging in size from a small puddle to a BIOME or larger.

biotechnology, *n*. the use of organisms, their parts or processes, for the manufacture or production of useful or commercial substances. The term denotes a wide range of processes, from using earthworms as a source of protein, to the genetic manipulation of bacteria to produce human gene products such as growth hormone.

biotic community, *n*. all the organisms living on and contributing to a specific region (a BIOTA).

biotic factor, *n*. the influences which occur within an environment (on both the physical environment and the organisms living there) as a result of the activities of living organisms. See EDAPHIC FACTOR.

biotic potential, *n*. the theoretical maximum rate of increase of a species of organism in the absence of any adverse environmental factors such as predators or disease.

biotin, *n*. a water-soluble vitamin of the B-COMPLEX present in many foods, including yeast, liver and fresh vegetables. Biotin acts as a COENZYME in amino acid and lipid METABOLISM. A deficiency (rare in humans) of biotin causes dermatitis and intestinal problems.

biotrophic, *adj*. (of microorganisms) feeding on living organisms, parasitic. Compare NECROTROPHIC.

biotype, *n*. a group of individuals that are genetically identical, forming a physiologically distinct race within a species.

bipedalism, *n*. a mode of locomotion found in many primates (particularly man) and birds, in which only the hind limbs are used in walking. True bipedalism (i.e. where locomotion is normally bipedal) has required evolutionary changes to the vertebral column and pelvis, with their associated musculature. A principal advantage of bipedalism would seem to be that the forelimb can become modified for a nonwalking function, e.g. tool handling in man, flight in birds.

bipolar cell, *n.* a NEURON cell with two axons emerging from opposite sides of the cell body. Such cells are found in the vertebrate retina.

biramous appendage, *n.* the forked appendage of CRUSTACEANS formed by the protopodite (coxopodite and basipodite) which is nearest the body, and the two branches, exopodite and endopodite, which may form pincers, mouthparts or legs. See Fig. 67.

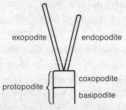

Fig. 67. **Biramous appendage.** A crustacean's biramous appendage.

bird, *n.* a vertebrate organism belonging to the class Aves and characterized by the presence of feathers, the forelimbs modified as wings (usually for flying), a bill or beak without teeth in the jaws, internal fertilization, calcareous shelled egg(s) incubated outside the body and the ability to control body temperature.

birth, *n.* the act of bringing forth young from the female animal, usually from the uterus of a female mammal, but also denoting the birth of any animal where eggs develop inside the adult and young are produced as independent organisms, e.g. seahorses.

birth control, *n.* any method used to limit the size of the human population, which usually involves the prevention of fertilization of the ovum by the sperm but can also include abortion of the foetus. Behavioural methods include (a) abstention from copulation, (b) the so-called 'rhythm method' which takes advantage of less-fertile phases of the menstrual cycle, (c) coitus interruptus. Other methods involve the use of *contraceptive* devices, hormonal treatment and sterilization.

Many countries have government-sponsored birth control programmes, initiated in an attempt to control the rapidly increasing human population. For example, China not only encourages one child per family but is also attempting to produce a longer period between generations (over 25 years) by favouring marriages at a late date.

birth rate, *n.* the average number of offspring produced by a female in a given time. This is calculated as the total number of new individuals added to the population during the time in question, divided by the number of mature breeding females in the population at that time.

bisexual *adj*. of or referring to (a) HERMAPHRODITE individuals, (b) a population that contains both males and females.

Biuret reaction, *n*. a chemical change used as a quantitative method for protein determination. The reaction takes place when dilute copper sulphate solution is added to a protein solution which is then made alkaline by the addition of sodium hydroxide. A copper hydroxide precipatate forms and a purple/violet colour is produced, the strength of which indicates the quantity of protein present. The reaction only works if at least two PEPTIDE BONDS are present in the protein. Thus single AMINO ACIDS and DIPEPTIDES with only one peptide bond will not give a positive Biuret reaction.

bivalent, *adj*. (of a pair of homologous CHROMOSOMES) being paired during prophase I of MEIOSIS. Compare MULTIVALENT.

bivalve, *n*. any marine or freshwater mollusc of the class Lamellibranchiata (Pelecypoda) having two hinged parts to its shell. BRACHIOPODS are also bivalves, in that there are two hinged parts to the shell, but the term is usually restricted to true molluscs.

blackwater fever, *n*. a rare tropical disease triggered by MALARIA PARASITE infestation, in which there is massive destruction of red blood cells producing dark red or black urine.

bladder, *n*. a hollow muscular bag situated in the lower abdominal cavity of mammals serving as a reservoir for urine from the kidneys. The bladder is composed of an internal epithelium surrounded by a coat of smooth muscle running in both circular and longitudinal directions, contraction of which causes complete collapse of the bag-like shape. The flow of urine down the URETERS from kidneys to the bladder is continuous, the amount depending on body fluid levels. When the bladder is empty the opening to the outside is closed by an internal SPHINCTER of smooth muscle which, like the bladder muscle, is controlled by the AUTONOMIC NERVOUS SYSTEM. When full, the internal sphincter relaxes under nervous control and urine enters the URETHRA (the duct to the outside), but is prevented from being voided by contraction of an external sphincter of striated muscle. Regulation of the sphincter (and thus of urine release or 'micturition') is under voluntary nervous control. See Fig. 68 and Fig. 69 on page 82.

bladder worm, *n*. the CYSTICERCUS larva of a tapeworm.

blastema, *n*. an undifferentiated mass of animal cells that later forms a structure or organ either embryologically or through regeneration, e.g. the head of a flatworm.

blasto -, *prefix*. denoting an embryo or bud or the process of budding.

blastocoel, *n*. the cavity within the BLASTULA.

blastocyst, *n*. the BLASTULA of mammals, differing from that of lower

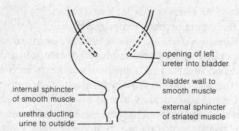

Fig. 68. **Bladder.** The mammalian bladder.

Bladder state	Autonomic system stimulation	Bladder muscle	Internal sphincter muscle
Empty	Sympathetic	Relaxed	Contracted
Full	Parasympathetic	Contracted	Relaxed

Fig. 69. **Bladder.** Summary of bladder control.

vertebrates in consisting of an outer wall of cells, the TROPHOBLAST, by which the embryo implants in the uterus, and in having an inner cell mass, the embryonic disc, from which the embryo is formed.

blastoderm, *n.* the layer of cells formed by cleavage of the fertilized egg in the presence of large amounts of yolk, e.g. in birds, so that the blastoderm forms on one side of the yolk mass, initially as a small blastodisc.

blastodisc, see BLASTODERM.

blastokinesis, *n.* the movement of an embryo within an egg during the course of development.

blastomere, *n.* any cell that occurs in the BLASTULA.

blastopore, *n.* the opening of the GASTRULA.

blastosphere, see BLASTULA.

blastozooid, see OOZOOID.

blastula or **blastophere,** *n.* a stage in the development of an embryo, where the embryo consists of a hollow ball of cells with a central cavity (a MORULA) formed by a process of CLEAVAGE. Invagination (in-pushing of the ball at one side) of cells into the blastula gives rise to the GASTRULA. See ARCHENTERON (Fig. 44). See Fig. 70.

blending inheritance, *n.* a pre-20th century theory that the male and female GAMETES each contained essences from the various parts of the parental body, which were mixed at fertilization to produce offspring showing some features of both parents. The theory was backed up by results from crosses involving quantitative characters (see POLYGENIC INHERITANCE) in which there was an apparent mixing of PHENOTYPES due

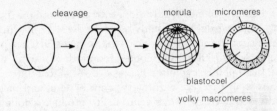

Fig. 70. **Blastula.** Formation of the blastula by cell cleavage.

to segregation of polygenes. However, the idea of blending was overtaken by the discoveries of MENDEL and his laws of particulate inheritance.

blepharoplast, *n*. a structure of unknown function which occurs at the base of the flagellum in some Protozoa.

blight, *n*. any of various plant diseases caused by microorganisms, where the entire plant is infected and soon dies.

blind spot, *n*. the point at which the optic nerve leaves the retina of the eye and from which light-sensitive nerve cells (rods and cones) are absent.

blood, *n*. a connective tissue with a liquid matrix called BLOOD PLASMA. Suspended in the plasma are three types of cell which form about 45% of total blood volume: (a) red blood cells or ERYTHROCYTES, (b) white blood cells or LEUCOCYTES and (c) cell fragments or PLATELETS. See Fig. 71.

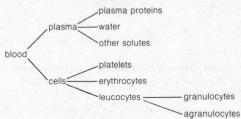

Fig. 71. **Blood.** The constituents of blood.

blood circulatory system, *n*. the mechanism by which blood is moved around the body. All vertebrates and a few invertebrates such as ANNELIDS have 'closed' blood circulatory systems which involve muscular tubes (BLOOD VESSELS) connecting the heart to all areas of the body. Other invertebrates (notably the ARTHROPODS) have an 'open' system in which a pool of blood is circulated by action of the heart.

The various vertebrate classes display a fascinating range of circulatory structures which is thought to relate to their evolutionary history (see HAEKEL'S LAW). The circulatory system of fish is regarded as primitive because blood from the heart goes straight from the respiratory organs (gills) to the body tissues of the systemic circulation. Such a single circulation through two CAPILLARY BEDS 'in series' produces a relatively sluggish flow of blood around the system, which is inefficient. In amphibia, evolution has progressed so that the circulation to the lungs is 'in parallel' with the systemic system, and the heart is partially divided (two atria, one ventricle) to produce a double route. The two separate circuits enable a higher blood pressure to be maintained and thus a more efficient flow of blood. However, a true 'double' circulation system did not occur until the birds and mammals evolved from the reptiles. In the mammal the heart is divided into two halves, the right half pumping blood to the lungs in the pulmonary circulation, the left half pumping blood to the body tissues in the systemic circulation. (See Fig. 72).

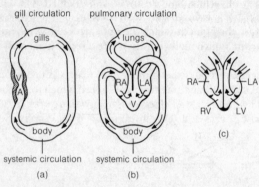

Fig. 72. **Blood circulatory system.** (a) Fish. (b) Amphibian. (c) Mammal (pulmonary and systemic circulation similar to amphibian). A = atrium, V = ventricle.

Although the circulatory system of the mammal is complex, it is possible to recognise several important features:

(a) in the systemic circulation, arteries carry oxygenated blood from the left ventricle of the heart, starting at the AORTA, while veins carry deoxygenated blood, collecting together to form the VENAE CAVAE which return blood to the right atrium of the heart.

(b) the pulmonary circulation is unusual in that the pulmonary artery carries deoxygenated blood from the right ventricle to the lungs, while the pulmonary vein carries oxygenated blood back to the left atrium of the heart.

(c) while all blood vessels are basically similar there are quite large differences between the various types. See ARTERY, VEIN and CAPILLARY.

(d) the blood supply from the gut does not feed back to the venae cavae directly, but via the liver in the HEPATIC PORTAL SYSTEM.

blood clotting, *n*. a condition where elements in the BLOOD PLASMA change consistency from a liquid to a gel-like structure, causing a 'clot' to form at the site of damage. The process of clotting is complex, but can be summarized as follows:

(a) damage to the blood circulatory system causes both BLOOD PLATELETS and the BLOOD VESSELS to secrete *thromboplastin* (see HAEMOPHILIA).

(b) thromboplastin, in the presence of calcium ions, causes *prothrombin* (a globulin type of plasma protein produced in the liver when vitamin K is present) to change to *thrombin* (see ANTICOAGULANT).

(c) thrombin acts as an ENZYME, causing FIBRINOGEN (another type of plasma protein) to be changed to fibrin which, when it contracts, forms a mesh of fibres in which blood cells become trapped. The final result of clotting is a hard lump which can plug the damaged area (if not too large) thus preventing further loss of blood and entry of microorganisms from outside. See Fig. 73.

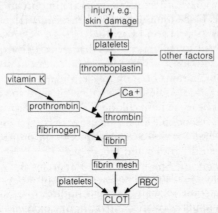

Fig. 73. **Blood clotting.** The process of clotting.

blood corpuscle, *n*. a white or red BLOOD cell suspended in a watery matrix, the plasma.

blood film, *n*. a smear of blood made on a microscope slide that is subsequently fixed and stained for examination.

blood fluke, *n. Schistosoma*, a close relative of the liver FLUKE which causes BILHARZIA.

blood grouping, *n*. a method in which blood is characterized by the presence or absence of particular ANTIGENS on the surface of red blood cells. The antigens can be recognised by their reaction to specific ANTIBODIES which causes AGGLUTINATION to occur. Two well-known blood groupings are the ABO BLOOD GROUP and RHESUS BLOOD GROUP systems.

blood islands, *n.* the beginning of the blood system in the vertebrate embryo in the form of cell masses which later form blood vessels.

blood pigments, *n.* the complex protein molecules found in blood and other tissues, which contain metal atoms in their structure. Such pigments have a high affinity for oxygen, functioning in oxygen transport, e.g. HAEMOGLOBIN in vertebrates (containing iron) and HAEMOCYANIN in molluscs and arthropods (with copper), and in oxygen storage within tissues, e.g. MYOGLOBIN (iron) found in muscle cells. See OXYGEN DISSOCIATION CURVE.

blood plasma, *n.* the watery matrix of BLOOD in which the blood cells are suspended. Plasma consists of about 90% water, acting as a solvent, and the following solutes;

(a) *plasma proteins.* The largest component (about 7% by weight), which is divided into three groups; albumin (55%), globulin (44.8%) and fibrinogen (0.2%). Plasma proteins are important in the maintenance of correct blood osmotic potential, the regulation of blood pH and (fibrinogen only) in BLOOD CLOTTING.

(b) NITROGENOUS WASTE substances that are carried from their site of production to the kidneys. These excretions consist mostly of UREA, but also have small amounts of ammonia and URIC ACID.

(c) *inorganic salts* of sodium, calcium, magnesium and potassium, the most common being sodium chloride. These salts form about 0.9% by weight and are responsible for maintenance of correct osmotic pressure and pH in the blood as well as maintenance of the proper physiological balance between the tissues and blood.

(d) *organic nutrients.* Among the most important are: (i) blood sugar, mainly glucose, derived from the breakdown of foods, either directly from the gut or from GLYCOGEN stored in the liver (glycogenolysis). The precise level of blood sugar is critical for the maintenance of homeostatis and is controlled by a negative FEEDBACK MECHANISM in which insulin plays a major part (see DIABETES). (ii) blood LIPIDS such as fats and cholesterol, derived from dietary intake or activity of the liver.

(e) hormones manufactured in the ENDOCRINE GLANDS.

(f) dissolved gases such as nitrogen (physiologically inert), small quantities of oxygen (mostly carried by haemoglobin in the ERYTHROCYTES) and carbon dioxide carried as bicarbonate (HCO_3^-) in the plasma (see CHLORIDE SHIFT).

blood platelets or **platelets** or **thrombocytes,** *n.* disc-like components of mammalian BLOOD, consisting of non-nucleated cytoplasmic fragments of large bone-marrow cells 3 μm in diameter called *megakaryocytes* that have entered the blood circulatory system. Platelets play an important part in BLOOD CLOTTING.

blood pressure, *n*. the force exerted by blood against the walls of the blood vessels, caused by heart contractions forcing a constant volume of blood round a closed system. Strong contraction of the left ventricle (SYSTOLE) ejects blood at high pressure into the AORTA, stretching the arterial walls. When the heart relaxes (DIASTOLE), force is no longer exerted on the arterial blood so pressure drops, although maintenance of pressure is helped by elastic recoil of the arterial walls. These oscillations of blood pressure are largest in the aorta, gradually diminishing as the blood flows along the arteries, becoming nonexistent in the CAPILLARIES.

The level of blood pressure also decreases from heart to tissue and back to the heart, these differences in pressure enabling the flow of blood around the system. See Fig. 74. Blood in the veins is prevented from moving backwards by the presence of one-way valves. Venous circulation is also enhanced by activity of the skeletal muscles, hence leg and arm movements aid blood flow back to the heart.

Note that, although the comments above refer to the systemic circulation, a similar situation also applies in the smaller pulmonary system of mammals (see BLOOD CIRCULATORY SYSTEM).

Several factors control the exact level of blood pressure: (a) heart action (rate of heartbeat, force per beat, volume per beat); (b) peripheral resistance to blood flow in the capillary beds, caused by friction; (c) elasticity of arteries; (d) total blood volume (the higher the volume the higher the pressure); (e) viscosity of blood (an increase in viscosity causes an increase in blood pressure but a decrease in flow rate).

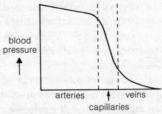

Fig. 74. **Blood pressure.** The variation of blood pressure in different vessels.

blood serum or **serum,** *n*. a fraction of the blood consisting of BLOOD PLASMA from which FIBRINOGEN (a clotting factor) has been removed. See Fig 75 on page 88.

blood sugar, see BLOOD PLASMA.

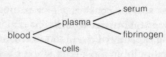

Fig. 75. **Blood serum.** The fractions of blood serum.

blood vessel, *n*. one of a number of muscular tubes found in higher invertebrates and all vertebrates which connect the heart to the tissues (via arteries and capillaries) and the tissues to the heart (via veins) forming a BLOOD CIRCULATORY SYSTEM.

bloom, *n*. an overabundant growth of algae, often resulting from nutrient enrichment.

blubber, *n*. the thick layer of fat deposited below the DERMIS which acts as a thermal insulation layer in some mammals, e.g. whales, seals.

blue baby, *n*. a rare condition in human infants where there is incomplete separation of oxygenated and deoxygenated blood in the heart. Some deoxygenated blood enters the aorta rather than the pulmonary artery, resulting in an inadequate oxygen supply to the body tissues and a 'bluish' appearance to the skin. The problem arises because of incomplete closure of two routes which cause bypassing of the lungs during foetal life: (a) DUCTUS ARTERIOSUS, a blood vessel connecting the pulmonary artery and the aorta and (b) FORAMEN OVALE, a hole in the septum which divides left and right atria. The condition is a serious one which can, however, be rectified by surgery.

blue-green algae, *n*. a division of algae-like organisms (Myxophyta or Cyanophyta) whose members are PROCARYOTES and live in marine or freshwater. They are usually blue-green in colour and propagate mainly by asexual reproduction. Some blue-green algae are filamentous (see FILAMENT sense 1), with no well-defined nucleus or chromatophores, and reproduce by fission, or fragmentation if filamentous. Some blue-greens live in very inhospitable environments, for example hot springs, where the temperature is in excess of 85°C. The pigments produced are chlorophyll (green) phycocyanin (blue) and phycoerythrin (red).

BMR, see BASAL METABOLIC RATE.

BOD, see BIOCHEMICAL OXYGEN DEMAND.

body cavity, *n*. the space within the body of most animals in which the gut and various organs are suspended. It normally contains fluid and is derived embryologically in different ways in different groups of animals. Absent in PLATYHELMINTHS and NEMERTINE worms, it is derived from the COELOM in vertebrates, the HAEMOCOEL in arthropods and molluscs and occurs as an intercellular space in nematode worms.

bog, *n*. an area of peat formation, typically in upland situations, which supports an extremely OLIGOTROPHIC vegetation. See FEN.

Bohr effect or **Bohr shift,** *n.* a phenomenon named after its discoverer, the Danish physiologist Christian Bohr (1855–1911), who showed that the oxygen-carrying capacity of blood HAEMOGLOBIN varies with PH. At high pH values haemoglobin has a high affinity for oxygen, but more acid conditions cause haemoglobin to release its oxygen, as in tissues with a high concentration of dissolved carbon dioxide. See OXYGEN DISSOCIATION CURVE.

bolting, *n.* an unusual lengthening of plant stems, due to elongation of cells, which can be induced by plant hormones called GIBBERELLINS producing a stem with long INTERNODES. Bolting can occur naturally, as when biennials such as cabbage and sugar beet produce elongated flower stalks in their first year rather than flowering in the second year. The problem arises when such plants are subjected to winter conditions in their first year (usually having been planted too early in the spring) which induces growth typical of the second year.

bolus, *n.* a soft mass of chewed food, suitable for swallowing, shaped by the tongue in the BUCCAL cavity.

Bombay blood type, *n.* a rare blood type in which individuals appear to be Group O (see ABO BLOOD GROUP) and yet can produce offspring of a type normally impossible for a Group O parent. It appears that Bombay types are homozygous (see HOMOZYGOTE) for a pair of recessive *h* alleles that inactivates the production of the ABO blood group antigen, even though they have the genes to manufacture these antigens, a form of EPISTASIS. See Fig. 76.

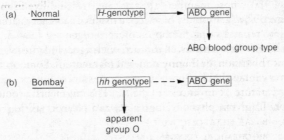

Fig. 76. **Bombay blood type.** (a) Normal and (b) Bombay blood group types.

bomb calorimeter, *n.* a CALORIMETER in the form of a thick-walled container in which organic material is ignited by electricity, burned, and the heat generated measured. The instrument is used to estimate the energy content of materials per unit weight.

bond, *n.* **1.** the force of mutual attraction that holds atoms together in molecules (see VAN DER WAALS INTERACTIONS and SULPHUR BRIDGE), such

as high-energy bonds in ATP, weak hydrogen bonds in DNA, PEPTIDE
BONDS and the disulphide bond of proteins. **2.** also called **pair bond**.
The attraction which maintains a male/female relationship, for purposes
of breeding, during the life cycle of some animals, mainly warm-
blooded vertebrates.

bone, *n.* the skeletal substance of vertebrate animals, consisting largely of
calcium and phosphate which make up 60% of the weight and gives it
hardness. This 'bone salt', together with large numbers of COLLAGEN
fibres, forms a matrix in which cells (OSTEOBLASTS) are distributed, and
these are connected by delicate channels (canaliculi). Larger channels
carry blood vessels and nerves (Haversian canals) and the cells are
arranged concentrically around them. Haversian bone is found in the
shafts of limb bones and is compact, whilst spongy bone is found at the
ephiphyses (ends of bone). See Fig. 77.

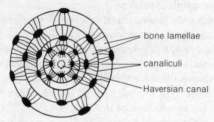

bone lamellae

canaliculi

Haversian canal

Fig. 77. **Bone.** The shaded areas are lacunae occupied by
osteoblasts.

bone marrow, *n.* a modified connective tissue of a vascular nature found
in long bones and some flat bones of vertebrates.

bony fish or **Osteichthyes,** *n.* the ACTINOPTERYGII and CHOENICHTHYES,
classes of fish that have a bony skeleton (as compared with cartilage in
ELASMOBRANCH fish).

bordered pit, *n.* a communicating channel between cells, found in the
VESSELS (2) and TRACHEIDS of higher vascular plants. See Fig. 78.

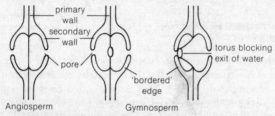

primary
wall

secondary
wall

pore

'bordered'
edge

torus blocking
exit of water

Angiosperm   Gymnosperm

Fig. 78. **Bordered pit.** Bordered pits in the tracheids of vascular
plants.

botany, *n.* the scientific study of the plant kingdom, usually including microorganisms.

botryoidal tissue, *n.* the MESENCHYME of leeches, consisting of dark-coloured, tubular cells containing blood-like fluid.

bottleneck, *n.* a period in the history of a population during which the number of individuals is reduced to a low number, perhaps by disease or extreme environmental conditions. As a result RANDOM GENETIC DRIFT may occur in the population. See also FOUNDER EFFECT.

botulism, *n.* a dangerous type of food poisoning, caused by toxins from the bacterium *Clostridium botulinum,* an obligate ANAEROBE that grows well in airtight containers (e.g. tins) which have not been properly sterilized before being filled. The toxin acts against the nervous system, particularly the CRANIAL NERVES, producing weakness and sometimes paralysis. The disease is named after the Latin *botulus* for sausage, a common source of the bacterium.

Bowman's capsule, *n.* the capsule attached to the end of the uriniferous tubule of the vertebrate kidney, which contains the GLOMERULUS. The capsule performs a process of ultrafiltration so that the composition of the capsular fluid contains all the constituents of the blood in the glomerulus minus the blood cells and plasma proteins. See Fig. 79.

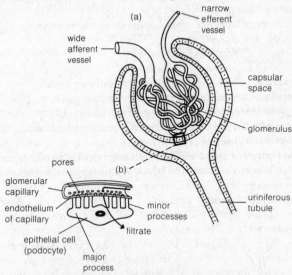

Fig. 79. **Bowman's capsule.** (a) The capsule is a small sac at the end of a kidney nephron. (b) An enlarged section of the uriniferous tubule.

Boyle's Law, *n.* the principle that the pressure of a gas varies inversely with its volume at a constant temperature. Named after Robert Boyle (1627–91).

BP, *abbrev. for* Before Present, a term used in GEOLOGICAL TIME scales to denote any time before the present.

brachi- or **brachio-,** *prefix.* denoting an arm.

brachial, *adj.* of or relating to the arm or to an armlike structure.

brachiation, *n.* the movement of an animal in its arborial habitat by swinging from tree to tree by means of the arms.

brachiopod or **lamp shell,** *n.* any marine invertebrate animal of the phylum Brachiopoda. They were the dominant marine forms of PALAEOZOIC and MESOZOIC times and a few species survive. See BIVALVE.

brachy -, *prefix.* denoting something short.

brachycephalic, *adj.* (of humans) having a short broad head, a characteristic common amongst the Mongolian people, where the breadth of the head is at least equivalent to the length. See CEPHALIC INDEX.

brachydactylic, *adj.* having abnormally short fingers or toes, an inherited characteristic in man (and some other animals), in which the phalanges are short and partially fused, and may even be reduced in number. The condition is controlled by a dominant autosomal gene, which was the first human Mendelian dominant gene discovered (see MENDELIAN GENETICS).

bract, *n.* a specialized leaf with a single flower or inflorescence growing in the axil. See also SPADIX.

bracteole, *n.* a small BRACT.

brady -, *prefix.* denoting slowness.

bradycardia, *n.* an abnormal reduction in heart rate.

bradykinin, *n.* a hormone formed from a kininogen in the blood plasma that has the effect of rapid VASODILATION in the skin.

bradymetabolism, *n.* the pattern of thermal physiology in which an animal possesses a relatively low BASAL METABOLIC RATE when measured at a body temperature of 37°C. In such animals a fall in core temperature simply results in a reduction of BMR. Although usually poikilothermic (see POIKILOTHERM) and ectothermic (see ECTOTHERM), many bradymetabolic forms are able to attain varying degrees of homoiothermy (see HOMOIOTHERM). All living lower vertebrates and invertebrates are bradymetabolic.

brain, *n.* the enlarged part of the CENTRAL NERVOUS SYSTEM beginning at the anterior end of bilaterally symmetrical animals (see BILATERAL SYMMETRY). The enlargement is associated with the aggregation of sense organs at the point which first contacts the changing environment. The brain, together with the rest of the central nervous system, coordinates the body functions. See also HEAD, CEPHALIZATION. See Fig. 80.

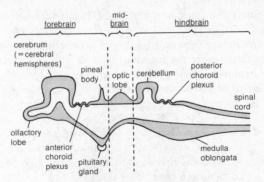

Fig. 80. **Brain.** The vertebrate brain.

brain stem, *n.* that part of the brain in vertebrates which excludes the CEREBELLUM and CEREBRAL HEMISPHERES.

branched pathway, *n.* a biochemical PATHWAY in which an intermediate substance serves as a precursor for more than one final product.

branchi-, *prefix.* denoting a gill.

branchial arches, *n.* the cartilagenous or bony masses supporting the gills of fish. There are usually five pairs of arches.

branchial chamber, *n.* a chamber or cavity containing the gills.

branchial clefts, *n.* the apertures in the walls of the PHARYNX of fish or young amphibia (four – seven pairs) which allow water entering by the mouth to pass out over the gills.

Branchiopoda, *n.* the most primitive class of Crustacea including the fairy shrimp *Chirocephalus* and the water flea *Daphnia*.

brassica, *n.* any member of the family Brassicaceae or Cruciferae, particularly members of the genus *Brassica*, e.g. cabbage, swede.

bread mould, *n.* the fungus *Mucor* or a related form (e.g. *Rhizopus*) which grows on bread. The term can also denote *Neurospora crassa*, a member of the class Pyrenomycetes which is associated with decomposing or burned vegetation. This organism was used by BEADLE and Tatum in their work on which the ONE GENE/ONE ENZYME HYPOTHESIS is based.

breast-feeding, *n.* (in humans) the direct feeding of an unweaned child on breast milk from the MAMMARY GLAND, as opposed to *bottle-feeding*.

breathing, *n.* a process in which air is taken into the lungs (*inspiration*) and then expelled from the lungs (*expiration*). In a mammal such as man the structures involved are as shown in Fig. 81 on page 94.

Inspiration occurs when the rib cage and the DIAPHRAGM alter shape to increase the volume and reduce the pressure in the thoracic cavity. The ribs are drawn up and outwards by contraction of the external INTERCOSTAL MUSCLES (chest breathing), while contraction of the

BREATHING

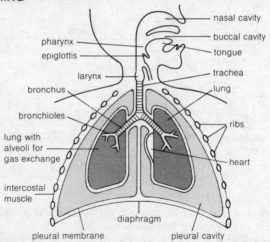

Fig. 81. **Breathing.** The human respiratory system.

diaphragm causes it to flatten downwards (abdominal breathing). Expiration can be simply a passive process in which the muscles relax, causing the ribs to drop down and inwards and the diaphragm to curve upwards. These movements reduce the thoracic volume which, along with elastic recoil of the lungs, results in a raising of the internal pressure so that air is expelled. Forced expiration can also occur where the abdominal muscles force the diaphragm further up into the thoracic cavity. See Fig. 82.

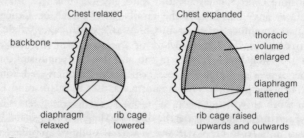

Fig. 82. **Breathing.** Expansion and relaxation of the chest during breathing.

Control of breathing is by a series of reflex actions so that, although the muscles involved are skeletal and therefore can be controlled voluntarily, respiration movements are automatic to a large extent. The main area of control is a *respiratory centre* in the hindbrain, located in the pons and the MEDULLA OBLONGATA. The steps in breathing control are as

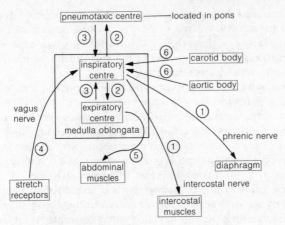

Fig. 83. **Breathing.** Steps in breathing control.

follows (see also Fig. 83, steps 1–6):

(a) spontaneous nerve impulses are generated in the *inspiratory centre* which cause contraction of the intercostal muscle (via intercostal nerve) and the diaphragm (via phrenic nerve). The thoracic volume increases.

(b) the *pneumotaxic* and *expiratory centres* are excited by impulses from the inspiratory centre.

(c) when sufficiently excited the pneumotaxic and expiratory centres send inhibitory signals to the inspiratory centre.

(d) at the same time as (c), stretch receptors in the walls of the lungs become excited and begin to send inhibitory signals (via the vagus nerve) to the inspiratory centre, which eventually stops stimulating the intercostal muscle and the diaphragm so that they now relax.

(e) the expiratory centre stimulates contraction of the abdominal muscles. Steps (d) and (e) cause reduction of thoracic volume so that the stretch receptors are no longer stimulated. As a result the breathing cycle begins again at step (a).

(f) finally, there is a chemical method of control. Excess carbon dioxide in the blood causes a lowering of pH which is detected in the AORTIC BODY and the CAROTID BODY as well as the respiratory centre itself. As a result, the inspiratory centre is stimulated to produce deeper breathing so as to reduce the level of CO_2 in the blood. This negative FEEDBACK MECHANISM thus relies not on oxygen levels but rather on CO_2 levels to control breathing rate. See Fig. 83.

breeding individual, *n.* an individual in a POPULATION that is involved in transferring its genes to the next generation via the process of SEXUAL REPRODUCTION. Thus a population will have two components: breeding individuals and those not breeding (juveniles and individuals that have undergone GENETIC DEATH).

breeding range, *n.* the geographical area within which a species breeds.

breeding season, *n.* the period of the year during which the female comes into reproductive condition.

breeding true, *n.* an organism that is not genetically variable for a particular character. Thus, crossing two parents (or self-pollinating one parent) of this type would produce offspring that were all identical with the parents. Such types are also known as 'pure-breeding' and were, for example, used by MENDEL in his experiments with the garden pea. In more precise terms, parents breed true when they are homozygous (see HOMOZYGOTE) for a pair of ALLELES of a particular *gene*.

breeding value, *n.* the (economic) worth of an individual's GENOTYPE, as judged by the average performance of its offspring. From the size of the breeding value it is possible to choose suitable parents for future crosses. See PEDIGREE ANALYSIS.

broad-sense heritability, *n.* the widest estimate of the role of genes in influencing total phenotypic variability (see HERITABILITY), measured by the ratio of total genotypic variance to total phenotypic variance. Compare NARROW-SENSE HERITABILITY.

bronchiole, *n.* one of the smaller tubes branching off the two main bronchii in the lungs of higher vertebrates. See BREATHING and Fig. 81.

bronchitis, *n.* an inflammation of the bronchial tubes.

bronchus, *n.* one of a pair of tubes (bronchii) linking the trachea to the lungs in mammals. Each bronchus consists mainly of connective tissue and a small amount of smooth muscle, the tubes becoming finely divided into bronchioles within the lungs, forming a 'bronchial tree'. See BREATHING and Fig. 81.

brood pouch, *n.* a pouch or cavity in certain animals in which eggs are contained and hatched.

brood spot or **patch,** *n.* a prolactin-induced (see LUTEOTROPHIC HORMONE) bare layer of skin in birds from which feathers are virtually absent and which receives a rich blood supply. Such patches are used to incubate the eggs.

brown earth soil, *n.* any mull soil typically found on loams or clays under deciduous forest. Most British agricultural soils fall into this group.

Brownian movement, *n.* a random movement of microscopic particles suspended in liquids or gases which results from the impact of molecules in the fluid around the particles. Such movements can be seen in COLLOIDS in a solid state or in a suspension of microorganisms. Named after Robert Brown (1773–1858).

brown fat, *n.* a special fat layer found between the neck and shoulders of some mammals, e.g. bats and squirrels, whose function is to enable the production of large amounts of heat, particularly after HIBERNATION. The fat is heavily vascularized (see VASCULAR) and has many mitochondria (see MITOCHONDRION), the latter giving it its brown colour due to the presence of mitochondrial cytochrome oxidase. Heat is released by very rapid fat metabolism (rather than the more normal fatty acid metabolism) and is rapidly transported away via the large vascular system.

brown rot, *n.* any fungal or bacterial disease of plants where browning and tissue decay occur.

brucellosis, *n.* a feverish disease caused by the bacterium *Brucella* that occurs commonly in cattle, sheep and goats. Infection of *B. abortus* in cattle can cause spontaneous abortion of calves and an attenuated live VACCINE has been developed to decrease the prevalence of the pathogen.

Brunner's glands, *n.* the specialized areas of the inner lining of the DUODENUM, lying at the bottom of the CRYPT OF LIEBERKUHN. The glands produce alkaline fluid and mucus but probably few or no digestive enzymes.

brush border, *n.* see MICROVILLUS.

bryophyte, *n.* any plant of the division Bryophta, a division containing the liverworts (see HEPATICA) and mosses (Musci). These plants do not possess a vascular system and there is always an ALTERNATION OF GENERATION with distinct SPOROPHYTE and GAMETOPHYTE generations.

bryozoan or **polyzoan** or **sea mat,** *n.* any aquatic invertebrate animal of the phylum Bryozoa. These are small animals superficially resembling hydroid COELENTERATES, because of their ciliated tentacles and horny or calcareous exoskeletons. There are two groupings, each sometimes referred to as a class though by many considered as separate phyla – the Ectoprocta and Endoprocta.

bubble respiration, *n.* respiration by means of a bubble attached to the rear end of an insect, e.g. a water beetle, which enables the insect to stay below water until the oxygen is used up, when the insect surfaces and takes on a new bubble.

buccal, *adj.* relating to the mouth cavity. For example, buccal epithelium lines the inside of the cheeks.

bud, *n.* an undeveloped embryonic shoot in a plant containing a meristematic area (see MERISTEM) for cell division, surrounded by leaf primordia (immature leaves) with often an outer protective layer of scales formed from modified leaves. The tip of a twig usually carries a terminal bud, while leaves generally have a lateral bud in their AXILS.

budding, *n.* a method of ASEXUAL REPRODUCTION common in some lower animal groups (e.g. COELENTERATES) in which part of the body wall bulges outwards and eventually forms a new individual which becomes detached from the parent. Budding can also take place in single-celled organisms such as yeasts, but here the process is more akin to MITOSIS with daughter cells of unequal size being produced. The production of 'plantlets' from the leaf margins of *Bryophyllum* is also called budding by some biologists.

buffer, *n*. a chemical substance which has the capacity to bond to H^+ ions, removing them from solution when their concentration begins to rise and releasing H^+ ions when their concentration begins to fall. In this way buffers stabilize the pH of biological solutions and are thus important in maintaining HOMEOSTASIS. HAEMOGLOBIN is an excellent example of a buffer, maintaining a stable pH in the ERYTHROCYTE.

bulb, *n*. a short, specialized underground stem in which nutrients (particularly sugars) are stored in fleshy scale leaves overlapping the stem. One or more buds are present in the AXILS of the scale leaves and in the spring these develop into new shoots, utilizing the stored food material. Examples of plants producing bulbs are onion, tulip and hyacinth. See Fig. 84.

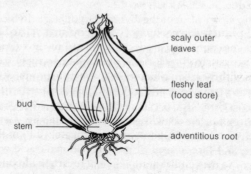

scaly outer leaves

fleshy leaf (food store)

bud

stem

adventitious root

Fig. 84. **Bulb.** Vertical section.

bulbil, *n*. a small BULB or tuber which arises on the aerial part of a plant in the axil of a leaf, or in an inflorescence.

bulla, *n*. a bony projection.

bundle end, *n*. the simplified ending of a VASCULAR BUNDLE in the mesophyll of a leaf.

bundle of His or **atrioventricular bundle,** *n*. a group of specialized muscle fibres in the mammalian heart carrying electrical signals from the ATRIOVENTRICULAR NODE down the septum between the ventricles and eventually dividing up into Purkinje fibres. See HEART and Fig. 183.

bundle sheath, *n.* a layer of PARENCHYMA that occurs round a VASCULAR BUNDLE.

bursa copulatrix, *n.* a depression around the genital aperture of insects which receives the male organ during copulation.

bursicon, *n.* an insect hormone produced in the central nervous system which tans and hardens the freshly moulted cuticle.

bush, *n.* a low woody perennial plant with branches at or near ground level.

butterfly, *n.* a diurnal insect of the order LEPIDOPTERA, which possesses clubbed antennae.

bypass vessel, *n.* a blood vessel which joins arteries and veins, thus bypassing capillaries. Constriction or dilation regulates the blood flow through the capillaries. See ARTERIOLE and Fig. 48.

byssus, *n.* the threads which attach certain molluscs to the SUBSTRATE (2) or the stalk in some fungi.

C

C$_3$ and C$_4$ plants, *n.* a grouping of higher plants, related to the carbon content of the compound produced when carbon dioxide is fixed during PHOTOSYNTHESIS. In C$_3$ plants (the most common type), carbon dioxide combines with 5-carbon ribulose diphosphate to produce two molecules of 3-carbon phosphoglyceric acid (PGA). In C$_4$ plants (e.g. maize), there is a special KRANZ ANATOMY in which mesophyll cells are adjacent to bundle sheath cells containing large chloroplasts. Carbon dioxide combines with 3-carbon phosphenol pyruvate (PEP) in the mesophyll cells to form 4-carbon oxaloacetic acid and malic acid which are then transported to the bundle sheath cells where carbon dioxide is released to go into the CALVIN CYCLE. The advantage of using PEP is that C$_4$ plants can fix carbon dioxide even when in low concentration in the atmosphere (e.g. in dense tropical vegetation). The system is also used in some XEROPHYTES as a CO_2 source during the day when the STOMA is closed. See Fig. 85 on page 100.

cactus, *n.* the common name for members of the Family Cactaceae, all of which, with one possible exception (*Rhipsalis*), are native to the continent of America. Most cacti are XEROPHYTES and succulents, found in deserts which have infrequent but heavy rainfall; cacti are absent from deserts with little or no rainfall. *Epiphylum* species and their relatives are found in rain forests and are chiefly EPIPHYTES. Cacti are distinguished

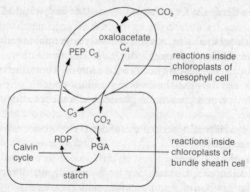

Fig. 85. **C₃ and C₄ plants.** Starch formation in C₄ plants.

from other succulents by having an *areole*, a pin-cushion type of structure from which wool, spines, new shoots and flowers develop. Other succulents do not possess an areole even though spines may be present.

cadophore, *n.* a projection from the dorsal side of TUNICATES to which individual newly separated offspring from the adult attach themselves.

caducous, *adj.* being shed at an early stage in development, as in sepals of the poppy, and stipules of the lime tree.

caecilian, see APODA.

caecum, *n.* **1.** (in animals) also called **hepatic caecum**. a blind-ending sac in the digestive system, which in mammals occurs at the junction of the small and large intestines. In herbivores the caecum contains bacteria that produce CELLULASE, enabling the breakdown of cellulose cell walls. See also APPENDIX. **2.** (in plants) a protrusion of the EMBRYO SAC into the endosperm tissue of seeds.

caeno- or **caino-,** *prefix.* denoting recent.

caenogenetic, *adj.* (of larval features) evolved because of use in the larval stage, being absent in the adult form.

caffeine, *n.* a bitter purine derivative found especially in coffee beans, tea leaves and cacao beans, that acts as a stimulant and a DIURETIC.

Cainozoic period, see CENOZOIC PERIOD.

Calamites, *n.* an extinct group of reed-like plants found in the upper Devonian, Carboniferous and Triassic periods. They are usually classified with EQUISETALES.

calcareous, *adj.* of, containing, or resembling calcium carbonate; chalky.

calcicole, *n.* a plant usually found on or limited to soils containing free calcium carbonate, such as *Fragaria vesca*, the wild strawberry.

calciferol or **vitamin D₂,** *n.* a compound that has vitamin D properties and is obtained by ultraviolet irradiation of ergosterol.

calcifuge or **oxylophyte,** *n.* a plant usually absent from soils containing free calcium carbonate, e.g. *Calluna*, the ling or heather.

calcitonin, *n.* a polypeptide hormone secreted by both the THYROID and PARATHYROID glands that lowers the calcium content of the blood.

calcium, *n.* an essential element to all animals and plants. Symbol: Ca.

callose, *n.* a thick CARBOHYDRATE plant material deposited in the pores of sieve plates in response to damage, disease or simply over a period of time, resulting in a blockage of the SIEVE TUBE. Callose is also present in the walls of fungal hyphae.

callus, *n.* a mass of immature plant cells which can differentiate into mature tissues, depending upon the relative concentrations of plant growth hormones present. Callus can develop from EXPLANTS in laboratory tissue culture experiments but also occurs naturally at the end of cut or wounded surfaces of shoots and roots.

calorie, *n.* the heat required to raise 1 g (1 cm³) of water through 1°C (i.e. from 14.5°C to 15.5°C). A Calorie (with a capital C) is used sometimes to denote a *kilocalorie*. Used as a unit of energy content or output, but now largely superseded by the S.I. unit joule (4.19 J = 1 cal).

calorimeter, *n.* an instrument for measuring quantities of heat, e.g. a BOMB CALORIMETER measures the heat output on burning a sample of material. Calorific values used in energetics are now measured in JOULES rather than CALORIES.

Calvin, Melvin (b. 1911) American biochemist best known for this work on CO_2 fixation in PHOTOSYNTHESIS. In his experiment he labelled CO_2 with ^{14}C and, using the 'Lollipop apparatus' (for exposing algal cells to light), and CHROMATOGRAPHY of cell extracts after varying time intervals, was able to isolate the various steps from CO_2 to starch, a sequence later called the CALVIN CYCLE. He was awarded a Nobel Prize for this work in 1961.

Calvin cycle, *n.* a series of chemical reactions, first described by Melvin CALVIN, which take place in the watery matrix of CHLOROPLASTS, where carbon dioxide is incorporated into more complex molecules and eventually carbohydrate. Energy for the reactions is supplied by ATP with NADPH (see NADP) acting as a reducing agent, both having been produced in the light reactions of PHOTOSYNTHESIS. Since light is not required for the Calvin cycle to continue (provided CO_2, ATP and NADPH are present) the steps are called the 'dark' reactions. See Fig. 86 on page 102. Every turn of the cycle fixes one molecule of carbon dioxide by producing two molecules of PGA and then two molecules of PGAL; three turns are necessary to release one molecule of PGAL (C_3) for

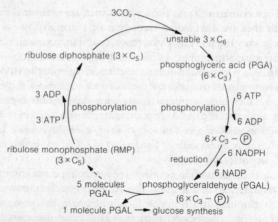

Fig. 86. **Calvin cycle.** The 'dark' reactions of the Calvin cycle.

the glucose pathway with the remaining five PGAL molecules remaining within the cycle. Thus six turns produce sufficient quantities of PGAL for the production of one molecule of glucose ($C_3 + C_3 = C_6$).

calyptera, *n.* a development of the archegonial wall (see ARCHEGONIUM) that in liverworts and mosses provides a hood-like covering to the capsule.

Calyptoblastea, *n.* an order of hydrozoan COELENTERATES in which the PERISARC surrounds the hydranths and gonothecae.

calyptrogen, *n.* a layer of cells covering the growing part of roots which gives rise to the root cap.

calyx, *n.* the group of SEPALS which usually form a whorl at the base of a flower.

CAM, see CRASSULACEAN ACID METABOLISM.

cambium or **fascicular cambium** or **lateral meristem,** *n.* a group of actively dividing cells found in the VASCULAR BUNDLES of roots and stems, whose function is to produce new plant tissue for lateral growth. INTERFASCICULAR CAMBIUM sometimes occurs between vascular bundles. Compare APICAL MERISTEM.

Cambrian period, *n.* a geological period which began about 515 million years ago and ended 445 million years ago. At this time the British Isles lay in the southern hemisphere and the Sahara was at the South Pole and subject to extensive glaciation. Trilobites and Brachiopods flourished and most invertebrate phyla had evolved. The first *graptolites* appeared in the mid-Cambrian, but few traces of plants have been found and these are largely limited to simple algae. See GEOLOGICAL TIME.

camel, *n.* the common name for a member of the mammalian family Camelidae that also includes dromedaries and llamas. They are even-toed UNGULATES and the RUMEN of the stomach has numerous pouches and diverticulae which are capable of storing large amounts of water.

camouflage, *n.* a disguise resulting from an organism having similar coloration to the background, or markings which cause breaking of the outline, so that the organism blends into the background and is hidden from predators. See CRYPTIC COLORATION.

campanulate, *adj.* (of plant structures) bell-shaped, as in the harebell flower, *Campanula rotundifolia*.

campylotropous, *adj.* (of a plant ovule) curved over so that the FUNICLE appears attached to the side between the MICROPYLE and the CHALAZA.

Canada balsam, *n.* a commonly used microscope slide mounting material for biological specimens. It is miscible with xylene.

canaliculi, *n.* fine channels, e.g. those which occur in bone.

canalization, genetical, *n.* the progression of a cell down a develop-mental pathway (see EPIGENETIC LANDSCAPE) as it becomes modified into its final, adult form (see CELL DIFFERENTIATION). Only at certain, fixed times in its development (see COMPETENCE) is the cell capable of being induced to move into another developmental pathway and thus become a different end type. The process of INDUCTION is thought to be due to chemicals secreted by tissues surrounding the developing cell.

canalizing selection, *n.* gene selection that stabilizes developmental pathways so that the phenotype is less affected by genetic and environmental disturbances.

cancer, *n.* a disease affecting the growth rate of affected tissues, in which the control mechanisms of cells become altered and the cells divide to form neoplastic growths or tumours. 'Benign' tumours consist of well-differentiated cells similar to those in the surrounding tissues and are usually harmless unless located in regions where no operation is possible. 'Malignant' tumours are dangerous and usually contain embryonic cells, which are capable of floating away and forming new malignant growths in other sites.

Little is known of the cause of cancer, though exposure to CARCINOGENS such as nicotine and mustard gas or the presence of certain microorganisms are possible causes. Radiation, surgery and chemo-therapy are all used in the treatment of different cancers.

canine tooth, *n.* one of four pointed teeth present in the jaw between the INCISORS and the PREMOLARS in mammals, being particularly prominent in the Carnivora and having a stabbing function.

canker, *n.* a plant disease giving a limited NECROSIS of affected tissue and caused by bacteria or fungi.

canopy, *n.* the branches and leaves of a woody plant, particularly trees, forming the uppermost light-restricting area some distance above the ground.

capacitance, *n.* the property of storing an electric charge by electrostatic means.

capacity, *n.* the ability to store an electric charge, measured in farads (Fd).

capillarity, *n.* the action by which the surface of a liquid (usually water) is elevated when in contact with a solid surface by attraction of molecules between the liquid and solid surfaces. When the liquid is in a narrow container (e.g. a capillary tube) the level of water will rise considerably, but capillarity can also occur in such structures as soils, causing a rise in the water table. Capillarity has been suggested as the explanation of water ascent in XYLEM vessels in plants, but is not considered to exert a significant effect. Compare COHESION/TENSION HYPOTHESIS, ROOT PRESSURE.

capillary, *n.* one of many minute blood vessels (5–20 μm diameter) which connect ARTERIOLES and VENULES in vertebrate tissues allowing a high level of exchange of materials to take place between blood and tissues via the interstitial fluid or LYMPH. See Fig. 87. Capillary walls are made up of a single layer of epithelial cells which are flexible, allowing changes in diameter with changing blood pressure. See Fig. 88.

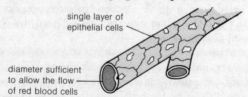

single layer of epithelial cells

diameter sufficient to allow the flow of red blood cells

Fig. 87. **Capillary.** Exchange route of materials between blood and tissues.

cells of tissue

interstitial fluid

capillary wall

capillary contents

Fig. 88. **Capillary.** The structure of a capillary vessel.

Capillaries are efficient because: (a) they are thin-walled, with narrow diameters giving a high surface-area to volume ratio, (b) they are very numerous, forming a CAPILLARY BED, and (c) blood flow is slow, allowing maximum time for exchange.

capillary bed. *n.* a fine network of blood CAPILLARIES which extends throughout each tissue, enabling efficient exchange of materials between cells and the blood. See ARTERIOLE and Fig. 48.

capillary soil water, *n.* water held between particles of soil by CAPILLARITY and taken up by plant roots.

capitate, *adj.* head-like.

capitellum, *n.* a rounded bone articulation. See CAPITULUM (1).

capitulum, *n.* **1.** (in animals) the rounded rib head which articulates with the centrum of the vertebra. **2.** (in plants) a head of flowers.

capsid, *n.* the protein coat of a virus.

capsule, *n.* a containing structure with a strong outer covering, found in many different groups, e.g.:

(a) the outer coat of some bacteria (referred to as *encapsulated*) which enhances resistance to the defences of the host (see TRANSFORMATION for genetic control).

(b) the sporangium of BRYOPHYTES (e.g. mosses) consisting of a hard outer layer inside which are developing spores.

(c) a type of fruit in ANGIOSPERMS that splits open when dry (DEHISCENT). Examples include poppy, willowherb, snapdragon.

(d) a spherical bony structure in some vertebrate skulls (e.g. the auditory capsule of dogfish).

(e) the blind-ending part of the kidney NEPHRON.

carapace, *n.* the part of the exoskeleton of a CRUSTACEAN that spreads over the head and thorax; in other animals, e.g. the tortoise, the carapace is the dorsal covering of bony plates.

carbamate, *n.* a type of insecticide, related to the organophosphates, the first of which (carbaryl) was introduced in 1956.

carbaminohaemoglobin, *n.* the compound formed by the reaction of HAEMOGLOBIN with carbon dioxide, which forms one of the methods of transportation of CO_2 in blood.

carbohydrase, *n.* an enzyme that catalyses the hydrolysis of carbohydrates.

carbohydrate, *n.* a family of organic molecules with the general formula $(CH_2O)_X$, ranging from simple sugars such as glucose and fructose to complex molecules such as starch and cellulose. All complex carbohydrates are built up from simple units called MONOSACCHARIDES which cannot be hydrolysed to a simpler structure.

The types of carbohydrate are described in detail under their own heading, but are summarized in Fig. 89 on page 106.

carbon, *n.* the element which is the basis of organic structure. Carbon has a valency of four, each atom forming four covalent bonds in its compounds. Long chains may be formed which give rise to the

Name	Type	Structure	Location
Glucose	Reducing monosacch.	$C_6H_{12}O_6$ (hexose)	sweet fruits
Fructose	Reducing monosacch.	$C_6H_{12}O_6$ (hexose)	honey, fruit juice
Lactose	Reducing disacch.	Glucose + Galactose	milk
Maltose	Reducing disacch.	Glucose + Glucose	germinating grain
Sucrose	Non-R. disacch.	Glucose + Fructose	sugar cane
Starch	Polysaccharide	repeated glucose units (linear)	potato tuber
Glycogen	Non-R. polysacch.	repeated glucose units (branched)	liver
Cellulose	Non-R. polysacch.	repeated glucose units (linear)	cell walls
Ribose	Reducing monosacch.	$C_5H_{10}O_5$ (pentose)	RNA
Deoxyribose	Reducing monosacch.	$C_5H_{10}O_5$ (pentose)	DNA

Fig. 89. **Carbohydrate.** The types of carbohydrate.

complexity of many organic compounds.

carbon-14, see CARBON DATING.

carbon cycle, n. the circulation of carbon by the metabolic processes of living organisms in an ECOSYSTEM so that it always returns to an arbitrary starting point. See Fig. 90.

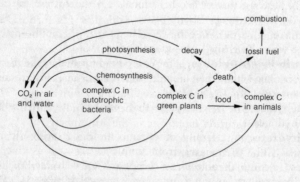

Fig. 90. **Carbon cycle.** The main steps.

carbon dating, n. the dating of organic remains by measuring the radioactive carbon content. Atmospheric carbon dioxide contains carbon atoms of two types, ordinary carbon ^{12}C and radioactive carbon ^{14}C. Like other radioactive isotopes, ^{14}C decays with age, so that the proportion of radioactive carbon present in, say, peat gives an indication of its age, assuming no more ^{14}C has been incorporated in it since it was a live plant, and also that the amount of ^{14}C in the atmosphere has remained constant. ^{14}C has a HALF-LIFE of 5,570 years and one carbon atom in every million million in the atmosphere is radioactive. Dating organic remains by the use of ^{14}C is a well-used and valuable technique, but not entirely accurate due to variations in atmospheric ^{14}C over long periods of time. Comparisons with tree ring dates (see DENDRO-CHRONOLOGY) have shown errors in the order of 900 years in 5,000.

carbon dioxide, *n.* a colourless, odourless gas, heavier than air, produced in respiration of organisms, and utilized to form sugars in PHOTOSYNTHESIS. Formula: CO_2.

carbon dioxide exchange, see GAS EXCHANGE.

carbonic acid, *n.* a compound formed by carbon dioxide and water.

carbonic anhydrase, *n.* an enzyme that accelerates the reaction between carbon dioxide and water to form carbonic acid in the red blood cells.

Carboniferous period, *n.* a geological period that began about 370 million years ago and ended 280 million years ago. It is often divided at about 325 million BP into Lower and Upper Carboniferous; the coal measures from which the period derives its name occurred mainly in the latter period. Club mosses, horsetails and ferns were the dominant plants during the period, and amphibians the commonest vertebrates, though this was also the time of the emergence of the reptiles. Britain crossed the equator.

carbon monoxide, *n.* a colourless, odourless gas formed by the incomplete oxidation of carbon; poisonous to animals. Formula: CO.

carboxyhaemoglobin, *n.* a stable compound produced when carbon monoxide combines irreversibly with HAEMOGLOBIN in the red blood cells, giving the blood a bright red colour. A consequence of the reaction is that haemoglobin is less able to combine freely with oxygen. This can lead to poisoning by lack of oxygen in the blood and eventually death if there is sufficient carbon monoxide in the atmosphere.

carboxylase, *n.* an enzyme that assists in the CARBOXYLATION or decarboxylation of a SUBSTRATE (3).

carboxylation, *n.* the addition of carboxyl (COOH) or carbon dioxide into a molecule as, for example, when carbon dioxide is added to ribulose diphosphate (C_5) in the CALVIN CYCLE to give an unstable C_6 product.

carboxyl group, *n.* the monovalent group –COOH, found in aldehydes and ketones.

carboxypeptidase, *n.* an exopeptidase that catalyses the hydrolysis of amino acids in polypeptide chains from the C-terminal.

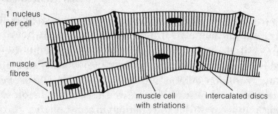

Fig. 91. **Cardiac muscle.** The intercalated discs enable the rapid transmission of excitatory waves across the tissue.

carcinogen, *n.* a substance which is a CANCER-causing agent.

carcinoma, *n.* a malignant tumour of epithelial tissue. See CANCER.

cardiac, *adj.* **1.** of or relating to the heart. **2.** see PYLORIC.

cardiac cycle, see HEART, CARDIAC CYCLE.

cardiac frequency, *n.* the rate at which the heart beats.

cardiac muscle, *n.* a type of vertebrate muscle found only in the HEART, which appears to be halfway between INVOLUNTARY MUSCLE and STRIATED MUSCLE in that its fibres are striated, but contain a single nucleus (see Fig. 91). The action of cardiac fibres is to produce strong and rhythmic contractions from within, even when removed from the body (see MYOGENIC CONTRACTION). Unlike striated muscle, cardiac muscle does not become fatigued even though it is repeatedly stimulated. The heartbeat is controlled by the AUTONOMIC NERVOUS SYSTEM.

cardiac output, *n.* the total volume of blood pumped out by the heart per unit time.

cardiac sphincter, *n.* the sphincter at the junction of the vertebrate oesophagus and stomach (the 'heart' end). Compare PYLORIC SPHINCTER.

cardinal veins, *n.* paired longitudinal vessels present in fish and vertebrate embryos which collect blood from most parts of the body and return it to the heart via the CUVIERIAN DUCTS.

cardiovascular centre, *n.* a group of nerve cells in the MEDULLA of the brain that controls heartbeat.

care of young, *n.* a feature of the behaviour of mammals and birds in which the young are tended by the adult(s) until such time as the young become independent. The phenomenon is less evident or absent in other groups of animals.

caridoid facies, *n.* the body arrangement found in shrimps, crayfish, lobsters, etc. (the MALACOSTRACA), where six body segments are present in the head, eight in the thorax and six in the abdomen. The animals have a CARAPACE, stalked compound eyes, SWIMMERETS on the abdomen and a tail fan.

caries, *n.* progressive decay of a bone or tooth.

carina or **keel,** *n.* **1.** the keel-shaped edge of a leguminous flower, consisting of two fused lower petals. This may have a role in pollination, acting, e.g., as a landing platform for bees. **2.** the breast-bone of a bird.

carnassial teeth, *n.* the last premolar of the upper jaw and first molar of the lower jaw, which form cutting edges in carnivorous mammals.

Carnivora, *n.* an order of EUTHERIAN mammals containing bears, weasels, wolves, cats and seals, most of which have well-developed incisors and canine teeth (for tearing flesh).

carnivore, *n.* any flesh-eating animal. The term is sometimes restricted to members of the CARNIVORA, although OMNIVORES also eat meat. See also CARNIVOROUS PLANTS.

carnivorous plants, *n.* plants which obtain at least some of their nutrients through trapping small animals, usually insects. In the UK such plants often grow in boggy places where nitrogen is difficult to obtain, extra nitrogen being extracted from the bodies of captured prey. The sundew *Drosera* and the bladderwort *Utricularia* are the commonest British carnivorous plants.

carotene, *n.* an orange plant pigment of the CAROTENOID group which is usually present in the CHLOROPLASTS, and sometimes occurs in pigment-containing structures called CHROMOPLASTS which are found in yellow/orange leaves, vegetables and fruits. Carotene is necessary for the production of vitamin A in man and has an ABSORPTION SPECTRUM of about 450 nm.

carotenoids, *n.* a group of yellow/orange pigments found in plants which are subdivided into CAROTENES (orange) and xanthophylls (yellow).

carotid artery, *n.* either of the two vessels carrying oxygenated blood from the AORTA to the head in the mammal and other vertebrates (see Fig. 92). Nervous receptor areas are located in the CAROTID SINUSES and the CAROTID BODIES.

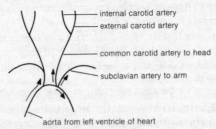

internal carotid artery
external carotid artery
common carotid artery to head
subclavian artery to arm
aorta from left ventricle of heart

Fig. 92. **Carotid artery.** Mammalian carotid artery.

carotid body, *n.* an area of gland-like tissue situated near the joining of the external CAROTID ARTERY to the common carotid which acts as a CHEMORECEPTOR (see Fig. 93 on page 110). When blood-oxygen

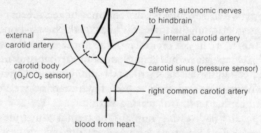

Fig. 93. **Carotid body.** Location.

concentration levels are low and/or the concentration of carbon dioxide is high, the carotid body is stimulated to produce nervous impulses which are transmitted to the respiratory centre in the hindbrain, thus influencing BREATHING rate.

carotid sinus, *n.* a small swelling at the base of the internal CAROTID ARTERIES which contains baroreceptors (pressure receptors) that act as STRETCH RECEPTORS (see CAROTID BODY and Fig. 93). When blood pressure drops below normal the sinus is stimulated to produce nervous impulses which are transmitted to the cardiovascular centre in the hindbrain (see HEART).

carpal, *n.* any bone present in amphibia, reptiles, birds and mammals in the part of the forelimb attached to the radius and ulna. They articulate distally (see DISTAL) with the METACARPALS. There are eight carpals in the human wrist. See Fig. 94.

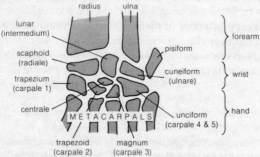

Fig. 94. **Carpal.** The human wrist and hand.

carpel or **pistil,** *n.* the flask-shaped female reproductive unit of a flower, composed of ovary, style and stigma. One or more carpels goes to make up the entire female structure, the GYNOECIUM.

carpogonium, *n.* the female sex organ of red algae which contains the egg in the swollen base. The male gamete enters the narrower mouth or TRICHOGYNE.

carpospore, *n*. the sexually produced spore of red algae.

carpus, *n*. that part of the forelimb of TETRAPODS which contains the CARPALS.

carrier, *n*. **1.** an individual plant or animal that is infected with pathogenic organisms internally or externally without showing signs of disease, and which is capable of transferring them to others, thus causing disease. **2.** an individual with a GENOTYPE containing a deleterious recessive gene that does not show in the PHENOTYPE. For example, a carrier of PHENYLKETONURIA.

carrier molecule, *n*. a molecule that is lipid–soluble and carries other molecules with a lower mobility within biological membranes. See ACTIVE TRANSPORT.

carrion feeder or **necrophagus feeder,** *n*. any organism that feeds on dead animals.

carrying capacity, *n*. the capacity of a particular habitat with reference to the maximum number of organisms which that habitat can normally support, e.g. the maximum number of wading birds an estuary can support.

cartilage or **gristle,** *n*. a form of connective or skeletal tissue characterized by the presence of rounded cartilage corpuscles (CELLS), surrounded by a matrix of mucopolysaccharide (CHONDRIN) in which, besides the cartilage cells, there are numerous collagen fibres. Cartilage forms the first parts of the skull, vertebrae and long bones of the developing embryo but in adult mammals (and many other vertebrates) is largely replaced by bone. It remains at the ends of bones, in joints, at the ventral ends of the ribs and in a few other places. The types of cartilage are as follows:

(a) *hyaline cartilage*, which is bluish–white and transluscent and contains some very fine collagen fibres. It is present at rib ends, in tracheal rings, in the nose, in the embryos of all vertebrates and the adult stages of cartilagenous fishes.

(b) *elastic cartilage*, containing yellow fibres and present in the ear and EUSTACHIAN TUBE.

(c) *fibrocartilage*, containing few cells and large numbers of fibres, and associated with joints subject to severe strains. It is present as discs between the vertebrae and in the pubic symphysis.

cartilage bone or **replacing bone,** *n*. any bone that develops from cartilage by means of OSTEOCLASTS breaking down the cartilage, and OSTEOBLASTS replacing it with bony material. For example, long bones.

cartilaginous fish, *n*. any fish of the classes Chondrichthyes or Elasmobranchii, including the sharks, skates and rays, whose skeleton is entirely cartilaginous.

caruncle, *n.* a wart-like growth present on the seeds of some ANGIOSPERMS.

caryopsis, *n.* an ACHENE in which the ovary wall and see coat are united.

casein, *n.* a phosphoprotein which is the principal protein in cheese.

Casparian strip, *n.* a waterproof thickening on the radial (side) and end walls of endodermal root cells (see ENDODERM) which is thought to influence the route by which water passes from the cortex into the VASCULAR BUNDLE of the STELE. See Fig. 95.

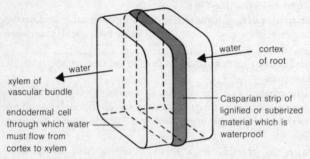

Fig. 95. **Casparian strip.** The strip surrounds the endodermal root cell.

caste, *n.* a specialized individual amongst social insects. For example, workers, and drones in a bee colony.

castration, SEE EMASCULATION (2).

casual, *n.* an introduced plant that has not become established in an area but which may occur away from cultivation.

catabolism or **katabolism,** *n.* a type of METABOLISM in which biochemical processes taking place in a cell result in the breaking down of complex compounds into simpler ones to release energy. Catabolism usually involves a series of step-by-step reactions, each catalysed by its own enzyme, for example, AEROBIC RESPIRATION.

catadromous, *adj.* (of fishes such as the eel) migrating down river to the sea to spawn. Compare ANADROMOUS.

catalase, *n.* an iron-containing ENZYME found in tissues such as liver and potato tubers whose function is to catalyse the breakdown of toxic hydrogen peroxide into water and oxygen:

$$H_2O_2 \rightarrow H_2O + \tfrac{1}{2}O_2$$

Catalase has a very high TURNOVER RATE and works by reducing the ACTIVATION ENERGY required from about 80 kJ to less than 10 kJ.

catalyst, *n.* a compound which is able to increase the speed of a reaction by lowering the ACTIVATION ENERGY needed for the reaction to start.

Protein catalysts are called ENZYMES and are found in all living cells.

cataphyll, *n.* a simplified leaf form, e.g. a bud scale, cotyledon, scale leaf.

catecholamine, *n.* any catechol–derived compound such as adrenalin or dopamine, which exert an action similar to that of the sympathetic nervous system (see AUTONOMIC NERVOUS SYSTEM).

Catharrhini, *n.* the primate group containing the Old World apes and monkeys and man. Members of this group have an internasal septum and a bony external ear passage. Compare PLATYRRHINI.

cathepsin or **kathepsin,** *n.* the intracellular, proteolytic enzymes that bring about AUTOLYSIS.

cathode, *n.* a negatively charged electrode to which positively charged ions move. Compare ANODE.

cathode–ray oscilloscope, *n.* an instrument that gives a video display of small electrical charges. The device is used to record electrical impulses from nerves and muscles.

catkin, *n.* an INFLORESCENCE usually in the form of a pendulous spike of unisexual, much–reduced flowers, e.g. a male catkin of hazel.

cattle, *n.* domesticated animals, usually oxen, but a term often extended to include sheep and pigs.

caudal, *adj.* relating to or in the position of the tail.

cauline, *adj.* (of leaves) borne on the aerial and mainly upper part of the stem of a plant.

C–banding, *n.* a technique of chromosomal staining in which chromosomes are exposed to alkaline and then acid conditions, in order to reveal bands of constitutive HETEROCHROMATIN that are identified with Giemsa stain.

cell, *n.* the structural unit of most organisms, consisting of a microscopic mass of protoplasm bounded by a membrane and containing one or more nuclei (in EUCARYOTES) or pieces of chromosome material (in PROCARYOTES). Cell size is roughly constant in all tissues and organisms, being limited by the physical restraints of unfavourable surface area/ volume ratios as size increases.

cell body, *n.* the enlarged region of a NEURON containing the nucleus.

cell cycle, *n.* the series of stages through which a cell progresses when it is actively dividing, consisting of three subdivisions of INTERPHASE (G1, S and G2) plus MITOSIS. See Fig. 96 on page 114.

cell differentiation, *n.* the process by which young cells develop into the various cell types making up the plant or animal structure. All young cells in an organism are thought to contain the same genetic information (DNA) and may become differentiated into any type of cell, the mature

CELL DIVISION

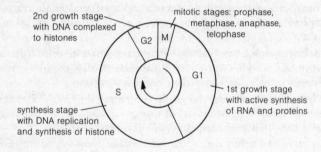

Fig. 96. **Cell cycle.** The interphase stages.

cell being produced by GENE SWITCHING rather than loss of DNA. Compare DEDIFFERENTIATION.

cell division, *n.* the division of a cell into two new cells during growth or reproduction. The new cells are called *daughter cells*. In PROCARYOTES, two identical cells are produced by BINARY FISSION. In EUCARYOTES, cell division is more complicated because the genetic material is located on CHROMOSOMES inside the nucleus, being composed of two distinct processes that usually occur together; (a) nuclear division (MITOSIS producing two identical nuclei, or MEIOSIS producing four nuclei with a halving of the genetic complement) and (b) cytoplasmic division (CYTOKINESIS).

cell fusion, *n.* the fusion of two cells in TISSUE CULTURE.

cell lineage, *n.* the developmental history of a cell from the ZYGOTE.

cell membrane or **plasma membrane,** *n.* the outer boundary of cells, the structure of which is visible only under the ELECTRON MICROSCOPE and which is still not clearly understood. Two major models have been proposed for membrane structure: the UNIT MEMBRANE MODEL and the FLUID–MOSAIC MODEL structures. The cell membrane gives shape and some protection to the cell, and also acts as a regulatory filter for transport of materials in and out of the cell (see ACTIVE TRANSPORT, DIFFUSION). Higher plants, fungi and bacteria have a CELL WALL outside the cell membrane.

cell plate, *n.* a membrane of differentially staining material in plant cells that appears at the equator of the spindle in TELOPHASE of MITOSIS, and is concerned with the formation of the MIDDLE LAMELLA between the two new cells. See CYTOKINESIS.

cell respiration, see CELLULAR RESPIRATION.

cell sap, *n.* a watery liquid contained in the VACUOLES of plant cells which is usually hypertonic (see HYPOTONIC) to the external medium. This causes the vacuoles to take in water by OSMOSIS, maintaining TURGOR in the cell.

cell structure, see CELL.

cell theory, *n.* the idea that all living things are composed of CELLS, first proposed by the German biologists Mattias Schleiden and Theodor Schwann in 1838/39. The cell theory is also used to suggest that all living cells come from pre-existing living cells and do not arise by SPONTANE-OUS GENERATION.

cellular respiration or **cell respiration,** *n.* a catabolic process (see CATABOLISM) occurring in cells where complex organic molecules are broken down to release energy for other cellular processes. Cell respiration usually occurs in the presence of oxygen (see AEROBIC RESPIRATION) but some organisms can respire without oxygen (see ANAEROBIC RESPIRATION).

cellulase, *n.* an ENZYME capable of splitting CELLULOSE into glucose, used particularly in the softening or digestion of plant cell walls. Most animals are not capable of producing cellulase and therefore of digesting plant material themselves, relying instead on a variety of gut micro-organisms to produce the enzyme and then absorbing the glucose product (see CAECUM). Cellulase is also produced in large quantities in the ABSCISSION layer formed in leaf stalks of higher plants, causing a weakening of cell walls prior to leaf fall.

cellulose, *n.* a type of unbranched polysaccharide carbohydrate composed of linked GLUCOSE units which can be hydrolysed by the enzyme CELLULASE. Cellulose is the main constituent of plant cell walls and is the most common organic compound on earth.

cell wall, *n.* a thick, rigid coat formed outside the CELL MEMBRANE of plants, fungi and bacteria, that is composed mainly of CELLULOSE secreted by the protoplasm of the cell. Older cells may also produce a secondary wall inside the primary wall which is thicker and contains LIGNIN for extra strength. Such cells often die after producing the secondary wall, e.g. XYLEM vessel cells.

cement, *n.* the spongy bone-like substance surrounding the roots of mammalian teeth, which assists in holding the teeth in sockets. Part of the enamel of the crown of the teeth in some mammals, e.g. ungulates, is also covered by cement.

cenospecies, *n.* all members of an ECOSPECIES related to each other by the ability to breed and exchange genes through hybridization, e.g. dogs and wolves.

Cenozoic or **Cainozoic period,** *n.* a geological era that includes the QUATERNARY PERIOD and TERTIARY PERIOD, beginning some 65 million years ago and extending up to the present time. See GEOLOGICAL TIME.

centi-, *prefix.* denoting a hundred.

central dogma, *n.* the hypothesis (based on WEISMANNISM) that genetical information flows only in one direction, from DNA to RNA to PROTEIN, and not in the opposite direction. Thus, in general, changes to protein structures produced by external forces are not inherited. See SOMATIC MUTATION.

central nervous system (CNS), *n.* the main mass of nervous material lying between the EFFECTOR and the RECEPTOR organs, coordinating the nervous impulses between the receptor and effector. The CNS is present in vertebrates as a dorsal tube which is modified anteriorly as the BRAIN and posteriorly as the SPINAL CORD; these are enclosed in the skull and backbone respectively. In invertebrates the CNS often consists of a few large cords of nervous tissue associated with enlargements called ganglia (see GANGLION). In some forms, i.e. COELENTERATES, the place of the CNS is taken by a diffuse nerve net. In addition to relaying messages for the sense organs, (in higher organisms at least) the CNS takes on an additional activity of its own in the form of memory which is the storage of past experiences. See also AUTONOMIC NERVOUS SYSTEM.

centric fission, *n.* the splitting of a chromosome at or near the CENTROMERE to produce two chromosomes.

centric fusion, *n.* the joining together of two ACENTRIC CHROMOSOMES to produce one METACENTRIC CHROMOSOME.

centric leaf, see LEAF.

centrifugation, *n.* a process in which particles in suspension are subjected to centrifugal forces by being spun in a CENTRIFUGE, so forming a sediment. Subsequent removal of the SUPERNATANT fluid from the sediment and subjecting it to higher speeds of centrifugation can enable particles of different sizes to be separated out and collected. Eventually the supernatant fluid will contain only soluble materials. See DIFFEREN-TIAL CENTRIFUGATION, ULTRACENTRIFUGE.

centrifuge, *n.* a rotating machine that separates liquids from solids, or dispersions of one liquid in another liquid, by the action of centrifugal force.

centriole, *n.* one of a pair of small ORGANELLES lying at right angles to each other just outside the nucleus of lower plants and all animals (see Fig. 97). Centrioles are self-replicating, dividing into two during the 'S' phase of the INTERPHASE of the CELL CYCLE and then separating into two pairs, one pair migrating to each pole of the future mitotic spindle, from which an ASTER forms. The role of centrioles in nuclear division is unclear, since they are absent from most plant cells and laser-beam irradiation of centrioles has no effect on division.

centrolecithal, *adj.* (of eggs) having the yolk aggregated in the centre.

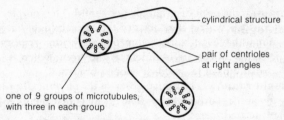

Fig. 97. **Centriole.** Structure and orientation.

centromere, *n.* a structure occurring along the length of a CHROMOSOME, often visible under the light microscope as a bump or a constriction whose location can help to identify the chromosome. The centromere contains a complex system of fibres called the *kinetochore* which becomes duplicated when the chromosomes divide into CHROMATIDS. The kinetochore attaches to SPINDLE microtubules during nuclear division. Damaged chromosomes without centromeres (ACENTRIC CHROMOSOMES) fail to move normally during nuclear division.

centrosome, *n.* an area of cell CYTOPLASM found near the nucleus, whose function is thought to be the organization of nuclear division since it is capable of assembling and disassembling MICROTUBULES. When nuclear division starts, the centrosome divides into two organizing centres which migrate to each pole (along with the CENTRIOLES, if present) and the spindle develops between them.

centrum, *n.* the central part of each VERTEBRA. The centrum lies ventral to the spinal cord and derives embryologically from the NOTOCHORD. Each centrum is joined to the centrum of the next vertebra by COLLAGEN fibres.

Cepaea, *n.* a genus of land snail that has been much studied because of its GENETIC POLYMORPHISM.

cephal- or **cephalo-,** *prefix.* denoting the head.

cephalic, *adj.* pertaining to the head.

cephalic index, *n.* an anthropological measurement obtained by dividing the maximum head width by the maximum length (back to front) and expressing it as a percentage.

dolichocephalic	$<75\%$	(long head)
mesocephalic	75–80%	
brachycephalic	$>80\%$	(short head)

The ratios of one index to another vary between populations and have been used in studies of race.

cephalization, *n.* the development of the head of an organism during evolution. The development is largely associated with a concentration of sense organs at the front of an animal (see BRAIN). The complexity of the head region depends on the habits of the animal; highly active, free-living animals have the best-developed cephalization requiring maximum detection of the environment at the front, whereas sedentary forms normally have less well-developed heads.

Cephalochordata, *n.* a primitive group of the phylum Chordata that is usually given the status of a class. The group contains only AMPHIOXUS and a few related forms.

cephalopod, *n.* any member of the class Cephalopoda containing molluscs such as squids (10 arms) and octopuses (8 arms). They have a well-developed brain, eyes very similar to vertebrate eyes (an example of CONVERGENCE), and are capable of rapidly changing colour through the possession of CHROMATOPHORES.

cephalothorax, *n.* the fused head and thorax which occurs in many ARTHROPODS, particularly CRUSTACEANS.

cerato, *prefix.* denoting horn or a hornlike part.

cerc- or **cerco-,** *prefix.* denoting the tail.

cercaria, *n.* the last larval stage of the liver FLUKE. It lives in the freshwater snail *Limnaea*, produces a cyst round itself and develops into the adult fluke after ingestion by a sheep or another primary host.

cerci, *n.* appendages which are often sensory and occur at the posterior tip of the abdomen in many insects. Cerci may be short, blunt or long, giving the insect the appearance of having a forked tail.

cere, *n.* that portion of the base of the upper mandible of the bill in birds which is swollen and appears waxy.

cereal, *n.* any member of the plant family Gramineae which produces edible fruits, e.g. rice, wheat, maize, oats and barley.

cerebellum, *n.* the anterior dorsal part of the HINDBRAIN which controls muscular coordination. It is best developed in birds and mammals; in the latter there is a cortex of grey matter and the surface is complexly folded. The folds are lined with PURKINJE CELLS. Removal of the cerebellum unbalances an animal and affects the accuracy of voluntary movements.

cerebral cortex, *n.* a layer of GREY MATTER forming the most superficial layer of the roof of the CEREBRAL HEMISPHERES in the forebrain of higher vertebrates. The cerebral cortex consists of millions of densely packed nerve cells and is rich in synapses. It is present in reptiles, birds and mammals and in some lower vertebrates. See BRAIN.

cerebral hemisphere, *n.* one of a pair of large lobes in the forebrain of vertebrates. In reptiles, birds and mammals the coordinating function is dominant and the cerebral hemispheres control most of the activities of the animals, whereas in lower vertebrates the hemispheres are associated mainly with the sense of smell. In mammals the enlargements of the cerebral hemispheres, the largest part of the BRAIN, are caused by the development of the neopallium which forms the entire roof and sides of the forebrain. The frontal lobes are particularly developed in humans and are the seat of memory, thought and a considerable part of what is considered to be intelligence. See Fig. 98.

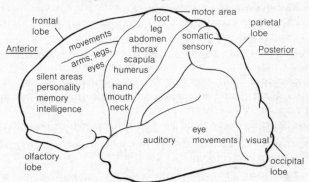

Fig. 98. **Cerebral hemispheres.** Localization of functions in the human brain.

cerebro-, *prefix.* denoting the brain.

cerebrospinal fluid (C.S.F.), *n.* a solution secreted by the CHOROID PLEXUSES (one in the roof of each of the four brain VENTRICLES (2) in man) which fills the cavity of the brain and SPINAL CORD and the space between the CENTRAL NERVOUS SYSTEM and its ensheathing membrane. It contains most of the small molecules found in blood, e.g. salts and glucose, but little protein and few cells, and serves as a nutritive medium. In humans the total volume is about 100 cm^3.

cerebrum, *n.* that part of the forebrain which expands to form the CEREBRAL HEMISPHERES, found in all vertebrates except fishes.

cerumen, *n.* the wax which is secreted in the ear of mammals.

cervic- or **cervico-,** *prefix.* denoting the neck or cervix.

cervical, *adj.* of or relating to the neck or cervix.

cervix, *n.* the neck of the UTERUS in female mammals that leads into the VAGINA. It contains numerous glands which supply MUCUS to the vagina.

Cestoda, *n.* a phylum of the Platyhelminthes containing the parasitic tapeworms, the adults of which are intestinal parasites of vertebrates. They have complex LIFE CYCLES usually involving intermediate hosts that are preyed upon by the primary host which thus becomes infected.

cetacean or **whale,** *n.* any aquatic EUTHERIAN mammal of the order Cetacea, comprising the porpoises, dolphins and whales. Cetaceans are completely aquatic, have lost their hind limbs during evolution and have fin-like forelimbs, often a dorsal fin and, in marked comparison with fishes, a blowhole and transverse tail fin. The young are large at birth and suckled like other mammals. Cetaceans possess an extensive outer layer of fat associated with their marine habit, have a large brain and are capable of complex communication.

chaet- or **chaeto-**, *prefix*. denoting hair.

chaeta or **seta**, *n.* a bristle-like structure, often chitinous (see CHITIN). The structures are found particularly in ANNELID worms, where they occur as projections on most segments, and serve as anchors for that segment during locomotion.

chaetognath, *n.* any small wormlike marine invertebrate of the phylum Chaetognatha, such as the arrowworms, which are INDICATOR SPECIES of water type.

Chaetopoda, *n.* the ANNELID worms of the classes Polychaeta or Oligochaeta, characterized by the possession of CHAETAE. See OLIGOCHAETE, POLYCHAETE.

chain reaction, *n.* a chemical or atomic reaction in which the products of each state promote a subsequent reaction. Initially, there is a slow induction period, but as the reaction progresses the reaction rate is accelerated.

chalaza, *n.* the basal part of the OVULE of a flowering plant to which the FUNICLE (1) is attached, and through which the pollen tube sometimes enters the ovule prior to fertilization.

chalice cell, see GOBLET CELL.

chalone, *n.* one of a group of chemicals (PEPTIDES and GLYCOPROTEINS) occurring in living animal tissues which specifically inhibit MITOSIS.

chamaeophyte, *n.* a plant that bears its perennating buds (see PERENNATION) between soil level and a height of 25 cm. See RAUNKIAER'S LIFE FORMS.

character, *n.* a genetic feature of an individual that can be assessed in some way. Such features often appear in various alternative 'forms', each controlled by different ALLELES. For example, the height of a pea plant is a 'character', with tall plants (2 m high) and dwarf plants (0.3 m high) as alternative forms (see QUALITATIVE INHERITANCE). Characters sometimes display a continuous range of forms, as in human height, which may be influenced strongly by environmental conditions (see POLYGENIC INHERITANCE). See also MULTIPLE ALLELISM.

character displacement, *n.* the divergence of characteristics (see CHARACTER) in sympatric species (see SYMPATRIC SPECIATION), as a result of the selective effects of competition.

character index, *n.* a value made up of the sum of the ratings of several

GENETIC CHARACTERS which is used in NUMERICAL TAXONOMY to indicate degrees of difference between related taxa. See KEY.

Charophyta, *n.* a group of the ALGAE, including the stoneworts, which inhabits ponds.

chartaceous, *adj.* (of plant parts) having a papery texture.

Chase, Martha, see HERSHEY.

chasmogamic, *adj.* (of a flower) opening before pollination.

checklist, *n.* a list of the organisms contained within a GROUP which can be used for quick reference in the arrangement of collections, for example, a list of the organisms' presence in various habitats.

cheese, *n.* a partially fermented, coagulated product of milk. Varieties of cheese are produced by the use of different microorganisms in the fermenting process (and by the use of different types of milk).

cheironym, *n.* an unpublished manuscript name for an organism.

Cheiroptera, see CHIROPTERA.

chel-, *prefix.* pertaining to claws.

chela, *n.* the large pincer-like claw of crabs and lobsters, formed from the modification of the 5th segment (dactylopodite) and 6th segment (propodite) of the limb.

chelation, *n.* the binding of a metal ion to an organic molecule from which it can later be released. In complex molecules, chelation results, for example, in zinc binding with amino acids in carboxypeptidase enzymes. The process also enables plants to take up metal ions such as iron that are not readily available in a free state.

Chelonia, *n.* the order of reptiles containing the tortoise and turtles.

chemiosmosis, see ELECTRON TRANSPORT SYSTEM.

chemoautotrophic or **chemotrophic,** *adj.* (of an organism) capable of manufacturing organic molecules from carbon dioxide and water by a process of CHEMOSYNTHESIS. For example, *Thiobacillus* oxidizes hydrogen sulphide to sulphur in order to produce the energy for chemosynthesis. Chemoautotrophic organisms are one form of AUTOTROPH, the other form displaying PHOTOTROPISM.

chemoheterotroph, *n.* a CHEMOAUTOROPHIC heterotroph.

chemoreceptor, *n.* a RECEPTOR that is stimulated by contact with molecules and is capable of reacting to and differentiating between different chemical stimuli, for example in taste and smell.

chemostat, *n.* a growth chamber designed to allow input of nutrients and output of cells on a controlled basis.

chemosynthesis, *n.* the process of obtaining energy and synthesizing organic compounds from simple inorganic reactions. This is brought about by special methods of respiration involving the oxidation of inorganic compounds such as iron, ammonia and hydrogen sulphate, and is carried out by several kinds of CHEMOAUTOTROPHIC bacteria. See AUTOTROPH.

chemotaxis, *n.* the orientation of an animal in relation to the presence of a particular chemical, the response being negative (moving away) or positive (moving towards). For example, the movement of a wasp towards an attractive odour such as beer would be positive chemotaxis.

chemotherapy, *n.* the use of chemical substances to combat disease caused by microorganisms. The term is often extended to include cancer treatment by chemicals.

chemotrophic, see CHEMOAUTOTROPHIC.

chemotropism, *n.* a growth response in plants in relation to a particular chemical. For example, growth of the pollen tube after pollination perhaps due to a chemical substance in the stigma.

chiasma, *n.* the cross-shaped configuration of chromatids produced during CROSSING OVER in meiosis.

chickenpox, *n.* an acute, contagious disease distinguished by slight fever and skin vesicles, and caused by the herpes varicella–zoster virus. The same virus can cause shingles in adults, a painful inflammation along the path of a major nerve, frequently across the back or sides.

chief cells, *n.* the gastric epithelial cells that release PEPSIN.

chilo-, *prefix.* pertaining to a lip.

chilopod, *n.* any arthropod of the class Chilopoda (centipedes), carnivorous MYRIAPODS distinguished by having one pair of legs on each segment.

chimaera, *n.* **1.** also called **graft-hybrid**. an organism, usually a cultivated plant, whose tissues are of more than one genetical type resulting from mutation or grafting. **2.** a smooth-skinned cartilaginous deep-sea fish ('king of the herrings') of the subclass Holocephali, which were common in the Jurassic.

Chiroptera or **Cheiroptera,** *n.* the order of mammals that includes the bats, the wings of which are formed from a membrane of skin, the patagium, stretched from the front to hind limbs and over the fingers (but not the thumb) of the forewing.

chi-squared test, *n.* a statistical routine which is a test of SIGNIFICANCE, comparing the observed results of an experiment or sample against the numbers expected from a theory or prediction. The test produces a value called chi-squared (X^2) which is:

$$X^2 = \text{sum of} \ \frac{\left(\begin{array}{c}\text{observed no.} \\ \text{in class}\end{array} - \begin{array}{c}\text{expected no.} \\ \text{in class}\end{array}\right)^2}{\text{expected no. in class}} \ \text{for each class}$$

The X^2 number is then converted to a probability value (P) using an X^2

table. If the P value is larger than 5% we can conclude that there is 'no significant difference' between the observed results and those expected, any deviation being due to chance. If, however, the probability is less than 5%, it must be concluded that there is a 'significant difference' between the observed results and those expected from theory. Note that the X^2 test can only be used with data that fall into discrete categories, e.g. heads or tails, long or short, yellow or orange.

chitin, *n*. a POLYSACCHARIDE containing nitrogen that has considerable strength because of its long fibrous molecules. It is resistant to chemicals and is found in the cuticle of insects and some other arthropods where the outer parts are impregnated with tanned proteins which gives it added strength. Many fungi contain a chitin-like substance in their cell walls.

chlamydo-, *prefix*. pertaining to a cloak.

Chlamydomonas, *n*. a flagellate, unicellular green alga now normally placed in the division CHLOROPHYTA but in many (older) texts regarded as a protozoan and placed in the class MASTIGOPHORA. It is characterized by the presence of two flagella (see FLAGELLUM) and, because of the presence of CHROMATOPHORES (1), is referred to as a phytoflagellate (often given subclass status: Phytoflagellata). Reproduction is by repeated fission or varies between species from an asexual method to a sexual method of CONJUGATION where ISOGAMETES rather than differentiated male and female gametes are produced. See Fig. 99.

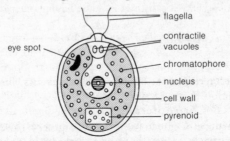

Fig. 99. **Chlamydomonas.** Generalized structure.

chlamydospore, *n*. an asexually produced, thick-walled fungal spore which is capable of surviving adverse conditions that are unfavourable to the main body of the fungus.

chloragogen cells, *n*. the yellow cells surrounding the gut of earthworms which break loose into the COELOM, absorb nitrogenous waste, break up, and are either excreted via nephridial tubes or deposited elsewhere as pigment.

chloramphenicol, *n*. a BACTERIOSTATIC antibiotic produced by a species of *Streptomyces* that inhibits protein synthesis in a variety of organisms.

CHLORELLA

Chlorella, *n.* a microscopic green alga of high protein content, used commercially as an additive in foodstuffs and confectionary.

chlorenchyma, *n.* plant cells (usually of the PARENCHYMA type) that contain CHLOROPLASTS in their cytoplasm. The cells are found in leaves (see MESOPHYLL) and other plant organs.

chloride secretory cells, *n.* cells present in the gills of TELEOST fish which actively extrude salts in marine conditions. In freshwater, the cells move salts into the body from the surrounding water.

chloride shift, *n.* the movement of chloride ions into a red blood cell (see ERYTHROCYTE) from the plasma. The carriage of carbon dioxide results in an accumulation of bicarbonate ions in the red blood cell, but because of the permeability of the cell membrane to negative ions they readily diffuse out into the plasma. The red blood cell thus develops a net positive charge (because of its retaining positive ions) which is neutralized by inward movement of negative chloride ions from the plasma. This ensures ionic and electrical stability during the transport of carbon dioxide. See Fig. 100.

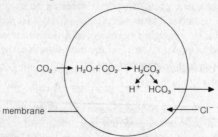

Fig. 100. **Chloride shift.** The entry of chloride ions (Cl^-) into a red blood cell from the surrounding plasma. Carbon dioxide is carried into the cell from respiring tissues.

chlorocruorin, *n.* a green respiratory pigment closely related to haemaglobin that is present in some polychaete worms.

chlorophyll, *n.* a group of pigments giving a green coloration to most plants that is found in any part of the plant that is exposed to sunlight. The pigments are usually contained in cell organelles called CHLOROPLASTS. Chlorophyll exists in several forms of which chlorophylls *a* and *b* are the most common, and has the vital function of absorbing light energy for PHOTOSYNTHESIS. See ACTION SPECTRUM.

Chlorophyta, *n.* the green algae, which constitute the largest division of algae, ranging from microscopic unicellular forms which are nonmobile or have flagella (see FLAGELLUM), to large forms with a flattened THALLUS. Reproduction may be asexual (see CELL DIVISION, FRAGMENTATION,

ZOOSPORE) or sexual (see ANISOGAMY, ISOGAMETE). They occur terrestrially in damp places such as tree trunks, in fresh water or in the marine environment.

chloroplast or **granum** *n.* a type of PLASTID containing CHLOROPHYLL found within the cells of plant leaves and stems. The chlorophyll is packed within granules called *quantasomes* which are located in the walls of flattened sacs called LAMELLAE. The lamellae lie closely together in certain areas to form grana and are more separate in the intergranal regions. LIGHT REACTIONS of photosynthesis are thought to take place in the quantasomes while DARK REACTIONS occur in the watery matrix surrounding the lamellae, the *stroma*. See Fig. 101.

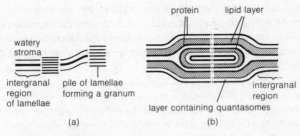

Fig. 101. **Chloroplast.** (a) Thylakoid arrangement. (b) An enlarged section of a granum.

chlorosis, *n.* a yellowing of plant leaves caused by lack of CHLOROPHYLL pigment due to mineral deficiency (e.g. magnesium, iron) or disease (e.g. virus yellows) which results in a decrease in photosynthetic rate.

choan- or **choano-,** *prefix.* pertaining to a funnel.

choanae, *n.* the internal nostrils which open into the roof of the mouth in all air-breathing vertebrates.

choanate, *n.* any vertebrate possessing internal nostrils.

Choanichthyes, *n.* the class of fishes closely related to the basic stock from which amphibians arose. Common in the Devonian and Carboniferous periods, the group includes present-day lungfish and COELOCANTHS.

choanocyte or **collar cell,** *n.* a cell in which the FLAGELLUM is surrounded by a protoplasmic sheath, occurring only in sponges (Porifera) and choanoflagellates.

choanoflagellate or **collar flagellate,** *n.* any stalked protozoan that occurs either singly or in branching colonies, possessing a FLAGELLUM surrounded by a cup-like structure into which food particles are wafted by flagellar movement. They are very similar to CHOANOCYTE cells.

choice chamber, *n.* a piece of apparatus used in experiments in animal behaviour in which it is possible for an organism to choose between, say, damp or dry, hot or cold conditions.

cholecystokinin-pancreozymin (CCK-PZ), *n.* a single hormone secreted by the wall of the duodenum in mammals when food enters the small intestine. CCK-PZ causes contraction of the gallbladder muscle, resulting in bile being pumped into the duodenum via the bile duct, and stimulates the pancreas to secrete pancreatic juice containing large quantities of digestive enzymes, which enter the duodenum via the lower part of the bile duct. At one time it was considered that cholecystokinin and pancreozymim were separate, different hormones. CCK-PZ causes VASODILATION of the intestinal blood vessels.

cholera, *n.* a serious human disease caused by gut infection of the bacterium *Vibrio cholerae* that results in severe diarrhoea, vomiting and abdominal cramps. Up to 15 litres per day of fluid may be lost from the gut, resulting in extreme dehydration and even death. Treatment is effected by replacement of the lost body fluids and salts. Cholera is endemic in certain Third World countries, occurring more often where natural resistance of the host is impaired, particularly by malnutrition.

cholesterol, *n.* a steroid that occurs in the cell membranes of animal cells, but not in plants. Cholesterol is produced in the liver and when in excess is excreted in the bile but is partly reabsorbed by the ileum. It may precipitate gallstones in the gallbladder or bile duct. Alternatively, if there is excess cholesterol in the blood, it may be deposited on the walls of the blood vessels, obstructing them and often leading to an intravascular clot which if it occurs in the region of the heart gives rise to a 'heart attack' or coronary thrombosis. Cholesterol is the precursor of animal steroid hormones and bile acids.

choline, *n.* an organic base which is a constituent of ACETYLCHOLINE.

cholinergic, *adj.* (of nerve fibres) secreting ACETYLCHOLINE as a TRANSMITTER SUBSTANCE; in vertebrates these include motor fibres to STRIATED MUSCLE, parasympathetic fibres to INVOLUNTARY MUSCLE and fibres to sympathetic ganglia from the central nervous system. Compare ADRENERGIC.

cholinesterase, *n.* an enzyme that hydrolyses and destroys excess ACETYLCHOLINE after it has been liberated and has produced its effect on specific sites on the postsympatic membrane. See NERVE IMPULSE.

chondr- or **chondro-,** *prefix.* pertaining to cartilage.

Chondrichthyes, *n.* a class of the phylum Chordata containing the cartilaginous fish, sharks, rays and chimaeras. In some classifications, the term is synonymous with ELASMOBRANCHII. They are the lowest vertebrates with complete and separate vertebrae, moveable jaws and paired appendages. All are predators and virtually all are marine. They are characterized by the absence of true bone and the presence of CLASPERS (1) in males, and DENTICLES.

chondrin, *n.* the white, translucent, matrix of CARTILAGE.

chondrioclast, see OSTEOCLAST.

chondriosome, see MITOCHONDRION.

chondroblast, *n.* a cell which gives rise to CARTILAGE.

chondrocranium, *n.* the cartilaginous original cranium surrounding the brain of the embryo, most of which is subsequently replaced by bone in further development.

chorda cells, *n.* cells of the embryo that give rise to the NOTOCHORD.

chordamesoderm, *n.* the MESODERM and NOTOCHORD of the embryo of vertebrates which originally form a mass of very similar cells and are conveniently referred to by a single term.

chordate, 1. *n.* any animal of the phylum Chordata, characterized by the presence of a notochord, hollow dorsal nerve cord and gill slits. The major subdivisions are the PROTOCHORDATES and the VERTEBRATES. **2.** *adj.* of or relating to the Chordata.

chordo-, *prefix.* pertaining to a string.

chordotonal receptors, *n.* the sense organs of insects that detect changes in muscle tension.

chorioallantoic grafting, *n.* an experimental process of culturing tissues in eggs, whereby the tissue is introduced to the surface of the CHORION (2) of a live chick embryo, the blood vessels of which enter the tissue and maintain it after sealing the egg shell.

chorion, *n.* **1.** the superficial outer coat of the insect egg which is non-cellular and secreted by the ovary around the ovum. **2.** an embryonic membrane of AMNIOTES formed from an outer ectodermal layer (see ECTODERM) and an inner mesodermal layer (see MESODERM). See AMNION. In birds and mammals it forms the outer membrane covering the amniotic cavity, and in mammals the mesodermal component of the chorion fuses with the allantois to form the chorioallantoic PLACENTA. Here the chorion develops finger-like outgrowths, the *chorionic villi*, which project into blood spaces in the uterus of the mother. See CHORIONIC BIOPSY.

chorionic biopsy, *n.* a technique that enables the prenatal assessment of a foetus between the 7th and 11th week of pregnancy, involving the sampling of cells from the chorionic villus. These villi are small, finger-like outgrowths of the CHORION(2) that, at the time of the sampling, are full of rapidly dividing foetal cells. A biopsy of the villus is taken with a small tube inserted through the vagina and cervix. No anaesthetic is needed and no amniotic fluid is collected. The large number of dividing foetal cells in the sample means that chromosomal and biochemical studies can be quickly completed, so that termination (if required) is usually before the 10th week, a significantly less traumatic period of gestation than termination after AMNIOCENTESIS.

choroid, 1. *n.* a layer behind the retina of the vertebrate eye which contains blood vessels and pigment. **2.** *adj.* resembling the CHORION.

choroid plexus, *n.* a non-nervous EPITHELIUM which projects into and forms part of the roof of the brain of vertebrates. It secretes the CEREBROSPINAL FLUID.

choroid rete, *n.* a countercurrent arrangement of arterioles and venules which occurs in the eyes of teleost fish, behind the retina.

chrom or **chromo-** or **chromato-,** *prefix.* pertaining to colour.

chromatid, *n.* one of a pair of duplicated CHROMOSOMES produced during the 'S' phase of the CELL CYCLE, which are joined together at the CENTROMERE. See Fig. 102. During nuclear division the centromere splits (in anaphase of mitosis, anaphase 2 of MEIOSIS) to produce two separate chromosomes.

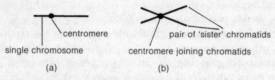

Fig. 102. **Chromatid.** (a) Before duplication. (b) After duplication.

chromatid interference, *n.* the restriction of CROSSING OVER between nonsister CHROMATIDS during prophase I of MEIOSIS, due to a prior CHIASMA having already formed. Such interference produces a reduction in the amount of RECOMBINATION and hence an underestimate of the map distance between genes linked on the same chromosomes.

chromatin, *n.* that part of the cell nucleus which becomes deeply stained with basic dyes. This is now known to be chromosomal material consisting of DNA together with HISTONE and nonhistone proteins.

chromatography, *n.* a technique designed to separate out the constituent parts of a mixture by taking advantage of the property of molecular substances to combine physically (for example, by electrical charges) with inert material. An inert substrate such as paper or silica gel supports the mixture over which a solvent flows. Components of the mixture will have different affinities for the solvent and are carried to different parts of the substrate and thus separated out. They can be subsequently identified by comparing with the migration pattern of known chemicals.

chromatophore, *n.* **1.** also called **chromoplast.** a pigmented PLASTID of plant cells which may be green due to the presence of chlorophyll or differently coloured because of the presence of other CAROTENOID pigments. CHROMATOPHORES are often CHLOROPLASTS in which the pigment has broken down, as in the ripening of fruit. **2.** (in animals) a cell with pigment in the cytoplasm which can be dispersed or concentrated so changing the colour of the animal as a whole, e.g. frog, chameleon, cephalopod. **3.** (in bacteria and blue-green algae) a LAMELLA carrying photosynthetic pigments.

chromomere, *n.* one of many densely staining bands across the SALIVARY GLAND CHROMOSOMES of certain insects, each of which are thought to represent the actual location of a different gene.

chromoplast, see CHROMATOPHORE (1).

chromosomal mutation, *n.* any genetic change in the structure of a chromosome, usually of a fairly major type (compare POINT MUTATION) and often causing severely adverse effects. Such mutations are classified into various types: (a) *inversion*, the rearrangement of a chromosomal segment so that it has the reverse sequence, e.g. ABCDE → ADCBE; (b) deletions, the removal of a segment; (c) *duplication*, the repeating of a segment either next door to the original or elsewhere along the chromosome; (d) *translocation*, the exchange of segments between nonhomologous chromosomes. Sometimes the loss or gain of whole chromosomes occurs, which is an extreme example of chromosomal mutation (see ANEUPLOIDY and EUPLOIDY).

chromosome, *n.* a coiled structure found in the nucleus of EUCARYOTE cells which contains DNA (the genetic material making up the genes), basic proteins called HISTONES, and nonhistone acidic proteins which may regulate the activity of DNA. Each species of organism has a typical number of chromosomes (e.g. 46 in man, 20 in maize) which come in identical HOMOLOGOUS pairs in DIPLOID (1) types, although lower types such as some fungi have only one chromosome of each type (see HAPLOID, sense 1). Chromosomes are not visible during the interphase parts of the CELL CYCLE, but during MITOSIS and MEIOSIS they shorten and thicken and, after suitable preparation, may be observed under the microscope. Individual chromosomes can be recognized by their overall length and the position of the CENTROMERES.

In PROCARYOTES, the chromosome consists of an intact DNA molecule lacking a centromere and is often circular. In viruses, the chromosomal material can be DNA or RNA.

chromosome map or **linkage map** or **genetic map,** *n.* a map on which the position of genes relative to each other is marked on lines representing the separate CHROMOSOMES, the estimation being carried out by GENETIC LINKAGE analysis.

chromosome puffs or **puffs,** *n*. cloudy areas visible on specific CHROMOMERES of SALIVERY GLAND CHROMOSOMES at certain times, which contain DNA and RNA and are thought to represent gene TRANSCRIPTION. This idea is reinforced by the fact that such puffs occur at regular intervals, suggesting that GENE SWITCHING is in operation.

chronic, *adj*. **1.** (of a disease) of long duration, developing slowly but not acute. **2.** (of a radiation dose) applied at a low level over a long time, rather than at an ACUTE dose level.

chrys- or **chryso-,** *prefix*. denoting the colour of gold.

chrysalis, *n*. the pupal stage which ENDOPTERYGOTES such as moths and butterflies enter into after the larval stage. It is an immobile, seemingly dormant, stage but one in which there is considerable internal activity, where larval structures are broken down and those of the IMAGO formed.

Chrysophyta, *n*. a division of the algae. The colouring is caused by the presence of CAROTENOID pigments that are present with the CHLOROPHYLL. This very diverse group of freshwater and marine forms includes the golden-brown algae (Chrysophyceae), yellow-green algae (Xanthophyceae) and the diatoms (Bacillariophyceae).

chyle, *n*. the form of food after passage through the mammalian SMALL INTESTINE, forming an alkaline, fluid emulsion.

chymase, see RENNIN.

chyme, *n*. food in the semifluid state after being subjected to the rhythmical contractions of the stomach.

chymotrypsin, *n*. an enzyme found in the pancreatic juice of mammals that functions as an endopeptidase, catalysing the hydrolysis of PEPTIDE BONDS. It attacks the carboxyl groups of specific amino acids (phenylalanine, tyrosine, leucine, tryptophan, and methionine) and so produces large peptides. The enzyme works in the alkaline medium of the small intestine and is secreted by the pancreas in an inactive form.

chymotrypsinogen, *n*. the inactive precursor of CHYMOTRYPSIN.

cichlid fish, *n*. a group of TELEOST fish that carry the young in the mouth.

cilia, see CILIUM.

cili- or **cilio-,** *prefix*. pertaining to, or being similar to, an eyelash.

ciliary body, *n*. the thickened rim of the CHOROID of the vertebrate eye which surrounds the lens and iris. It contains the CILIARY MUSCLES and secretes the AQUEOUS HUMOUR.

ciliary feeding, *n*. a method of obtaining food used by animals such as molluscs, where cilia (see CILIUM) beat to cause currents that carry food to the gut opening of the organism concerned.

ciliary movement *n*. locomotion brought about by the beating of cilia (see CILIUM) as found in some protozoans (ciliates) and free-living flatworms (Turbellaria). Numerous cilia projecting from the body beat in relays giving the effect of waves. This is called METACHRONAL RHYTHM. Cilia are rigid on the back stroke, pushing the organism forward, but bend on the forward stroke.

ciliary muscle, *n.* any muscle contained within the CILIARY BODY surrounding the lens of the vertebrate eye. Through attachment to the lens they change its shape and bring about ACCOMMODATION (1).

ciliate, *n.* any PROTOZOAN of the class Ciliata (usually placed in the subphylum Ciliophora), members of which possess cilia (see CILIUM) during some part of their life cycle for locomotion and/or food capture. Ciliates are said to be the most specialized of the protozoans and have various organelles to perform particular processes. There are more than 5,500 species and they are common in both fresh and salt water.

ciliated epithelium, *n.* a sheet of cells (see EPITHELIUM) that carry cilia (see CILIUM) on the exposed surface. These cilia exhibit METACHRONAL RHYTHM and commonly move fluids about the bodies of animals, for example, in the respiratory passages of terrestrial vertebrates expelling dust or other foreign matter in microphagous animals. Ciliated epithelium is usually columnar.

Ciliophora, *n.* a very large subphylum of PROTOZOANS, including the Ciliata and Suctoria, that at some time during the life cycle possess cilia (see CILIUM) and usually a micro- and meganucleus.

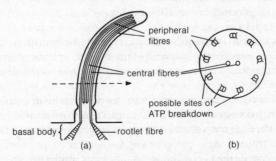

Fig. 103. **Cilium.** (a) Vertical section. (b) Transverse section.

cilium, *n.* (*pl.* cilia) a fine cytoplasmic structure in the form of a thread which projects from the surface of cells. Each cell may bear numerous cilia which beat constantly in one direction, either moving liquid over the surface of the cells concerned or moving the cell(s) in relation to the liquid, for example, locomotion in PROTOZOANS. They are also present in all Metazoa except the ARTHROPODS and NEMATODES, but in only a few plants, for example, cycads. The cilium has a similar structure to the FLAGELLUM, consisting of an outer membrane surrounding a matrix containing two central MICROTUBULES around which is a ring of nine more microtubules (a '9+2' structure). See Fig. 103.

circadian rhythm, *n.* the basic rhythm with a periodicity of approximately 24 hours that organisms undergo when isolated from the daily rhythmical changes of the environment, for example, when kept entirely in the dark. This rhythm demonstrates the ability of the organs to measure time, but the physiological basis of this is unknown. See DIURNAL RHYTHM, BIOLOGICAL CLOCK.

circular overlap, *n.* a naturally occurring arrangement of contiguous and intergrading populations which interbreed, curving round geographically until the beginning and end overlap and behave as *good species*. A species that behaves in this way is called a *ring species*.

circulatory system, see BLOOD CIRCULATORY SYSTEM.

circum-, *prefix.* denoting surrounding.

circumnutation, *n.* the spiral path followed by a plant organ as it grows, as seen, for example, in the twining stem of *Convolvulus*.

circumscissile, *adj.* (of a DEHISCENT spore or seed capule) exhibiting transverse dehiscence, the top opening like a lid.

cirripede or **cirriped,** *n.* any marine arthropod of the class Cirripedia, such as the barnacles and the parasite *Sacculina*.

cis-phase, *n.* an isomer (see ISOMERISM) having similar atoms or groups on the same side of the molecule.

cisterna, *n.* (*pl.* cisternae) a cellular space enclosed by a membrane, as in the GOLGI APPARATUS.

cis-trans isomerization, *n.* the conversion of the configuration of a 'cis' isomer (see ISOMERISM), with similar atoms or groups on the same side of the molecule to a 'trans' form, with the similar atoms or groups on opposide sides.

cis-trans test, *n.* a genetic COMPLEMENTATION TEST to determine if two mutations have occurred within the same CISTRON or in adjacent cistrons. The test depends upon the two mutations being introduced into a cell at the same time. If they 'complement' each other by producing a WILD TYPE phenotype the mutations are in different cistrons (i.e. nonallelic). If a mutant phenotype is produced this shows the two mutations to be noncomplementary and therefore to be located in the same cistron (i.e. allelic). Two arrangements are possible in the cis-trans test (see Fig. 104) which can be carried out in a double HETEROZYGOTE individual (a HETEROKARYON in a haploid type). The more important is the transconfiguration, which is diagnostic, since two normal cistrons are present if the mutations are nonallelic (giving wild type) whereas the same cistron is mutant on both chromosomes if the mutations are allelic (giving a mutant phenotype).

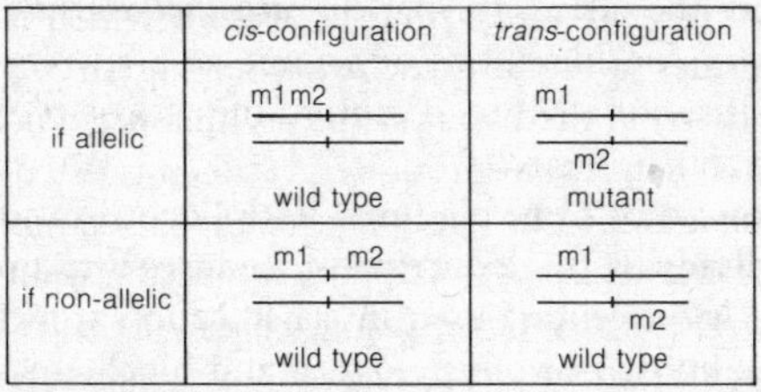

	cis-configuration	*trans*-configuration
if allelic	m1 m2 —————— ———— wild type	m1 ———— m2 —————— mutant
if non-allelic	m1 m2 —————— ———— wild type	m1 ———— m2 —————— wild type

Fig. 104. **Cis-trans test.** The cis-trans test configurations. m1 and m2 are mutations. The two lines represent two homologous chromosomes.

cistron or **functional gene,** *n.* a portion of DNA coding for one POLYPEPTIDE CHAIN. The one gene/one enzyme hypothesis thus becomes the 'one cistron/one polypeptide' hypothesis.

citric–acid cycle, see KREBS CYCLE.

clad– or **clado–,** *prefix.* pertaining to a twig.

cladistics, *n.* an approach to CLASSIFICATION by which organisms are ordered and ranked entirely on a basis which reflects recent origin from a common ancestor, i.e. like a family tree. The system is concerned simply with the branching of the tree and not with the degree of difference. The latter is the concern of evolutionary taxonomists who oppose the cladistic approach.

cladoceran, *n.* any minute crustacean of the order Cladocera, with a laterally compressed carapace and large forked antennae used in swimming.

cladode, *n.* a green flattened stem that functions as a leaf. For example, the lateral branches of butcher's broom *Ruscus aculeatus* are cladodes and bear flower buds.

cladogram, *n.* a geneological DENDROGRAM based on the principles of CLADISTICS in which rates of evolutionary divergence are ignored.

clasmatocyte, see HISTOCYTE.

claspers, *n.* **1.** the rod-like projections between the pelvic fins of male ELASMOBRANCH fish that are utilized in copulation. **2.** the processes at the tip of the male abdomen of insects used to hold the female during copulation.

class, *n.* a TAXON below the level of phylum and above order; related classes make up a phylum just as related orders make up a class. See CLASSIFICATION.

classification, *n.* the ordering of organisms into groups on the basis of their relationships. The groups are referred to as TAXONS, for example, kingdom, phylum (in animals), division (in plants), class, order, family, genus, species. *Natural classification* based on overall evolutionary relationships is the usual form, but *artificial classification* based on nonevolutionary considerations or on one or a few characters is often used in identification. More recently, the CLADISTICS approach has gained favour in some circles.

Classification is not to be confused with 'identification' which is the placing of individuals by deductive procedures into previously established groups. Most criteria used in classification are structural, but as more becomes known of physiological and biochemical criteria these will be used in classification, as they are already with many microorganisms. See NUMERICAL TAXONOMY, TAXONOMY.

clavate, *adj*. (of plant or animal structures) club-shaped.

clavicle, *n*. a bone associated with the ventral side of the shoulder girdle on each side of many vertebrates. In humans it is the collar bone. See Fig. 105.

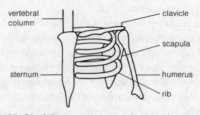

Fig. 105. **Clavicle.** Location on left side of a primate.

clay soil, *n*. a soil formed on a clay substrate.

clearing, *n*. a process used in the preparation of materials for microscopic examination when, after dehydration using alcohols, clearing in benzene or xylene is carried out as these are MISCIBLE with the mountant and alcohols whilst the alcohols are immiscible with the mountant.

cleavage, *n*. the division of the cytoplasm during nuclear division (see MEIOSIS, MITOSIS) following FERTILIZATION of the egg to form the ZYGOTE. *Holoblastic cleavage* occurs in animals where there is little yolk, and here the entire zygote is involved; *meroblastic cleavage* occurs where the yolky part of the zygote fails to divide, only part of the zygote undergoing cleavage. BILATERAL CLEAVAGE gives rise to a bilaterally symmetrical arrangement of blastomeres, as opposed to spiral cleavage which gives rise to a spiral arrangement of blastomeres. Bilateral cleavage occurs in chordates, echinoderms and a few smaller groups, indicating their common origins whereas most other invertebrate phyla have spiral cleavage. See Fig. 106.

cleidoic egg, *n*. an egg that possesses a protective shell.

cleistocarp, *n*. the fruiting body of some ASCOMYCETE fungi that is completely closed and must be ruptured to release the spores.

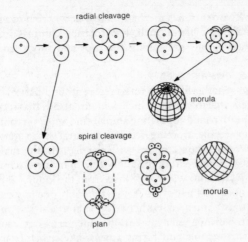

radial cleavage

morula

spiral cleavage

plan

morula

Fig. 106. **Cleavage.** Radial and spiral cleavage.

cleistogamy, *n.* the production of small, inconspicuous flowers in which self-fertilization occurs before the flowers open. For example, the violet forms large, showy, cross-pollinating flowers, but also forms cleistogamic flowers late in the summer.

climacteric, *n.* **1.** any critical event. **2.** the MENOPAUSE.

climacteric phase, *n.* a stage preceding sensescence in certain fully grown fruits, in which there is a marked increase in respiration rate. Climacteric fruits (for example apples and bananas) can be prevented from undergoing this respiratory increase by treatment with high levels of carbon dioxide and/or exposure to low temperatures. Exposure to ethylene may then be used to hasten the climacteric phase, so that the fruit is ready for sale. Many fruits such as citrus types do not exhibit the climacteric phase.

climate, *n.* the general meteorological conditions prevailing in a given area.

climax, *n.* a plant community that has reached stability and is in equilibrium with the climatic conditions appertaining at the time, for example, oak woodland in Britain. A succession of climaxes resulting from climatic change is called a *clisere*.

cline, *n.* a gradual and continuous (or nearly continuous) change in a character (such as size or colour) in relation to its geographical or ecological distribution. For example, there is an increase in the percentage of bridled forms in the guillemot population with more northerly latitudes.

clinostat or **klinostat,** *n.* a mechanism that slowly rotates a pot or other holder containing a plant. Such slow rotation eliminates the effect of gravity on the plant, preventing the root and shoot from showing GEOTROPISM.

clisere, *n.* see CLIMAX.

clitellum, *n.* a saddle-like structure present around part of the anterior half of some ANNELID worms which contains glands that secrete mucus, forming a sheath round copulating animals and a cocoon round the eggs.

clitoris, *n.* an erectile structure occurring ventral to the uterus in female mammals which is homologous with the PENIS in the male.

cloaca, *n.* the terminal part of the gut system of most vertebrates (except higher mammals) into which the ducts from the kidney and reproductive system open. In these types there is thus only one posterior aperture to the body as compared with two in mammals, the anus and the opening of the urinogenital system. In some vertebrates, such as birds, the cloaca is reversible and forms a penis-like structure in the males during copulation.

clock, internal, see BIOLOGICAL CLOCK.

clone, *n.* **1.** any of two or more individuals with identical genetic makeup produced from one parent by ASEXUAL REPRODUCTION. For example, daughter plants produced by strawberry RUNNERS, whole plants produced by tissue culture. **2.** either of the identical individuals produced by the splitting of a young embryo.

closed community, *n.* a plant community that is so dense on the ground that new species cannot colonize.

closed population, *n.* a population into which there is no gene input from outside, i.e. the only possible genetic change is through MUTATION.

cloven–footed herbivore, *n.* a member of the mammalian order ARTIODACTYLA with two functional toes (numbers three and four) on each foot. While all in this group are herbivorous, some are RUMINANTS (camels, deer, sheep and cattle) while others are not (pigs and hippopotamuses).

club moss, *n.* any mosslike pteridophyte plant of the order Lycopodiales, having erect or creeping stems covered with small overlapping leaves. The group existed as far back as the Palaeozoic era. *Lycopodium* and *Selaginella* are present-day forms.

clustering methods, *n.* the means by which groups of related or similar species are arranged in species groups or into higher taxonomic groupings.

clypeus, *n.* the cuticular plate on the head of an insect above the LABIUM.

Cnidaria, *n.* a subphylum of the phylum Coelenterata in some classifications, where the Ctenophora (sea gooseberries) form the other subphylum (see COELENTERATE). However, in more modern classifications the Ctenophora are given the status of a phylum, so that the organisms classified in the Cnidaria (hydroids, jellyfishes, sea anemones and corals) are the only organisms in the new phylum Coelenterata, thus making the term Cnidaria obsolete.

cnido-, *prefix.* pertaining to a stinging nettle.

cnidoblast or **nematoblast,** *n.* a stinging *thread cell.* See NEMATOCYST.

cnidocil, *n.* the trigger on a CNIDOBLAST.

CNS, see CENTRAL NERVOUS SYSTEM.

co- or **con-** or **com-,** *prefix.* denoting with, together.

CO_2 acceptor, *n.* a molecule (ribulose diphosphate) whose function is to combine with carbon dioxide in the CALVIN CYCLE.

coacervate theory, *n.* a theory expressed by the Russian biochemist A.I. Oparin in 1936 suggesting that the origin of life was preceded by the formation of mixed colloidal units called 'coacervates'. These are particles composed of two or more colloids which might be protein, lipid or nucleic acid. Oparin proposed that whilst these molecules were not living, they behaved like biological systems in the ancient seas. They were subject to natural selection in terms of constant size and chemical properties, there was a selective accumulation of material and they reproduced by fragmentation. Subsequent work by the American biochemists Stanley Miller and Harold Urey shows that such organic materials can be formed from inorganic substances under the conditions prevailing on the prebiological earth. They synthesized amino acids by passing a spark through a mixture of simple gases in a closed system.

coagulation, *n.* the separation or precipitation of suspended particles from a dispersed state.

coalescent, *adj.* (of plant structures) joining together.

cobalamin or **cyanocobalamin** or **vitamin B_{12},** *n.* one of the B-COMPLEX of water-soluble vitamins, containing cobalt and required for red blood cell formation. A deficiency of B_{12} can lead to pernicious anaemia especially amongst old people. The vitamin is synthesized by bacteria in the gut rather than being obtained from the diet.

coccidiosis, *n.* a disease caused by SPOROZOAN parasites that occurs in rabbits and poultry.

coccus, *n.* a bacterium having a spherical or globular form.

coccyx, *n.* the fusion of the posterior vertebrae to form a single unit, the coccyx. In humans three to five vertebrae are involved and these form the remnant of a tail.

cochlea, *n.* a part of the inner ear which is concerned with the detection of the pitch of the sound received by the ear. A projection of the SACCULE (2), it occurs in some reptiles, birds and mammals. In the mammal it is a coiled tube consisting of three parallel canals and contains the *organ of Corti*, the part which actually responds to sound. See Fig. 107.

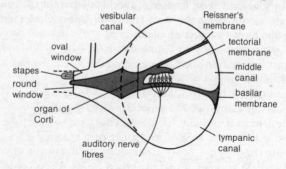

Fig. 107. **Cochlea.** Cross section.

cocoon, *n.* a protective covering of eggs or larvae found in several invertebrate groups. In some cases, e.g. ANNELIDS, a cocoon is produced by the adults to contain the eggs and in others, e.g. insects, by the larvae to protect the pupa during the course of development.

codominance, *n.* the relationship between two different ALLELES of a gene in the HETEROZYGOTE, producing a PHENOTYPE containing both parental forms of the character. For example, allele *A* of the ABO blood group is codominant to allele *B*, so that an individual with both alleles (*A/B*) produces both *A* and *B* antigens. Compare INCOMPLETE DOMINANCE. See DOMINANCE (1).

codon or **triplet,** *n.* a group of three DNA nucleotide bases that codes for a specific amino acid in the related protein. See PROTEIN SYNTHESIS, GENETIC CODE.

coefficient of the difference, *n.* the difference of the means divided by the sum of the STANDARD DEVIATION.

coefficient of dispersion, *n.* the variance divided by the mean. Where this value is equal to 1 there is a random or POISSON DISTRIBUTION; greater than 1 aggregation or under-dispersion, and a value smaller than 1, indicates an even distribution or over-dispersion (e.g. territoriality in birds).

coefficient of inbreeding, *n.* the probability that any two gametes fusing to form a zygote will carry the same allele of a gene from a COMMON ANCESTOR. Symbol: F.

coefficient of variability, *n.* the STANDARD DEVIATION as a percentage of the mean.

coel-, *prefix.* pertaining to a hollow.

coelacanth, *n.* a primitive bony fish with lobed fins (as distinct from ray fins). Almost all its relatives were freshwater fossil forms dating from the DEVONIAN PERIOD. *Latimeria* is a living marine form found round the Comoro Islands off Southeast Africa and collected for the first time in 1938. It is remarkable in that it has remained almost unchanged from its ancestors in the lower CARBONIFEROUS PERIOD, that existed more than 300 million years ago. See Fig. 108.

Fig. 108. **Coelacanth.** *Latimeria chalumnae.* Length: 1.5 m.

coelenterate, *n.* any invertebrate of the phylum Coelenterata, including the hydroids (Hydrozoa), jellyfish (Scyphozoa), sea anemones and corals (Actinozoa). In some classifications the sea gooseberries (Ctenophora) are included as a separate subphylum. See CNIDARIA.

coelenteron, *n.* the digestive cavity in COELENTERATES which has only one opening, the mouth.

coelom, *n.* the main body cavity of most animals with three body layers, being a fluid-filled space between inner and outer layers of MESODERM. The coelom often has tubes connecting to the outside called COELOMODUCTS. See Fig. 109.

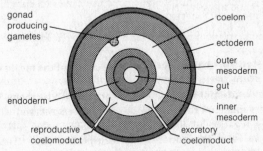

Fig. 109. **Coelom.** Location.

coelomoduct, *n.* any duct leading from the COELOM of an animal to the exterior that can carry either excretory materials or gametes. For example, the oviduct in vertebrates is thought to have evolved as a coelomoduct.

coen- or **coeno,** *prefix.* denoting common.

coenobium, *n.* a colony of algal cells, where the cells are arranged in a constant form and number and are surrounded usually by a gelatinous matrix. They are interconnected, act as a single functional unit and are thus regarded as being a single organism, e.g. *Volvox*.

coenocyte, *n.* a multinucleate mass of protoplasm formed when nuclei divide without the division of the cytoplasm of the original cell, as in many fungi.

coenospecies, *n.* a group of species which may be capable of forming hybrids with each other.

coenzyme, *n.* an organic COFACTOR molecule smaller than protein that bonds with a specific ENZYME while the reaction is being catalysed. Like enzymes, coenzymes are not altered or used up in the reaction and can be used many times, but a minimal quantity is required for normal level of enzyme function and thus normal health. This explains why VITAMINS, which often act as coenzymes, are so essential. See also ACETYLCO-ENZYME A.

coenzyme A, see ACETYLCOENZYME A.

coevolution, *n.* the evolution of unrelated organisms that has taken place together because of the special link between them, e.g. insects and the flowers they pollinate (see ENTOMOPHILY), parasites and their host, members of a symbiotic relationship (see SYMBIOSIS). The ARUM LILY is a notable example.

cofactor, *n.* a substance that is essential for the catalytic activity of some enzymes, binding to the enzyme only during the reaction. Cofactors can be metallic ions, or nonprotein organic molecules (coenzymes) such as vitamins in the B–COMPLEX.

coherent, *adj.* (of plant structures) united at some point but appearing free.

cohesion–tension hypothesis, *n.* a hypothesis that explains the ascent of water from roots to leaves in a plant as due to a combination of upward pull created by TRANSPIRATION losses producing a tension on the xylem vessels and cohesion of water molecules to each other, aided by the adhesion of water molecules to the sides of the narrow vessels. Evidence for the hypothesis comes from: (a) the long rise in mercury in a column attached to a leafy shoot as compared with mercury rise in a vacuum; (b) the decrease in tree trunk diameter during the day (due possibly to tension) when transpiration is at its highest level. See DENDROMETER.

coincidence, coefficient of, *n.* the proportion of double recombinant types observed in a progeny as compared to the number expected. A value less than 1.0 indicates that CHROMATID INTERFERENCE has taken place.

coitus, *n.* the act of copulation between male and female animals during which sperm is transferred from male to female.

colchicine, *n.* a poisonous alkaloid extracted from the corms of *Colchicum autumnale* that acts as a spindle inhibitor during NUCLEAR DIVISION and can thus be used to produce cells with double sets of chromosomes, due to NONDISJUNCTION.

cold receptor, *n.* a sensory structure that responds particularly to cold and sometimes to pressure. Such receptors occur in the skin of vertebrates, and in humans are more abundant and occur more superficially than warm receptors. Fibres from cold receptors are active between 10° and 40°C, with a maximum firing frequency between 20° and 34°.

Coleoptera, *n.* an order of insects, including beetles and weevils. The forewing is thick, leathery and veinless, and is called an *elytrum*. When closed, elytra meet along the midline and protect the membranous hindwings, which fold forward. Some of the approximately 280,000 species of Coleoptera are wingless, however. There is a complete METAMORPHOSIS. See Fig. 110.

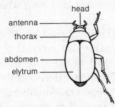

Fig. 110. **Coleoptera.** Generalized structure.

coleoptile, *n.* a nonchlorophyllous covering over the growing shoot of young plant seedlings in certain MONOCOTYLEDONS (e.g. oats) that is the first structure to break through the soil into the air after germination. As growth proceeds the coleoptile is ruptured by the first foliage leaf of the enclosed shoot. Oat coleoptiles have been used extensivly in experiments with AUXINS.

coleorhiza, *n.* a structure similar to the COLEOPTILE but located around the radicle of young seedlings.

coliform, *adj.* (of Gram-negative rod bacteria) normally inhabiting the colon, e.g. *Escherichia coli, Enterobacter aerogenes, Klebsiella.* See GRAM'S STAIN.

colinearity, *n.* the linear relationship between a piece of DNA coding (a CISTRON) and the POLYPEPTIDE CHAIN (see Fig. 111 on page 142). Therefore:

$$\frac{\text{A to X}}{\text{A to B}} = \frac{\text{C to Y}}{\text{C to D}}.$$

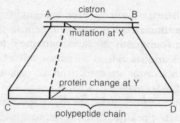

Fig. 111. **Colinearity.** The DNA bases are transcribed from left to right.

collagen, *n.* a fibrous protein that forms the white fibres of vertebrate CONNECTIVE TISSUE. These have a high tensile strength, e.g. tendons, but are not elastic. Collagen tissues consist of a glycoprotein matrix containing densely packed collagen fibres.

collagenoblast, *n.* a type of FIBROBLAST giving rise to COLLAGEN.

collar cell, see CHOANOCYTE.

collar flagellate, see CHOANOFLAGELLATE.

collateral, *n.* a minor side branch of a blood vessel or nerve.

collateral bud, *n.* an accessory bud on a plant stem, lying beside the axillary bud.

collateral bundle, *n.* a plant vascular bundle in which the PHLOEM is on the same radius as the XYLEM and external to it.

collecting duct, *n.* that part of the renal tubule in a kidney in which water absorption takes place under the control of ADH, producing a urine of variable concentration depending on overall water levels in the body.

Collembola, *n.* an order of small primitive wingless insects comprising the springtails, most of which possess a springing organ (furcula). They live in soil and leaf litter. See VENTRAL TUBE.

collenchyma, *n.* a type of plant tissue in which the cells are similar to PARENCHYMA but have cellulose wall thickenings, particularly at the angles when seen in transverse section. Collenchyma acts as a supporting tissue, especially in young stems and leaves. See Fig. 112.

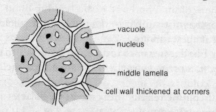

Fig. 112. **Collenchyma.** Transverse section.

colloid, *n.* a mixture of two substances which are immiscible (see

MISCIBLE), but where the particles of one are too small to settle out, and so remain suspended indefinitely. Glue is a colloid of animal gelatin in water; the water is defined as the matrix and the gelatin as the inclusion. Colloid particles measure 1×10^{-4} to 1×10^{-6} mm in diameter, forming either a SOL or GEL structure which does not diffuse through cell membranes. Colloids are common in cells, where their large surfaces are important for chemical changes constantly in progress there.

colon, *n*. a part of the large intestine of mammals, consisting of a wide tube with folded walls between the ILEUM of the small intestine and the RECTUM. The main function of the colon is the reabsorption of water from the FAECES.

colon bacillus, *n*. the bacillus *Escherichia coli* found in the colon.

colonial animal, *n*. an association of animals in which the individuals are connected, as in the Hydrozoa or Polyzoa. The term is not usually attributed to higher animals, such as birds, that live in colonies.

colony, *n*. **1.** an aggregated group of separate organisms which have come together for specific purposes, e.g. breeding in birds. **2.** a group of incompletely separated individuals organised in associations, as in some hydrozoan COELENTRATES and polyzoans. **3.** a localized population of microorganisms, e.g. bacteria, derived from a single cell grown in culture.

colostrum, *n*. a yellowish, watery secretion expressed from the breast nipples of female mammals when in late pregnancy and for a few days after birth. Colostrum has a high protein content and is rich in vitamin A and ANTIBODIES which give the baby an immediate, short-term, passive immunity to foreign ANTIGENS.

colour blindness, *n*. a condition in which certain colours cannot be distinguished, due to a lack of one or more colour-absorbing pigments in the retina. In humans the most common example is red/green colour-blindness which can be considered as being controlled by a single locus on the X-CHROMOSOME (although at least two, closely linked loci are known to influence the character). The 'normal vision' ALLELE is dominant to the 'colour blindness' allele; colour-blind females must therefore be HOMOZYGOUS for the colour-blindness allele, while males will be affected if they carry only one colour-blindness allele. Consequently, colour blindness is more common in males than females.

colour change, *n*. a process in which certain animals (e.g. squid, sole) can alter their colours or shade by changing the concentration of pigment in their pigment cells (see CHROMATOPHORES sense 1) under nervous or hormonal control.

colour vision, *n*. the ability of some animal groups to detect colour in an object, due in vertebrates to activity of CONE CELLS in the retina of the eye. Examples of animals with colour vision include primates, many fishes, most birds and the insects. Animals without colour vision probably see objects in shades of grey (monochrome vision).

columella, *n*. **1.** the sterile central column in the sporangia of mosses and liverworts. **2.** the domed structure forming a pillar supporting the sporangium of some fungi, e.g. *Mucor*. See Fig. 113. **3.** the central area of a root cap which contains STATOLITHS (1). **4.** the central axis of a fruit.

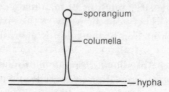

Fig. 113. **Columella (2).** The sporangium of *Mucor*.

columella auris, *n*. a bony or cartiliginous rod which connects the ear drum to the inner ear in reptiles, birds and some amphibians. It is homologous with the STAPES of mammals and the HYOMANDIBULA of fishes.

columnar epithelium, see EPITHELIUM.

coma, *n*. a tuft of hairs attached to the testa of a seed and used in wind dispersal.

commensal, *adj*. (of an organism) living in close association with another organism of a different species where neither has an obvious effect on the other, e.g. some POLYCHAETE worms will live in the tubes of others. See SYMBIOSIS.

Commission, *n*. short for the International Commission on Zoological Nomenclature.

commisure, *n*. **1.** (in invertebrate animals) a bundle of nerve fibres joining ganglia; in vertebrate animals, a similar bundle of nerve fibres joining left and right sides of the brain or spinal cord. **2.** (in plants) the faces by which two CARPELS are joined.

common ancestor, *n*. a relative occurring in the family trees of both parents in a cross. For example, cousin matings have grandparents as common ancestors.

communication, *n*. the transmission of information from one individual to another of the same or different animal species, which receives and understands it. Any of the five senses can be used in communication. Many higher organisms use sound, but posturing in specific displays, e.g. courtship in birds and primates, dancing in bees, touch and smell (see PHEROMONE) are other important means of conveying information.

community, *n*. a naturally occurring group of different species of organisms that lives together and interacts as a selfcontained unit, relatively independent of inputs and outputs from adjacent communities. Ideally, it is selfcontained in terms of food relationships, and usually the only energy required from outside is that of the sun.

companion cell, see SIEVE TUBE.

compatibility, *n*. a state where two organisms or parts of organisms can be mixed. For example, when two plants can cross-fertilize, or when two tissues can be added together, as in transplant surgery or blood transfusion. Animal compatibility depends on suitable matching of ANTIGENS and ANTIBODIES. See also SELF-INCOMPATIBILITY.

compensation light intensity, *n*. the level of illumination at which photosynthesis exactly balances respiration in plants. Only at levels greater than this can organic compounds accumulate to give a net gain of plant material.

compensation period, *n*. the time taken when a plant is removed from darkness to reach the *compensation point* where photosynthesis and respiration proceed at the same rate and there is no net gain or loss in carbohydrate. The compensation period is shorter in shade plants, which are very efficient at utilization of low levels of light.

compensation point, see COMPENSATION PERIOD.

compensatory hypertrophy, *n*. the enlargement of a tissue or organ as a result of removal or damage, e.g. if one kidney is removed from a human, the other enlarges.

competence, *n*. **1.** a period when a differentiating cell or tissue is capable of switching to an alternative developmental PATHWAY. See INDUCTION, CELL DIFFERENTIATION, GENE SWITCHING, CANALIZATION. **2.** a state in bacteria when they are able to receive DNA from other bacteria in a process called TRANSFORMATION.

competition, *n*. the interaction between organisms striving for the same end. Such interaction may adversely affect their growth and survival. This includes competition for resources (food, living space), mates, etc. See INTERSPECIFIC COMPETITION, INTRASPECIFIC COMPETITION.

competitive exclusion, *n*. the principle that two species cannot coexist if they have identical ecological requirements.

competitive inhibition, *n*. a form of enzyme control in which an inhibitor molecule very similar in structure to the normal SUBSTRATE of an enzyme becomes reversibly bound to the ACTIVE SITE, thus reducing the quantity of enzyme available. However, if excess substrate is present the inhibitor can be forced out by the substrate molecule which takes its place and the reaction proceeds. Compare NONCOMPETITIVE INHIBITION.

complement, *n*. the protein components of blood serum that can bind to antigen/antibody groups already formed on the surface of cells, thus enhancing destruction of the foreign body by PHAGOCYTES.

complemental males, *n.* those that live attached to their females which are very much larger, e.g. angler fish.

complementation test, *n.* a genetical test to determine the precise location of mutations in one area of DNA. See CIS-TRANS TEST.

complementary base pairing, *n.* the arrangement of NUCLEOTIDE BASES so that purines bond with pyrimidines, as occurs in the two polynucleotide chains of DNA (adenine with thymine, guanine with cytosine) and between DNA and MESSENGER RNA nucleotides during TRANSCRIPTION. See Fig. 130.

compost heap, *n.* a pile of vegetable material which is allowed to break down through decomposition by microorganisms, and is subsequently used to improve the fertility of garden soils.

compound, *adj.* (of plant structures) made up of several similar parts, as in a leaf compound of several leaflets. A *simple* structure is one *not* divided into similar parts.

compound microscope, see MICROSCOPE.

compressed, *adj.* flattened.

concentration gradient, *n.* a system that is set up where the concentration of any SUBSTRATE (usually in the form of ions or molecules) is different with respect to the concentration elsewhere in the system. For example, ionic concentration across cell membranes, solutes in sap.

concentric bundle, *n.* a vascular bundle in which one tissue surrounds another, e.g. phloem around xylem (AMPHICRIBRAL BUNDLE) or xylem around phloem (AMPHIVASAL BUNDLE).

conceptacle, *n.* the cavity which contains the sex organs in some species of brown algae, e.g. *Fucus*.

conchiolin, *n.* a fibrous insoluble protein that forms the basic structure of molluscs.

concolorous, *adj.* having the same colour throughout the structure concerned, e.g. petals, leaves.

concordance, *n.* any similar traits occurring in MONOZYGOTIC TWINS or the possessors of similar genes.

condensation reaction, *n.* a reaction involving the removal of water (dehydration) from two or more small molecules to form a new larger compound. For example, the disaccharide carbohydrate sucrose is formed by a condensation reaction between glucose and fructose molecules. Compare HYDROLYSIS.

conditional lethal, *n.* a mutation that causes death in certain circumstances (e.g. high temperature) but not in others (e.g. low temperature).

conditional reflex, *n.* a reflex action which is modified by experience, so that whilst the response remains the same the stimulus is replaced by one different from the original. The classic example is the experiment by the Russian physiologist Ivan Pavlov (1849–1936) with dogs, where the smell, sight and taste of food causes them to salivate. Pavlov rang a bell immediately before producing food and, after considerable repetition, salivation could be brought about by ringing the bell in the absence of food (a *Pavlovian response*).

conductance, *n.* **1.** the measure of the ability of a conductor to carry a current, measured in siemens (S) (reciprocal of the ohm [Ω]). **2.** the ability of heat to flow by conduction across an object under a temperature gradient.

conduction, *n.* the transmission of an electrical current by a conductor.

conductivity, *n.* **1.** the property of conducting an electric current. **2.** the passage of a physiological disturbance through tissue or a cell, as in a NERVE IMPULSE.

condyle, *n.* the protruberance of a bone which fits into the socket of another bone, thus forming a joint.

cone, *n.* **1.** (in plants) a reproductive structure in the form of a conical mass of scale-like sporophylls surrounding a central axis, found particularly in GYMNOSPERMS but also in other plant groups, e.g. horsetails (*Equisetales*). **2.** (in animals) a light-sensitive structure in the vertebrate eye. See CONE CELL.

cone cell, *n.* a light-sensitive nerve cell containing the photochemical pigment IODOPSIN which is not readily bleached by bright light. Cones are concerned with the discrimination of colour and are found in the retina of most vertebrate eyes. The highest density of cones is in the FOVEA of the eye, although they are absent in many animals which live in the dark. There are now thought to be three types of cone cells, each giving maximum response when stimulated by the blue (450 nm), green (525 nm), and red (550 nm) parts of the spectrum. Colour vision is a result of differential stimulation of three types of cone so that the colour seen results from the differential excitation of the three types of cone. This is in accord with the trichromatic theory of colour vision which suggests that all colours can be produced by the mixing of blue, green and red. In general, cones have more direct connections to the brain than RODS, and in the centre of the fovea each cone has its own optic nerve fibre whereas groups of rods converge on a single fibre (retinal convergence). Away from the fovea there are more nerve fibres per cone than per rod, so increasing the precision of cones and the sensitivity of rods.

congeneric, *adj.* (of species) in the same genus.

congenital, *adj.* existing at or before birth and referring especially to defects and diseases that are environmental in origin and not inherited.

conidiophore, see CONIDIUM.

conidum, *n*. a sexual SPORE found in certain fungi and borne on a *conidiophore* which is a specialized HYPHA.

conifer, *n*. any GYMNOSPERM tree or shrub of the group Coniferae, including pines and spruces, which are more northerly forms, and cedars, yews and larches, which are the more temperate forms. Most are tall, forest trees, usually evergreen, but some, e.g. larches, are deciduous. Usually they are MONOECIOUS and possess separate male and female cones.

conjugant, *n*. one of a pair of gametes in CONJUGATION.

conjugation, *n*. a method of sexual reproduction involving the fusion of similar GAMETES which are not freed from the parent forms, as in ciliates where two individuals fuse together, exchange micronuclear material, and then separate. Conjugation also occurs in some algae, e.g. *Spirogyra* and fungi, e.g. *Mucor*. The partial exchange of genetic material between two bacterial cells may also be regarded as 'conjugation'.

conjunctiva, *n*. a protective covering over the front of the vertebrate eye and extending to line the eyelids, formed of a layer of MUCUS-secreting epithelium and underlying connective tissue.

connate, *adj*. (of plant organs) originally distinct, but growing together and joining.

connective tissue, *n*. an animal tissue in which the intercellular matrix forms the major part. Such tissues fall into three main groups: (a) 'true' connective tissue (ADIPOSE TISSUE, AREOLAR TISSUE, LIGAMENTS and TENDONS), (b) CARTILAGE, (c) blood, the last differing from the other two in containing no fibres in the matrix (BLOOD PLASMA).

connivent, *adj*. (of plant parts) originally widely separate at the base but growing towards each other apically (see APIX).

consanguineous mating, *n*. mating between closely related individuals.

conservation, *n*. the preservation, protection and management of an environment which takes into account recreational and aesthetic needs, in addition to preserving as much as possible of the natural fauna and flora and allowing for the harvesting of natural resources and agriculture. This necessitates the sensible planning of what is taken from the environment in terms of the yield of plants, animals and materials, whilst at the same time maintaining as much natural habitat as possible, and thus the largest possible GENE POOL.

conservative replication model, *n*. a model of DNA synthesis, now disproved, in which both chains of the original molecule act as a template for a new daughter molecule. The SEMICONSERVATIVE REPLICATION MODEL is now accepted as the correct method of DNA replication.

consociation, *n*. a climax vegetation in which one species is dominant, e.g. oakwood, as opposed to ASSOCIATION where there are several dominant species.

conspecific, *adj.* (of animals or plants) belonging to the same species.

constant (C) region, *n.* an area of an IMMUNOGLOBULIN molecule that shows little variation between molecules with different specificities. Compare VARIABLE REGION.

constitutive enzyme, *n.* an enzyme that is made all the time at a constant rate, unaffected by inducers.

consumer, *n.* any organism which consumes other organisms, e.g. HETEROTROPHS or SAPROBIONTS. Macroconsumers (*phagotrophs*) are chiefly animals which ingest other organisms or particulate organic matter (detritus). Microconsumers are chiefly bacteria and fungi which break down dead organic material, absorb some of the decomposition products, and release inorganic nutrients (which are then re-available to the PRODUCERS) and organic materials which may provide energy sources to other organisms.

contact insecticide, *n.* a chemical (e.g. DDT, Malathion) lethal to insects when they alight on a surface coated with the poison. Compare SYSTEMIC BIOCIDE.

contagious, *adj.* (of infections) transmitted by contact.

contaminant, *n.* an organism, usually a microorganism, introduced into a pure culture.

contiguous, *adj.* (of plant parts) touching at the edges.

continental drift, *n.* the theory originally put forward in 1912 by Alfred Wegener that continental masses were continuously moving over the surface of the globe as though floating on a sea of molten rock. Doubted extensively for many years, the theory is now generally accepted and supported by evidence from geomagnetic studies and the study of PLATE TECTONICS. 200 million years ago the present continents were joined in a universal land mass called *Pangaea*. Since that time, the continents have drifted apart and the study of plate tectonics shows that they are still moving. Over a period of 400 million years Britain has drifted from south of the tropic of Capricorn to its present position – over 80° of latitude. Pangaea was probably formed by previously separate continents coming together, and the line of the Appalachian, Caledonian, Greenland and Scandinavian mountains was probably formed by North America, Europe and Africa coming into contact. Previous separation would account for the position of the equator and South Pole 440 million years ago. See Fig. 114 on page 150.

contingency table, *n.* a method of presenting results so as to show relationships between two characters which can then be tested statistically by a modified CHI-SQUARED TEST (the *heterogeneity chi-square test*). For

Fig. 114. **Continental drift.** The continent of Pangaea 200 million years ago.

example, eye colour and hair colour might be measured in a sample population and the results presented in a contingency table. See Fig. 115.

		Eye colour			
		Blue	Grey	Brown	total:
Hair colour	Blonde	96	42	37	175
	Brown	58	61	107	226
	Black	17	32	93	142
	total:	171	135	237	543

Fig. 115. **Contingency table.** A contingency table for eye and hair colours in a sample population.

continuity, *n.* the principle in taxonomy that continued usage of a particular scientific name should have priority over date of publication. Normally in scientific classification the date of publication has priority.

continuous variation, *n.* a feature of certain CHARACTERS, e.g. height, in which there is a complete spread of forms across a range, often producing a bell–shaped, NORMAL DISTRIBUTION CURVE when sampling in a population. See also POLYGENIC INHERITANCE.

contorted, *adj.* (of PERIANTH lobes) overlapping and appearing to be twisted when in bud.

contour feathers, *n.* the vaned feathers possessing barbules that chiefly cover the body of a bird and give it a streamlined form. Those which extend beyond the body on the wings and serve for flight are termed *flight feathers.*

contraceptive, see BIRTH CONTROL.

contractile vacuole, *n.* a membrane-surrounded vacuole that periodically fills with liquid, expands and then voids the contents to the outside of the organism. Particularly found in protozoans and freshwater sponges, the contractile vacuole appears to have an osmoregulatory function. This theory is supported by the relative absence of such vacuoles in marine amoebae, whereas they are common in freshwater forms that take in water along an osmotive gradient because of the higher (less negative) osmotic pressure.

control, *n.* **1.** an experiment carried out to afford a standard of comparison for other experiments, e.g. where the effect of a nutrient substance on a plant is being tested, control plants are grown in exactly similar conditions but without the addition of the nutrient substance. **2.** also called **population control.** the limitation by man of numbers of harmful plants or animals by artificial means, e.g. spraying with chemicals, poisoning, shooting or by seminatural means such as BIOLOGICAL CONTROL. See also REGULATION (2).

controlling factor, *n.* any factor which has an increasing effect as the population density increases. See DENSITY–DEPENDENT FACTORS.

conus arteriosus, *n.* the projection from which the pulmonary artery arises on the right ventricle of the heart.

convection, *n.* the propagation of heat through liquids and gases by the movement of the heated particles.

convergence or **convergent evolution** or **parallelism,** *n.* a form of evolution which results in unrelated organisms independently producing similarities of form, usually because they become adapted to living in similar types of environment. For example, fish and cetaceans have evolved similar streamlined body shapes and fins.

converging, *adj.* (of plant parts) having apices closer together than their bases.

convolute, *adj.* coiled or rolled together.

Cooley's disease, see THALASSAEMIA.

cooling, see TEMPERATURE REGULATION.

copepod, *n.* any minute free-living or parasitic crustacean of the subclass Copepoda. Copepods lack a carapace and are extremely common in both freshwater and marine plankton, where they constitute an important food source for larger animals such as fish.

coprodaeum, *n.* the part of the CLOACA into which the rectum opens in birds.

coprophage, *n.* an animal that feeds on the faeces of other animals.

copulation or **coition,** *n.* the act of coupling of male and female animals in sexual intercourse. In mammals the male organ (penis) penetrates the female and introduces the male GAMETES. MATING is a term usually restricted to the behavioural aspects of pair formation and this occurs normally before copulation whilst fertilization takes place only after copulation.

coracoid, *n.* one of a pair of bones that form the ventral part of the pectoral girdle of many vertebrates. In most mammals they are reduced to small processes on the SCAPULA and their role taken over by the CLAVICLES.

coral, *n.* **1.** the calcareous skeleton of certain COELENTERATES. **2.** the skeleton together with the animals which live in it and secrete it, the whole structure often forming a reef. The animals, usually anthozoans and hydrozans, have a POLYP structure.

Cordaitales, *n.* an extinct order of carboniferous GYMNOSPERMS having the form of tall, slender trees at the top of which was a crown of branches bearing large, elongated leaves.

cordate, *adj.* heart-shaped.

corepressor, *n.* a substance of low molecular weight which unites with an APOREPRESSOR and reduces the activity of particular structural genes (see OPERON MODEL).

core temperature (T_c), *n.* the mean temperature of the tissues of organisms at a depth below that directly affected by a change in the AMBIENT temperature. T_c cannot be measured accurately and is generally represented by a specified body temperature, e.g. rectal and cloacal temperatures.

coriaceous, *adj.* (of plant structures) having a leathery appearance.

corium, see DERMIS.

cork, *n.* a plant tissue made up of cells with thick walls impregnated with SUBERIN. Cork cells are dead when mature, forming an outer layer in stems and roots of woody plants that is impervious to water and air. The cork oak *Quercus suber* produces very large quantities of cork which can be removed and used commercially.

cork cambium, *n.* a specific CAMBIUM that contributes towards the production of bark.

corm, *n.* a modified underground stem found in some MONOCOTYLEDONS that is usually broader than high and contains no fleshy scale leaves (as in the BULB) but rather has its food reserves in the stem. For example, crocus and meadow saffron. See Fig. 116.

cormidium, *n.* an aggregation of polyps in a SIPHONOPHORE.

corn, *n.* any of various cereal plants. 'Corn' usually denotes the predominant cereal crop of a region, e.g. wheat in England, oats in Scotland and Ireland, and maize in North and South America.

cornea, *n.* that part of the SCLERA at the front of the eye of vertebrates overlaying the iris and lens. It is a transparent layer of EPITHELIUM and

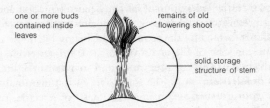

Fig. 116. **Corm.** Generalized structure.

CONNECTIVE TISSUE through which light enters the eye and is refracted so that the lens can then focus it on the retina.

cornification, *n.* the formation of KERATIN in the epidermis from epidermal cells, as occurs in the skin when subjected to abrasive stresses.

corolla, *n.* a collective term for the petals of a flower.

corona, *n.* a trumpet-shaped outgrowth of the PERIANTH, as in daffodil flowers.

coronary thrombosis, *n.* the formation of a blood clot (*thrombus*) in a coronary artery.

coronary vessels, *n.* the arteries and veins supplying the muscles of the heart.

corpus, *n.* a body.

corpus allatum, *n.* a gland present in the head of insects that is responsible for the production of *juvenile hormone* and maintains the larval characteristics at each moult until the adult metamorphosis takes place. See Fig. 117.

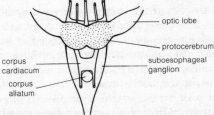

Fig. 117. **Corpus allatum.** The brain of *Rhodnius* (the kissing bug).

corpus callosum, *n.* a band of nervous tissue connecting the cerebral hemispheres in the higher mammals.

corpus cavernosum, *n.* one of a pair of laterally situated masses of erectile tissue in the mammalian penis which, together with the CORPUS SPONGIOSUM, causes the erection of the penis when filled with blood.

corpus luteum, *n.* an endocrine organ formed in mammals in the ruptured GRAAFIAN FOLLICLE after the process of OVULATION. The corpus luteum is responsible for secreting PROGESTERONE under the influence of

LH and LTH. If fertilization of the egg occurs the corpus luteum persists during pregnancy, otherwise it degenerates at the end of the OESTRUS CYCLE. It is formed by the action of luteinizing hormone produced by the anterior lobe of the PITUITARY GLAND. Progesterone prepares the reproductive organs for pregnancy and maintains the uterine lining.

corpus quadrigemina, *n.* the four lobes of the mammalian brain.

corpus spongiosum, *n.* a median mass of erectile tissue in the mammalian penis. See CORPUS CAVERNOSUM.

correlated response, *n.* a change in the PHENOTYPE resulting from SELECTION for a completely different character, e.g. selection for more bristles, which in some insects can also cause sterility.

correlation, *n.* a statistical association between two variables, calculated as the correlation coefficient r. The coefficient can range from $r = 1.0$ (a perfect positive correlation) to $r = -1.0$ (a perfect negative correlation), with an r value of 0 indicating no relationship between the two variables. Height and weight in humans are positively correlated (as values for height increase so do values for weight), whereas other variables give a negative correlation, e.g. as human age increases so mental agility tends to decrease.

cortex, *n.* an outer zone of any organ or part, as in the mammalian kidney and brain, or the layer of plant tissue outside the VASCULAR BUNDLES but inside the epidermis.

corticotrophin, see ADRENOCORTICOTROPIC HORMONE.

cortisol or **hydrocortisone,** *n.* an adrenocortical steroid with effects similar to CORTISONE.

cortisone, *n.* a GLUCOCORTICOID hormone secreted by the adrenal cortex, whose function is to combat stress. It causes shrinkage of lymph nodes and lowers the white blood cell count, reduces inflammation, promotes healing and stimulates GLUCONEOGENESIS. Cortisone controls its own production, which is triggered by the ADRENOCORTICOTROPIC HORMONE in a negative FEEDBACK MECHANISM.

corymb, *n.* a type of RACEME in which the pedicals shorten towards the apex, as a result of which flowers occur at approximately the same level and produce a flat top.

corymbose cyme, *n.* a flat-topped CYMOSE inflorescence that resembles a CORYMB but is not produced in the same way, in that the oldest parts are in the centre (top) of the inflorescence.

cosmic rays, *n.* a stream of atomic particles entering the earth's atmosphere from outer space at nearly the speed of light. Cosmic rays are thought to be a cause of SPONTANEOUS MUTATIONS.

cosmoid, *adj.* (of the scales of coelacanths and lungfish) consisting of an outer layer of *cosmin* (which resembles dentine), a mid-layer of spongy bone and a lower layer of bony lamellae. Compare GANOID, PLACOID.

costal, *adj.* of the ribs.

cotyledon, *n.* **1.** a part of the plant embryo in the form of a specialized seed leaf that acts as a storage organ, absorbing food from the endosperm and functioning as a leaf after EPIGEAL germination. Some ANGIOSPERMS have one cotyledon per seed (MONOCOTYLEDONS) while others have two (DICOTYLEDONS). **2.** a part of the mammalian placenta on which a tuft of villi occurs, particularly in ruminants.

cotype, see SYNTYPE.

Coulter counter, *n.* an electronic particle analyser, in which a pair of electrodes detects the effect of a passing particle on the impedance. The counter is used to measure the total cell count in a microbiological sample.

countercurrent exchange, *n.* a biological mechanism designed to enable maximum exchange between two fluids. The mechanism's effect is dependent on the two fluids flowing in opposite directions, and having a concentration gradient between them.

| fluid ONE | high concentration | → | low concentration |
| fluid TWO | high concentration | ← | low concentration |

Such mechanisms occur, for example, in oxygen exchange between water and the blood vessels in fish gills, and in the ascending and descending tubules of the LOOP OF HENLE in the mammalian kidney.

countercurrent multiplier, *n.* the amount of cross-transport in a COUNTERCURRENT-EXCHANGE system per unit distance, being a function (= multiplication) of the total distance over which the exchange takes place.

counterflow system, *n.* the flow of water and blood in opposite directions across the gills of fish which ensures that blood meets water with the highest possible oxygen content. See COUNTERCURRENT EXCHANGE.

coupling, *n.* an arrangement in a double HETEROZYGOTE where WILD TYPE alleles of two genes are located on one chromosome with MUTANT alleles of these two genes on the other HOMOLOGOUS CHROMOSOME. Such an arrangement of linked genes is usually referred to as *in coupling* (see Fig. 118). Such arrangements are important when considering the implications of GENETIC LINKAGE. Compare REPULSION.

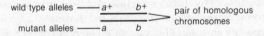

Fig. 118. **Coupling.** The arrangement of genes in coupling.

courtship, *n*. those aspects of behaviour which involve displays and posturings in order to bring about copulation.

coverslip, *n*. a very thin piece of glass used to cover a preparation on a microscope slide for microscopical examination.

coxa, *n*. the basal segment of the leg of an insect.

crab, *n*. a decapod CRUSTACEAN (suborder Brachyura) characterized by a large flattened, laterally expanded carapace below which the abdomen is curled forward.

cranial nerve, *n*. one of 10 (in ANAMNIOTES) or 12 (in AMNIOTES) peripheral nerves that emerge from the vertebrate brain as distinct from the spinal cord.

Craniata, see VERTEBRATE.

cranium, *n*. the skull of vertebrates.

crassulacean acid metabolism (CAM), *n*. a method of PHOTOSYNTHESIS found in certain succulent plants (members of the family Crassulaceae) that close their stomata during the day to avoid excessive TRANSPIRATION losses and open them at night. During the night CO_2 is taken in and stored as organic acids (e.g. malic acid); during the day the CO_2 is released from the organic acids and used in the CALVIN CYCLE.

crayfish, *n*. a decapod crustacean (suborder Macrura) that possesses an elongate body of the CARIDOID FACIES type.

creatine phosphate, *n*. see PHOSPHAGEN.

creatinine, *n*. the nitrogenous waste material of muscle creatine.

crenate, *adj*. (of plant leaves) having a notched edge.

Cretaceous period, *n*. the geological period which began about 135 million years BP and ended 65 million years BP. It is usually divided into Lower (135–95 million years) and Upper (94–65 million years) periods. The dominance of flowering plants began during this period and large reptiles (e.g. dinosaurs) and ammonites were extinct by its end. By the end of the period, London was at 40°N and Britain was still moving northwards. See GEOLOGICAL TIME.

cretin, *n*. a person suffering or who has suffered from serious reduction in THYROID GLAND activity during development. A cretin shows slow growth, pot belly, gross intellectual deficiency and retarded sexual development.

Crick, Francis, (1916–   ) English biochemist best known for his work with James WATSON and Maurice WILKINS on the structure of the DNA molecule, for which they shared a Nobel Prize in 1962. They were able to suggest a model for DNA structure and how it replicates which is widely accepted today. Crick also proposed the *wobble hypothesis* which refers to the fact that the third base of each codon often can be varied without changing the AMINO ACID coded for by that codon.

crinoid, *n.* any echinoderm of the class Crinoidea, including the present-day feather stars and sea lilies. They are common fossils from the CAMBRIAN PERIOD throughout geological history, and are mainly sedentary. *Antodon* is free-living in British waters.

crisped, *adj.* (of plant structures) curled.

crista, *n.* (*pl.* cristae) a fold in the inner membrane of a MITOCHONDRION on which the electron-transport reactions of AEROBIC RESPIRATION take place.

critical group, *n.* a taxonomic group in which it is difficult to recognize taxonomic differences.

CRM (cross-reacting material), *n.* a protein, usually without an enzymic function, that can be recognized in serological tests.

crocodile, *n.* any large, tropical, aquatic lizard-like reptile of the order Crocodilia, possessing a four-chambered heart that is unique in reptiles.

crop, *n.* **1.** in vertebrates, particularly some birds, an expanded part of the oesophagus where food is stored. **2.** in invertebrates, an expansion of the anterior part of the gut system where food is either digested or stored. **3.** the agricultural or commercial fishery yield. **4.** in ecological terms, the difference between gross annual production and the net production – i.e. the material eaten by predators (or herbivores where the food is a vegetable), including that taken by man, and that consumed by organisms responsible for decay. See STANDING CROP.

crop growth rate, *n.* the total dry matter production of a COMMUNITY per unit land area after a known time span.

crop milk, *n.* pigeon's milk that is secreted by the epithelium of the CROP (1) and which is fed to nestlings.

crop pruning, *n.* the removal of old growth to promote new growth and better CROP (3) production.

crop rotation, *n.* the growing of CROPS (3) in a regular sequence over a number of seasons so as not to exhaust the soil. A simple example is: root crop, oats, leguminous plants and wheat. The leguminious crop is of particular importance in returning nitrogen to the soil from the atmosphere through the nitrogen–fixing bacteria (see NITROGEN FIX-ATION) in their root nodules.

crop spraying, *n.* the treatment of CROPS (3) with chemical sprays to prevent or limit attack by insects and plant pathogens.

cross, *n.* **1.** a mating between a male and a female of a plant or animal species, from which one or more offspring are produced. **2.** a hybrid produced by mating two unlike parents.

crossbreed, see INTERBREED (2).

cross–bridge link, *n.* the connection of the globular head of the MYOSIN cross-bridge with the myosin thick filament in skeletal muscle fibrils.

cross-fertilization or **allogamy,** *n.* the production of an offspring by the female GAMETE being fused with a male gamete from a different plant. Compare SELF–FERTILIZATION.

crossing over, *n.* a process occurring in the DIPLOTENE stage of Prophase 1 of MEIOSIS in which there is exchange of nonsister CHROMATIDS, producing genetic RECOMBINATION. (See Fig. 119). See also UNEQUAL CROSSING OVER.

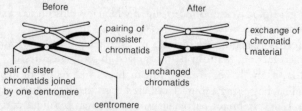

Fig. 119. **Crossing over.** The exchange of chromatid material.

Crossopterygii, *n.* a subclass of the class Choanichthyes, an ancient group of mainly freshwater fish, including many fossil forms but only a single living species, the COELACANTH *Latimeria*.

cross–over value, *n.* an estimation of the number of gametes that contain a gene combination produced by CROSSING OVER during MEIOSIS, as a proportion of all gene combinations in the gametes produced. The frequency produced by this calculation is used as a measure of GENETIC LINKAGE, one percent cross–over value being equivalent to one map unit.

cross–pollination, *n.* the transfer of pollen from the anthers of one flower to the stigma of another by the action of wind, insects, etc., with the subsequent formation of pollen tubes. Compare SELF–POLLINATION. See POLLINATION.

crustacean, *n.* a member of the class Crustacea in the phylum Arthropoda. The class includes shrimps, lobsters, crabs and water fleas. Most forms are aquatic, although a few are terrestrial, living in damp places, e.g. woodlice.

cryoturbation, *n.* the movement of frozen earth sediments as a result of ice formation.

cryoturbation, see additional entries.

cryophytes, *n.* plants that grow on ice or snow – usually algae, mosses, fungi or bacteria.

crypt- or **crypto-,** *prefix.* denoting hidden.

cryptic coloration, *n.* any coloration which conceals animals in their normal habitats by its resemblance to that of the surroundings (see CAMOUFLAGE).

cryptic species, *n.* a sibling species.

crypt of Lieberkuhn, *n.* a narrow pit at the base of a villus in the lining of the DUODENUM and ILEUM that passes secretions from BRUNNER'S GLANDS, PANETH CELLS and GOBLET CELLS. See Fig. 120.

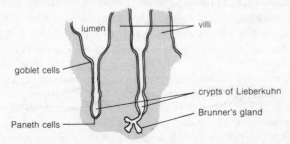

Fig. 120. **Crypt of Lieberkuhn.** A transverse section of the wall of the duodenum.

cryptogam, *n.* (in obsolete classification schemes) any plant that does not produce seeds, i.e. plants placed in the groups Thallophyta, Bryophyta and Pteridophyta. Whereas in conifers and flowering plants (Phanerogamia) the organs of reproduction are prominent, they are not so in the Cryptogamia – hence the name ('secret marriage').

cryptophyte or **geophyte,** *n.* a hidden plant, one in which the renewal bud is buried.

cryptozoa, *n.* **1.** any animals that live in litter, dark places and the upper layers of soil. **2.** any animals which *may* exist, such as the Yeti, but whose existence has yet to be positively demonstrated.

C.S.F. see CEREBROSPINAL FLUID.

ctenidium, *n.* (*pl.* ctenidia) one of the comb-like gills present in the mantle of many molluscs, particularly Lamellibranchia.

ctenophore, *n.* any marine invertebrate of the phylum Ctenophora, including the sea gooseberries, that moves by means of comb rows of cilia. They are sometimes classified with the COELENTERATES but more usually given the status of a phylum.

cubical epithelium, see EPITHELIUM.

cud, *n.* the food from the first stomach of a ruminate which is regurgitated into the mouth and rechewed.

cull, *vb.* **1.** to take out (an animal, especially an inferior one) from a herd. **2.** to reduce the numbers of a population of animals by killing them, e.g. seal culls.

culmen, *n.* the median longitudinal ridge of the upper mandibles of the bill of a bird.

cultivar, *n.* a cultivated variety of a plant or animal species that contains unique features and can be propagated sexually or asexually. In horticulture the use of cultivar names is common and is written after the species name, e.g. the common ivy cultivar *Hedera helix* (Goldheart).

cultivation, *n.* **1.** the preparation of ground for crops. **2.** the planting, care and harvesting of crops.

culture medium, see MEDIUM.

cuneate, *adj.* (of leaves) wedge-shaped.

cuniform, *adj.* (of leaves) wedge-shaped with the thin end of the wedge at the base.

cupula, *n.* **1.** any cup-shaped structure. **2.** a structure in the AMPULLA of the ear containing sensory end organs.

curare, *n.* a paralysing poison originally extracted from the root of *Strychnos toxifera* by South American Indians and used on arrowheads. Now it is a valuable source of drugs. Its action in paralysis is to prevent ACETYLCHOLINE depolarizing the postsynaptic membrane, particularly at nerve/muscle junctions, thus preventing the passage of the nerve impulse and so rendering the victim immobile.

cusp, *n.* one of a series of projections on the biting surface of a molar tooth in mammals.

cutaneous respiration, *n.* the exchange of gases through the skin. Many lower organisms in the animal kingdom are able to take up oxygen and lose CO_2 entirely by diffusion provided they are small enough. Organisms up to 1 mm in diameter can get all their oxygen supply through mere diffusion on the assumption that they consume about 0.05 $cm^3g^{-1}h^{-1}O_2$. If their metabolic rate was considerably lower and required less O_2, they could have a larger diameter, but lower O_2 consumptions than this are rare. Thus, animals exceeding 1 mm in diameter must have a special oxygen-transport system to supply internal tissues. However, even organisms which possess such a system may also carry out gaseous exchange through the skin. Indeed, amphibians have a higher mean-oxygen uptake through the skin than through the lungs in temperatures up to $22°C$.

cuticle, *n.* **1.** in both animals and plants, a thin noncellular layer secreted by the epidermis. In higher plants it covers all the exterior, except where stomata and lenticels occur, and prevents water loss (*cuticularization*). In many arthropods, it forms part of the EXOSKELETON and is composed of CHITIN and PROTEIN, with a calcareous element in CRUSTACEANS. **2.** the epidermal cells of vertebrates (the *stratum corneum*) that have been converted to KERATIN (as in hair, feathers, scales, nails, claws, hooves, horns, etc.), giving proofing against water loss, bacterial entry and ultraviolet light.

cutinization, *n.* the formation of CUTICLE in plants by the impregnation of the cell wall with cutin, a fatty substance with waterproofing qualities.

cutting, *n.* a method of artificial propagation in plants where a small stem, usually with attached leaves, is removed from the parent plant (e.g. *Begonia, Geranium*) and placed in water or moist sand. Once ADVENTITIOUS (1) roots have developed the cuttings are transferred to soil.

Cuvierian duct or **ductus Cuvieri,** *n.* one of pair of blood vessels which lead from the sinuses and blood vessels to the sinus venosus of the heart in fish. They are homologous with the venae cavae of higher vertebrates.

cyanide, *n.* any salt of hydrocyanic acid. Potassium cyanide (KCN) and hydrogen cyanide (HCN) are the commonest examples of cyanides, all of which are extremely poisonous. They combine with cytochrome enzymes (e.g. CYTOCHROME OXIDASE), which transfer hydrogen atoms in CELLULAR RESPIRATION, and thus block the production of energy in the cells.

cyanocobalamin, see COBALAMIN.

cyanophyte, see BLUE-GREEN ALGAE.

cybernetics, *n.* the study of the comparison of control in the workings of the living body with man-made mechanical systems such as are used in robots.

cycad, *n.* any tropical or subtropical GYMNOSPERM of the order Cycadales. Cycads date from the MESOZOIC PERIOD. Present-day forms grow to 20 m in height and have a crown of fern-like leaves. They live for up to a thousand years.

Cycadofilicales, *n.* an order of extinct GYMNOSPERMS whose members are of interest from a phylogenetic point of view. They reproduce by seeds but have a fern-like vascular system and develop secondary wood.

cycl- or **cyclo-,** *prefix.* denoting a circle.

cyclic GMP (guanosine monophosphate), *n.* a nucleotide that usually produces cell responses opposite to those of CYCLIC AMP. Analogous with cAMP, it is present in much lower concentrations.

cyclic AMP (cAMP, adenine monophosphate), *n.* a molecule thought to act as an intermediary between a hormone and the biochemical process of its target cell. The process is thought to be (a) the hormone arrives at the target cell and becomes complexed to receptor sites in the cell membrane; (b) the adenyl cyclase enzyme is activated, enabling conversion of ATP to cAMP; (c) specific cellular enzymes are activated by cAMP starting a chain reaction. See also CYCLIC GMP.

cyclic phosphorylation, see LIGHT REACTIONS.

cycling matter, *n.* any material recirculated within a system.

cyclopia, *n.* the possession of a single median eye.

cyclosis, see CYTOPLASMIC STREAMING.

cyclostome, *n.* **1.** any member of the order Cyclostomata, a group consisting of lampreys and lungfish, characterized by the lack of jaws and the presence of a suctorial mouth. **2.** a suborder of Polyzoa possessing a tubular body wall and lacking an operculum.

cyme, *n.* a CYMOSE inflorescence.

cymose, *adj.* (of an inflorescence) having growing parts that end in flowers, as a result of which the combined growth depends on the production of lateral growing points; the oldest part of the inflorescence is thus at the apex.

cyphonautes larva, *n.* the larva of a polyzoan, similar to a TROCHOSPHERE, with a bivalve shell.

cypris larva, *n.* the larva of barnacles.

cypsela, *n.* the fruit of members of the Compositae, similar to an ACHENE and formed from two carpels.

cyst, *n.* **1.** a bladder or bag-like structure that may contain the resting stage of an organism. Many groups or organisms have encysted stages, e.g. protozoans, nematodes, flukes, tapeworms. **2.** a MORBID structure arising as an outgrowth of the skin, e.g. a sebaceous cyst, usually arising from a blocked duct, or an internal growth, e.g. ovarian cyst.

cysteine, *n.* one of 20 AMINO ACIDS common in proteins that has a polar 'R' structure and is water-soluble. See Fig. 121. The ISOELECTRIC POINT of cysteine is 5.1.

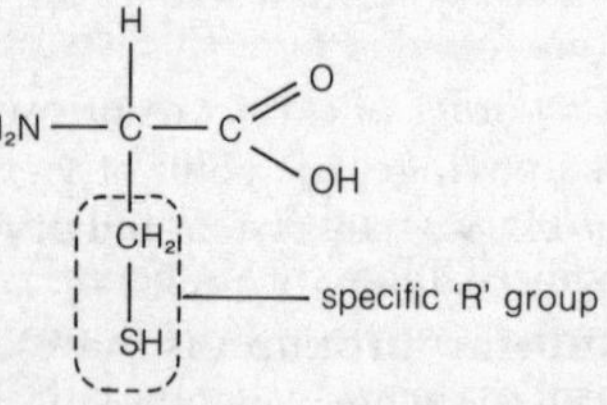

Fig. 121. **Cysteine.** Molecular structure.

cysticercoid, *n.* a larva of tapeworms that is similar to a CYSTICERCUS but possesses only a small bladder.

cysticercus, *n.* the larval form of a tapeworm which grows into the adult when eaten by the primary host, and consists of a SCOLEX inverted into a large bladder. When ingested the outer cyst wall is digested, the scolex everts and the bladder disappears. The scolex attaches to the intestinal wall of the host and a new tapeworm forms with the growth of proglottides. See Fig. 122.

cystic fibrosis, *n.* a rare disease of humans (affecting one child per 2,000

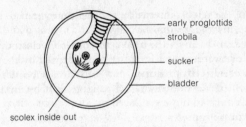

early proglottids

strobila

sucker

bladder

scolex inside out

Fig. 122. **Cysticercus.** Generalized structure.

live births) in which various MUCUS-secreting glands, particularly in the pancreas and in the lungs, become fibrous, producing an abnormally thick mucus. Symptoms of the disease include intestinal blockage and pneumonia which, despite modern medicines, can still be fatal. The condition is inherited as an autosomal recessive gene on the long arm of chromosome 7, so that affected children are homozygous recessive, a MUTANT allele having been passed on by both parents. At present there is no cure for the disease, but in 1986 a gene probe prenatal test was announced involving CHORIONIC BIOPSY of the foetus at 9–11 weeks. If the result of the test is positive, the parents have a choice of whether to terminate the pregnancy or not.

cystitis, *n.* a painful inflammation of the bladder caused by infection with the gut bacterium *Escherichia coli.*

cystocarp, *n.* a structure produced from the fertilized CARPOGONIUM of a red alga which irritates the CARPOSPORES.

cystolith, *n.* a structure found in some plants, e.g. nettles, formed by an ingrowth of the cell wall, and carrying a grain of calcium carbonate at its tip.

cytochrome, *n.* a protein pigment containing iron that is capable of being alternately oxidized and reduced, acting as an electron carrier in the ELECTRON TRANSPORT SYSTEM of a MITOCHONDRION.

cytochrome oxidase, *n.* an enzyme acting as the last hydrogen electron carrier in the ELECTRON TRANSPORT SYSTEM, receiving an electron from CYTOCHROME and passing it on to oxygen, with the formation of water.

cytogenetics, *n.* the study of the inheritance of cells and their chromosomes. Cytogenetic analysis of foetal cells (see AMNIOCENTESIS and CYSTIC FIBROSIS) is used to assess potential risks to the baby.

cytokinesis, *n.* a part of CELL DIVISION that usually occurs during TELOPHASE of nuclear division. The cytoplasm becomes divided into two parts which are sometimes unequal, e.g. the production of ova and polar bodies during OOGENESIS. The mode of splitting differs between organisms. In higher plants, a special membrane-producing area (the

CELL PLATE) develops internally and becomes joined to the outer membrane; in lower plants, the cell wall and plasma membrane grow in from outside; in animals, the plasma membrane forms a deep cleft from the outside inwards which eventually divides the cell.

cytokinin or **kinetin** or **kinin,** *n.* a PLANT HORMONE which promotes cell division (CYTOKINESIS) and cell enlargement by swelling. Several effects are similar to those produced by AUXIN, but others are different, e.g. cytokinin promotes lateral bud development and growth while auxin supresses lateral buds (see APICAL DOMINANCE).

cytology, *n.* the study of cells.

cytolysis, *n.* the breakdown of cells.

cytoplasm, *n.* that part of the cellular PROTOPLASM not located in the nucleus. The range of organelles contained in the cytoplasm varies widely between PROCARYOTE and EUCARYOTE cells.

cytoplasmic inheritance or **extranuclear inheritance,** *n.* the control of certain characters by genetic factors located in the cytoplasm of EUCARYOTES. These cytoplasmic mechanisms can show themselves as general 'maternal' influences (since the female gamete contains more cytoplasm than the male) as in the control of shell coiling in the snail *Limnaea.* DNA has also been located in several cytoplasmic organelles, such as a MITOCHONDRION or a CHLOROPLAST, which can replicate and function independently of the nucleus.

cytoplasmic streaming or **cyclosis,** *n.* the movement of cytoplasm from one region of a cell to another, often in definite currents, thought to be controlled by MICROFILAMENTS composed of protein similar to ACTIN. Such streaming has several possible functions: (a) transport of substances from one part of the cell to another (e.g. during TRANSLOCATION in SIEVE TUBES); (b) cellular movement (e.g. in PSEUDOPODIA of white blood cells); (c) maintenance of optimal temperature; (d) provision of optimal light conditions for chloroplasts (e.g. in MESOPHYLL cells of leaf).

cytosine, *n.* one of four types of nitrogenous bases found in DNA, having the single-ring structure of a class known as PYRIMIDINES. Cytosine forms part of a DNA unit called a NUCLEOTIDE and always forms complementary pairs with a DNA purine base called a GUANINE. Cytosine also occurs in RNA molecules. See Fig. 123.

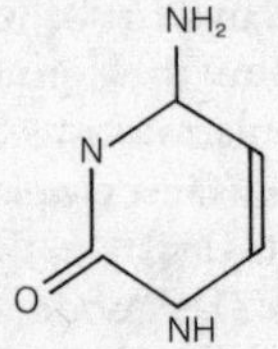

Fig. 123. **Cytosine.** Molecular structure.

cytoskeleton, *n.* a network of MICROTUBULES and MICROFILAMENTS in the cytoplasm of cells which is thought to give the cell its characteristic shape. The network enables the movement of specific organelles within the cytoplasm (as in vesicles produced by the GOLGI APPARATUS), and the production of general CYTOPLASMIC STREAMING.

cytosol, *n.* the soluble portion of the cytoplasm including dissolved solutes but excluding any particulate matter.

cytostome, *n.* a mouth-like opening present in many unicellular organisms.

cytotaxonomy, *n.* a method of classification based on the characteristics of CHROMOSOMES.

D

dactylo-, *prefix.* denoting finger or toe.

dactylozooid, *n.* a stinging COELENTERATE polyp, usually possessing long tentacles that serve to collect prey in colonial forms.

damping off, *n.* a disease caused usually by the fungus *Pythium* in which young seedlings are attacked at ground level when in overcrowded, damp conditions causing them to rot and fall over.

damselfly, *n.* any insect of the suborder Zygoptera, similar to but smaller than dragonflies.

Danielli, James, see UNIT–MEMBRANE MODEL.

dark adaptation, *n.* the increase in light sensitivity of an eye as a result of remaining in the dark.

dark-field microscopy, *n.* a microscope technique that involves creating a dark background to the subject to be observed under the microscope and viewing this by reflected light; the object thus appears light on a dark background.

dark reactions, *n.* those processes in PHOTOSYNTHESIS that do not require light energy but which can occur in both light and dark conditions; the reactions involve the reduction of carbon dioxide to carbohydrate in the CALVIN CYCLE, using ATP and $NADPH_2$ formed in the LIGHT REACTIONS.

dark repair, *n.* repair of DNA by a mechanism that does not require light.

dart sac, *n.* a sac branching from the vagina of snails which contains a dart that is shot out prior to copulation between two snails, so that the darts from each penetrate the tissues of the other.

Darwin, Charles Robert (1809–92) English naturalist. His works include: *Voyage of the Beagle* (1839), *On the Origin of Species by means of Natural Selection* (1859), and *The Descent of Man* (1871). He was with

WALLACE, the originator of modern evolutionary theory. See DARWINISM.

Darwinism, *n.* the theory of evolution formulated by Charles DARWIN that holds that different species of plants and animals have arisen by a process of slow and gradual changes over successive generations, brought about by NATURAL SELECTION. The essential points of Darwin's theory are:

(a) in organisms that reproduce sexually there is a wide range of variability, both within and between species.

(b) all living forms have the potential for a rapid rise in numbers, increasing at a geometric rate.

(c) the fact that populations usually remain within a limited size must indicate a 'struggle for existence' in which those individuals unsuited to the particular conditions operating at that time are eliminated or fail to breed as successfully as others (see FITNESS).

(d) the struggle for existence results in natural selection that favours the survival of the best-adapted individuals, a process described by Herbert Spencer (1820–93) as the 'survival of the fittest' in his *Principles of Biology* (1865).

Darwin's finches, *n.* a group of finches that occurs on the Galapagos Islands in the Pacific. The islands are oceanic and were colonized by an ancestral type which has speciated and provides an excellent example of ADAPTIVE RADIATION.

daughter cell, see CELL DIVISION.

Davson, H., see UNIT–MEMBRANE MODEL.

day–neutral plant, *n.* a plant which flowers after a period of vegetative growth, regardless of PHOTOPERIOD. Examples include dandelion, tomato, sunflower. Compare SHORT–DAY PLANT, LONG–DAY PLANT.

day sleep, *n.* the folding of the leaflets of a compound leaf in conditions of bright light so that stomatal surfaces come together, thus reducing water loss.

DDT, *abbrev. for.* a chlorinated hydrocarbon (*d*ichloro*d*iphenyl-*t*richloroethane) which acts as a powerful insecticide with long-lasting effects. DDT was the first major insecticide in use. Although DDT is cheap to manufacture, it has produced adverse ecological consequences. Its lack of biodegradability and the fact that it tends to accumulate in fatty tissues has resulted in its transfer from one consumer to another up the FOOD CHAIN, becoming concentrated at each step. One effect of this has been to endanger the top carnivorous birds whose eggshells have become paper-thin because DDT has prevented the mobilization of calcium in the oviduct, so reducing the reproductive potential of many rare species. While these processes have been occurring the target insects have been subjected to strong SELECTION pressure from the DDT, with the result that highly resistant populations now exist, making the insecticide useless in many parts of the world.

de-, *prefix.* denoting from, out or away.

dealation, *n.* the loss of wings in ants and other insects. For example, the queen ant removes her wings prior to laying eggs after fertilization.

deamination, *n.* the removal of the amino group (NH_2) from a molecule, as in the release of ammonia (NH_3) from AMINO ACIDS, the residue of which can then enter the KREBS CYCLE, usually via ACETYL-COENZYME A. The amino group then enters the ORNITHINE CYCLE.

death rate, see MORTALITY RATE.

deca-, *prefix.* denoting ten.

decapod, *n.* a taxonomic name unusually applied to two completely different groups of organisms. **1.** any crustacean of the largely marine order Decapoda, having five pairs of walking limbs, including the crabs, lobsters, shrimps, prawns and crayfish. **2.** any cephalopod mollusc of the order Decapoda, having a ring of eight short tentacles and two long ones, including the squids and cuttlefish.

decarboxylation, *n.* the removal of carbon dioxide from a molecule, as in the conversion of oxalosuccinic acid (C_6) to -ketoglutaric acid (C_5) in the KREBS CYCLE.

decay, *n.* the decomposition of dead tissue.

decerebrate rigidity, *n.* a state in which the limb extensor muscles go into tonic contraction and the limbs are held rigidly out from the body. This occurs if the brain is transected immediately behind the ganglionic centre (red nucleus) of the midbrain.

deceribration, *n.* an experimental cutting off of cerebral activity either by sectioning the brain stem or by interrupting the blood supply to the brain.

decidua, *n.* that part of the lining of the uterus of mammals which, after thickening during pregnancy, is shed as AFTERBIRTH.

deciduous teeth or **milk teeth,** *n.* the first set of teeth, present in most mammals. These are replaced by a second, permanent set of teeth together with several molars that are absent as deciduous teeth.

deciduous tree, *n.* any tree in which there is leaf-fall in the autumn with new leaves produced in the spring after a winter spent in dormancy. Such trees are generally ANGIOSPERMS (e.g. oak, ash, beech) although some CONIFERS are also deciduous (e.g. the larch).

decomposer, *n.* any saprophytic organism such as bacteria and fungi, which breaks organic materials down into simpler compounds and eventually into inorganic materials. The latter can then be utilized by PRODUCERS in the synthesis of organic compounds.

decomposition, *n.* **1.** the break-up of a chemical substance into two or more simpler substances. **2.** the breakdown of organic material by microorganisms.

decumbent, *adj.* (of stems) lying on the ground and growing upwards only at the tip.

decussate, *adj.* (of leaves) at right angles to each other.

dedifferentiation, *n.* a process in which tissues that have undergone CELL DIFFERENTIATION can be made to reverse the process so as to become a primordial cell again (see GURDON). In theory, all cells should possess this ability since the mature cell does not lose DNA (see TOTIPOTENCY), but reversal has been demonstrated in plants much more easily than in animal cells.

deer, *n.* any of the ruminant quadrupeds included in the family Cervidae.

defecation, *n.* the elimination of faeces from the alimentary canal via the anus, a process requiring rhythmic contraction of the smooth muscle of the gut wall (under autonomic control) and relaxation of the striated muscle of the anal SPHINCTER under voluntary control.

deficiency disease, *n.* any condition exhibiting abnormalities produced by lack of a particular component in the diet. Examples include BERI-BERI (vitamin B_1), SCURVY (vitamin C), KWASHIORKOR (protein deficiency), RICKETS (vitamin D). Plants can also exhibit conditions brought about by deficiency of minerals. For example, magnesium is required in the synthesis of CHLOROPHYLL, a lack of the mineral producing CHLOROSIS of the leaves.

deflexed, *adj.* (of bent structures) bent sharply downwards.

degeneracy, *n.* a situation in the GENETIC CODE where most amino acids are coded by more than one triplet of DNA NUCLEOTIDE bases. For example, ARGININE is coded by CGU, CGC, CGA and CGG CODONS. See also CRICK.

degeneration, *n.* the loss or reduction in size of an organ during the lifetime of an organism, or during the course of evolution. The latter process may give rise to a vestigial organ.

deglutition, *n.* the act of swallowing which is brought about by a complex series of reflexes initiated by stimulation of the pharynx.

degrees of freedom (d.f.), *n.* the number of unrestricted variables in a frequency distribution, a factor that is of great importance in statistical testing. For example, in a simple CHI-SQUARED TEST the number of degrees of freedom is one less than the number of classes (types) of individuals, one d.f. having been lost due to the assumption that a certain proportion of each class is expected. Thus in testing a 9:3:3:1 ratio there are three degrees of freedom.

dehiscent, *adj.* (of plant structures) spontaneously bursting open to liberate the enclosed contents. The two main dehiscent structures are ANTHERS releasing pollen grains and certain dry fruits that split open to release seeds, for example, peas, beans, wallflowers, delphiniums. See also CIRCUMSCISSILE.

dehydration, *n.* the process by which water is removed from any substance. It is utilized in freeze-drying for the preservation of materials, and in the removal of water from microscopical preparations where it is necessary to use substances which are immiscible with water.

dehydrogenase, *n.* an enzyme, such as any of the respiratory enzymes, that catalyses a reaction in which hydrogen is removed from a molecule (often to be taken up by the ELECTRON TRANSPORT SYSTEM). For example, in the KREBS CYCLE the conversion of succinic acid to fumaric acid is catalysed by succinate dehydrogenase, the hydrogen released being passed on to FAD.

dehydrogenation, *n.* the removal of hydrogen atoms from a donor molecule, which is oxidized, to a receptor molecule which is reduced, a reaction involving DEHYDROGENASES.

deleterious gene, *n.* an ALLELE of a gene whose effects on the PHENOTYPE are likely to result in a reduced FITNESS. Such genes are often recessive, so that they can be transmitted through families without being detected unless two occur together in the same individual.

deletion mapping, *n.* the use of overlapping deletions (see DELETION MUTATION) to locate the position of a gene on a CHROMOSOME MAP.

deletion mutation, *n.* a type of gene MUTATION in which genetic material is removed from chromosomes (see CHROMOSOMAL MUTATION, POINT MUTATION). The deletion can be as small as a single DNA base (which can cause a misreading of the base sequence during PROTEIN SYNTHESIS, see FRAMESHIFT) to a large piece of chromosome containing many GENES.

delimitation, *n.* the formal statement in taxonomy that defines and limits the characters of a TAXON.

deltoid, *adj.* triangular.

deme or **local population,** *n.* a group of individuals within a species that is largely isolated genetically from other individuals of the species, with clearly definable genetical, cytological and other characteristics.

demersal, *adj.* (of organisms) living on or near the sea bottom, beyond the tidal zone.

demi-, *prefix.* denoting half.

demographic transition, *n.* a theory of demography which states that, as a nation industrializes, it goes through a series of populational changes, starting with a decline in infant and adult mortality and followed later by a reduction in birth rate. The time lag between the decline in deaths and births produces a rapid population growth in 'developing' nations. In 'developed' nations births and deaths become approximately equal, giving a stable population structure.

demography, *n.* the study of human populations.

denaturation, *n.* the breakdown of the bonds forming the quaternary, tertiary and secondary bonds of PROTEINS, a process that can be caused by various agents, such as excess heat, strong acids or alkalis, organic solvents and ultrasonic vibration. Usually denaturation is irreversible as in, for example, the heating of egg albumen to give a solid egg white.

dendr- or **dendro-,** *prefix.* denoting a tree.

dendrite or **dendron,** *n.* the projections from the nerve cell which branch and conduct impulses to the cell body from other axons with which they have SYNAPSES. See NEURON and Fig. 222.

dendrochronology, *n.* the study of ANNUAL RINGS in timber so as to discover the age of the wood (often very precisely) and the climatic situation at the time of growth.

dendrogram, *n.* a diagram illustrating taxonomic relationships in the manner of a family tree.

dendrometer, *n.* a device for monitoring variations in the girth of a tree trunk, often mentioned in relation to evidence for the COHESION-TENSION HYPOTHESIS.

dendron, see DENDRITE.

denitrification, *n.* the process by which nitrogenous compounds are degraded and nitrogen is returned to the air in gaseous form, e.g. the breaking down of nitrates and nitrites to gaseous nitrogen, carried out in the absence of oxygen by soil bacteria. See NITROGEN CYCLE.

denitrifying bacteria, *n.* the bacteria responsible for DENITRIFICATION.

density-dependent factor, *n.* any factor that regulates the size of a population under natural circumstances by acting more severely on a population when it is large than when it is small. Such factors can affect either the birth rate or the mortality, but the latter is more usual. At high densities of populations some organisms have fewer young, or the mortality rate (brought about by predation, disease or food shortage) might be higher than at low densities. The factors tend to cause population numbers to be maintained at a relatively constant level over long periods of time. See CONTROL, REGULATION (2). Compare DENSITY-INDEPENDENT FACTOR.

density-gradient centrifugation, *n.* the separation of the components of an homogenate by a process of CENTRIFUGATION where each component is moved to a level equivalent to its density in the solute gradient used. See DIFFERENTIAL CENTRIFUGATION, ULTRACENTRIFUGE, EQUILIBRIUM CENTRIFUGATION.

density-independent factor, *n*. any factor, such as climate, that is not dependent on the density of population. Compare DENSITY-DEPENDENT FACTOR.

dent-, *prefix*. denoting a tooth.

dental formula, *n*. a formula derived from numbering the incisors, canines, premolars and molars on one side of the mouth. The formula displays the upper jaw numbers over the lower jaw numbers, e.g. the dental formula in humans is:

$$\frac{2.1.2.3}{2.1.2.3}$$

dentate, *adj*. (of plant structures) having teeth or tooth-like projections.

dentary, *n*. the lower jawbone of mammals.

denticle, see PLACOID.

dentine, *n*. the calcified tissue surrounding the pulp cavity of a tooth. Dentine is acellular, but like bone in structure. It is penetrated by cellular processes from the central cavity of the tooth which secrete the enamel.

dentition, *n*. the arrangement, type and number of teeth in vertebrates.

deoxyribonucleic acid, *n*. see DNA.

deoxyribose, *n*. a pentose sugar which is the carbohydrate constituent of deoxyribonucleic acid (see DNA).

dependent differentiation, *n*. the differentiation of embryonic tissue initiated from other tissue by ORGANIZERS. For example, the eye lens on a frog develops under the influence of organizers from the optic cup. See CELL DIFFERENTIATION.

depolarization, *n*. the process of reversing the charge across a cell membrane (usually a NEURON), so causing an ACTION POTENTIAL. In depolarization, the inside of the membrane, which is normally negatively charged, becomes positive and the outside negative. This is brought about by positive sodium ions rapidly passing into the axon. The RESTING POTENTIAL is restored by the SODIUM PUMP mechanism. See Fig. 124 on page 172.

depressed, *adj*. (of plant structures) **1.** sunken, producing a concavity. **2.** flattened.

depressor muscle, see FLIGHT.

depth distribution, *n*. the vertical distribution of organisms, e.g. in soil or water.

derived character, *n*. a characteristic that is considerably altered from the ancestral condition.

derm-, *prefix*. denoting the skin.

dermal bone, *n*. any bone deposited in the lower layer of skin that does

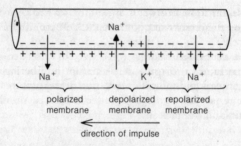

Fig. 124. **Depolarization.** Depolarization of a nerve fibre.

not replace existing cartilage. Examples include the clavicle, and bones of the mammalian cranium.

Dermaptera, *n*. the order of insects that contains the earwigs.

dermatogen, *n*. a superficial layer of cells in the growing point of a plant stem or shoot thought to give rise to the epidermis.

dermis or **corium,** *n*. the deeper portion of the SKIN of vertebrates that is derived embryologically from the MESODERM, and lies beneath the EPIDERMIS which is of ectodermal origin. The dermis contains nerves, blood vessels, muscles and connective tissue.

Dermoptera, *n*. the order of insectivorous mammals containing the flying lemurs that have a membrane stretching between wrist, ankle and tail.

des– or **desmo–,** *prefix*. denoting a bond or chain.

description, *n*. the formal statement of the characters of a TAXON with particular emphasis on those distinguishing it from other closely related forms.

desert, *n*. a terrestrial environment in which there is poor soil, climatic extremes of high and low temperatures and low rainfall, and consequently a scarcity of living organisms, all of which must be specially adapted. See XEROPHYTE, METABOLIC WATER.

design of experiments, *n*. the planning of an experiment so that the results will provide evidence for or against the particular hypothesis which is being tested. For example, any experiment which involves sampling should be planned in such a way that the results are sufficient to allow proper statistical testing; thus, in designing the experiment, it is best to first consult a statistician.

desiccation, *n*. the process by which a substance is dried out and the moisture removed. This is often carried out in a desiccator, which contains a substance which will take up water, e.g. calcium chloride.

desmognathous, *adj*. (of jaws) having large, spongy maxillopalatine bones, usually united ventral to the vomer. Such jaws are found in swans, geese, ducks, parrots and some birds of prey.

desmosome, *n.* a thickened zone in the cell membrane of adjacent eucaryote cells.

detergent, *n.* a substance that when dissolved in water acts as a cleansing agent for the removal of grease by altering the interfacial tension of water with other liquids or solids. Powerful detergents are used to break up oil spillages at sea.

determination, *n.* the process by which embryonic tissues become able to produce only one particular sort of adult organ or tissue.

detoxification, *n.* the process by which poisonous substances are rendered less harmful, e.g. the liver converts ammonia into the less toxic compound urea (see ORNITHINE CYCLE), and hydrogen peroxide is split into water and oxygen by the enzyme CATALASE.

detritivore, *n.* an organism feeding on DETRITUS.

detritus, *n.* any organic debris.

deuto- or **deutero-,** *prefix.* denoting secondary.

deutoplasm, *n.* the yolk or food material of an egg.

Deuteromycotina, see FUNGI IMPERFECTI.

development, *n.* the proceeding towards maturity of eggs, embryos or young organisms.

Devonian period, *n.* a geological period lasting from about 415 to 370 million years ago. Many freshwater fish existed during this period and plants became established on land. True seed plants evolved and lycopods reached the size of trees. Amphibians evolved, the first known insects, spiders and mites occurred and most graptolites became extinct. Britain still lay south of the equator with London at 10°S. See GEOLOGICAL TIME.

de Vriesianism, *n.* the hypothesis that evolution has taken place as a result of drastic mutations, particularly at the species level. The hypothesis was suggested *c.* 1900 by the Dutch botanist, H. de Vries (1848–1935), who was also responsible for the rediscovery of MENDELIAN GENETICS.

dextrin, *n.* a polysaccharide carbohydrate that may form an intermediate step in the hydrolysis of insoluble starch to soluble glucose, ready for CELL RESPIRATION, TRANSLOCATION or further synthesis. See Fig. 125 on page 174.

dextrorotatory, *adj.* (of a crystal, liquid or solution) having the property of rotating a plane of polarized light to the right (e.g. glucose). Compare LAEVOROTATORY.

dextrose, see GLUCOSE.

di-, *prefix.* denoting two, twice.

starch ⟶ dextrin ⟶ maltose ⟶ glucose

Fig. 125. **Dextrin.** The hydrolysis of insoluble starch to soluble glucose.

dia-, *prefix*. denoting through or across.

diabetes, *n*. a metabolic disorder, either where there is an increase in the amount of urine excreted (*diabetes insipidus*), or where excess sugar appears, in the blood and urine, associated with thirst and loss of weight (*diabetes mellitus*). The former is caused by the failure of the pituitary to secrete ADH, and the latter by the failure of the ISLETS OF LANGERHANS to produce sufficient insulin.

diadelphous, *adj*. (of stamens) having filaments united and in two groups, or having one separate and the rest united.

diageotropism, *n*. the phenomenon of growing horizontally in response to gravity, e.g. a RHIZOME.

diagnosis, *n*. a statement that distinguishes the particular TAXON in question from other similar taxa, in terms of the most important characters.

diakinesis, *n*. the final stage in Prophase I of MEIOSIS, where the chromosomes reach maximum contraction with the HOMOLOGOUS CHROMOSOMES tending to separate from each other. The NUCLEOLUS disappears, the nuclear membrane degenerates and a spindle is formed of microtubules produced by the CENTROSOMES.

dialysis, *n*. a process by which small molecules can be separated from larger ones using a fine membrane, e.g. collodion, to contain the larger, but which allows the smaller molecules to pass through into the excess water on the other side. The kidney functions by means of this principle, which is also the basis for kidney machines used in cases of kidney disease or failure.

diapause, *n*. a period of arrested growth and development in insects which is under the control of the endocrine system. Diapause is an adaptation to avoid adverse conditions, but does not automatically end with the termination of the adverse conditions as it is genetically determined. However, diapause can be 'broken' by an appropriate environmental change, or artificially by temperature shocks or chemical stimulation.

diapedesis, *n*. the passage of blood cells through the unruptured wall of a blood vessel by changing in shape.

diaphragm, *n*. a sheet of tissue present only in mammals, that separates the thorax from the abdomen. It consists mainly of muscle and tendons, and its flattening from a convex position projecting into the thorax is an important aspect of the expansion of the lungs as inspiration takes place. See BREATHING.

diaphysis, *n.* the shaft of a long bone. Compare EPIPHYSIS.

diapsid, *n.* **1.** a vertebrate skull having two fossae, one above the other, behind the orbit. **2.** (in some classifications) a reptile of the subclass Diapsida.

diastase, *n.* an enzyme mixture common in seeds such as barley, that is responsible for starch hydrolysis. The mixture contains amylases for conversion of starch to MALTOSE (sometimes via DEXTRIN) and MALTASE for conversion of maltose to glucose.

diastema, *n.* a gap in the teeth.

diastole, see HEART, CARDIAC CYCLE.

diatom, *n.* a member of the BACILLARIOPHYCEAE.

dicarboxylic acid, *n.* an organic acid containing two carboxyl $(-COOH)$ groups.

dicentric, *n.* a chromosome with two CENTROMERES.

dichasium, *n.* a CYME where the branches are opposite and approximately equal.

dichlamydeous, *adj.* having two whorls on the perianth segments of flowers.

dichogamy, *n.* the maturation of anthers and ovules in a flower at different times.

dichotomous, *adj.* **1.** (of a plant) branching by repeated division into two equal parts. **2.** (of characters) choosing between two possibilities. See KEY, IDENTIFICATION.

dicotyledon, *n.* any flowering plant of the subclass Dicotyledonae, class Angiospermae. Other ANGIOSPERMS belong to the subclass Monocotyledonae (see MONOCOTYLEDON). See Figs. 126 and 166.

Feature	Dicotyledons	Monocotyledons
1. No. cotyledons/embryo.	two.	one.
2. Leaf vein structure.	pinnate or palmate.	parallel.
3. Stem vascular bundles.	cylindrically arranged.	scattered.
4. Root system.	typically a taproot.	typically fibrous.
5. Cambium in vascular bundle?	Yes.	No.
6. Floral parts.	in fours, fives or multiples.	in threes or multiples

Fig. 126. **Dicotyledon.** The differences between dicotyledons and monocotyledons.

dictyosome, see GOLGI APPARATUS.

dictyostele, *n.* a type of STELE found in some forms of plant, with phloem internal and external to the XYLEM, which is broken up into distinct strands each surrounded by endodermis.

Didelphia, see MARSUPIAL.

didymous, *adj.* (of plant structures) formed of two similar parts which are attached at some point.

didynamous, *adj.* (of stamens) being in two pairs, one pair being larger.

dieback, *n.* shoot NECROSIS starting at the plant apex and moving downwards.

dieldrin, *n.* a chlorinated hydrocarbon that is a powerful insecticide, with long-lasting effects. However, it has become ecologically damaging and is banned from many countries. See also DDT.

differential centrifugation, *n.* a technique in which cell organelles can be separated by spinning an homogenate at various speeds, each speed causing sedimentation of a specific cell fraction. The smaller the organelle, the higher the speed required for sedimentation. See DENSITY-GRADIENT CENTRIFUGATION, ULTRACENTRIFUGE, CENTRIFUGATION.

differential permeability, see SELECTIVE PERMEABILITY.

differentiation, see CELL DIFFERENTIATION.

diffusion or **passive transport** *n.* the movement of molecules of a particular substance from regions of high concentration to regions of low concentration of that substance, i.e. down a CONCENTRATION GRADIENT. The rate of diffusion in gases is much greater than in liquids, due to differences in their molecular structure. For example, carbon dioxide will diffuse 10,000 times more rapidly in air than in water.

diffusion pressure deficit (DPD), see WATER POTENTIAL.

digastric, 1. *adj.* (of muscle) having two swollen parts, or bellies, interconnected by a tendon. **2.** *n.* a muscle concerned with the swallowing reflex in the human neck.

Digenea, *n.* a subclass or order of trematodes containing the liver FLUKES. Compare MONOGENEA.

digestion, *n.* a process requiring enzymes in which complex food molecules are broken down by HYDROLYSIS into a state in which they can be absorbed. In some organisms digestion is 'extracellular', the enzymes being released outside the body and the digested products absorbed (for example, all fungi), but in the majority of HETEROTROPHS digestion takes place in a specialized cavity or internal tube, the ALIMENTARY CANAL, which is usually muscular, with food being squeezed through the DIGESTIVE SYSTEM by a process of PERISTALSIS.

digestive system, *n.* the ALIMENTARY CANAL together with associated digestive glands (see DIGESTION). Taking mammals as an example, the main areas of digestive activity are:

(a) MOUTH cavity. starch digestion occurs during chewing, which increases the surface area of food using teeth modified for particular foodstuffs. The pH in the mouth ranges from slightly acid to slightly alkaline.

(b) STOMACH. the site of the start of protein digestion in acid conditions with strong muscular churning movements. Bones are partly digested.

(c) SMALL INTESTINE. the main area of digestion for all types of food, with enzymes secreted by pancreas and intestinal glands. Strong alkaline conditions.

digit, *n*. **1.** a primate finger or toe. **2.** any part of the PENTADACTYL LIMB of vertebrates which contains a linear group of PHALANGES.

digitate, *adj*. having finger-like processes.

digitgrade, *adj*. walking on the toes or fingers, as in most fast-running animals such as dogs and cats.

dihybrid, 1. *n*. an organism that carries two different ALLELES of one gene and two different alleles of another gene. For example, character *A* (controlled by alleles *A*1 and *A*2) and character *B* (alleles *B*1 and *B*2). The dihybrid would have the genotype *A*1*A*2 *B*1*B*2, i.e. a double HETEROZYGOTE. **2.** *adj*. (of a cross) having two parents that have been mated to produce dihybrids that are then also mated. A single dihybrid plant is self-fertilized (see SELF–FERTILIZATION). MENDEL formulated his law of INDEPENDENT ASSORTMENT from the results of a dihybrid cross involving pairs of contrasting characters in the pea plant.

dikaryon, *n*. **1.** a fungal hypha with two types of nuclei in an unfused state in the same cells, so that cytologically they are $n + n$. When the nuclei are from different genetic strains the individual or cell is called a HETEROKARYON. **2.** a single animal cell with two nuclei, each from a different species, e.g. mouse and man.

dilute, *vb*. to make a substance less concentrated by the addition of water.

dimer, *n*. a molecule made by the joining of two molecules of the same kind, i.e. two MONOMERS. Ultraviolet light can induce thymine dimers in DNA, where two thymine bases on one polyneucleotide chain form into a dimer that displaces their position in the helix, preventing the normal hydrogen bonding with adenine bases on the opposite polynucleotide chain and so inactivating the DNA molecule.

dimictic lake, *n*. a lake that has two annual periods of free mixing of water.

dimorphism, *n*. the occurrence of an organism in two forms, e.g. the male and female (see SEXUAL DIMORPHISM). Dimorphism can occur in body form or in colour phases, e.g. the two–spotted ladybird which has a brown form with four red spots, and a red form with two dark spots. See GENETIC POLYMORPHISM.

dino-, *prefix.* denoting **1.** fearful (e.g. dinosaur). **2.** whirling (e.g. dinoflagellate).

dinoflagellate, *n.* any of a group of unicellular biflagellate aquatic organisms forming a constitutent of plankton now normally placed in the division Pyrrhophyta. They can be regarded as an order of protozoans (Dinoflagellata) or a class of algae (Dinophyceae).

Dinornithidae, *n.* the moas, large, extinct flightless birds (taller than 3 m) of New Zealand, that probably were hunted to death by humans.

dinosaur, *n.* any extinct reptile of the orders Saurischia and Ornithischia. Dinosaurs (the name means 'fearful lizard') were the dominant form of land vertebrates during the JURASSIC and CRETACEOUS periods.

dinucleotide, *n.* a compound formed of two NUCLEOTIDES.

dioecious, *adj.* (of plants) the occurrence of male flowers in one individual and female flowers in another individual, for example, willow. The term literally means 'two homes'. Compare MONOECIOUS.

dioxin, *n.* a chemical byproduct of the manufacture of certain herbicides and bactericides, particularly tetrachlorodibenzo-paradioxin (TCDD), which is extremely toxic.

dipeptide, *n.* the product of joining together two AMINO ACIDS by a CONDENSATION REACTION forming a PEPTIDE BOND. See Fig. 127.

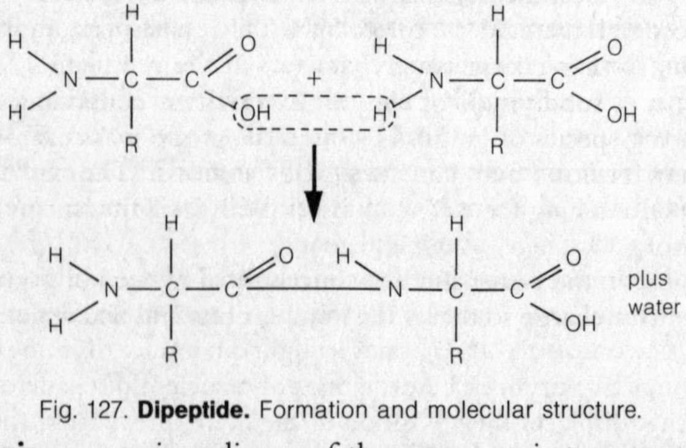

Fig. 127. **Dipeptide.** Formation and molecular structure.

diphtheria, *n.* a serious disease of the upper respiratory tract in man caused by toxins produced by the bacterium *Corynebacterium diphtheriae.* The toxins cause NECROSIS of epithelial cells in the throat, resulting in the production of a greyish EXUDATE that gradually forms a membrane on the tonsils and can spread upwards into the nasal passages, or downwards into the larynx causing suffocation if not treated.

diplanetism, *n.* the presence in the life cycles of some fungi of two ZOOSPORE stages.

dipleurula, *n.* a bilaterally symmetrical larva of ECHINODERMS.

diplo-, *prefix.* denoting double.

diploblastic, *adj.* (of animals) having a body wall of two layers of cells, the ECTODERM and ENDODERM, an arrangement found only in the COELENTERATES.

diploid, 1. *adj.* (of a cell nucleus) containing two of each type of chromosome in homolgous pairs and formed as a result of sexual reproduction. **2.** *n.* an organism in which the main life stage has cell nuclei with two of each type of chromosome, written as 2n. Diploid stages occur in all EUCARYOTES, apart from certain fungi, and allow a greater degree of genetic variability in individuals than the HAPLOID (2) state (n).

diplonema, *n.* chromosome structure during the DIPLOTENE stage of meiosis.

diplont, *n.* an organism in the DIPLOID stage.

diplopod, *n.* any arthropod of the subclass Diplopoda (class Myriapoda), the millipedes. They have a cylindrical body and two pairs of limbs on each abdominal segment.

diplospondyly, *n.* the state in which organisms have two vertebrae in each body segment, as in the tails of some fish.

diplotene, *n.* a stage of MEIOSIS near the end of Prophase 1, in which the CHROMATIDS become separated (except at the CHIASMATA), and the chromosomal material is contracting. The chiasmata indicate that CROSSING OVER is taking place.

Dipnoi, *n.* a subdivision of the CHOANICHTHYES containing the only three living species of lungfish, though the group was common in the DEVONIAN PERIODS and CARBONIFEROUS PERIODS. The air bladder is developed as a lung for air breathing and this facilitates survival during dry seasons.

dipolarity, *n.* the possession by a molecule of regions of both positive and negative charge so that it is positive at one end and negative at the other.

Fig. 128. **Dipteran.** A fly of the family Statiomyidae.

dipteran, *n.* any two-winged fly of the order Diptera in the class Insecta.

DIRECTIONAL SELECTION

Dipterans are ENDOPTERYGOTES and have the hind pair of wings very much reduced to form HALTERES which are used for balance in flight. See Fig. 128.

directional selection, *n.* the SELECTION of a character in one direction, e.g. towards larger size, that may occur as a result in a change in environment. For example, if another ice age was imminent then temperature would fall and the overall size of warm-blooded vertebrates living in a particular region would probably increase. See BERGMANN'S RULE.

disaccharide, *n.* a 'double' sugar, such as MALTOSE or LACTOSE, with the general formula $C_{12}H_{22}O_{11}$, synthesized by a condensation reaction joining together two MONOSACCHARIDE sugars. See Fig. 129.

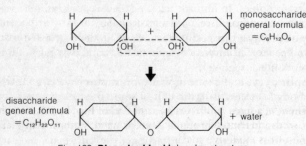

Fig. 129. **Disaccharide.** Molecular structure.

disc or **disk,** *n.* that part of the receptacle surrounding the plant ovary which is fleshy and sometimes nectar-secreting.

disco-, *prefix.* denoting a disc.

Discomedusae, *n.* an order of the class Scyphozoa (COELENTERATES) that contains most of the common disc-shaped jelly fish.

discontinuous distribution, *n.* a distribution in which populations of related organisms are found in widely separated parts of the world, e.g. lung flukes in Australia, Africa and South America. Such a distribution is thought to indicate the great age of the group, with intermediate populations having become extinct.

discontinuous variation, *n.* a variation in which different forms of a CHARACTER fall clearly into a particular grouping without overlapping each other, and are often controlled by a small number of major genes. For example, white and red eye colours in fruitflies. See also QUALITATIVE INHERITANCE.

discordance, *n.* any difference in a character between individuals due to genetic differences, as may occur in DIZYGOTIC TWINS, but not usually in MONOZYGOTIC TWINS.

disease, *n.* an abnormality of an animal or plant caused by a pathogenic organism that affects performance of the vital functions and usually gives diagnostic symptoms.

disease resistance, *n.* the ability of some organisms to withstand the attack of PATHOGENS and remain virtually unaffected. The resistance can occur naturally but may be enhanced in a population by ARTIFICIAL SELECTION, as in plant and animal breeding. For example, the breeding of varieties of tomatoes resistant to the fungal disease that produces brownroot rot.

disjunction, *n.* the separation of HOMOLOGOUS CHROMOSOMES to opposite poles during anaphase of nuclear division, particularly MEIOSIS. Abnormalities in disjunction (see NONDISJUNCTION) can lead to a chromosomal mutation in which daughter cells possess too many or too few chromosomes. In humans extra AUTOSOMES can result in severely abnormal individuals (see DOWN'S SYNDROME, PATAU SYNDROME, EDWARDS' SYNDROME). See also TURNER'S SYNDROME and KLINEFELTER'S SYNDROME for disjunction abnormalities affecting the SEX CHROMOSOMES.

dispersal, *n.* the act of disseminating or scattering the seeds of plants, or the larvae of animals (particularly important in sessile animals), or any movement of adults. Animal larvae may be mobile, but the dispersal of spores, seeds and fruits is brought about by wind, water and animals, and in some plants explosive mechanisms are present.

Many mechanisms exist to aid dispersal: some fruits are hooked, e.g. teasel, burdock, becoming attached to the bodies of animals; the use of vectors, e.g. mosquito in the case of malarial parasite, to carry parasitic organisms between hosts; and some larvae of marine animals move to different layers of water where current direction may be different from the level at which the adults live. Dispersal techniques include EMIGRATION, IMMIGRATION and MIGRATION.

dispersion, *n.* the distribution of individual organisms once any DISPERSAL has taken place. For example, organisms may be randomly dispersed, under-dispersed (aggregated) or over-dispersed (as in territorial animals). Dispersion should not be confused with DISTRIBUTION which normally refers to a species as a whole and not to individuals.

displacement activity, *n.* any animal behaviour that takes place outside its normal context, usually in a stress situation. It has been interpreted as being a result of aggression and fleeing drives occurring together. Examples include: false brooding in birds; birds picking up nesting material from the ground and throwing it away on being disturbed from the nest; sticklebacks suddenly digging in the vertical position during a boundary clash. See VACUUM ACTIVITY.

display, *n.* any ritualized behaviour, including posturing, vocalization, and movements that elicit specific reactions in other organisms. Courtship displays, particularly in birds, are often complex, but displays may also be concerned with threat, DISTRACTION, etc.

disruptive selection, *n*. the selection in favour of phenotypic extremes in a population. Disruptive selection produces divergence between subpopulations. Compare STABILIZING SELECTION.

dissociation, *n*. a process in which a chemical combination breaks up into component parts, as with haemoglobin and oxygen. See OXYGEN-DISSOCIATION CURVE.

distal, *adj*. furthermost from the body in any structure, or furthest from the centre of the system concerned. For example, the finger is at the distal end of the human arm. Compare PROXIMAL.

distal-convoluted tubule, see KIDNEY.

distance receptor, *n*. a RECEPTOR that detects stimuli from an animal's environment and thus allows the animal to orientate with respect to the stimuli.

distichous, *adj*. (of plant structures) arranged in two rows that are diametrically opposite.

distraction, *n*. a display by animals, generally in response to a predator that threatens eggs or young. For example, feigning injury, behaviour directed at a predator with the aim (often effective) of diverting attention.

distribution, *n*. the occurrence of a species over the total area in which it occurs, i.e. its range or geographical distribution. In aquatic organisms or soil organisms, or even organisms living on mountains, vertical distribution is also important. In some organisms vertical distribution may vary seasonally, as does geographical distribution, particularly in migratory forms. See also FREQUENCY DISTRIBUTION, DISPERSION.

disulphide bridge, *n*. the covalent bond between two sulphur atoms, particularly in peptides and proteins.

diuresis, *n*. the increased output of watery urine by the kidneys. See ADH.

diuretic, *n*. a substance that enhances DIURESIS.

diurnal rhythm, *n*. a pattern of activity based upon a 24-hour cycle, in which there are regular light and dark periods. See also CIRCADIAN RHYTHM.

divaricate, *adj*. diverging widely.

divergent evolution, *n*. EVOLUTION from a common ancestral form.

diversity, *n*. the variety of species richness of a COMMUNITY.

diversity (D) region, a sequence of amino acids that is unique to particular IMMUNOGLOBULINS, enabling the molecule to act as an ANTIBODY against specific ANTIGENS.

diversity index, *n*. a single integrated measure that expresses the diversity of a particular community of organisms, e.g. freshwater invertebrates. The index can be related to the level of pollution in an ecosystem, and the greater the diversity, the less the pollution.

diverticulum, *n.* any sac or pouch formed by herniation of the wall of a tubular organ or part, especially the intestines.

division, *n.* **1,** a major grouping in plant classification (see PLANT KINGDOM). **2.** the process of the formation of daughter cells from a parent cell (see CELL DIVISION).

dizygotic twins or **fraternal twins,** *n.* a pair of children born at the same time to the same mother but each developing from a different egg. Such twins are therefore nonidentical, being equivalent to normal SIBLINGS, and can be of the same or different sexes. Compare MONOZYGOTIC TWINS.

DNA (deoxyribonucleic acid), *n.* a complex NUCLEIC ACID molecule found in the chromosomes of almost all organisms, which acts as the primary genetical material, controlling the structure of proteins and hence influencing all enzyme-driven reactions.

(a) *structure.* The model proposed by WATSON and CRICK in 1953 has now become universally accepted. DNA is considered to consist of two POLYNUCLEOTIDE CHAINS joined together by hydrogen bonds between NUCLEOTIDE bases, with COMPLEMENTARY BASE PAIRING between specific bases ensuring a parallel-sided, stable structure: ADENINE pairing with THYMINE (2H bonds) and CYTOSINE with GUANINE (3H bonds). See Fig. 130. The two polynucleotide chains each have an opposite polarity due to the way the phosphates are attached to the sugar, and the whole molecule is twisted into a *double helix* shape, with a complete turn every 10th base. See Fig. 131 on page 184.

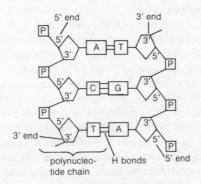

Fig. 130. **DNA.** Complementary base pairing.

DNA

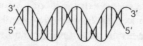

Fig. 131. **DNA.** The 'double helix' shape produced by coiling.

(b) *replication* (see SEMICONSERVATIVE REPLICATION MODEL). The molecule 'unzips' through enzymic breakage of the hydrogen bonds between polynucleotide chains. Each chain acts as a 'template' for attachment of free nucleotides under the influence of various DNA POLYMERASES working in the 5' to 3' direction only. See Fig. 132. DNA replication occurs in the 'S' phase of the CELL CYCLE prior to nuclear division.

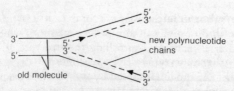

Fig. 132. **DNA.** Replication.

(c) *location*. DNA is found in all chromosomes except those of certain viruses (e.g. tobacco mosaic virus, TMV), where the heritable material is RNA. In PROCARYOTES, DNA is in the form of a single coiled molecule in a continuous loop and may also occur as extra-chromosomal material in the cytoplasm. In EUCARYOTES, the DNA is also highly coiled but is complexed with basic and acidic proteins. There is probably only one very long DNA molecule per chromosome. DNA is also found in CHLOROPLASTS and MITOCHONDRIA of eucaryote cytoplasm (see CYTOPLASMIC INHERITANCE).

(d) *DNA as a genetic material*. There are several pieces of evidence to suggest the role of DNA in inheritance: (i) TRANSFORMATION experiments with *Pneumococcus* in 1928, by F. GRIFFITH; (ii) the identification of the 'transforming principle' as DNA by AVERY, MacLeod and McCarty in 1944; (iii) the fact that the wavelength of ultraviolet light which causes most mutations in various procaryotes and eucaryotes matches the ABSORPTION SPECTRUM of nucleic acids (260 nm); (iv) HERSHEY and Chase's experiment with labelled BACTERIOPHAGE. DNA has several features which make it an ideal genetic material: great stability (see *structure* above); accurate replication so that all cells contain an identical copy of information; four nucleotide bases providing storage of coded information (see GENETIC CODE); it is capable of mutation by altering the base sequence; it may be broken and rejoined to form new genetic combinations (see RECOMBINATION); stored information can be accurately 'read' by other cell molecules (see TRANSCRIPTION).

DNA ligase, *n.* an enzyme that catalyses the formation of covalent bonds between adjacent parts of a broken DNA POLYNUCLEOTIDE CHAIN (5′ to 3′), thus helping in repair.

DNA polymerase, *n.* a group of three types of enzymes that are responsible for the attachement of free NUCLEOTIDES onto the unzipped DNA molecule during DNA replication. The first DNA polymerase was discovered by A. Kornberg in 1957 (for which work he received the Nobel Prize), although it now appears that this enzyme is not the main one in the group, but rather helps to repair DNA after damage.

dogfish, *n.* any small cartilaginous fish of the family Scyllidae; a small shark.

dolicho-, *prefix.* denoting long.

dolichocephalic, *adj.* (of humans) having a long head. See CEPHALIC INDEX.

Dollo's law of irreversibility, a principle formulated by the Belgian palaeontologist L. Dollo, stating that a structure once lost in evolution cannot be regained.

dominance, *n.* **1.** a genetic interaction where one ALLELE of a gene masks the expression of an alternative allele in the HETEROZYGOTE, so that the PHENOTYPE is of that form controlled by the dominant allele. For example, gene A has two alleles $A1$ and $A2$:

$$A1/A1 = \text{black phenotype}$$
$$A2/A2 = \text{white phenotype}$$

but when heterozygous, $A1/A2 =$ black phenotype. Thus $A1$ is dominant to $A2$, with $A2$ 'recessive' to $A1$. In molecular terms, the $A1$ allele is coding for a protein of such quality and quantity to allow the normal amount of black pigment to be produced, even though the $A2$ allele is not coding for a normal enzyme. See also CODOMINANCE, INCOMPLETE DOMINANCE. **2.** the preponderance of one species within an ecological COMMUNITY, for example, oaks in oak woods.

dominance hierarchy, *n.* a phenomenon in animal societies where some members are subordinate to others, giving a reproductive advantage to the less subordinate members. See PECKING ORDER.

dominant epistasis, *n.* a form of EPISTASIS in which dominant ALLELES of one gene can cause a masking effect on the expression of alleles at another locus. Such an interaction would produce a DIHYBRID (1) ratio of 12:3:1 instead of the more normal 9:3:3:1. Compare RECESSIVE EPISTASIS.

dominant species, *n.* the commonest species in a COMMUNITY.

Donnan equilibrium, *n.* an electromechanical equilibrium that is set up when two solutions are separated by a membrane that is impermeable to some of the ions in solution.

donor, *n.* an individual supplying tissue (e.g. blood), to a recipient. See COMPATIBILITY, ABO BLOOD GROUP, UNIVERSAL DONOR/RECIPIENT.

DOPA (dihydroxy phenylalanine), *n.* a precursor in the biochemical PATHWAY leading to MELANIN formation in animals. DOPA is not metabolized in individuals with ALBINISM.

dopamine, *n.* the decarboxylation product of DOPA. Formula: $C_8H_{11}O_2N$.

dormancy, *n.* a state in which seeds and other structures (such as underground stems) reduce their metabolic activities to a minimum level during unfavourable conditions (e.g. low temperature, drought) so as to survive until conditions improve. In fungal spores dormancy can be EXOGENOUS, requiring an external stimulus for germination (e.g. presence of nutrients) or ENDOGENOUS, where dormancy is controlled by an innate property of the spore and is unaffected by the ambient environment when in this state. See GERMINATION for events that occur when seed dormancy is broken.

dormin, *n.* a growth-inhibiting substance found in buds of plants undergoing DORMANCY, with a chemical structure identical to ABSCISIC ACID.

dorsal, *adj.* **1.** (of an animal) the part that normally occurs uppermost. The back of an animal is called the *dorsal surface*. The dorsal side is normally directed upwards (dorsal fin) but backwards in primates in the upright position. **2.** (of a plant) of, or situated on the side of an organ that is directed away from the axis. See also DORSIVENTRAL LEAF.

dorsal fin, *n.* any unpaired median fin on the backs of fishes and some other aquatic vertebrates. The fin maintains balance during locomotion.

dorsal lip, *n.* that part of the rim of the BLASTOPORE which in further development becomes the dorsal side.

dorsal root, *n.* a nerve trunk containing only sensory axons, that enters the SPINAL CORD near the dorsal surface.

dorsifixed, *adj.* (of ANTHERS) attached at the back.

dorsiventral leaf, *n.* the leaf type, found mainly in DICOTYLEDONS, with a structure that changes from dorsal (upper) surface to ventral (lower) surface. The major areas are: an upper EPIDERMIS with no stomata; a palisade MESOPHYLL layer below which is spongy mesophyll with large intercellular spaces; a lower epidermis with irregularly spaced STOMATA. Such leaves are generally held horizontally.

dosage compensation, *n.* a genetical mechanism in which ALLELES of a gene automatically regulate the amount of a useful product produced, so that homozygous–dominant GENOTYPES produce the same amount of gene product as heterozygotes. In this way the two genotypes cannot be distinguished. Dosage compensation occurs regularly in female mammals where one X-chromosome in each cell is thought to become inactivated, leading to equal amounts of gene products in males and females. See INACTIVE-X HYPOTHESIS.

dose, *n.* the known amount of chemical or other treatment received by an organism.

double circulation, *n.* the type of circulation found in mammals and birds where the pulmonary and systemic circulations are completely separate. See BLOOD CIRCULATORY SYSTEM.

double cross, *n.* a method of plant breeding in which four different inbred lines are crossed together (A × B and C × D) and the progenies crossed again ((A × B) × (C × D)) to produce a four-way hybrid seed with good vigour.

double crossover, *n.* two separate CROSSING–OVER events occurring between CHROMATIDS. In a TEST CROSS involving three genes, progeny that have carried out this process can be identified and usually from the least frequent type of offspring.

double fertilization, *n.* the situation in flowering plants where a male nucleus combines with the egg nucleus and another male nucleus combines with the primary endosperm to form an endosperm nucleus. See EMBRYO SAC.

double helix, see DNA.

double recessive, *n.* a GENOTYPE in which both ALLELES of a gene are of the recessive type, i.e. HOMOZYGOUS. See DOMINANCE (1).

doubling time, *n.* (in microbiology) the time in which the number of cells in a population doubles.

down, *n.* the soft, small feathers lacking BARBULES that cover the body of a bird and aid heat retention.

Down's syndrome or **mongolism,** *n.* a human abnormality caused by the presence of an extra AUTOSOME (number 21) in the chromosome complement of each cell (see ANEUPLOIDY). The main features of the syndrome are: mental retardation, short stature with stubby fingers, characteristic slanting eyes (hence 'mongolism') and heart defects. The abnormality arises in the mother's egg as a rule, with older mothers having a far higher chance of producing abnormal ova than younger mothers. This may be because the mother's eggs have been part-way through Prophase 1 of MEIOSIS since birth, so that eggs of older mothers have been at risk of NONDISJUNCTION much longer than those of younger mothers. Down's syndrome is the most common chromosomal abnormality in humans, with an incidence of 1 in 600 live births.

drainpipe cells, *n.* any epithelial cells (see EPITHELIUM) with an intracellular lumen in the form of a drainpipe joined end to end, as in the TRACHEOLES of insects and excretory ducts of flatworms.

drift, see RANDOM GENETIC DRIFT.

Drosophila or **fruit fly,** *n.* any small dipterous fly (see DIPTERAN) of the genus *Drosophila*, which has been used extensively in genetical investigations. *D. melanogaster* is probably the best-understood animal in terms of inheritance studies.

drought tolerance, *n.* the ability of plants to resist drought. See XEROPHYTE.

drug, *n.* **1.** any substance used as an ingredient in medical preparations. **2.** any substance that affects the normal body functions.

drupe, *n.* a type of succulent, fleshy fruit produced by some plants in which the seed is enclosed in a hard, woody 'stone'. The fruit is formed from the swollen tissue of the PERICARP. Examples include peach, plum, cherry, olive, apricot.

dryopithecine, *n.* any extinct Old World ape of the genus *Dryopithecus*.

dry rot, *n.* **1.** a rotting of timber caused by the fungus *Serpula lacrymans* in which infected timber shows characteristic cubical cracking. Only damp wood is attacked initially although other, drier wood can be attacked later via long water-conducting structures. **2.** any general symptoms of plant disease in contrast to WET ROT, e.g. dry rot of gladioli.

dry weight, *n.* the weight or mass of biological material dried at 105°C until no further water loss takes place. Because water content varies considerably between individuals, dry weight is the most commonly used method of assessing weight in plants and animals. See BIOMASS.

ductless gland, see ENDOCRINE GLAND.

ductus arteriosus, *n.* a blood vessel found in foetal mammals which connects the pulmonary artery to the AORTA and thus creates a bypass to the lungs. See Fig. 133. The vessel closes soon after birth so that all blood is now sent through the lungs for oxygenation. Occasionally the vessel does not close completely, which may cause a condition called BLUE BABY.

ductus Cuvieri, see CUVIERIAN DUCT.

Duffy blood group, *n.* a red blood cell ANTIGEN found in certain human racial groups, controlled by a single autosomal gene on chromosome 1. Around 65% of Caucasians and 99% of Chinese are positive for the antigen being either homozygous or heterozygous for the allele Fy$^+$, while 92% of West Africans are not, being homozygous for Fy$^-$ alleles.

duodenum, *n.* that part of the SMALL INTESTINE connecting the stomach

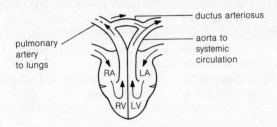

Fig. 133. **Ductus arteriosus.** Location in mammals.

to the ileum. It is about 25 cm long in man. The wall is highly folded internally with microscopic projections called VILLI, which increase the surface area for digestion and absorption. Within the wall are BRUNNER'S GLANDS and PANETH CELLS which, together with secretions from the pancreas entering the duodenum via the bile duct, produce a whole range of enzymes to complete digestion.

duplication, chromosomal, *n.* a type of CHROMOSOMAL MUTATION in which part of a chromosome is replicated, so producing extra copies of those genes contained in the duplicated segment. For example, the 'Bar-eye' mutation in *Drosophila* is due to a duplication of a segment along the X-CHROMOSOME.

dura mater, *n.* the connective tissue which covers the brain and spinal cord of vertebrates. See MENINGES.

dwarfism, *n.* a form of body malfunction in which the adult individual does not reach the normal height and may sometimes have other abnormalities. Such conditions can be due to a deficiency of GROWTH HORMONE secreted by the anterior pituitary, or to cartilage abnormalities due to genetical defects (see ACHONDROPLASIA). Compare GIGANTISM.

dyad, *n.* a chromosome consisting of two sister CHROMATIDS joined at their centromere which is produced after DISJUNCTION.

dynamic equilibrium, *n.* a balanced state of continual change, for example, water and water vapour, where particles are constantly passing from one phase to the other.

dysentery, *n.* a severe disorder of the ileum and colon caused by the bacterium *Shigella dysenteriae* (and several other species), resulting in abdominal cramps, diarrhoea and fever. The disease is spread by 'food, faeces, fingers and flies', and can be controlled by sanitary precautions.

dysgenic, *adj.* of, or relating to a deleterious genetic change. See also TURNER'S SYNDROME.

dyslexia or **word blindness,** *n.* impairment of the ability to read, due to a brain disorder.

dysplasia, *n*. abnormal growth or development, as may occur in organs or cells.

dyspnea, *n*. laboured breathing, with breathlessness.

E

e- or **ex-,** *prefix*. denoting out of, without.

ear, *n*. the sense organ of vertebrates concerned with reception of sound (hearing), BALANCE (detecting position with respect to gravity) and acceleration. The external ear is absent in amphibia and some reptiles, where the eardrum is at the skin surface; in other forms the external ear consists of an AUDITORY CANAL and the pinna, a projection of skin and cartilage. The *middle ear* or *tympanic cavity* (not present in some amphibians and some reptiles) lies between the ear drum and the auditory capsule. The EUSTACHIAN TUBE connects the middle ear to the pharynx; it contains the ear ossicles and lies within the bulla which is a projection of the skull. The inner ear or *membranous labyrinth* is contained in the auditory capsule; the utricle gives rise to the semicircular canals (for balance), and from the saccule the hearing organ arises in the form of the COCHLEA in some tetrapods. See Fig. 134.

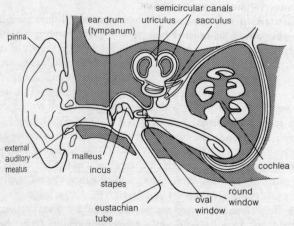

Fig. 134. **Ear.** The human ear.

Hearing results from sound waves striking the tympanic membrane and causing it to vibrate. The vibrations are transmitted to the oval

190

window by means of the ear ossicles which magnify them. This disturbs the fluid in the vestibular canal of the cochlea and causes movement in REISSNER'S MEMBRANE, which then results in the fluid of the middle canal being displaced. This moves the basilar membrane and then disturbs the fluid in the tympanic canal which stretches the membrane covering the round window. Movement of the basilar membrane stimulates the organ of Corti (see COCHLEA) and impulses are fired in the auditory nerve. Loud sounds cause greater movement of the basilar membrane and a higher frequency of impulses from the organ of Corti. Pitch of a sound determines the frequency of movement of the basilar membrane.

eardrum, see TYMPANIC MEMBRANE.

ear ossicle, *n.* the COLUMELLA AURIS of lower vertebrates and the MALLEUS, INCUS or STAPES of mammals. See EAR.

earthworm, *n.* any ANNELID of the order Oligochaeta.

ec-, *prefix.* denoting out of, off.

ecad, *n.* a plant form that results from the habitat in which the plant occurs and is not brought about genetically. See PHENOTYPIC PLASTICITY.

ecdysis, *n.* the process of moulting the cuticle in insects, usually in the preadult stage. The old cuticle is split and cast off to reveal a new, soft cuticle underneath; the insect increases in size, often by intake of air, and the new cuticle hardens. Each larval stage is referred to as an INSTAR, so that the first instar is terminated by the first ecdysis, the second instar by the second ecdysis. Ecdysis is initiated by the MOULTING HORMONE.

ecdysone, see MOULTING HORMONE.

ecesis, *n.* the germination and subsequent establishment of a plant colonizing a new habitat.

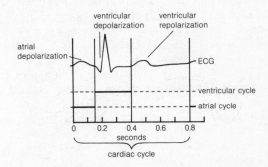

Fig. 135. **ECG.** A tracing of the electric currents that initiate the heartbeat. The broken line indicates diastole, the solid line systole.

ECG (electrocardiogram), *n.* a recording of the electrical changes occurring as the heart beats (see HEART, CARDIAC CYCLE) that can be used in the diagnosis of heart malfunction. To obtain an ECG, electrodes are attached to various parts of the body surface, usually both arms and the left leg. See Fig. 135.

echidna, *n.* any of the spine-covered MONOTREME mammals of the family Tachyglossidae, found in Australia and New Guinea. They have a long snout and claws for digging out ants.

echino-, *prefix.* denoting spines or prickles.

echinoderm, *n.* any member of the phylum Echinodermata, including the sea urchins, starfish, brittle-stars, sea cucumbers, feather stars and sea lillies. The phylum is characterized by a pentaradiate structure (a five-sided RADIAL SYMMETRY) and the presence of tube feet in most forms.

echinoid, *n.* any of the ECHINODERMS of the class Echinoidea, including the sea urchins.

eclipse, 1. *n.* the period during which a virus exists as a free nucleic acid in the host cell. **2.** *adj.* (of plumage in birds) occurring for a short time after the breeding plumage is moulted, particularly in ducks.

ecogeographical rules, see BERGMANN'S RULE, ALLEN'S RULE, GLOGER'S RULE.

ecological equivalent, *n.* one of two species that have arisen from the same ancestral stock which have evolved in similar environments and have the same adaptive characters.

ecological isolation, *n.* the separation of organisms (usually) within the same geographical region because of their preference for different habitat types. For example, the toads *Bufo fowleri* and *B. americanus* both live in the same areas but breed in different places, the former in large, still bodies of water such as ponds, the latter in puddles, or pools in brooks. Compare GEOGRAPHICAL ISOLATION. See also SPECIATION.

ecological niche, *n.* **1.** the physical space occupied by an organism. **2.** the organism's functional role in the community (e.g. TROPHIC LEVEL). **3.** other conditions of the organism's existence, such as preferred temperature, moisture and pH.

ecological pyramid, *n.* a graphical means of illustrating the trophic structure or trophic function of communities. They are of three general types:

(a) the PYRAMID OF NUMBERS, which illustrates the numbers of organisms at each TROPHIC LEVEL (see Fig. 256).

(b) the pyramid of BIOMASS, based on DRY WEIGHT (or occasionally live weight) at each level.

(c) the pyramid of energy which shows the energy flow between each level. See Fig. 136. See FOOD CHAIN.

ecological race, *n.* a localized form that has evolved characteristics as a result of the selective effect of a specific environment.

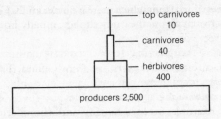

Fig. 136. **Ecological pyramid.** A hypothetical ecological pyramid for standing crop on tropical grassland. Figures in $Kcal/m^{-2}/Y_r^{-1}$ per year.

ecology or **bionomics,** *n.* the study of plants and animals in relation to their total environment.

ecophenotype, *n.* a modification of the phenotype of an organism, resulting from environmental influences, that is not heritable. See PHENOTYPIC PLASTICITY.

ecospecies, *n.* a species of plant or animal that can be divided into several ECOTYPES (see CENOSPECIES) which are able to exchange genes freely without loss of fertility or vigour in the offspring.

ecosphere, *n.* those parts of the earth and its atmosphere where life can exist.

ecosystem, *n.* an ecological system that includes all the organisms and their environment within which they occur naturally.

ecotone, *n.* the transition zone between two major ecological communities where one does not merge gradually into the other, for example, that between grassland and woodland. Such steep gradients between communities are usually man-made.

ecotype, *n.* the product arising as a result of the response of the GENOTYPE of an organism to the particular habitat in which it lives, for example, *Plantago maritima* has a height of about 17.5 cm in waterlogged mud and 56 cm in fertile meadow. See also PHENOTYPIC PLASTICITY.

ecto- or **ect-,** *prefix.* denoting outside.

ectoblast, see ECTODERM.

ectoderm or **ectoblast,** *n.* the germ layer lying on the outside of the developing embryo that eventually gives rise largely to the EPIDERMIS, but also to nervous tissue and, where present, nephridia (see NEPHRIDIUM). Compare ENDODERM.

ectoenzymes, *n.* enzymes secreted across the body wall of a SAPROPHYTE, into the materials which they have invaded, so facilitating absorption.

ectoparasite, see PARASITE.

ectoplasm, *n.* the outer layer of cytoplasm as distinct from the ENDOPLASM of a cell. It is often much more gel-like (see PLASMA GEL) than the liquid endoplasm, from which there is no clear line of distinction. It is important in the movement of unicellular animals such as *Amoeba*.

Ectoprocta, see BRYOZOAN.

ectotherm, see POIKILOTHERM.

ectotroph, *n.* any HETEROTROPH that absorbs food materials directly from outside the body. Examples include bread mould and tapeworms. In contrast, an *endotroph* is a heterotroph that absorbs food material from a gut cavity. Examples include humans and *Hydra*.

eczema, *n.* a blistery skin rash usually due to an allergy.

edaphic factor, *n.* any characteristic of the environment resulting from the physical, chemical or biotic components of the soil.

edaphic race, *n.* an ECOPHENOTYPE resulting from the properties of the SUBSTRATE rather than other environmental factors.

edema, see OEDEMA.

edentate, *n.* any of the placental mammals that constitute the order Edentata. Edentates are primitive mammals, lacking teeth or having only peg-like teeth from which enamel is absent, for example, sloths, anteaters, armadillos.

edentulous, *adj.* without teeth.

Edwards' syndrome, *n.* a human genetical abnormality in which there are multiple congenital malformations: elongated skull, low-set ears, webbed neck, severe mental retardation. The condition is caused by TRISOMY of chromosome 18 and, like DOWN'S SYNDROME, is related to maternal age. 90% of cases die in the first six months after birth.

EEG (electroencephalogram), *n.* a recording of the electrical changes occurring in the brain, produced by placing electrodes on the scalp and amplifying the electrical potential developed. The EEG shows three main types of wave called alpha, beta and delta, that differ in their rates of production. Delta waves are the slowest and are found normally only during sleep.

eel, *n.* any TELEOST fish of the Anguilliformes, having a smooth, shiny skin, a long, snakelike body and reduced fins.

eelworm, *n.* any plant-parasitic or free-living NEMATODE.

effector, *n.* a structure or organ that brings about an action as a result of a stimulus received through a RECEPTOR which can come from the CNS or from a hormone.

efferent neurone, *n.* a nerve fibre carrying impulses away from the CNS to effector cells. Compare AFFERENT NEURONE.

egestion, *n.* the evacuation of faeces or unused food substances from the body.

egg, *n.* **1.** see OVUM. **2.** a structure produced by insects, birds and reptiles whose function is to enable embryonic development outside the female on land without the use of water as a growth medium. The vertebrate egg consists of an outer shell (hard in birds, leathery in reptiles), four types of embryonic membranes, a food supply in the yolk sac and surrounding albumen ('egg white') and the embryo which develops from an OVUM fertilized before the shell is deposited. Domestic birds can produce unfertilized eggs in which the embryo does not develop.

egg cell, see OVUM.

egg membrane, *n.* any membrane, shell or jelly-like layer surrounding and protecting the EGG, including primary mechanisms secreted by the OVUM or OOCYTE, secondary membranes secreted by cells of the ovary and tertiary membranes secreted by glands of the oviduct, for example, albumen and egg shell.

egg tooth, *n.* a small horny projection at the tip of the skull in birds used to pierce the egg shell.

eglandular, *adj.* without glands.

ejaculation, *n.* the process by which SEMEN is expelled from the penis by strong muscular contractions of the urethral wall.

elaioplast, *n.* a colourless PLASTID that stores oil, found commonly in MONOCOTYLEDONS and LIVERWORTS.

elasmobranch, *n.* any cartilaginous fish of the subclass Elasmobranchii (class CHONDRICHTHYES) including fish such as sharks, dogfish, skate and rays – all the class except the chimaeras. In some classifications the term is synonymous with Chondrichthyes.

elastic cartilage, see CARTILAGE.

elastic fibre, *n.* any of the highly extensible fibres that are found scattered in CONNECTIVE TISSUE in vertebrates, occurring particularly in the lungs and artery walls.

elastin, *n.* the protein found in ELASTIC FIBRES.

elater, *n.* **1.** a cell in which the wall is reinforced by spiral bands of thickening whose function is to assist in the discharge of spores in liverworts in response to high humidity. **2.** an appendage of horsetail spores, whose function is similar to **1.**

elaterid, *n.* any COLEOPTERAN of the family Elateridae, including the click beetles, fireflies and wireworms.

electric organ, *n.* an organ that is capable of giving an electric shock to any organism coming in contact with it, and occurring particularly in ELASMOBRANCH fish such as rays.

electrocardiogram, see ECG.

electroencephalogram, see EEG.

electromagnetic spectrum, *n.* the entire range of wavelengths of electromagnetic radiation, most of which are not detectable by the human eye except in the *visible spectrum* from about 400–700 nm wavelength. Wavelengths shorter than the visible spectrum contain large quantities of energy which can be harmful to living material. See Fig. 137. See X-RAY, GAMMA RADIATION, ULTRAVIOLET LIGHT.

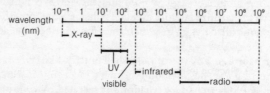

Fig. 137. **Electromagnetic spectrum.** The wavelength ranges of the electromagnetic spectrum.

electron microscope (EM), *n.* a microscope that produces high-resolution images by the interaction of electrons with the specimen, the electrons being guided by electromagnetic lenses. There are two major types of EM:

(a) *transmission (TEM).* The beam of electrons passes through the specimen (e.g. thinly-sectioned tissues) and is focused onto a flurorescent screen or a photographic film. Magnifications well in excess of × 250,000 are possible, with a RESOLUTION of less than 1 nm.

(b) *scanning (SEM).* The specimen (which can be whole cells or tissues) is bombarded with high-energy electrons causing generation of low-power, secondary electrons from the specimen surface, which are collected to form an image of the surface. Magnifications in excess of × 100,000 are possible, with a resolution of about 5 nm. See Fig. 138.

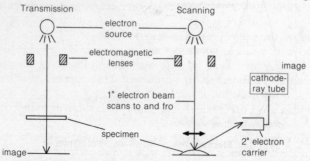

Fig. 138. **Electron microscope.** The movement of electrons through the transmission electron microscope (TEM) and the scanning electron microscope (SEM).

electron transport system (ETS), *n.* a series of biochemical steps by which energy is transferred in steps from a higher to a lower level. Each step involves a specific electron carrier which has a particular energy level (or REDOX POTENTIAL), with the carriers organized in a sequence of decreasing energy. See Fig. 139.

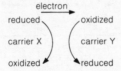

Fig. 139. **Electron transport system.** The role of electron carriers.

Thus the oxidation of X releases more energy than is required to reduce Y. The released energy can be used to produce ATP, in MITOCHONDRIA and CHLOROPLASTS, probably by a process of *chemiosmosis*. This mechanism, proposed by P. Mitchell (b. 1920, Nobel prize-winner in 1978), involves hydrogen ions (H^+) being pumped in or out of the membranes containing the ETS, generating a *proton motive force* that enables ATP synthesis from ADP and P.

Electron transport is vital in both PHOTOSYNTHESIS and AEROBIC RESPIRATION:

(a) *photosynthesis*. Two ET systems are utilized during the LIGHT REACTIONS in the grana of the chloroplasts. One ETS enables the production of ATP by PHOTOPHOSPHORYLATION and the other enables the production of reduced NADP. See Fig. 140.

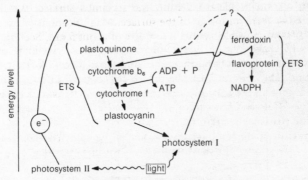

Fig. 140. **Electron transport system.** Photosynthesis.

(b) *aerobic respiration*. A molecule of NADH from GLYCOLYSIS or the KREBS CYCLE is passed to the cristae of mitochondria where it is oxidized in the *respiratory ETS*, the final products being water and three molecules of ATP (= OXIDATIVE PHOSPHORYLATION). See Fig. 141. The free energy of NADH is approximately 220 kJ. Of this, about 102 kJ are stored in the three ATP molecules (3×34 kJ) with the remaining energy being lost as

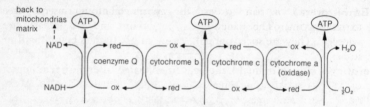

Fig. 141. **Electron transport system.** Aerobic respiration.

heat, a conversion efficiency of about 46%. FADH enters the ETS at a lower energy level than NADH, yielding only two ATP molecules, not three.

electro–osmosis, *n.* a type of OSMOSIS in which the SEMIPERMEABLE MEMBRANE has a potential difference across it so that charged solvents move from the negative to the positive side.

electrophoresis, *n.* a method for separating particles with different electrical charges, for example, proteins. The apparatus consists of a supporting medium soaked in a suitable buffer with an electrical field set up across it. The mixture to be separated (e.g. blood proteins) is placed on the supporting medium. The components with different charges then separate from each other and their eventual position is compared with the position of known standards.

element, *n.* a substance that cannot be destroyed by normally available heat or electrical energy.

elephantiasis, *n.* an abnormality of the human lymphatic system in which blockage of the lymph ducts causes gross swellings of the surrounding tissues, often resulting in limbs that appear like those of an elephant. The condition can be caused by bacterial infection, but in tropical regions is caused most often by nematode FILARIAL WORMS, for example, *Wucheria bancrofti.*

elevator muscle, see FLIGHT.

elimination, *n.* the removal of waste and undigested materials from the body, defacation and exhalation of CO_2 during breathing.

elytron, see COLEOPTERA.

emarginate, *adj.* (of plant structures) notched shallowly towards the apex.

emasculation, *n.* **1.** (in plants) the removal of stamens from hermaphrodite flowers before pollen is liberated, thus enabling artificial CROSS-POLLINATION. **2.** (in animals) also called **castration** or **gelding**. the removal of the testicles.

embedding, *n.* the process of sealing a specimen in wax which is to be sectioned, usually with a microtome.

Embioptera, *n.* an order of hemimetabolous insects (see EXOPTERYGOTA). The males have four wings and the females are wingless. They are social animals, living in silk tunnels below bark or stones.

embryo, *n.* **1.** (in animals), the stage immediately after the beginning of CLEAVAGE up to the time when the developing animal hatches, or breaks out of egg membranes, or in higher animals, is born. **2.** (in plants), the partly developed SPOROPHYTE, which in ANGIOSPERMS is protected within a seed. At one end of the embryo axis is the RADICLE or ROOT, and at the other the apical MERISTEM, or PLUMULE in some forms, and one or two young leaves (COTYLEDONS).

embryology, *n.* the study of the developing EMBRYO in animals or plants.

embryonic membrane or **extraembryonic membrane,** *n.* any of the protective membranes surrounding the developing embryo in animals, that are involved in the respiration and nutrition of the developing organism. They derive from the zygote, lie outside the embryo proper, and form the CHORION, ALLANTOIS, AMNION and YOLK SAC.

embryophyte, *n.* any plant of the nontaxonomic group Embryophyta, possessing an embryo and multicellular sex organs. Examples include mosses, liverworts, ferns and seed plants.

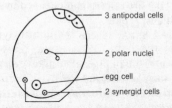

3 antipodal cells

2 polar nuclei

egg cell

2 synergid cells

Fig. 142. **Embryo sac.** Vertical section.

embryo sac, *n.* the female GAMETOPHYTE of flowering plants, consisting of a sac-like structure inside the OVULE in which are found six HAPLOID cells (without cell walls) and two haploid nuclei. See Fig. 142. The name 'embyo sac' comes from the fact that the plant embryo develops from within the sac. Two MALE GAMETE NUCLEI enter the embryo sac from the POLLEN TUBE and carry out a 'double' fertilization. One male nucleus fuses with the egg cell to give a DIPLOID zygote (the future plant), the other male nucleus fuses with two polar nuclei giving a triploid (3n) ENDOSPERM nucleus (the future food supply in some species).

emendation, *n.* the correction of a previously published misspelt scientific name.

emigration, *n.* the movement of animals away from a specific area. Compare IMMIGRATION.

emphysema, *n.* a pulmonary disorder involving overdistention and destruction of the air spaces in the lungs.

emulsification, *n.* a process in which an emulsion is formed, an emulsion being a liquid containing fine droplets of another liquid that do not together form a solution, for example, fats in milk. BILE is an important emulsifier in the mammalian digestive system.

en- or **em-,** *prefix.* denoting in, into.

enamel, *n.* a substance consisting mainly of a calcium phosphate-carbonate salt, bound together by KERATIN, found on the crowns of teeth and the denticles of fish. It is formed from the EPITHELIUM of the mouth. See TOOTH.

enation, *n.* an outgrowth on a leaf produced by a virus infection resulting in the multiplication of leaf cells.

encephalin, see ENKEPHALIN.

encephalitis or **sleeping sickness,** *n.* an inflammatory viral disease of the human CENTRAL NERVOUS SYSTEM that is ENDEMIC in parts of North America. Encephalitis is not to be confused with AFRICAN SLEEPING SICKNESS, which is caused by a trypanosome protozoan.

encystment, *n.* a process found in the life stages of some organisms in which they surround themselves with a CYST. Encystment is usually a protective stage in the life history and occurs, for example, in *Euglena* and in a larval stage of FLUKES.

endangered species, *n.* any species at risk of extinction as a result of human activity.

endemic, *adj.* (of organisms or disease) occurrence limited to a particular geographical area such as an island.

endergonic reaction or **endothermic reaction,** *n.* a reaction requiring free energy (e.g. heat) from an outside source to begin the reaction. See also ACTIVATION ENERGY.

endo-, *prefix.* denoting inside.

endoblast, see ENDODERM.

endocardium, *n.* the inner lining of the HEART.

endocrine gland or **ductless gland,** *n.* the glands of internal secretion, which shed their secretions (HORMONES) directly into the blood system.

endocrinology, *n.* the study of ENDOCRINE GLANDS and their secretions.

endocytosis, *n.* an active process by which some cells can enclose a smaller body (e.g. food particles) forming a membrane-bound vesicle. Compare EXOCYTOSIS. See PHAGOCYTOSIS, PINOCYTOSIS,

endoderm or **endoblast,** *n.* the embryological germ layer in animals that gives rise to the gut system and its association organs. It arises initially as a result of GASTRULATION from cells which have moved in from the surface of the BLASTULA. Compare ECTODERM.

endodermis, *n.* a one-cell layer of tissue found outside the vascular areas in many ANGIOSPERMS, which is particularly important in the roots where endodermal cells are thickened with CASPARIAN STRIPS for control of water transport.

endogamy, *n.* pollination of a flower by another flower on the same plant. Compare EXOGAMY.

endogenous, *adj.* (of growth or production) from within the body, for example, the development of new roots from the PERICYCLE of the old ROOT. Compare EXOGENOUS.

endolymph, *n.* the fluid contained in the membranous labyrinth of the vertebrate ear.

endometrium, *n.* the lining of the uterus in female mammals, being a glandular MUCUS membrane that undergoes changes during the OESTRUS CYCLE during which it builds up in preparation to receive the fertilized egg. Where fertilization does not occur, it regresses, or in the case of humans, anthropoid apes and Old-World monkeys, suddenly breaks down, producing bleeding. See MENSTRUAL CYCLE.

endomitosis, *n.* the doubling of the chromosome number without subsequent cell division, so producing a POLYPLOID.

endonuclease restriction enzymes, *n.* bacterial ENZYMES capable of cleaving DNA at points where specific NUCLEOTIDE sequences occur, a function thought to be important in defending an organism against invasion by foreign DNA. Such enzymes have been used extensively to map the base sequences of DNA controlling, for example, the structure of HAEMOGLOBIN molecules.

endoparasite, see PARASITE.

endopeptidase or **proteinase,** *n.* a type of protein-splitting enzyme that hydrolyses peptide bonds between amino acids located *inside* the chain, but not at the ends. There are three major endopeptidases in the mammalian gut: PEPSIN (stomach); TRYPSIN and CHYMOTRYPSIN (pancreas). Such enzymes are responsible for the first stage of protein digestion; other proteases called EXOPEPTIDASES complete the digestion of protein in the ILEUM.

endoplasm, *n.* any cytoplasm present within the plasma membrane and ECTOPLASM of a cell. It is often more liquid (see PLASMA SOL) than the ectoplasm and is important in locomotion of some PROTOZOANS. It contains more granules than the ectoplasm, from which it is difficult to distinguish, as there is no distinct boundary between the two.

endoplasmic reticulum (ER), *n.* a series of interconnected, flattened cavities lined with a thin membrane about 4 nm thick which is continuous with the NUCLEAR MEMBRANE. It can be covered with RIBOSOMES, when it is referred to as *rough ER*, or ribosomes may be absent in which case the ER is known as *smooth ER*, and gives rise to the GOLGI APPARATUS.

Endoprocta, see BRYOZOAN.

Endopterygota or **Holometabola,** *n.* a subclass of the class Insecta, containing those insects that have a marked METAMORPHOSIS in which a larval form pupates and gives rise to an adult (imago) with a very different body form from the larva. Examples include DIPTERANS and LEPIDOPTERA.

end organ, *n.* a single or multicellular organ situated at the end of a fibre of the peripheral nervous system (outside the CNS). It is either a receptor or a means of transferring a nerve impulse to an effector, as in a motor endplate (see ENDPLATE, MOTOR).

endorphin, *n.* a small protein produced in the nervous system of vertebrates exhibiting actions similar to morphine.

endoskeleton, *n.* a skeleton present within the body of an organism, for example, the vertebrate skeleton. Compare EXOSKELETON.

endosperm, *n.* a TRIPLOID (1) tissue found in many angiosperm seeds (e.g. castor oil), serving as a food source for the embryo which develops within it. Nonendospermic seeds (e.g. runner bean) store their food substances within the cotyledons. See EMBRYO SAC for origin.

endospore, *n.* a very resistant bacterial SPORE that develops by a process called SPORULATION, from a vegetative cell.

endostyle, *n.* a ciliated groove on the ventral wall of the pharynx of urochordates, cephalochordates and the ammocoete larva of the lamprey, that passes food by ciliary action backwards to the gullet.

endothelium, *n.* a layer of flattened cells, one cell thick, that lines the heart, blood vessels and lymph vessels of vertebrates.

endothermic reaction, see ENDERGONIC REACTION.

endothermy, *n.* the ability of an organism to produce sufficient metabolic heat to raise its CORE TEMPERATURE above its surroundings. It may be maintained continually or for limited periods only, such as during activity.

endotroph, see ECTOTROPH.

endplate, motor, *n.* the modification of the muscle fibre membrane at a nerve muscle junction to which the DENDRITE is attached, forming an END ORGAN. ACETYLCHOLINE released from the nerve ending diffuses across the junction and depolarizes the endplate, so giving rise to an endplate potential which, as it builds up, may give rise to an ACTION POTENTIAL and initiate muscle contraction. See Fig. 143.

end-product inhibition, *n.* a type of FEEDBACK MECHANISM in which

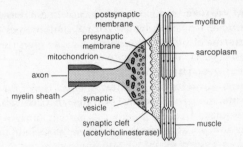

Fig. 143. **Endplate.** Vertical section.

the movement of substances along a biosynthetic PATHWAY is prevented by the end product of the pathway, resulting in self-regulation.

energy, *n.* the capacity of a body or system to do work. The most important energy forms, as far as living organisms are concerned, are heat, radiant, chemical and mechanical energy. Energy units of importance are:

1 calorie = heat to raise 1 g of water through 1°C

$$1 \text{ erg} = \frac{\text{energy to raise 1g through 1cm against gravity}}{981}$$

1 joule $= 10^7$ ergs
1 Calorie $= 1$ kcal ($= 1000$ cals)
1 calorie $= 4.2 \times 10^7$ ergs
1 calorie $= 4.2$ joules

The quantity of solar energy entering the earth's atmosphere is $64.3 \times 10^8 \text{ J/m}^{-2}\text{yr}^{-1}$. The amount of solar energy available to plants in Britain is $10.5 \times 10^8 \text{ J/m}^{-2} \text{ yr}^{-1}$. The SI UNIT of energy is the Joule (J). In plants and animals, energy is stored in ATP (short-term storage), and starch and FAT (long-term storage).

energy acceptor, *n.* a molecule (e.g. CYTOCHROME) capable of receiving energy (usually in the form of electrons) and of passing it on to another acceptor, as in an ELECTRON TRANSPORT SYSTEM.

energy donor, *n.* the molecules that give up the energy to drive an ENDERGONIC REACTION.

energy flow, *n.* the movement of energy through an ECOSYSTEM. The energy is usually first trapped as *radiant energy* (from the sun) by PRODUCERS, which are then eaten by HERBIVORES which, in turn, might be consumed by PREDATORS, with considerable loss of energy (both material energy and heat) back into the ecosystem (and out of it) at each level.

encephalin, see ENKEPHALIN.

enrichment culture, *n.* a technique for isolating an organism from a mixed culture by manipulating the growth conditions in its favour and to the detriment of other organisms.

ent- or **ento-,** *prefix.* denoting within.

enteric canal, see ALIMENTARY CANAL.

entero-, *prefix.* denoting the intestine.

enterocrinine, *n.* a gastrointestinal hormone that controls the secretion of intestinal juice.

enterogastrone, *n.* a hormone secreted by the MUCOSA of the duodenum that decreases gastric secretions and movement in response to the ingestion of fat.

enterokinase, *n.* an ENZYME secreted by the wall of the small intestine, whose function is to catalyse the conversion of inactive trypsinogen in the pancreatic juice to active TRYPSIN.

enteron, see ALIMENTARY CANAL.

Enteropneusta, see HEMICHORDATA.

entire, *adj.* (of plant structure) not toothed or cut.

entomo-, *prefix.* denoting insects.

entomogenous, *adj.* (of fungi) parasitic on insects.

entomology, *n.* the study of insects.

entomophily, *n.* the POLLINATION of plants by insects. Such animal pollinators are one of the two main mechanisms for the transport of pollen to the stigma, the other being ANEMOPHILY. In entomophilous flowers the colours are adapted to their pollinators, for example, moths are mainly active at dusk and at night and they visit flowers that are mostly white; bees cannot see red and will visit mainly blue or yellow flowers. Many flowers have patterns visible only with ULTRAVIOLET LIGHT which insects (but not mammals) can detect. Deep flowers are pollinated by insects with long mouthparts, and short flowers by insects with short mouthparts, an example of COEVOLUTION of plants and insects.

entropy, *n.* the amount of disorder or the degree of randomness of a system. For example, when a protein is denatured by heat (see DENATURATION), the molecule (which has a definite shape) uncoils and takes up a random shape, producing a large change in entropy.

envelope, *n.* any enclosing structure, such as a membrane or skin. In bacteria, it is the part of the cell enclosing the cytoplasm, i.e. the cytoplasmic membrane cell wall and capsule. In VIRUSES, it is the outer lipid-containing layer of some virions.

environment, *n.* the surroundings of any organism, including the MEDIUM, SUBSTRATE, climatic conditions, other organisms (see BIOTIC FACTORS), light and pH.

environmental resistance, *n.* the collective effect on population growth of predators and competition for food and space.

environmental temperature, *n.* the temperature at which an inanimate body of the same shape and size as a given organism will come to equilibrium with its surroundings when placed at the same point in space as the organism. The temperature includes radiative and convective influences on the organism.

environmental variation, see PHENOTYPIC PLASTICITY.

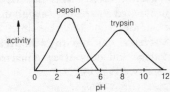

Fig. 144. **Enzyme.** The effect of pH activity.

enzyme, *n.* a protein molecule that catalyses a biochemical reaction by lowering the ACTIVATION ENERGY required for the reaction to proceed. Enzymes are usually specific to particular substrates (see ACTIVE SITE) and are sensitive to environmental conditions such as pH and temperature (see Fig. 144, and Q_{10}). ALLOSTERIC ENZYMES exist in inactive and active forms, while others can be inhibited by nonsubstrate molecules (see COMPETITIVE INHIBITION, NONCOMPETITIVE INHIBITION). Protein-splitting enzymes (PROTEASES) are produced in nonactive forms in the mammalian digestive system to minimize the risk of self-digestion. For example, TRYPSIN is produced as inactive trypsinogen.

enzyme induction, *n.* the process in which a 'structural' gene is activated by a substrate binding with a repressor substance, thus enabling the production of an enzyme that will catalyse the metabolism of the substrate. See OPERON MODEL.

enzyme inhibitor, *n.* a molecule that prevents an enzyme from catalysing a reaction. Such inhibitors can compete with the normal substrate (see COMPETITIVE INHIBITION) or can block the active site, preventing entry of the substrate (see NONCOMPETITIVE INHIBITION). Enzyme inhibitors often form part of a FEEDBACK MECHANISM to regulate a biochemical pathway.

eobiont, *n.* a hypothetical chemical precursor of a living cell.

Eocene, *n.* a geological epoch lasting from 54 to 38 million years ago; a subdivision of the TERTIARY PERIOD. During this time extensive planktonic populations of Foraminifera laid down rock beds (from which the pyramids of Egypt were built), and many groups of mammals appeared for the first time, for example, rodents, whales, carnivores.

Britain was still moving northwards. See GEOLOGICAL TIME.

eosinophil leucocyte, *n*. a type of white blood cell (of the GRANULOCYTE group) which takes up the stain of an acid dye. Eosinophils make up about 4% of the adult human leucocytes. They are produced in the bone marrow, and their function is probably connected with the IMMUNE RESPONSE, particularly ALLERGIES.

ep- or **epi-** or **eph-,** *prefix*. denoting upon.

ephemeral, *adj*. (of organisms) having a very short life cycle. The term is used specifically to describe those plants that have more than one generation a year, as opposed to ANNUAL, BIENNIAL, PERENNIAL.

Ephemeroptera, *n*. an order of the class Insecta, containing the mayflies. They are EXOPTERGYOTES, and spend the great majority of their life in the nymphal stage and only a few minutes or hours in the nonfeeding, adult stage. See Fig. 145.

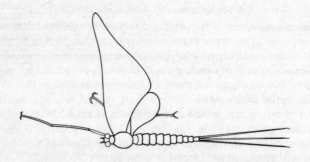

Fig. 145. **Ephemeroptera.** *Ephemera sp.*

ephyra, *n*. the free-swimming larva of a COELENTERATE jellyfish that results from STROBILATION of a scyphistoma larva (see SCYPHOZOAN).

epi-, *prefix*. denoting outside.

epiblem, *n*. the outermost layer of PARENCHYMA cells in the root that becomes the functional outer layer when the piliferous layer sloughs off as roots increase in age.

epiboly, *n*. the process by which a BLASTULA is converted into a GASTRULA by growth of ectoderm over the endoderm.

epicalyx, *n*. a CALYX-like structure that occurs outside and around the true calyx, as in the strawberry flower.

epicardium, *n*. the external covering of the HEART wall.

epicotyl, *n*. the axis (or stem) of the shoot in a young germinating seedling, located above the COTYLEDON. See GERMINATION and Fig. 166.

epidemic, *n*. the occurrence of many cases of a disease within an area.

epidemiology, *n.* the study of the incidence, distribution and control of an EPIDEMIC disease in a population.

epidermis, *n.* **1.** (in plants) the thin tissue, usually one cell thick, that surrounds young roots, stems and leaves. In stems and leaves the epidermal cells secrete a CUTICLE (1), in roots they do not. In older roots and stems the epidermis is often replaced by CORK tissue. **2.** (in animals) the outer layer of the skin derived from embryonic ECTODERM. In vertebrates, the epidermal layer is usually made up of stratified EPITHELIUM with an outer layer of dead cells which become 'keratinized' (see KERATIN) forming a protective layer. The invertebrate epidermis is normally one cell thick and often forms a protective cuticle.

epididymis, *n.* the long, coiled narrow tube running from the TESTIS to the VAS DEFERENS in the higher vertebrates; it functions to store sperm.

epifauna, *n.* the part of the BENTHOS living on the mud surface.

epigamic character, *n.* any secondary sexual character brought into existence by hormone action and used in courtship display.

epigeal, *adj.* of or relating to seed GERMINATION in which the COTYLEDONS are carried above ground and form the first green foliage leaves of the plant, for example, french beans (*Phaseolus vulgaris*).

epigenesis, *n.* the formation of entirely new structures during the development of the EMBRYO.

epigenetic landscape, *n.* a concept related to the developmental PATHWAYS along which a cell passes during differentiation. See CANALIZATION.

epiglottis, *n.* a thin flexible structure, made of CARTILAGE, with a leaf-like shape that guards the entrance to the larynx (the *glottis*) and prevents food material entering the TRACHEA during swallowing. See Fig. 146.

soft palate
buccal cavity
pharynx
tongue
epiglottis
glottis
larynx
oesophagus
trachea

Fig. 146. **Epiglottis.** The left-side cervical region in humans.

epigynous, see GYNOECIUM.

epilepsy, *n.* a nervous condition due to abnormalities in the brain cortex that results in seizures ranging from a sense of numbness in certain body areas (*petit mal*) to extreme muscular convulsions and fits (*grand mal*). Epileptics exhibit large, abnormal brain waves, which can be detected on an EEG.

epimere, *n.* the dorsal part of the mesoderm of·a vertebrate embryo, consisting of a series of SOMITES.

epinasty, *n.* the increased growth of the upper surface of a structure, such as a leaf, resulting in its being downcurved.

epinephrine, see ADRENALINE.

epipelagic zone, see EUPHOTIC ZONE.

epipetalous, *adj.* (of plant structures) inserted upon the COROLLA, as in the stamens of the primrose.

epiphysis, *n.* the ossified part of the end of a mammalian limb bone or vertebra which, during growth, is separated by a plate of cartilage from the rest of the ossified bone. When growth is complete the epiphysis fuses with the rest of the bone. Compare DIAPHYSIS.

epiphyte, *n.* a plant that grows attached to another and uses it solely for purposes of support, for example, some mosses and some orchids, there being no parasitic association.

epiphytotic, *adj.* (of plant diseases and parasites) affecting plants over a wide geographical region.

episome, *n.* a circular DNA molecule found in bacterial cells that can exist independently in the cell or can become integrated into the main CHROMOSOME. In recent times, episomes have been added to a general group of extrachromosomal factors called PLASMIDS.

epistasis, *n.* a form of genetic interaction in which one gene interferes with the expression of another gene, for example, if genes A and B code for enzymes active in the same PATHWAY. See Fig. 147. If both ALLELES of gene A code for a nonfunctional version of enzyme A, then the pathway will shut down, irrespective of which B alleles are present, i.e. gene A is epistatic to gene B. Compare DOMINANT EPISTASIS, RECESSIVE EPISTASIS.

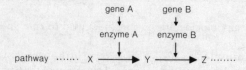

Fig. 147. **Epistasis.** The interaction of genes A and B.

epithelium, *n.* **1.** (in animals) a layer of covering cells that is normally one cell thick and usually covers connective tissue embryologically derived from the ECTODERM. The cells often have a secretory function, and their shape gives rise to names descriptive of the cells, e.g. *columnar, cubical, squamous* (see Fig. 148). Where the epithelium is more than one cell thick it is described as *stratified*. Similar cells can be derived from MESODERM and are referred to as *mesothelium* when lining the COELOM, and as ENDOTHELIUM when lining blood vessels. **2.** (in plants) a layer of

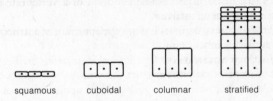

Fig. 148. **Epithelium (1).** Types of epithelium.

cells lining cavities and secretory canals, for example, resin canals.

epizoite, *n.* any animal that lives attached to another and uses it for protection or means of locomotion, there being no parasitic relationship.

epizootic, *adj.* (of a disease) suddenly and temporarily affecting a large animal population.

equal weighting, *n.* a taxonomic procedure where all characters are regarded as of equal importance. The procedure is commonly used in NUMERICAL TAXONOMY.

equatorial plate, *n.* an arrangement of chromosomes lying in one plane at the equator of the spindle in METAPHASE of MITOSIS or MEIOSIS.

equilibrium centrifugation, *n.* a process of DENSITY–GRADIENT CENTRIFUGATION that continues until there is no net movement of the molecules, each molecule having come to rest when its density equals that of the solution.

equilibrium of population, see GENETIC EQUILIBRIUM.

Equisetum, *n.* any pteridophyte plant of the genus *Equisetum,* comprising the horsetails. The plant is herbaceous, has rhizomes and a vertical stem with whorls of scale leaves, and is often found growing in water and damp conditions. It is the last living representative of the order Equisetales.

eradicant, *n.* any biocide (e.g. a fungicide) used to cure an established infection.

erector-pili muscle, *n.* a small muscle found in the DERMIS of a mammal that is connected to a hair and controls the position of the hair in relation to the skin surface for temperature regulation.

erection, penis, *n.* the hardening and lengthening of the mammalian penis in order to bring it into condition for copulation. An erection is brought about by engorgement of blood in the CORPUS SPONGIOSUM and CORPUS CAVERNOSUM, causing them to become turgid.

erepsin, *n.* a mixture of PROTEOLYTIC ENZYMES secreted by the small intestine of mammals.

ergot, *n.* **1.** a disease caused by the ascomycete *Claviceps purpurea* in cereals and grasses. **2.** a SCLEROTIUM forming in place of a grain in a diseased grain head. Ergots contain substances that are poisonous and sometimes used medicinally.

erose, *adj.* (of plant structures) having the appearance of being gnawed, with irregular notches.

erosion, *n.* the wearing away of geological formations such as rock, soil, etc. For example, deafforestation or the removal of hedges causes soil erosion.

erythr- or **erythro-,** *prefix.* denoting red.

erythroblastosis foetalis, *n.* a haemolytic disease of newborn children brought about when a rhesus negative mother produces a rhesus negative child (see RHESUS BLOOD GROUP). In late pregnancy, Rh⁻ cells may pass into the mother who produces antigens which pass back across the placenta and adversely affect the child.

erythrocyte or **red blood cell (RBC),** *n.* a vertebrate cell that carries some carbon dioxide as HCO_3- from tissues to lungs (see also CHLORIDE SHIFT), and contains HAEMOGLOBIN pigment for oxygen transport from lungs to tissues. Unlike other vertebrate cells, mammalian RBCs are non-nucleated and have definite biconcave shape. See Fig. 149. Compare LEUCOCYTE.

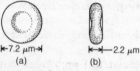

$\leftarrow$7.2 μm$\rightarrow$ $\leftarrow$2.2 μm
(a)   (b)

Fig. 149. **Erythrocyte.** (a) Surface view. (b) Vertical section.

erythromysin, *n.* an antibiotic produced by a strain of STREPTOMYCIN that includes protein synthesis, particularly in organisms which are Gram-positive (see GRAM'S STAIN).

escape, *n.* **1.** any cultivated plant growing in the wild that is not well naturalized. **2.** any organism that is normally captive but which has been freed into the natural environment.

escape response, *n.* any flight reaction elicited in an animal as a result of a threat.

Escherichia coli (E. coli), *n.* a bacterium common in the human gut which has been used extensively in biochemical and genetical studies.

eserine, *n.* a plant alkaloid that is capable of blocking CHOLINESTERASE.

essential amino acid, *n.* an AMINO ACID that must be taken in via the diet, as distinct from nonessential amino acids, which can be synthesized by the organism itself, in a process called TRANSAMINATION, using the essential amino acids as a source. In humans there are ten essential amino acids: ARGININE, TRYPTOPHAN, ISOLEUCINE, METHIONINE, THREONINE, LYSINE, LEUCINE, VALINE, HISTIDINE and PHENYLALANINE.

essential element or **mineral element,** *n.* an element without which normal growth and reproduction cannot take place. In plants, there are seven major essential elements: nitrogen, phosphorus, sulphur, potassium, calcium, magnesium and iron. There are also TRACE ELEMENTS required in much smaller quantities, for example, manganese, boron, chlorine. Animals also have requirements for elements, the list being quite similar to that for plants.

essential fatty acid, see LINOLEIC ACID.

ester, *n.* a compound formed from an alcohol and an acid.

esterification, *n.* the formation of an ESTER.

estrogen, see OESTROGEN.

estuary, *n.* the point at which a river meets the sea. There is thus a mixture of saline and freshwater conditions, often with areas of tidal mudflats and salt marsh. Because of the varying salinity and tidal cover there is often a specific flora and fauna associated with such areas.

etaerio, *n.* an aggregation of ACHENES or DRUPES.

ethanoic acid, see ACETIC ACID.

ethene, see ETHYLENE.

ethanol, *n.* an alcohol produced in ALCOHOLIC FERMENTATION.

ethology, *n.* the study of animal behaviour in the natural habitat of the animals concerned.

ethylene or **ethene,** *n.* a simple hydrocarbon with the formula $CH_2 = CH_2$ that can act as a PLANT HORMONE even when present in very low concentrations (down to 1 ppm.). Ethylene inhibits elongation in most growing tissues and promotes leaf ABSCISSION and fruit ripening in some plants. Plant cells produce ethylene from the amino acid METHIONINE.

etiolation, *n.* the range of symptoms developed by plants when grown in the dark. Examples include: pale yellow or white colour due to lack of chlorophyll, long internodes, small and rudimentary leaves, poor development of lignified tissue.

etiology, see AETIOLOGY.

ETS, see ELECTRON TRANSPORT SYSTEM.

eu-, *prefix.* denoting true, well or good.

eucarpic, *adj.* (of an adult fungal THALLUS) having clear distinction between vegetative and reproductive parts.

eucaryote or **eukaryote,** *n.* any member of a group of organisms that contains all plants and animals apart from bacteria and blue-green algae (which are PROCARYOTES). Eucaryotes are distinguished by the fact that their cells possess a membrane-bound nucleus containing the genetic material, but there are also other differences from the procaryotes. See Fig. 150.

Procaryotes	Eucaryotes
1. No true nucleus.	Nucleus with NUCLEAR MEMBRANE.
2. Single chromosome made up of nucleic acid.	Several CHROMOSOMES with nucleic acid complexed with protein.
3. Cell organelles absent.	GOLGI APPARATUS, ENDOPLASMIC RETICULUM, LYSOSOMES, MITOCHONDRIA present.
4. If present, chlorophyll not in chloroplasts.	If present, CHLOROPHYLL in chloroplasts.
5. Flagella lack 9+2 structure.	FLAGELLA with 9+2 structure.
6. Cell division by BINARY FISSION.	Cell division by MITOSIS AND MEIOSIS.

Fig. 150. **Eucaryote.** A comparison of procaryotes and eucaryotes.

euchromatin, *n*. a chromosome material that stains heavily in META-PHASE but hardly at all in INTERPHASE when it is relatively uncoiled and undergoes TRANSCRIPTION. Compare HETEROCHROMATIN.

eugenics, *n*. the study of ways of improving the hereditary qualities of a population (especially the human population) by the application of social controls, guided by genetical principles.

Euglena, *n*. a genus of large, green flagellates (division Euglenophyta) that is common in both fresh and salt waters. Euglenoids produce a polysaccharide storage molecule called PARAMYLUM and, (unlike other flagellates), do not possess a cell wall, but produce a flexible PELLICLE instead.

Euglenophyta, *n*. a division of the algae (sometimes classified by zoologists in the class Flagellata). They normally possess chlorophyll *a* and *b*, but this may be lost; colourless forms do not regain their chlorophyll and no members of the division are completely AUTOTROPHIC.

Eumycota or (formerly) **Mycophyta,** *n*. the 'true fungi' (see FUNGUS), a division of the kingdom Mycota.

euphausiid, *n*. any small, pelagic, shrimp-like CRUSTACEAN of the order Euphausiacea. Euphausiids are an important constituent of *Krill*, the main food of whalebone whales.

euphotic zone or **photic zone** or **epipelagic zone,** *n*. the top 100 m of the sea into which light can penetrate and in which PHOTOSYNTHESIS takes place. See SEA ZONATION.

euploidy, *n*. the condition of a cell, tissue or organism that has one or more multiples of a chromosome set, for example, a TRIPLOID (3n). Euploids with more than two sets of chromosomes are POLYPLOIDS.

eupnaea or **eupnea,** *n*. the normal, quiet breathing of an animal at rest.

euryhaline, *adj.* (of organisms) capable of tolerating a wide variety of salinity in the surrounding medium.

eurypterid, *n.* any large extinct scorpion-like aquatic arthropod of the order Eurypterida, found in the SILURIAN PERIOD.

eurythermous, *adj.* (of organisms) able to tolerate a wide range of environmental temperatures.

eusporangiate, *adj.* (of sporangia) arising from a group of parent cells and having a wall consisting of at least two layers of cells.

eustachian tube, *n.* a tube passing from the pharynx to the middle ear in higher vertebrates, serving to equalize the pressure on either side of the TYMPANIC MEMBRANE.

euthanasia, *n.* the act of painless killing to relieve human suffering from an incurable disease.

eutherian, *n.* any placental mammal of the subclass Eutheria, characterized by the embryo developing in the uterus of the female where it obtains food and exchanges gases through the attached PLACENTA.

eutrophic, *adj.* (of a body of water) being rich in organic and mineral nutrients, either naturally or by fertilization.

eutrophication. *n.* a process by which pollutants cause a body of water to become over-rich in organic and mineral nutrients, so that algae grow rapidly and deplete the oxygen supply.

e value, *n.* the solar energy present on other planets expressed as a percentage of the earth's solar energy.

E_0 values, *n.* a numerical series indicative of the REDOX POTENTIAL of molecules. Protons are accepted by a molecule from any other molecule with a more positive E_0 value.

evaporation, *n.* the physical change when a liquid becomes a gas. Since such a change usually requires heat as an energy source, heat is drawn from the immediate environment, which produces a significant cooling effect. The size of the cooling effect depends on the *latent heat of evaporation* of the liquid. Evaporation of water is used by mammals in temperature regulation (sweat) and occurs in plants from the surface of the mesophyll cells during TRANSPIRATION.

evergreen, *n.* a type of tree or shrub that possesses leaves in all seasons. Most CONIFERS are evergreen, as are many ANGIOSPERMS (e.g. laurel, privet).

eversible, *adj.* (of structures) being capable of protruding from an organism by being turned inside out. Examples include the eversible proboscis in some worms, and the eversible cloaca in some birds that can be protruded to form a penis-like structure used in copulation.

evocation, *n.* the induction of embryonic tissue by a chemical stimulus, for example, ECTODERM produces neutral material in the vertebrate embryo due to an evocator (see ORGANIZER REGION) from the underlying CHORDAMESODERM.

evocator region, see ORGANIZER REGION.

evolution, *n.* an explanation of the way in which present-day organisms have been produced, involving changes taking place in the genetic make-up of populations that have been passed on to successive generations. According to DARWINISM, evolutionary MUTATIONS have given rise to changes that have, through NATURAL SELECTION, either survived in better adapted organisms (see ADAPTATION, GENETIC), or died out. Evolution is now generally accepted as the means which gives rise to new species (as opposed to SPECIAL CREATION) but there is still debate about exactly how it has taken place and how rapidly changes can take place. See LAMARCKISM.

evolutionary tree, *n.* a diagram showing the relationships between a group of organisms and their evolution from an ancestral stock.

ex-, *prefix.* denoting outside.

excision, *n.* the removal of a DNA fragment from a chromosome.

excision repair, *n.* the removal of damaged polynucleotide segments from DNA (e.g. thymidine DIMERS produced by ultraviolet light) and replacement with a correct segment using DNA POLYMERASE and DNA LIGASE.

excitability, *n.* any change in membrane conductance as a response to stimulation.

excitation, *n.* the process by which the electrical stimulation of (a) a surface membrane results in contraction of the muscle, or (b) brings about secretion of a transmitter substance at a nerve ending.

excitatory postsynaptic potential (EPSP), *n.* a reduction in the RESTING POTENTIAL of a postsynaptic cell caused by the arrival of TRANSMITTER SUBSTANCE from the presynaptic cell. The reduction takes the membrane potential close to the THRESHOLD and, therefore, nearer to itself forming an ACTION POTENTIAL. See FACILITATION.

excitor neuron, *n.* a neuron that directly excites a muscle or other organ.

exclusion principle, *n.* the principle that two species with the same ecological requirements cannot exist together.

excretion, *n.* any elimination from an organism of unwanted materials, for example, carbon dioxide and nitrogenous substances produced in METABOLISM. It is worth noting that excretory materials must be *produced* by the organism rather than just pass through it; thus faeces in mammals contain a mixture of excreta (e.g. bile pigments) and undigested gut contents.

exergonic reaction or **exothermic reaction,** *n.* a reaction that releases free energy, as in a spontaneous change.

exine, *n.* the outer coat of a POLLEN GRAIN, that often has protuberances, spines, etc. The exine also has one or more areas where the wall is thin, forming a pore through which the pollen tube will protrude.

exocytosis, *n.* an active process in which vesicles containing excretory or secretory materials are actively carried to the periphery of the cell, and released to the outside when the vesicle membrane fuses with the cell membrane. Compare ENDOCYTOSIS. See PHAGOCYTOSIS, PINOCYTOSIS.

exodermis, *n.* a layer of cortical cells, with suburized walls, that replaces the piliferous layer in the older parts of roots.

exogamy, *n.* mating between unrelated individuals. Compare ENDOGAMY.

exogenous, *adj.* **1.** originating from or due to external causes. **2.** developing near the surface of an organism, as in the development of axillary buds in plants.

exobiology, *n.* the study of possible extraterrestrial organisms.

exon or **extron,** *n.* the DNA segments of a EUCARYOTE gene that are transcribed into mRNA and then into protein. Exons occur along the length of the gene, and are separated by segments called INTRONS whose sequences are also transcribed into mRNA. The intron mRNA segments are then excised, leaving behind the exon mRNA segments that join up to form a functional piece of mRNA, a process called RNA SPLICING.

exonuclease, *n.* an enzyme that removes a terminal NUCLEOTIDE ($3'$ or $5'$) in a POLYNUCLEOTIDE CHAIN. Exonucleases remove the nucleotides in a successive way, one by one, and are highly specific in their action.

exopeptidase, *n.* a type of protein-splitting ENZYME that hydrolyses the terminal PEPTIDE BONDS rather than those bonds within the chain. There are three main types of exopeptidase in the mammalian gut, each attacking a particular area of the protein: *carboxypeptidase* attacks the carboxyl end of the chain; *aminopeptidase* attacks the amino end of the chain; *dipeptidase* breaks the bond between DIPEPTIDES. Such enzymes complete the digestion of protein prior to absorption into the blood stream. Compare ENDOPEPTIDASE.

Exopterygota or **Heterometabola** or **Hemimetabola,** *n.* a subclass of the class Insecta, including those insects that do not have a marked METAMORPHOSIS, and in which nymphal stages gradually approach the form of the imago at each moult. Examples include EPHEMEROPTERA, ODONATA.

exopthalmic goitre, *n.* a human condition resulting from the over-production and release of THYROXINE. The thyroid gland becomes swollen, producing a goitre. There is elevated metabolism, nervousness, irritability, loss of weight and the eyes are prominent and staring, with protrusion of the eyeballs. The abnormality can be treated with anti-thyroid drugs and/or surgical excision of part of the gland.

exoskeleton, *n*. a skeleton present on the outside of an organism as in ARTHROPODS or MOLLUSCS. Some vertebrates possess an exoskeleton in addition to an ENDOSKELETON, for example, armadillos and turtles. The exoskeleton may lie outside the EPIDERMIS, as in the arthropods, or inside, as in vertebrates such as scaly fish, tortoises, etc.

exothermic reaction, see EXERGONIC REACTION.

expiration, see BREATHING.

expiratory centre, see BREATHING.

experimental cytology, *n*. the study of cells, utilizing microscopy and electron microscopy in conjunction with biochemical and biophysical experimental techniques.

explant, *n*. any actively dividing plant tissue that can be induced to produce CALLUS tissue in tissue culture.

exploitation, *n*. the situation in which one organism gains at the expense of another.

exposure, *n*. **1.** the aspect of a particular location with respect to the parts of the compass, for example, some garden plants prefer a southern exposure, as with the peach tree in England. **2.** a rock outcrop. **3.** a soil section.

exponent, *n*. a number or quantity placed as a superscript to the right of another number or quantity, indicating how many times the number is to be multiplied by itself. For example, 10^6.

expressivity, *n*. the degree to which a particular gene exhibits itself in the PHENOTYPE of an organism, once it has undergone PENETRANCE. Thus, for example, a penetrant baldness gene in man can have a wide range of expressivity, from thinning hair to complete lack of hair.

exserted, *adj*. (of plant structures) protruding.

exsiccata, *n*. the preserved material of a herbarium.

extensor, *n*. a muscle that extends or straightens a limb.

extero-, *prefix*. denoting outside.

exteroceptor, *n*. any sensory organ that detects stimuli from outside the organism.

extinct, *adj*. (of an animal or plant species) having died out, no longer present in the world population.

extinction, *n*. **1.** the act of making EXTINCT or the state of being extinct. **2.** the elimination of an allele of a gene in a population, due to RANDOM GENETIC DRIFT or to adverse SELECTION pressures. **3.** any periodical, catastrophic event resulting in a species or larger taxonomic group dying out abruptly at a particular point in geological history. Such extinctions are thought to be cyclical, occurring every 28.4 million years, and have been attributed to cosmic activity such as showers of large asteroids or comets, though neither the periodicity nor its causes are at present universally accepted.

exstipulate, *adj*. lacking STIPULES.

extra-, *prefix*. denoting outside.

extracellular, *adj*. situated or occurring outside the cell, as in extracellular digestion, where cells secrete enzymes to break down food material which is then absorbed.

extraembryonic, *adj*. lying outside the embryo, e.g. EMBRYONIC MEMBRANE.

extranuclear inheritance, see CYTOPLASMIC INHERITANCE.

extrapolation, *n*. the estimation of a value beyond a given series, for example, the extension of the line of a graph beyond the calculated points.

extrorse, *adj*. (of anthers) opening towards the outside of the flower.

exudate, *n*. the material that comes from a cut pore or break in the surface of an organism, such as sweat or cellular debris.

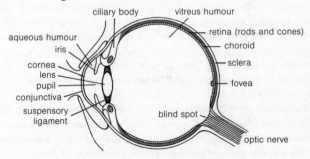

Fig. 151. **Eye.** Vertical section of the mammalian eye.

eye, *n*. the light-receptor organ of animals. Such organs range in complexity from the *ocellus* with a single lens, found in insects and some other invertebrates, to the vertebrate eye, where light is focused by a lens onto a retina consisting of light sensitive cells in the form of RODS and CONES (2). See Fig. 151. The front of the vertebrate eye is covered by the CORNEA behind which is the IRIS which controls the size of the pupil, thus determining the amount of light entering the eye. The shape (and thus the focus) of the lens is controlled by the INVOLUNTARY MUSCLES of the ciliary body under the control of the AUTONOMIC NERVOUS SYSTEM. Light is focused on the retina. See COLOUR VISION, ACCOMMODATION, BINOCULAR VISION and EYE, COMPOUND.

eye, compound, *n*. a type of light-receptor organ, found particularly in

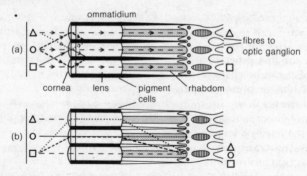

Fig. 152. **Eye, compound.** (a) Apposition image. (b) Superposition image.

insects and crustaceans, that is formed from numerous OMMATIDIA, each of which can form a separate image. Movement of the pigment between the ommatidia forms either (a) an apposition image, where light enters each ommatidium parallel to its long axis, so forming a mosaic image; or (b) a superposition image, where pigment is withdrawn, allowing in light to the sides of the ommatidia and giving a brighter, less sharp and overlapping series of images. See Fig. 152. The latter occurs in nocturnal insects, the former in diurnal insects, and change from one to the other gives rise to dark adaptation. In many insects the image is either apposition or superposition and there is no dark adaptation.

Ommatidia are larger than rods and cones, so fewer can be packed into the same space. Consequently the detail of the image is less good than in the vertebrate eye. For example, the honeybee has a visual acuity that is 1% of human capacity, and most other arthropods are worse than this. However, the compound eye of insects is capable of detecting movement over a large field, and since the reaction time is small, insects are capable of reacting to movement rapidly.

eye muscle, *n.* **1.** one of six muscles that move the eyeball, the so-called extrinsic eye muscles, a pair of oblique muscles situated anteriorly and four rectus muscles situated posteriorly. **2.** one of the intrinsic eye muscles inside the eyeball that are associated with the iris, lens and ciliary body. See EYE.

eye spot, *n.* **1.** a light-sensitive organelle present in unicellular organisms, green algae, zoospores and some gametes. **2.** a pattern on the wings of some insects, particularly moths, which appears eyelike (see WARNING COLORATION).

F

F₁, *n.* the first generation of filial offspring of a particular cross. The number can be changed to indicate the generation, for example, F_3 would be the great-grandchildren of a particular pair of individuals.

facial nerve, *n.* the 7th cranial nerve, a dorsal root that in mammals is mainly motor in function, supplying facial muscles, the salivary glands and the front-of-tongue taste buds.

facilitated transport, *n.* any carrier-mediated transport (see CARRIER MOLECULE) in a membrane where diffusion is aided by the molecule to enhance the mobility of the diffusing substance, but in which there is no ACTIVE TRANSPORT.

facilitation, *n.* a residual effect that enables (facilitates) a second impulse to be transmitted across a junction between a nerve cell or effector cell where a first impulse has failed to cross it. There is thus an increase in responsiveness, resulting from the SUMMATION of impulses, and this is called 'facilitation'.

factor, genetical, *n.* an obsolete term that was used by early geneticists (e.g. R.C. PUNNETT) to describe an ALLELE of a gene.

factorial experiment, *n.* an experiment in which all treatments are varied together rather than one at a time, so the effect of each or combinations of several can be isolated and measured.

facultative, *adj.* (of an organism) being capable of adopting an alternative life style from the normal one. In this context, the word 'facultative' is followed by the *unusual* life style. Thus, a facultative PARASITE is one that is normally saprophytic but is occasionally parasitic; a facultative SAPROPHYTE is one that is normally parasitic but can become saprophytic; a facultative ANAEROBE is an organism that is normally aerobic but can exist in anaerobic conditions. Compare OBLIGATE.

FAD (flavin adenine dinucleotide), *n.* an electron carrier similar in action to NAD, picking up hydrogen from succinic acid in the KREBS CYCLE. The hydrogen is transported to the mitochondrial lamellae where it enters an ELECTRON TRANSPORT SYSTEM at a lower point than NAD, with the release of only two molecules of ATP (rather than three ATP molecules when NAD is the carrier).

faeces, *n.* the bodily waste material that is formed in the large intestine and eliminated via the anus. Faeces contain a mixture of excretory material from the liver (e.g.bilirubin which gives the faeces their characteristic colour, see BILE), food material which has passed straight through the gut, dead bacteria, dead cells and mucus.

falcate, *adj.* sickle-shaped.

Fallopian tube, *n.* a tube forming part of the OVIDUCT, present on either side of a female mammal, that transmits eggs from the peritoneal cavity to the uterus and is the usual site of FERTILIZATION. Ciliary action is involved in the egg movement, and muscular action facilitates the upward movement of sperms from the vaginal tract where they have been deposited after copulation.

family, *n.* the TAXON between ORDER and GENUS that normally contains more than one genus. Family names of animals usually end in –idae, and of plants in –ceae, for example, Ursidae, the bear family; Rosaceae, the rose family.

family planning, see BIRTH CONTROL.

farmer's lung, *n.* a human disease in which lesions of minor blood vessels develop in the lungs. The disease is produced as a result of an IMMUNE RESPONSE to the presence of fungal spores in hay. Similar symptoms have been recorded in other occupations, for example, malt workers, pigeon breeders, cheese washers.

fascia, *n.* sheets of CONNECTIVE TISSUE.

fasciation, *n.* the growing together of branches or stems to form abnormally thick growths.

fascicular, see CAMBIUM.

$$\text{glycerol} + 3 \text{ fatty acids} \xrightleftharpoons[\text{hydrolysis}]{\text{condensation}} \text{fat} + H_2O$$

Fig. 153. **Fat.** The formation of a triglyceride.

fat, *n.* a type of simple LIPID found in almost all organisms, which is an important energy-storage molecule (containing twice as much energy as carbohydrates per gram) that can also aid in heat insulation, cushioning and protection. Fats are produced by a combination of one glycerol molecule, and three fatty acid molecules (which need not all be the same) forming a *triglyceride*. See Fig. 153.

Fats are abundant in plant seeds, and are also found in roots, stems and leaves, forming about 5% of the total dry weight. In animals, fats are stored in specialized cells making up ADIPOSE TISSUE. See also BROWN FAT.

fate map, *n.* a map of an embryo at an early stage of development, showing the various regions where future structures will form.

fat body, *n.* **1.** (in amphibians and lizards) a structure in the form of finger-like growths immediately in front of the gonads. The fat body stores fat as ADIPOSE TISSUE and is largest just before hibernation. It is of special importance in males where little or no food is taken in the breeding season, after which it is greatly reduced in size. **2.** (in insects) a loose network of tissue in spaces between organs and around the gut, storing fat, proteins, glycogen and uric acid. It is found especially in juvenile insects before METAMORPHOSIS.

fatigue, *n.* exhaustion in muscles resulting from exertion or over-stimulation following a period of activity.

fat-soluble vitamin, *n.* any of several vitamins, including A, D, E and K, that are soluble in organic solvents but insoluble in water.

fatty acids, *n.* a range of molecules with the general formula $C_nH_{2n+1}COOH$ that occur naturally in many organisms, often combined with glycerol to form FATS. Fatty acids are of two main types: *unsaturated*, with at least one carbon-to-carbon double bond, and *saturated*, with no such bonds. The greater the proportion of unsaturated fatty acids in a fat the lower its melting point, with many unsaturated fats being liquid oils at room termperature. There is evidence that excess consumption of saturated fats can lead to hardening of the arteries, but the facts are disputed, particularly by butter manufacturers.

fauna, *n.* the grouping of animals present in any one place or at any one time in geological history.

favism, *n.* a human disease characterized by the destruction of red blood cells, resulting in severe anaemia. The disease is triggered by the consumption of raw broad bean (*Vicia faba*), inhalation of broad bean, pollen or several other chemicals such as naphthalene (found in moth balls). The condition is due to a deficiency of the enzyme glucose-6-phosphate dehydrogenase found in red blood cells, the trait being controlled by an X-linked gene which is rare in most caucasian populations but more common in black populations. Since the condition is sex-linked it is more prevalent in males, although heterozygous females can be shown to have a deficiency of dehydrogenase enzyme.

F^+ cell, *n.* a bacterial cell having an F FACTOR.

F^- cell, *n.* a bacterial cell that receives the F FACTOR from an F^+ cell during conjugation, becoming F^+ in status.

feather, *n.* any of the flat light waterproof epidermal structures forming the plumage of birds, several types of which form the body covering of birds. The principal types of feather are: *remige* (wing feather); *rectrice* (tail feather); CONTOUR FEATHER (covering the outside of the bird); DOWN (the soft covering to the body); *filoplume* (hairlike feathers occurring between the contour feathers). Feathers consist of a central RACHIS that supports BARBS which, except in down feathers, are connected to form a lamella by means of BARBULES.

Fechner's law, see WEBER–FECHNER LAW.

fecundity, *n.* the numbers of young produced by an organism during the course of its life. Compare FERTILITY (3). See REPRODUCTIVE POTENTIAL.

feedback mechanism, *n.* a mechanism by which the products of a process can act as regulators of that process. Many biochemical processes are controlled by *negative feedback* mechanisms. See Fig. 154.

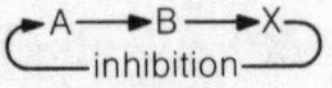

Fig. 154. **Feedback mechanism.** Excess production of X causes a shutdown of the process by inhibition of the first step from A to B.

Mammals utilize negative feedback mechanisms to maintain HOMEO-STASIS in several systems, for example blood thyroxine levels, body temperature, blood osmotic pressure. Occasionally *positive feedback* occurs, which is a disruptive process where products of a process cause further ACTIVATION, for example, excess core heat in mammals will in turn encourage a higher metabolic rate which produces yet more heat and eventually death.

feeding phase, *n.* that phase of the life history during which food is taken in. For example, the nymphal stage in mayflies.

Fehling's test, *n.* a procedure used to detect the presence of REDUCING SUGARS in an unknown solution. Two solutions are mixed: Fehling's solution A containing copper tartrate (Cu(II)), and Fehling's solution B containing sodium hydroxide. The mixture is then added to the test solution and boiled. An orange-red precipitate of copper oxide (CU(I)) indicates the presence of a reducing sugar.

femur, *n.* **1.** the thigh bone of TETRAPODS. **2.** the insect leg joint between the trochanter (2nd segment) and tibia (4th segment).

fen, *n.* a plant community on alkaline, neutral or slightly acid peat that is wet and usually low-lying.

fenestra ovalis and rotunda, *n.* the membrane covered 'windows' leading from the middle ear to the inner ear. See EAR.

fenestration, *n.* the presence in an organism of window-like openings, as in the palate of marsupials.

feral, *adj.* (of plants and animals) existing in a wild state, outside human cultivation or habitation.

fermentation, 1. see ALCOHOLIC FERMENTATION. **2.** any industrial process involving the large-scale culturing of cells in either aerobic or anaerobic conditions, using fermenters.

fern, *n.* any pteridophyte plant of the class Filicinae, subdivision Pteropsida of the division Tracheophyta, at one time classified in the division Pteridophyta. The ferns constitute the great majority of species in the division and possess large, conspicuous aerial DIPLOID (2) stems.

Sporangia are borne on the underside of leaves and the HAPLOID (2) spores usually give rise to homosporous prothalli which carry both ANTHERIDIA and ARCHEGONIA. There is a small group of heterosporous aquatic ferns (see HETEROSPORY).

ferredoxin, *n*. an important iron-containing protein acting as an electron carrier in the ELECTRON TRANSPORT SYSTEM that operates in the LIGHT REACTIONS of PHOTOSYNTHESIS, particularly in NONCYCLIC PHOTOPHOSPHORYLATION.

ferritin, *n*. a conjugated, electron-dense protein concerned in the absorption of iron through the intestinal mucosa. It serves as a storage protein for iron in the liver and spleen.

fertilization, *n*. the fusion of male and female GAMETES to give rise to a ZYGOTE which then subsequently develops into a new organism. See ACROSOME for further details of animal fertilization. See EMBRYO SAC for details of the 'double' fertilization of flowering plants.

fertilization membrane, *n*. a membrane that appears at the surface of the egg after FERTILIZATION. It is effectively a thickened VITELLINE MEMBRANE that may separate from the surface of the egg; it prevents the entry of additional sperms.

fertilizer, *n*. a fertilizing agent added to agricultural soils, in the form of an added chemical, or manure.

fertility, *n*. **1.** the readiness with which the gamete of an organism may fuse with a GAMETE of the other sex. **2.** the capability of a sperm or ovum of, in the case of the egg, being fertilized, or, in the case of the sperm, of fertilizing an egg and so giving rise to a viable ZYGOTE. **3.** the capability of an organism, particularly a hybrid (see HETEROSIS), of producing more fertilized ova than others. Where such organisms give rise to more young, they are also more fecund (see FECUNDITY), a term with which fertility is often confused. **4.** a measure of the productiveness of soil.

fertility factor, see F FACTOR.

Feulgen, Robert (1884–1955) German physiologist and chemist best known for his discovery of a selective staining technique for DNA, the *Feulgen reaction*, which is based upon the reaction of an aldehyde group with leucofuchsin. The intensity of the purple colour in the cell nucleus is used as a measure for the quantitative estimation of nuclear DNA.

Feulgen reagent, *n*. a deep purple dye that stains the deoxyribose sugar of DNA in cell nuclei.

F factor or **fertility factor** or **sex factor,** *n*. an EPISOME that confers donor status on an F^+ CELL, so that the F^+ cell transfers a copy of the F factor and a small piece of bacterial DNA to an F^- CELL during CONJUGATION, the recipient becoming F^+.

fibre, *n*. a slender fibre, in the form of an element of SCLERENCHYMA in plants, or of COLLAGEN, RETICULIN or ELASTIN in animals.

fibril, *n.* a small FIBRE.

fibrillar muscle, see FLIGHT.

fibrillation or **ventricular fibrillation,** *n.* very rapid, irregular contractions of the ventricular muscle of the HEART that cause blood circulation to stop immediately. Under suitable circumstances, ventricular fibrillation can be stopped by passing an electric current through the heart, using a cardiac defibrillator.

fibrin, see FIBRINOGEN.

fibrinogen or **fibrin,** *n.* a large, soluble protein found in BLOOD PLASMA that is formed in the liver and is converted to insoluble fibrin during the process of BLOOD CLOTTING.

fibroblast, *n.* a connective tissue cell which may differentiate into CHONDROBLASTS, COLLAGENOBLASTS or OSTEOBLASTS.

fibrocartilage, see CARTILAGE.

fibrous protein, see PROTEIN.

fibrous root system, *n.* a root system composed mainly of branches rather than of one principal root. Such fibrous systems are found in many herbaceous PERENNIALS, especially the grasses.

fibula, *n.* a bone present in the posterior limb of TETRAPODS lying slightly posterior to and parallel with the tibia. See PENTADACTYL LIMB.

fight–or–flight reaction, *n.* a defence reaction or alerting response in higher animals, involving an increase in blood pressure and heart rate and a redistribution of the blood away from the viscera towards the STRIATED MUSCLE. These changes result from integrated nerve pathways in the brain known as defence centres, and from the secretion of ADRENALINE.

filament, *n.* **1.** the stalk of a STAMEN bearing the ANTHER at its apex. **2.** a type of cellular organization consisting of a threadlike row of cells, as found in certain algae, for example, *Spirogyra*.

filarial worm, *n.* a NEMATODE worm that is parasitic on vertebrate animals and which has an intermediate ARTHROPOD host. For example, adults of the nematode worm *Wucheria bancrofti* obstruct lymph vessels in man and cause the disease ELEPHANTIASIS, with nocturnal mosquitoes as the intermediate host.

fili– or **filo–,** *prefix.* denoting a thread.

Filicales, *n.* the order that includes the FERNS in the old classification of PTERIDOPHYTES.

Filicinae, see FERNS.

filiform, *adj.* thread–like.

filter feeder, *n.* any marine or freshwater animal that feeds on microscopic organisms and creates currents, usually by ciliary action, so that food particles are carried into the body cavity, as occurs, for example in the sea-squirt. In other filter feeders, food particles are carried across the gills where they are trapped in MUCUS which is carried, again by ciliary action, into the entrance to the gut system, for example, the freshwater clam *Anodonta*. Filter feeders are described as MICROPHAGOUS.

fimbriate, *adj.* (of plant structures) having a fringe on the margin.

fin, *n.* a flattened limb found in aquatic animals, and used for locomotion.

fin rays, *n.* the skeletal structures that give rise to the shape of fins in fish. They may be cartilaginous, bony or fibrous.

first-division segregation, *n.* the segregation of a pair of different alleles of a gene into different nuclei at the first division or MEIOSIS. In TETRAD ANALYSIS of an ASCUS such segregation would be detected by four adjacent ascospores carrying one allele type and the other four ascospores carrying the other type.

first law of thermodynamics, see THERMODYNAMICS.

first-order reaction, *n.* a reaction in which the rate of reaction is directly proportional to the concentration of one of the reactants, either product or substrate.

fish, *n.* any of a large group of cold-blooded, finned aquatic vertebrates. Fish are generally scaled, and respire by passing water over gills. Fish were formerly placed in a single grouping, class Pisces. It is now recognised that there are four distinct classes, ACTINOPTERYGII, (ray-finned fishes), CHOANICHTHYES (fins with central skeletal axis – collectively sometimes classed as Osteichthyes, see BONY FISH) CHONDRYCHTHYES (sharks) and APHETOHYIDEAN (extinct, primitive, jawed fish).

fission, see BINARY FISSION.

fitness, *n.* the ability of an organism to transfer its genes to the next generation. Organisms favoured by SELECTION (natural or artificial) have a high fitness, while those subjected to adverse selection pressure have a low fitness. Thus, under conditions of insecticide treatment, resistant members of an insect population will have a high fitness and produce more offspring as compared to susceptible individuals that have a low fitness.

fixation, *n.* a genetical situation where all members of a population are HOMOZYGOUS for one particular ALLELE of a gene, so that no alternative alleles of that LOCUS exist in the population. Fixation often occurs in small populations. See also EXTINCTION (2).

Flagellata, see MASTIGOPHOHARA.

flagellate, *n.* any organism carrying a FLAGELLUM.

flagellum, *n.* (*pl.* flagella) a fine, hair-like process of a cell, associated with locomotion in unicellular organisms. It is similar in structure to the

CILIUM; however, it can be distinguished from cilia by its occurrence in smaller numbers and by generally being longer. Flagella occur in the flagellates, in most motile gametes, in ZOOSPORES, and occasionally in METAZOANS, as in the ENDODERM of some COELENTERATES. Its structure is similar to that of the CILIUM (see Fig. 103).

flame cell, *n.* a specialized cell that contains a central cavity with several beating cilia. These create a current in the tubules to which they are connected. The tubules usually open to the exterior and flame cells control the water content of the body. The flame cell and tubule is known as a *protonephridium* and is found in PLATYHELMINTHS, NEMERTINES, rotifers, ANNELIDS, the larvae of molluscs and AMPHIOXUS. See Fig. 155.

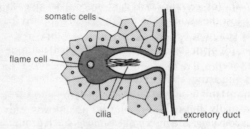

Fig. 155. **Flame cell.** General structure.

flatworm, see PLATYHELMINTH.

flavin adenine dinucleotide, see FAD.

flavin mononucleotide, see FMN.

flavoprotein, *n.* a protein that is combined with flavin prosthetic groups and acts as an intermediate carrier in respiratory chains between dehydrogenases and cytochromes. See ELECTRON TRANSPORT SYSTEM.

flea, *n.* any small wingless parasitic blood-sucking insect of the ENDOPTERYGOTE order Aphaniptera (Siphonaptera).

Fleming, Sir Alexander (1881–1955) Scottish bacteriologist and Nobel prizewinner (1945) who discovered the ANTIBIOTIC penicillin (1928).

flexor, *n.* any muscle that brings about the bending of a limb.

flexuous, *adj.* (of a stem) bending or wavy.

flight, *n.* any locomotion through air, either active or passive (gliding). Active flight is brought about by the movement of wings by muscles as in birds and insects; gliding involves a minimum of muscular effort and is found only in some larger birds and certain mammals adapted for flight, such as the flying lemur or flying fox.

In birds, muscles are attached directly to the wings and are of two main types: *depressor muscles* which produce the downstroke and run from the humerus to the STERNUM, and *elevator muscles* which produce the upstroke and are attached to the upper surface of the HUMERUS by a tendon which runs through the pectoral girdle to the sternum.

In insects such as bees, wasps, flies, beetles and bugs, the muscles raising and lowering the wing are attached to the walls of the thorax (indirect flight muscles) and not to the wings and are called *asynchronous fibrillar muscles*. Direct flight muscles attached to the wings alter the angle and adjust the wings to the resting position. In other insects, for example, the dragonflies, the flight muscles are called *synchronous muscles*, being attached directly to the wings. Asynchronous wing beats are much slower than synchronous ones.

flight feathers, see CONTOUR FEATHERS.

flightless, *adj.* (of certain birds and insects) unable to fly, having secondarily lost the ability. The term is usually applied to birds such as RATITES and penguins.

floccose, *adj.* (of plant structures) covered in small hairs giving a down-like appearance.

flocking, *n.* the active coming together of animals to form a flock. In most organisms this usually takes place outside the breeding season. The term is normally limited to birds and mammals. Fish are usually described as *schooling*, and insects as *swarming*. Such grouping may serve as a defence against predators, or as a means of transferring information on food supplies, where one member of the group may follow another to a known good feeding area.

flora, *n.* **1.** the plant life characteristic of a particular geographical area. **2.** a botanical manual from which plants can be identified by the use of KEYS.

floral diagram, *n.* a diagram illustrating the number and position of the parts in each set of organs which goes to make up the flower. See FLORAL FORMULA.

floral formula, *n.* an expression giving the information set out in the FLORAL DIAGRAM. For example, a buttercup has the formula $K_5C_5A\infty\bar{G}\infty$, indicating five sepals in the calyx (K), five petals in the corolla (C), and the androecium (A) and gynoecium ($\bar{G}$) of an indefinite number of stamens and carpels. $\bar{G}$ indicates a superior GYNOECIUM.

florigen, *n.* a hypothetical plant hormone that may be produced in leaves after suitable light treatment (see PHOTOPERIODISM), and moves to the buds to stimulate flowering.

floristics, *n.* the branch of botany concerned with the study of vegetation in terms of the number of different species present in the flora.

flower, *n.* the sexual reproductive structure of ANGIOSPERMS, consisting usually of four types of organs: SEPALS, PETALS, STAMENS and one or more CARPELS. See Fig. 156. Some plants have unisexual flowers. When these are found on the same plant, the plant is described as MONOECIOUS; when on different plants, the plants are called DIOECIOUS. Many plants produce their flowers in clusters called INFLORESCENCES.

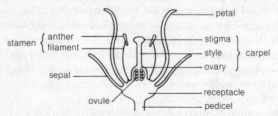

Fig. 156. **Flower.** A vertical section of a typical flower.

fluid-mosaic model, *n.* a hypothetical model of the structure of CELL MEMBRANES as observed with the ELECTRON MICROSCOPE. The model proposes two phospholipid layers with proteins sandwiched in the middle and sometimes projecting into the outer layer in a haphazard 'mosaic' pattern. See Fig. 157. The membrane structure is not static: lipid molecules can move laterally (due to weak bonds between molecules) as can the protein molecules, although to a lesser extent. Some membrane proteins function as carriers in ACTIVE TRANSPORT. See UNIT-MEMBRANE MODEL.

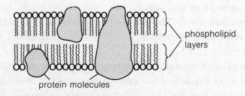

Fig. 157. **Fluid-mosaic model.** Cell membrane structure.

fluke, *n.* any parasitic flatworm, such as the BLOOD FLUKE or liver fluke, that inhabits the organs of vertebrates and in some cases causes serious disease. See BILHARZIA.

fluorescence, *n.* the property of giving out light when molecules are excited by incident light. Emitted light is always of a shorter wavelength than the incident light.

fluorescent antibody technique or **immunofluorescence,** *n.* a technique used to show up the presence of a particular ANTIGEN, in which an antibody is labelled with an ultraviolet fluorescent substance that combines with the antigen and is subsequently located by its FLUORESCENCE.

fluoridation, *n*. the addition of a fluoride, usually sodium fluoride, to drinking water in a concentration of about 1 ppm, in order to reduce the decay of teeth. Teeth may also be treated directly with a fluoride gel by the dentist, and this is usually undertaken as a treatment for children.

fluoride, *n*. a compound of fluorine that replaces hydroxyl groups in teeth and bones and reduces the tendency to tooth decay. Its therapeutic use was discovered accidentally at Bauxite, Arkansas, when water containing fluoride was replaced by water lacking fluoride, resulting in an increase of dental cavities in children. See FLUORIDATION.

flush, *n*. wet ground (often on hillsides) that is typified by the presence of *Sphagnum* moss, where water comes to the surface but does not form a stream bed.

flux, *n*. the rate of flow of matter or energy.

fly, see DIPTERAN.

flying fish, *n*. a species of fish (*Cypselurus californicus*) in which the pectoral fins are enlarged and act as aerofoils when the fish leaps out of the water, enabling it to glide up to 50 m, possibly to escape predators.

FMN (flavin mononucleotide), *n*. a COENZYME produced when RIBOFLAVIN is phosphorylated. It is necessary in the biosynthesis of fats.

focus or **primary focus,** *n*. an area containing a high concentration of diseased plants or animals and from which the disease probably spreads.

foetal membranes, *n*. the EXTRAEMBRYONIC MEMBRANES of the mammalian FOETUS.

foetus, *n*. the EMBRYO of mammals at the time when it achieves the main features of the adult form. This is usually at the time of the formation of the AMNION which is after about eight weeks of pregnancy in women. See AMNIOCENTISIS and Fig. 29.

foliar feeding, *n*. a method of supplying plants with nutrients by spraying an aqueous solution of ESSENTIAL ELEMENTS and other nutrients (e.g., urea) onto the leaves where they are absorbed through the CUTICLE.

folic acid or **vitamin M or vitamin B$_C$,** *n*. a member of the B-COMPLEX group of vitamins that is synthesized by microorganisms in the mammalian gut, but is also required in the normal diet. Folic acid is involved in the synthesis of NUCLEIC ACIDS as well as red blood cells, and a deficiency causes reduced growth and anaemia.

follicle, *n*. **1.** any small cavity or sac (see GRAAFIAN FOLLICLE, HAIR FOLLICLE, OVARIAN FOLLICLE). **2.** a dry fruit formed by a single carpel splitting along a line, usually ventral, to liberate its seeds.

follicle–stimulating hormone (FSH), *n*. a glycoprotein secreted by the anterior lobe of the PITUITARY GLAND of vertebrates. The hormone stimulates growth of the OVARIAN FOLLICLES and OOCYTES in the OVARY, and SPERMATOGENESIS in the seminiferous tubules of the TESTIS.

follicular phase, *n.* that part of the OESTROUS CYCLE in which GRAAFIAN FOLLICLES are formed and their secretions start.

fomite, *n.* any inanimate object via which pathogenic organisms may be transferred, although it does not support their growth, for example, a book.

fontanelle, *n.* a gap in the skull bone where the brain is covered only by skin. Human babies possess a parietal fontanelle between frontal and parietal bones on top of the head.

food, *n.* a substance containing or consisting of chemicals which can be used in the body of an organism to build structures and provide energy to sustain life. Most plants are AUTOTROPHS and require only ESSENTIAL ELEMENTS in their food, but animals are HETEROTROPHS and their food must contain certain carbohydrates, fats and proteins as well as vitamins and essential elements.

food chain, *n.* a sequence of organisms arranged in such a way that the second grouping, e.g., HERBIVORES, feeds on the first, e.g., PRIMARY PRODUCERS, and the third, e.g., CARNIVORES, feeds on the second. See Fig. 158.

primary producer $\longrightarrow$ primary producer $\longrightarrow$ secondary consumer
(P_1)   (herbivore: H_1)   (carnivore: C_1)

Fig. 158. **Food chain.** A simple food chain.

Each organism in the chain feeds and derives energy from its predecessor in the chain, the relative energy value at each TROPHIC LEVEL often being assessed by the construction of ECOLOGICAL PYRAMIDS. In practice, such simple systems rarely exist and the nutritional relationships between organisms normally constitute a *food web*. See Fig. 159.

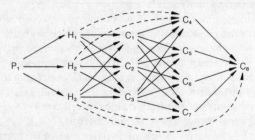

Fig. 159. **Food chain.** A complex *food web* of producer (P), herbivore (H) and carnivore (C).

food pollen, *n.* infertile pollen (produced by special ANTHERS), that attracts insects to the flower where they also collect fertile pollen.

food production, *n.* the growth of animal and plant material, and subsequent cropping for human consumption.

food test, *n.* a method by which certain components of a food mixture can be detected. For example, FEHLINGS TEST for reducing sugars.

food poisoning, *n.* an ACUTE (2) disorder of the gut caused by food contaminated with bacteria or their toxins (e.g. BOTULISM) or by some chemical.

food vacuole, *n.* a fluid-filled space in the cytoplasm of a cell. It is formed by the plasma membrane of the cells invaginating to form a flask-like depression around a food particle and then the neck of the flask closing off, so sealing the food within the cell – a process known as PHAGOCYTOSIS.

food web, see FOOD CHAIN.

foot, *n.* **1.** the part of the leg of vertebrates that contacts the ground in the standing position. **2.** the locomotive organ in invertebrates, for example, the foot of a mollusc, or the tube feet of an echinoderm.

foramen, *n.* an opening.

foramen ovale, *n.* an embryonic opening in the septum that divides the left and right atria of the mammalian heart, providing a bypass to the lung circulation. Failure to close the aperture at birth can lead to a BLUE BABY condition.

Foraminifera, *n.* amoeboid PROTOZOANS that possess chitinous, calcareous or siliceous shells that are usually many-chambered. Calcareous Foraminifera are the main constituent of chalk.

forb, *n.* any herbaceous plant other than grass.

Ford, E.B. (1901–) English entomologist famous for his work on the genetics of LEPIDOPTERA and GENETIC POLYMORPHISM.

forebrain, *n.* that part of the brain which gives rise to the CEREBRAL HEMISPHERES, olfactory lobes, the PINEAL GLAND, PITUITARY GLAND and the OPTIC CHIASMA (which carries nerve fibres from the eyes to the midbrain). It is marked by a constriction in the developing embryo where the brain appears to be three-lobed, the forebrain, MIDBRAIN and HINDBRAIN.

foreskin, *n.* the glandular and vascular fold of skin surrounding the tip of the mammalian penis.

forespore, *n.* a stage in the process of SPORULATION that can be identified as a refractile body not yet resistant to heat.

forest, *n.* an extensive wood or tract of wooded country.

form, *n.* **1.** a neutral term for (a) a uniform sample of individuals or (b) any taxon. **2.** the place in which hares lie and may give birth to young.

formation, *n.* any of the main natural vegetation types extending over a large area that is created by the nature of the climate. Examples include tundra, steppe, rainforest and coniferous forest.

formenkreiss, *n.* an aggregation of ALLOPATRIC SPECIES or SUB-SPECIES.

forward mutation, *n.* a genetic change in which a WILD TYPE piece of DNA is altered to a MUTANT sequence. Compare REVERSE MUTATION.

fossa, *n.* a depression, pit or cavity.

fossil, *n.* the remains of a once-living organism preserved on the rock strata.

fossil record, *n.* the remains of organisms, or traces of their existence (such as footprints), present in the rock strata and forming a history of the development of life from its origins on earth. The record is very largely incomplete due to the comparative rarity with which fossils are formed, but provides evidence of evolution having taken place, particularly where long series of a particular form can be traced over an extended period of time.

founder effect, *n.* the result of starting a new population with a low number of individuals (founders), so that their GENE POOL may not contain the same proportions of ALLELES for a particular LOCUS as in the original population. For example, instead of containing three alleles of the ABO BLOOD GROUP locus, Australian aborigines contain no B alleles and thus no Group B or Group AB individuals are produced, a situation probably caused by a 'founder effect'. Such small founder populations are subject to RANDOM GENETIC DRIFT.

fovea, *n.* an area in the centre of the retina of the eye in which CONES are concentrated and RODS are absent (see Fig. 151). The fovea takes the form of a shallow pit about 1 mm in diameter in man, there being no layer of nerve fibres over it as there is in the rest of the retina. It occurs in primates, diurnal birds and lizards. It is an area of acute vision, and in BINOCULAR VISION the image of the object observed is focused on both foveae.

Fox, Sidney W. (b. 1912) American biochemist and educator, famous for his work on the origin of life. His work was mainly on proteins, their structure, evolution and synthesis, particularly in connection with the role they played in the initiation of life on earth.

fragmentation, *n.* a type of ASEXUAL REPRODUCTION found in some lower plants (such as *Spirogyra* or blue-green algae) where there is separation of the threadlike FILAMENTS (2), each segment of which is capable of growing into a mature organism. Fragmentation is caused by such mechanical forces as wind, wave action, rain, etc.

frameshift, *n.* an altered reading of the GENETIC CODE during the TRANSLATION stage of PROTEIN SYNTHESIS caused by a change in the nucleotide base sequence of DNA and RNA due to an insertion or deletion in a POINT MUTATION. Fig. 160 shows an example of a DNA base insertion between bases 4' and 5'. When the sequence is read from left to right the first amino acid is unchanged, but the second and subsequent amino acids will be altered, producing a mutant protein.

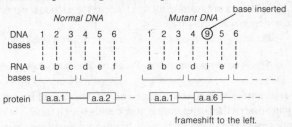

Fig. 160. **Frameshift.** The effect of inserting an extra DNA base into the nucleotide base sequence.

fraternal twins, see DIZYGOTIC TWINS.

free energy, *n.* the amount of energy that is available for work when released in a chemical reaction. For example, when a molecule of ATP is hydrolysed to ADP + P, the free energy released is about 34 kJ.

freemartin, *n.* a female twin born with a male, in which normal female sexual development has been adversely affected by hormones produced by the male twin.

freeze–drying, *n.* a process by which biological material can be preserved by removal of water under a vacuum while the material is frozen.

freeze–etching, *n.* the process by which a piece of tissue is frozen and then sectioned for examination under the ELECTRON MICROSCOPE.

frequency-dependent selection, *n.* any SELECTION in which the FITNESS of genotypes is directly related to the proportions of the various PHENOTYPES present in a population, so that the frequency of the more common types is decreased and the less common types is increased. Such selection pressure often produces a stable GENETIC POLYMORPHISM and when it involves predation is referred to as APOSTATIC SELECTION.

frequency distribution, *n.* an arrangement of statistical data in order of the frequency of each size of the variable. For example, the numbers: 2, 3, 5, 3, 4, 2, 1, 3, 4 would have the frequency distribution shown in Fig. 161 on page 234. Data from a large sample often produces a NORMAL DISTRIBUTION CURVE.

freshwater, *n.* an aquatic environment such as streams, rivers, lakes, etc. where there is little dissolved mineral matter and which results directly from precipitation (rain), as opposed to saltwater (the seas and oceans).

Frisch, Carl von (1886–) Austrian behaviourist famous for his researches on the language, orientation and direction finding of bees. He

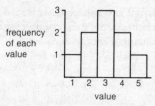

Fig. 161. **Frequency distribution.** Based on 9 numbers (see main entry).

also carried out extensive research on the colour sense of bees and fishes and their senses of hearing, smell and taste.

frog, see ANURAN.

frond, *n*. **1.** the megaphyllous 'leaf' of a fern, which may be subdivided into pinnae and pinnules. **2.** the thallus of a seaweed or a lichen.

frontal bone, *n*. one of a pair of dermal bones present at the front of the head, such as the human forehead.

frontal lobe, *n*. the anterior part of each cerebral hemisphere in mammals.

fructose, *n*. a carbohydrate that is considered to be an atypical KETOSE sugar because it acts as a reducing sugar in FEHLING'S TEST. See MONOSACCHARIDE and Fig. 218.

fruit, *n*. a plant structure consisting of one or more ripened ovaries (with or without seeds) together with any flower parts which may be associated with the ovaries. Many fruits such as the berry and drupe are succulent, but often they contain no fleshy tissue (e.g. ACHENE, NUT). See also PERICARP.

fruit fly, see DROSOPHILA.

fruiting body, *n*. a special fungal structure that bears sexually or asexually derived spores. The HYPHAE and the spores borne by them constitute the structure that has a specific name in different groups of fungi. Examples include ascocarp, conidiophore, coremium, pycnidium.

frustule, *n*. the hard, silica-containing wall of a DIATOM.

FSH, see FOLLICLE-STIMULATING HORMONE.

fucoxanthin, *n*. a brown CAROTENOID pigment found in brown algae and diatoms that is present together with CHLOROPHYLL *c*.

Fucus, *n*. a genus of algae, typically having greenish-brown slimy FRONDS (2), such as *Fucus vesiculosus*, the bladderwrack.

fugacious, *adj*. (of plant organs) withering and falling quickly.

fungicide, *n*. any substance such as captan or benomyl that kills FUNGI.

Fungi Imperfecti or **imperfect fungi** or **Deuteromycotina,** *n*. an artificial grouping of fungi with no known perfect (sexual) stages. The group apparently do not produce reproductive structures of any kind, but merely fragment. Imperfect fungi contain stages of some ASCOMYCETES and BASIDIOMYCETES.

fungistasis, *n.* the inhibition of further fungal growth.

fungus, *n.* (*pl.* fungi) an organism that may be unicellular or made up of tubular filaments (HYPHAE) and lacks CHLOROPHYLL. Fungi live entirely as SAPROPHYTES or PARASITES. There are two main divisions: the EUMYCOTA, the 'true fungi' and the MYXOMYCOTA, the slime moulds, each of which is now given the status of division, though in older classifications both were grouped with algae in the Thallophyta.

funicle, *n.* **1.** the stalk of an OVULE by which it is attached to the placenta in plants. **2.** the joints of the antenna of some HYMENOPTERA between the second joint and the distal joint or club.

furanose ring, *n.* a monosaccharide having a five-membered ring structure.

furca, *n.* any fork-like structure, such as that formed by the last two abdominal segments in some arthropods or which supports the two spongy pads on the proboscis of some insects.

furcula, *n.* **1.** the wishbone (clavicle) of birds. **2.** the forked springing organ of COLLEMBOLA.

fusiform, *adj.* (of plant structures) spindle-shaped.

G

G1 stage, *n.* the first growth phase of the CELL CYCLE.

G2 stage, *n.* the second growth phase of the CELL CYCLE.

gain, *n.* **1.** (in physiological experiments) an increase in signal as a result of amplification. **2.** a shift in the progeny mean relative to the mean of the original population brought about by DIRECTIONAL SELECTION. Gain is used to estimate the HERITABILITY of characters.

galactosaemia, *n.* a rare INBORN ERROR OF METABOLISM in which the breast-fed human infant is literally poisoned by the mother's milk. The affected individuals are unable to metabolize the milk sugar GALACTOSE, which normally is converted to glucose ready for oxidation and the release of energy. Instead, affected infants store the galactose in various tissues including the brain, resulting in severe malnutrition along with mental retardation.

Galactosaemia is due to blockage of the step galactose-1-phosphate to glucose-1-phosphate because the enzyme uridyltransferase is absent or inactive. The condition is controlled by an autosomal gene probably on chromosome 9, affected individuals being homozygous for the recessive alleles. The transferase enzyme can be detected in foetal cells from AMNIOCENTESIS before birth and, if newborn infants are given a special diet, development will be normal.

galactose, *n.* a monosaccharide CARBOHYDRATE that does not occur freely in nature, but is combined with GLUCOSE to form LACTOSE, a disaccharide sugar found in milk.

galactosidase, *n.* the inducible enzyme that catalyses the hydrolysis of galactosides (see GLYCOSIDE) formed from the reaction of GALACTOSE with alcohol.

Galapagos Islands, *n.* a group of 15 islands in the Pacific Ocean situated on the equator some 900 miles west of Ecuador. In 1835, Charles DARWIN visited the islands during his voyage on the Beagle and the fauna particularly influenced his views on evolution. According to his theory, GEOGRAPHICAL ISOLATION has influenced divergence of different forms so that, for example, DARWIN'S FINCHES have all probably evolved from a common ancestor, and now occupy numerous niches on the islands due to the absence of competition from other birds. Marine iguanas and giant tortoises are other interesting animals found there, and in the absence of palms and conifers, the normally small prickly-pear cactus has reached tree-like proportions.

galea *n.* the outer lobe of the MIXILLA in insects.

gall, *n.* an abnormal growth of plant tissue caused by insects, mites, eel worms or fungi.

gall bladder, *n.* a bag-like reservoir (of about 50 cm³ capacity in man) that lies at the edge of the liver closest to the gut and whose function is to store bile produced by the liver. The contents of the gall bladder are squirted into a gut lumen under the influence of the hormone CHOLECYSTOKININ–PANCREOZYMIN. See BILE and Fig. 64.

gallstones, *n.* calcareous concretions formed in the GALL BLADDER.

gametangium, *n.* the organ in which the GAMETES are produced in lower plants.

gamete or **germ cell,** *n.* a specialized HAPLOID cell that fuses with a gamete from the opposite sex (or mating type) to form a diploid ZYGOTE. In simple organisms, the process is called isogamy (see ISOGAMETE), and OOGAMY in more complex organisms. In animals where oogamy occurs male gametes are called sperm, the female gametes eggs. The situation in higher plants is more complicated, but essentially the male gamete is the generative nucleus found in the POLLEN GRAIN while the female gamete is the egg cell found within the EMBRYO SAC.

gameto- or **gamo-,** *prefix.* denoting germ cells.

gametocyte, *n.* an animal or plant cell that develops into gametes by meiosis. See GAMETOGENESIS.

gametogenesis, *n.* the process in which haploid GAMETES are produced from diploid cells by meiotic division (MEIOSIS). See Fig. 162 for animal cells and Fig. 163 for plant cells.

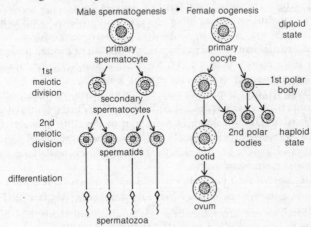

Fig. 162. **Gametogenesis.** In animals.

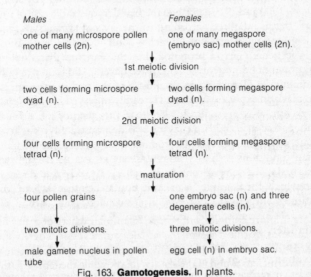

Fig. 163. **Gamotogenesis.** In plants.

gametophyte, *n.* that part of the life cycle of plants which has HAPLOID nuclei and gives rise to the sex cells which on fusing produce a DIPLOID stage, usually the SPOROPHYTE. See ALTERNATION OF GENERATIONS.

gametopetalous or **sympetalous,** *adj.* (of flowers) having a structure in which the petals are fused to form a tube–like shape, as in the primrose and convolvulus. Other flowers, such as the buttercup and tulip, have petals which are completely free, i.e. *polypetalous*.

gamma globulin, *n.* a member of a group of proteins found in BLOOD PLASMA that may act as an ANTIBODY. See IMMUNOGLOBULIN.

gamma radiation, *n.* electromagnetic radiation of shorter wavelength and higher energy than X-RAYS. See ELECTROMAGNETIC SPECTRUM.

gamosepalous, *adj.* (of flowers) having united SEPALS.

ganglion, *n.* (*pl.* ganglia) a structure composed of the CELL BODIES of neurones that in vertebrates is usually found outside the central nervous system (CNS), but which may form part of the CNS in invertebrates where such structures are usually well-developed in the head region. In vertebrates, ganglia are found in the PERIPHERAL nervous system and the AUTONOMIC NERVOUS SYSTEM.

ganoid, *adj.* (of fish) possessing ganoid scales, typified by a hard shiny layer of ganoine (an enamel–like substance) and the fact that they increase in thickness by adding layers all round, ganoine above, laminated from below. Compare COSMOID, PLACOID.

Garrod, Archibald, E. (1857–1936) English hospital physician who between 1900 and 1910 described the first demonstrable case of a human disease that is inherited according to the laws of MENDELIAN GENETICS. The condition is *alkaptonuria*, which is controlled by a single autosomal recessive gene. Garrod proposed that the enzyme that catalyses the breakdown of homogentisic acid to acetoacetic acid is nonfunctional in alkaptonurics. This results in a build–up of the acid in the urine, which turns black on exposure to air, and is readily observed in infant's nappies. It was not, however, until 1958 that the absence of a functional homogentisic–acid oxidase enzyme was demonstrated in the liver of a patient with alkaptonuria. Garrod went on to explain several other human conditions such as ALBINISM as INBORN ERRORS OF METABOLISM.

gas analysis, *n.* a comparison of atmospheric air and expired respiratory air, in which, for example, in a resting human being at sea level, oxygen constitutes only 16.4% (20.95% in atmospheric air), carbon dioxide 4.1% (0.04 in atmosphere) and nitrogen 79.5% (79% in atmosphere).

gas bladder, see AIR BLADDER.

gas carriage, *n.* see GAS EXCHANGE.

gas exchange or **gas carriage,** *n.* the transfer of gases between an organism and the environment. In RESPIRATION, oxygen is taken in and carbon dioxide given out. Photosynthesis in plants complicates this system in that during the process carbon dioxide is required by the plant and oxygen given off (see COMPENSATION PERIOD). In plants and small animals such as PROTOZOANS and PLATYHELMINTHS, gas exchange occurs by DIFFUSION. In higher animals, special respiratory surfaces have been developed, for example, internal and external gills, lungs and trachea.

gas loading, see GAS EXCHANGE, OXYGEN DISSOCIATION CURVE.

gasteropod or **gastropod,** *n.* any member of the class Gasteropoda in the phylum Mollusca, including MOLLUSCS such as slugs, snails, pteropods, limpets, winkles, whelks and sea slugs. Some forms lack a shell, but where a shell is present it is in the form of a single valve, often spiral. Marine, freshwater and terrestrial forms occur, and there is usually a distinct head, bearing a pair of tentacles and eyes.

gastric, *adj.* of, or relating to, the stomach.

gastric gland, *n.* any gland in the stomach wall that produces components of the GASTRIC JUICE.

gastric juice, *n.* the fluid secreted by glands of the stomach, containing PEPSIN, RENIN, and hydrochloric acid.

gastric mill, *n.* a structure in the proventriculus or stomach of CRUSTACEANS, formed of a series of cuticular teeth which assist in breaking down food.

gastrin, *n.* a hormone produced by gastrin cells of the pyloric gland, which induces gastric secretion.

gastroenteritis, *n.* an inflammation of the intestinal tract, resulting in diarrhoea, vomiting and nausea.

gastrointestinal tract, see ALIMENTARY CANAL.

gastrolith, *n.* a mass of CALCAREOUS material occasionally found in the proventriculus of crustaceans. It is probably formed as a result of calcium withdrawal from the exoskeleton prior to moulting.

gastropod, see GASTEROPOD.

gastrotrich, *n.* any minute aquatic multicellur animal of the phylum Gastrotricha, comprising unsegmented, worm-like organisms whose locomotion is brought about by epidermal cilia. They have some affinities with ROTIFERS and NEMATODES.

gastrozooid, *n.* a feeding polyp in colonial COELENTRATES.

gastrula, *n.* a stage in embryonic development in which the BLASTULA has invaginated, so giving rise to a two-layered embryo by a process of gastrulation. See ARCHENTERON.

gastrulation, *n.* the process by which the BLASTULA forms the GASTRULA.

Gause's Law, *n.* a law stating that no two species with identical ecology can exist together in the same environment. Named after the German anatomist G.F. Gause.

Gaussian curve, see NORMAL DISTRIBUTION CURVE.

G banding, *n.* a method of treating chromosomes with Giemsa stain to show areas of the chromosomes with light and heavy staining. Such patterns are different for each chromosome type and thus are most useful when arranging chromosomes in KARYOTYPE analysis.

gel, *n.* a semi–rigid COLLOID as distinct from the more liquid SOL.

gelding, see EMASCULATION.

gemma, *n.* (*pl.* gemmae) a small group of cells that serves as a means of vegetative reproduction in some mosses and liverworts. These cells become detached from the main THALLUS and often occur in cup–like structures referred to as *gemma-cups*. Each gemma is capable of developing into a new plant.

gemmation, *n.* **1.** the process of forming a GEMMA or gemmae. **2.** any budding (as in *Hydra*), where a new organism is produced as an outgrowth of the parent.

gemmule, *n.* a bud formed in sponges as an internal group of cells that gives rise to a new sponge after overwintering (in freshwater forms) and the decay of the parent.

gene, *n.* the fundamental physical unit of heredity that transmits information from one cell to another and hence one generation to another. Genes consist of specific sequences of DNA nucleotides which can code for the structure of polypeptide chains (see CISTRON), tRNA molecules and rRNA molecules (see TRANSCRIPTION, TRANSLATION). Individual genes can be recognized through the existence of variable forms, or ALLELES, that form the basis of GENETIC VARIABILITY. Organisms can be considered as gene carriers, being controlled by genes in such a way as to maximize the changes of survival of the genetic material.

gene amplification, *n.* a process in which many copies are made of some genes at one time, while other genes are not replicated. The replicated genes enable enhanced manufacture of product in a short time. For example, rRNA genes are amplified about 4,000 times in oocytes of the clawed toad *Xenopus laevis* to assist in protein production during egg development.

gene bank, *n.* **1.** a collection of clones containing all the genes of a particular organism, such as *E. coli.* **2.** a collection of many lines of a particular crop plant, used as a genetic resource by plant breeders.

gene bridge, *n.* the means by which a plant pathogen can survive between the main host-growing seasons by living on host plants grown out of the normal season. Thus, for example, a cereal pathogen can often transfer in September from a mature, spring-sown host to newly sown winter cereal plants.

gene cloning, *n.* the technique of genetic engineering in which specific genes are excised from host DNA, inserted into a vector plasmid and introduced into a host cell, which then divides to produce many copies (clones) of the transferred gene.

genecology, *n.* the study of plant populations in relation to habitat and genetical structure.

gene exchange, *n.* new combinations of genes in offspring resulting from normal sexual reproduction within a breeding population where genes are contributed by both parents.

gene flow or **gene migration,** *n.* the movement of genes (via GAMETES) from one population to another.

gene-for-gene concept, *n.* the theory that genes for PATHOGEN resistance in a host plant are matched by corresponding virulence genes in the attacking pathogen.

gene frequency, *n.* see ALLELE FREQUENCY.

gene induction, *n.* the activation of an inactive gene so that it can carry out TRANSCRIPTION. See OPERON MODEL.

gene locus, see LOCUS.

gene migration, see GENE FLOW.

gene mutation, see POINT MUTATION.

gene pool, *n.* the sum total of all the genes of all BREEDING INDIVIDUALS in a population at a particular time, represented by their GAMETES. Note that those individuals which are too young to breed (juveniles) or those that have undergone GENETIC DEATH do not contribute to the gene pool.

generation, *n.* **1.** a group of organisms of approximately the same age, usually derived from the same parents, referred to as the first filial generation (F_1). **2.** the act of propagating a species or bringing about the formation of new individuals.

generation time, *n.* **1.** also called **division time**. the time required for a population of cells to double in number. **2.** the time required for a generation of individuals to be born, reach sexual maturity and reproduce. The generation time is a critical factor in population growth rates and efforts are being made (particularly in China) to extend the human generation time so as to slow down the rate of population increase.

generative nucleus, *n.* one of two HAPLOID nuclei found within the POLLEN GRAINS of flowering plants, which enters the pollen tube when it is produced, divides by MITOSIS, and becomes the male gamete nucleus which fuses with the female egg cell at FERTILIZATION.

generator potential, *n.* the nonconducted electrical change that is developed when the sensitive part of a RECEPTOR is stimulated. The magnitude of the potential depends on the intensity of the stimulus, and when it reaches a certain threshold it may give rise to an ACTION POTENTIAL.

gene repression, *n*. the deactivation of an active gene that causes a shutdown of TRANSCRIPTION. See OPERON MODEL.

generic, *adj*. of or belonging to a genus.

gene switching, *n*. the process in which genes are activated or deactivated during development, so that there is a sequential output of gene products, some taking part in current biochemical pathways others acting as inducers and/or repressors of other genes (see JACOB, F). see also CHROMOSOME PUFFS for physical evidence of gene switching.

genetic, *adj*. of or relating to genes.

genetic background, *n*. **1.** any genes other than those of special interest in a study. **2.** any effects in an experiment which can alter the main results in different crosses.

genetic code, *n*. a collection of CODONS of DNA and RNA that contains information about POLYPEPTIDE CHAIN sequencing during TRANSLATION, each triplet usually coding for one AMINO ACID. See FRAMESHIFT and Fig. 160. A triplet code with four letters (Cytosine, Guanine, Adenine and Thymine bases) gives $4^3 = 64$ combinations. Each codon has been shown to have a specific function. See Fig. 164. Note that the code shows DEGENERACY, and that there is one triplet for initiating polypeptide chain production (AUG) and three triplets for terminating production (UUA, UAG and UGA). The code is more or less universal, the same proteins being produced from a particular base sequence in both *E. coli* and mammals.

genetic death, *n*. a sexually mature individual that is alive but unable (or unwilling) to breed and thus transmit his or her genes to the next generation via the GENE POOL. Genetic death is fairly rare in most populations of plants and animals, but in developed countries the amount of genetic death amongst humans is increasing, not only in terms of individuals surviving to old age but also because of those who opt to practice BIRTH CONTROL.

genetic dictionary, *n*. a construction showing the relationship between the triplets in MESSENGER RNA and each amino acid. See GENETIC CODE.

genetic drift, see RANDOM GENETIC DRIFT.

genetic engineering, *n*. the modification of the genetic complement of an organism to the benefit of man. The term applies to a wide range of genetical techniques, for example, plant and animal breeding, to improve physiological performance by SELECTION, but recently the term has become associated with the deliberate transfer of genetic material from one organism to another where it is not normally found. For example, a gene can be removed from human cells and transferred to

Second base

	U	C	A	G	
U	UUU ⎤ Phe UUC ⎦ UUA ⎤ Leu UUG ⎦	UCU ⎤ UCC ⎥ Ser UCA ⎥ UCG ⎦	UAU ⎤ Tyr UAC ⎦ UAA ⎤ Stop UAG ⎦	UGU ⎤ Cys UGC ⎦ UGA Stop UGG Trp	U C A G
C	CUU ⎤ CUC ⎥ Leu CUA ⎥ CUG ⎦	CCU ⎤ CCC ⎥ Pro CCA ⎥ CCG ⎦	CAU ⎤ His CAC ⎦ CAA ⎤ Gln CAG ⎦	CGU ⎤ CGC ⎥ Arg CGA ⎥ CGG ⎦	U C A G
A	AUU ⎤ AUC ⎥ Ileu AUA ⎥ AUG Met (start)	ACU ⎤ ACC ⎥ Thr ACA ⎥ ACG ⎦	AAU ⎤ Asn AAC ⎦ AAA ⎤ Lys AAG ⎦	AGU ⎤ Ser AGC ⎦ AGA ⎤ Arg AGG ⎦	U C A G
G	GUU ⎤ GUC ⎥ Val GUA ⎥ GUG ⎦	GCU ⎤ GCC ⎥ Ala GCA ⎥ GCG ⎦	GAU ⎤ Asp GAC ⎦ GAA ⎤ Glu GAG ⎦	GGU ⎤ GGC ⎥ Gly GGA ⎥ GGG ⎦	U C A G

First base (5′ end) — Third base (3′ end)

Fig. 164. **Genetic code.** The mRNA codons of the genetic code arranged in a GENETIC DICTIONARY (see Fig. 27 for the AMINO ACID abbreviations).

microbial cells (using BACTERIOPHAGE or PLASMID vectors) where the 'foreign' gene can control the formation of useful products. Industrial production of INSULIN and INTERFERON is now possible using such methods. In the near future it may be possible to transfer pieces of microbial genetic material to higher organisms. For example, genes from nitrogen-fixing bacteria could be introduced into the genomes of nonleguminous crop plants to supplement their nutrition.

genetic equilibrium or **equilibrium of population,** *n.* a state that occurs when GENE FREQUENCIES are constant in a population for several generations. Such equilibrium may also be found where SELECTION is operating to produce a stable GENETIC POLYMORPHISM. See also HARDY-WEINBERG LAW.

genetic homeostasis, *n.* the property possessed by a population for resisting sudden change and maintaining a steady genetic composition.

genetic isolation, *n.* a form of REPRODUCTIVE ISOLATION in which hybrids from two different species are unable to produce fertile GAMETES and thus are sterile. For example, a horse crossed with a donkey produces a sterile mule. See ISOLATION.

genetic linkage, *n.* the association between genes located (linked) on the same chromosome, thus producing proportions of gametes that are not those expected by INDEPENDENT ASSORTMENT although, unless there is very close linkage, the same TYPES of gamete will be produced. Instead,

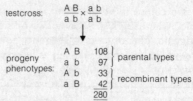

testcross: $\dfrac{A\ B}{a\ b} \times \dfrac{a\ b}{a\ b}$

progeny phenotypes:

A B	108	}	parental types
a b	97		
A b	33	}	recombinant types
a B	42		
	280		

Fig. 165. **Genetic linkage.** The phenotypes produced by a testcross.

there tends to be association of particular alleles together in the gametes (the 'parental' types, in the same combination as the parent), with other combinations being less frequent (see RECOMBINATION and CROSSING OVER).

By convention the amount of recombination between linked genes is a direct measure of their distance apart along the chromosome, 1% recombination being equivalent to the MAP UNIT. The amount of recombination occuring is best measured in a TESTCROSS calculated from the total number of recombinant types in the testcross progeny as a proportion of the total progeny. For example, a testcross with genes in COUPLING produced the PHENOTYPES in Fig. 165 where the amount of recombination would be:

$$\frac{33+42}{280} \times 100 = 26.8\%$$

and the genes A and B could be represented on a GENETIC MAP as:

$$\underset{26.8}{\overline{A \qquad\qquad\qquad B}}$$

Had the two genes been independently assorting (i.e., on different chromosomes) roughly equal numbers of each phenotype would have been expected in the progeny (about 70). Similarly, since the maximum expected amount of recombination between two genes is 50%, if two genes are located more than 50 map units apart on the same chromosome they will appear to be independent of each other.

genetic load, *n.* a measurement of the amount of deleterious genes in a population, calculated as the average number of lethal equivalents per individual.

genetic map, see CHROMOSOME MAP.

genetic polymorphism or **polymorphism,** *n.* the presence in a population of two or more MORPHS, produced when different alleles of a gene occur in the same population and the rarest allele is not maintained merely by repeated MUTATION (i.e. has a frequency higher than, say, 0.05%). Such a definition excludes continuously variable characters such as height or skin colour in humans, but the human blood groups are classic examples, where single genes have two or more alleles, producing different antigenic phenotypes. A genetic polymorphism can be maintained by several mechanisms such as heterozygous advantage or FREQUENCY–DEPENDENT SELECTION, and can be stable over several generations (a BALANCED POLYMORPHISM) or may become 'transient' as when the environment changes, see, for example, SICKLE–CELL ANAEMIA.

genetic population, see POPULATION.

genetic recombination, see RECOMBINATION.

genetics, *n.* the science that investigates patterns of inheritance between generations, together with how genes express themselves within the lifetime of individual organisms.

genetic variability, *n.* a range of phenotypes for a particular CHARACTER produced by alternative alleles of one or more genes, the range containing discrete groups (see QUALITATIVE INHERITANCE) or a continuous spectrum of types (see POLYGENIC INHERITANCE). Genetic variability arises initially by MUTATION and is maintained by sexual reproduction involving CROSSING OVER in MEIOSIS. Such variation is the raw material for NATURAL SELECTION to act upon, ensuring that the best-adapted variants are most likely to reproduce.

geniculate, *adj.* (of plant structures) bent abruptly, knee-like.

genitalia, *n.* the reproductive organs and their accessory parts.

genome, *n.* the complete complement of genetic material in a cell, or carried by an individual.

genotype, *n.* the genetic constitution of an individual, usually referring to specific CHARACTERS under consideration. Thus, the two alleles of the human albino gene can be written A and a, with three possible genotypes: a/a, A/a and A/A. See DOMINANCE (1) for the expression of the genotype in the PHENOTYPE.

genotype frequency, *n.* the proportion of a particular genotype amongst all the individuals in a population. Thus, if a population sample contained 5 albinos and 95 normally-pigmented individuals, the genotype frequency or albinos (a/a) would be 0.05 or 5%. Note, however, that it is not possible to be certain of the frequency of A/a and A/A genotypes among the normal individuals, since both genotypes have the same PHENOTYPE due to dominance, although their frequencies can be predicted using the HARDY–WEINBERG LAW.

genus, *n.* a TAXON immediately above that of a SPECIES and containing usually a group of closely related species. The name of a genus (generic name) is written or printed with an initial capital letter and in BINOMIAL NOMENCLATURE precedes the specific name, which invariably has a small initial letter.

geobotany, *n.* the part of botany related to ecology and the geography of plants.

Geoffroyism, *n.* the belief that there is an adaptive response of the genotype to the demands of the environment, i.e. an environmental induction of appropriate genetic changes. Named after the French naturalist Hilaire Geoffroy (1772–1844).

geographical distribution, see DISTRIBUTION.

geographical isolation, *n.* the effective isolation of a GENE POOL by geographical barriers. For example, oceanic islands are isolated because of the difficulty of the sea-crossing for many organisms. The evolution of new species takes place through the prevention or limitation of GENE FLOW and various isolating mechanisms may play a part in this, one of the most important being geographical isolation. Compare ECOLOGICAL ISOLATION. See SPECIATION.

geographical speciation, see SPECIATION.

geographical variation, *n.* any variation that occurs in spatially separated populations of species.

geological time, *n.* the period beginning with the formation of the earth and ending in the present day. It is often tabulated beginning with the Cambrian period, since the majority of fossil forms date from then. Precambrian time began with the formation of the earth about 4,600 million years ago. Dating of rocks is usually carried out by the measurement of radioactive decay. The older the rock the less radioactive it is. Organic remains are dated by CARBON DATING. See Appendix A.

geophyte, see CRYPTOPHYTE.

geotaxis, *n.* a type of animal movement (not growth) in which the body becomes orientated in relation to the force of gravity. For example, fruit flies exhibit negative geotaxis, moving upwards in the culture bottle, away from the gravitational force.

geotropism, *n.* a plant growth movement that occurs in response to gravity. Thus roots are positively geotropic (growing downwards), while shoots are generally negatively geotropic (growing upwards).

Gephyrea, *n.* a class of large, marine ANNELID worms that possess little or no segmentation and few or no chaetae, e.g. *Echiuroidea* and *Sipunculoidea*.

German measles or **rubella,** *n.* a mild human disease whose main feature is a rash on the body occuring about 18 days after infection. Caused by the rubella virus, the condition is highly contagious, being spread by direct contact via nasal secretions. The virus can be transmitted from mother to foetus and may cause serious damage in many foetal tissues and even death. The major period of risk to the child is the first 3 months of pregnancy when there is at least a 30% risk of abnormality.

For this reason many Western countries have a routine immunization programme for young females, either as babies or between the ages of 10–14 years.

germ cell, see GAMETE.

germicide, *n.* a substance that kills microorganisms.

germinal epithelium, *n.* a cell layer found on the outer surface of the vertebrate OVARY and lining the seminferous tubules of the TESTIS, which gives rise to follicle cells in females and to spermatogonia in males. In males it also produces SERTOLI CELLS which nourish the developing sperms.

germinal vesicle, *n.* the meiotic PROPHASE nucleus of the oocyte during the growth of the cytoplasm in animals.

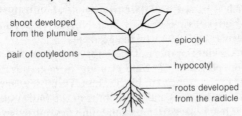

Fig. 166. **Germination.** A typical seedling.

germination, *n.* the beginning of the growth of a seed, spore or other structure that is dormant. Seed DORMANCY can be broken by several factors, depending on the species:

(a) the removal of a germination inhibitor which can be leached out by water.

(b) a period of cold temperature (STRATIFICATION).

(c) exposure to the correct wavelength of light to stimulate PHYTOCHROMES in the seed, for example, lettuce seeds require red light for germination and are inhibited by far-red light.

(d) rupture of a thick testa by (i) microbial breakdown, (ii) abrasive action of soil (as in desert plants), (iii) heat from bush fires, (iv) the effect of digestive juices (as when eaten by birds) by the softening action of water.

Once dormancy is broken a regular sequence of events takes place: water is imbibed which hydrates the tissues increasing enzymic action in the ALEURONE layer (when present), a process aided by the release of GIBBERELLIN from the embryo; food stores are mobilized in the ENDOSPERM or within the COTYLEDONS; AUXINS and CYTOKININS are formed which promote cell division and enlargement, causing the embryo to grow and burst through the testa. See Fig. 166.

There are two main types of germination, depending on whether the seed cotyledons are carried above the soil (EPIGEAL) or remain below ground (HYPOGEAL).

germ layer, *n.* any of the three embryonic layers of cells, ECTODERM, ENDODERM or MESODERM, that can be distinguished during development of the GASTRULA and for a short period after its formation.

germ line, *n.* a group of cells that give rise to the GONADS, becoming differentiated from the 'somatic' cells early in embryonic development. Only germ-line cells have the potential to undergo MEIOSIS, and a MUTATION event in these cells may well be transmitted to the offspring, unlike genetic change in somatic cells (see SOMATIC MUTATION).

germ plasm theory, *n.* a concept put forward by August Weismann (1834–1914) who suggested that the sex cells are produced by the 'germ plasm', which is passed on intact from generation to generation and gives rise to body cells in each individual. Nowadays we consider DNA to be the molecular equivalent of the germ plasm.

gestation, *n.* the development of the embryo in the uterus of mammals. The gestation period is the time elapsing from fertilization and the implantation of the embryo to birth. This takes 60 days in the domestic cat, 9 months in humans and 18 months in the Indian elephant.

giant fibre, *n.* a very large nerve axon found in some invertebrates such as ARTHROPODS and ANNELIDS. Such fibres are capable of transmitting impulses faster than normal fibres due to a reduction in internal longitudinal resistance.

gibberellic acid, *n.* a crystalline acid that occurs in plants having similar growth-promoting effects to GIBBERELLINS.

gibberellin, *n.* one of a range of growth regulators in plants that stimulate stem elongation but, unlike AUXINS, do not inhibit root growth. Gibberellins can be used to induce BOLTING in immature plants and to remove dwarfness in mature plants. Unlike auxins, gibberellins can move freely around the plant. Compare ANTIGIBBERELIN.

gibbon, *n.* a long-armed anthropoid ape of the genus *Hylobatea*.

gibbous, *adj.* (of a solid object) having a projection in the form of a rounded swelling.

gigantism, *n.* a rare human condition in which excess production of GROWTH HORMONE by the anterior PITUITARY GLAND during childhood and adolescence causes over-elongation of bones, producing a *pituitary giant*. Compare DWARFISM.

gill, *n.* **1.** the respiratory organ of aquatic animals. External gills, as in tadpoles, are produced by the embryonic ECTODERM; internal gills, as in fish, are developed from the pharynx and are thus endodermal (see ENDODERM). Gills are usually well supplied with blood vessels, and interchange of oxygen and carbon dioxide takes place across the extensive surface area (see COUNTERCURRENT EXCHANGE). Gills also occur in many invertebrates, for example, in insects such as the caddis fly larva and molluscs such as oysters. Occasionally, unusual structures act as gills, for example, the walls of the rectum in certain dragonfly nymphs, water being pumped in and out via the anus. **2.** the spore-carrying lamellae in basidiomycete fungi, located underneath the cap or 'pileus'.

gill bar, *n.* the tissue that separates the GILL CLEFTS, consisting of skeletal material, nervous tissue, blood vessels, etc.

gill book, *n.* the respiratory organs of aquatic ARACHNIDS such as the king crab (*Limulus*). Such organisms possess numerous leaves (hence 'book') through which blood circulates and over which oxygenated water is circulated.

gill cleft or **gill slit,** *n.* an inpushing of the epidermis in the region of the GILLS in developing CHORDATE embryos including humans. The gill clefts normally meet similar outpushings from the endoderm of the pharynx, and in fish and occasionally in terrestrial vertebrates such as amphibians, break through to the exterior to form the gill slits.

gill fungus, *n.* a BASIDIOMYCETE, the fruiting bodies of which possess a series of lamellae on the pileus that carries the spore-bearing hymenium.

gill pouch, *n.* outgrowths of the pharynx in early embryos that give rise to gills in fish and in amphibian larvae. In higher vertebrates, they usually disappear in further development.

gill slit, see GILL CLEFT.

gingivitis, *n.* an inflammatory disease of the gums associated with an accumulation of plaque containing bacteria of several types, for example, *Streptomyces sanguis*. Whilst painful and unpleasant, the condition usually does not lead to tooth loss.

Ginkoales, *n.* an order of the GYMNOSPERMS that is now represented by only a single species *Ginko biloba*, the maidenhair tree. Its ancestors flourished during the MESOZOIC PERIOD.

girdle, *n.* a bony structure to which the limbs of vertebrates are attached. See PECTORAL GIRDLE, PELVIC GIRDLE.

gizzard, *n.* a portion of the anterior part of the alimentary canal where food materials are broken up prior to digestion. Usually the gizzard possesses strong muscular walls and in invertebrates has internal projections. In some birds, such as the curlew, the gizzard lining is regularly renewed and the old lining cast out in the form of a pellet.

glabrous, *adj.* (of plant structures) without hairs.

glaciation, *n.* one of the so-called ice ages; a period of ice cover.

gland, *n.* an organ producing substances which are then secreted to the outside of the gland, sometimes by means of a duct, as in the exocrine glands, for example, the salivary, mammary, lachrymal glands, but also in the case of ENDOCRINE GLANDS directly into the blood system. Occasionally, individual cells act as glands, for example, gland cells of *Hydra* producing digestive enzymes.

glandular cells, *n.* cells that produce a secretion.

glandular epithelium, *n.* a layer of cells with a secretory function, the secreted products entering the space lined by the epithelium, for example, the lining of the vertebrate intestine.

glandular fever or **infective mononucleosis,** *n.* an ACUTE infectious disease, probably caused by the Epstein–Barr virus, that affects primarily the lymphoid tissue throughout the body, resulting in abnormal blood lymphocytes (the 'mononucleosis' of the title), enlarged lymph nodes and spleen, and sometimes fever and sore throat. The disease affects mainly teenagers from an affluent background and is prevalent therefore amongst students. No control measure is available at present.

glaucous, *adj.* (of plant structures) bluish.

glenoid cavity, *n.* **1.** the hollow in the scapula into which the humerus fits. **2.** the depression of the squamosal bone into which the joint of the mammalian lower jaw fits.

glia or **neuroglia,** *n.* the web of undifferentiated cells that packs and supports nerve cells in the brain and spinal cord.

globigerina ooze, *n.* the CALCAREOUS mud that occurs on the sea bed as a result of the aggregation of the shells of dead Foraminifera, *Globigerina* being the commonest genus of shell present.

globular protein, *n.* a type of PROTEIN.

globulin, *n.* a group of proteins that are soluble in salt solution and coagulated by heat. They occur in blood plasma and antibodies, and are the main proteins of plant seeds.

glochidium, *n.* a lamellibranch larva possessing a small bivalve shell and a tentacle-like sucker for attachment to a fish.

Gloger's Rule, *n.* a biological rule stating that populations of species which occur in warm, humid areas are more heavily pigmented than those in cool, dry areas.

glomerular filtrate, *n.* a filtrate that has passed through the GLOMERULUS in the vertebrate kidney.

glomerular filtration rate, *n.* the total amount of filtrate produced per minute by glomeruli of all nephrons in both kidneys of an individual.

glomerulus, *n.* a bunch of capillaries found in the vertebrate kidney, enclosed in the BOWMAN'S CAPSULE. Substances such as water and dissolved materials filter from the blood, across the endothelium of the capillaries and the epithelium of the capsule, into the tubules leading to the LOOP OF HENLE.

glossopharyngeal nerve, *n.* the 9th CRANIAL NERVE of vertebrates; a dorsal root nerve. It is concerned with the swallowing reflex and back of the tongue taste buds in mammals.

glottis, see EPIGLOTTIS.

glucagon, *n.* a polypeptide of 29 amino acids produced by the cells in the ISLETS OF LANGERHANS of the pancreas of vertebrates. Glucagon acts as a hormone, having the opposite effect to INSULIN, in causing the breakdown of liver GLYCOGEN and the release of glucose into the blood.

glucocorticoid, *n.* a steroid endocrine secretion produced by the adrenal cortex, influencing the metabolism of carbohydrates and proteins, e.g. cortisol (Hydrocortisone), cortico-sterone.

gluconeogenesis, *n.* the formation of GLUCOSE or GLYCOGEN in the liver from noncarbohydrate precursors such as certain AMINO ACIDS, GLYCEROL and LACTIC ACID.

glucose or **dextrose,** *n.* an important hexose sugar with an ALDOSE structure that occurs in two forms, alpha and beta; it has the general formula $C_6H_{12}O_6$ and is found in sweet fruits, especially ripe grapes. Glucose is formed in the CALVIN CYCLE of PHOTOSYNTHESIS and acts as a primary energy supply for both plant and animal cells, although usually it is converted to an insoluble form for long-term storage: STARCH in plants, GLYCOGEN in animals. See MONOSACCHARIDE for structure.

glucose-6-phosphate dehydrogenase (G-6-P.D.), *n.* an important enzyme in the PENTOSE PHOSPHATE PATHWAY for carbohydrate metabolism, catalysing the oxidation of glucose 6-phosphate in the presence of NADP coenzyme, giving 6-phosphogluconolactone and NADPH. G-6-P.D. of blood erythrocytes is of particular interest because a decrease in its activity can lead to the condition called FAVISM.

glume, *n.* the chaffy scale that encloses the spikelets of grass or the flowers of sedges.

glutamate, *n.* the dissociated form of the amino acid GLUTAMIC ACID.

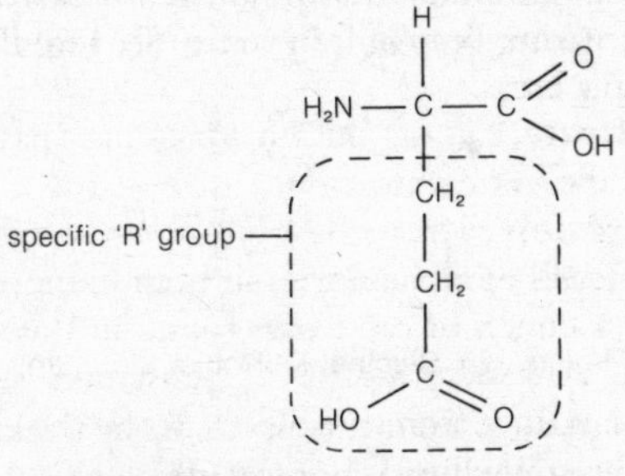

Fig. 167. **Glutamic acid.** Molecular structure.

glutamic acid, *n.* one of 20 AMINO ACIDS common in proteins that has an extra carboxyl group and is acidic in solution. See Fig. 167. The ISOELECTRIC POINT of glutamic acid is 3.2.

glutamine, *n.* one of 20 AMINO ACIDS common in proteins that has a polar 'R' structure and is soluble in water. See Fig. 168.

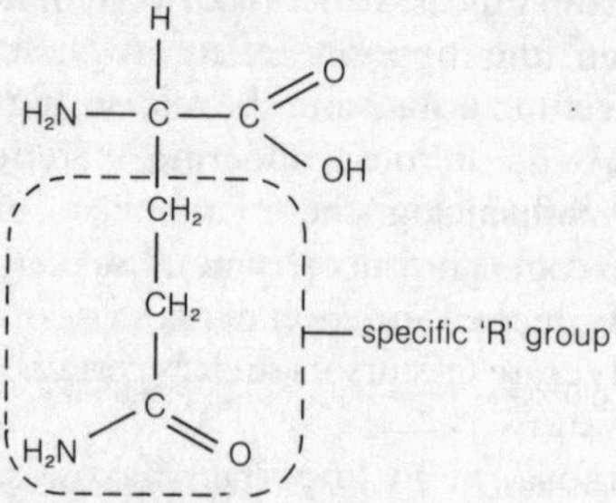

Fig. 168. **Glutamine.** Molecular structure.

glyceraldehyde phosphate, see PGAL.
glyceraldehyde 3-phosphate (GALP), see PGAL.
glycerate 3-phosphate (GP), see PGA.
glyceric acid phosphate, see PGA.
glycerol or **glycerin,** *n.* a simple LIPID that is a basic component of fats. See Fig. 169. Glycerol contains high amounts of energy which can be released in metabolism. See GLYCOLYSIS.

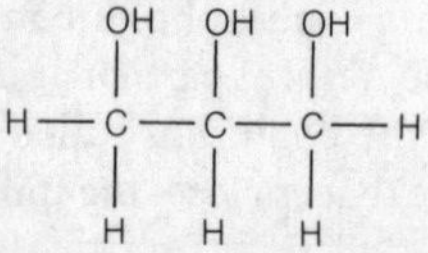

Fig. 169. **Glycerol.** Molecular structure.

glycine, *n.* one of 20 amino acids common in proteins that although it has a nonpolar 'R' structure, is soluble in water. See Fig. 170. The isoelectric point of glycine is 6.0.

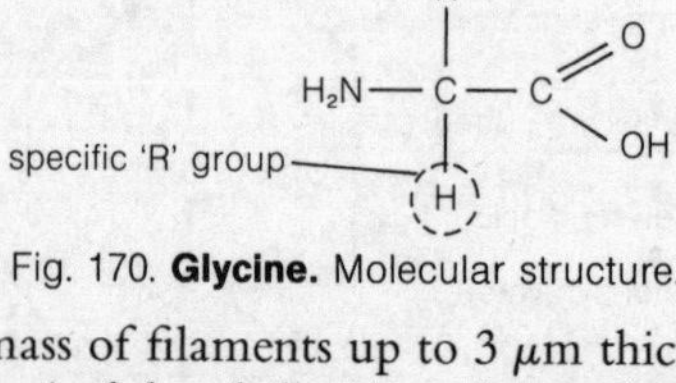

Fig. 170. **Glycine.** Molecular structure.

glycocalyx, *n.* a mass of filaments up to 3 μm thick produced by the membrane of intestinal brush-border microvilli, consisting of acid mucopolysaccharide and GLYCOPROTEIN, and thought to be associated with the digestion of small food molecules. Other animal cells also have a glycocalyx, on their cell coat, providing a mechanism that enables cells to recognise each other, an important process in embryonic development.

glycogen, *n.* the principal carbohydrate storage molecule of animals, being produced from glucose in the mammalian liver (see PHOSPHATASE) and muscles when blood sugar levels are too high, a process called *glycogenesis* which is under the influence of INSULIN. Glycogen in the liver can be broken down to glucose when blood sugar levels are low, a process called *glycogenolysis* which is under the influence of GLUCAGON. See Fig. 171. Glycogen in the muscle, however, is broken down to LACTIC ACID (not glucose) in GLYCOLYSIS.

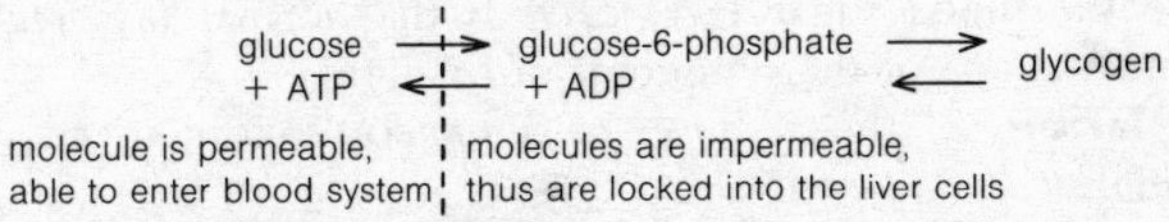

glucose ⟶ glucose-6-phosphate ⟶ glycogen
+ ATP ⟵ + ADP ⟵

molecule is permeable, | molecules are impermeable,
able to enter blood system | thus are locked into the liver cells

Fig. 171. **Glycogen.** Formation and breakdown of glycogen in the liver.

glycogenesis, see GLYCOGEN.

glycogenolysis, see GLYCOGEN.

glycolipid, *n.* any of a group of LIPIDS containing a carbohydrate.

glycolysis, *n.* the first stage of CELLULAR RESPIRATION, occurring with or without the presence of oxygen, in which glucose is converted to two molecules of pyruvic acid. See Fig. 172. See also AEROBIC RESPIRATION.

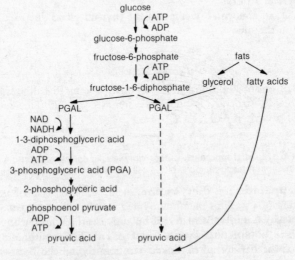

Fig. 172. **Glycolysis.** The individual steps of glycolysis.

glycophorin, *n.* a glycopolypeptide found in human erythrocytes.

glycoprotein, *n.* any PROTEIN that contains sugars as part of the molecule.

glycoside, *n.* an acetal derivative of a sugar that, on hydrolysis by enzymes or acids, gives rise to a sugar. Glycosides containing glucose are called *glucosides*, those with galactose are called *galactosides*. They render unwanted substances chemically inert or form food reserves, for example, GLYCOGEN.

glycosidic bond or **glycosidic link,** *n.* a bond between the anomeric carbon of a carbohydrate and another group or molecule.

glycosuria, *n.* the condition where glucose is excreted in the urine because the blood sugar level exceeds the normal (*hyperglycemia*). Glycosuria is one of the symptoms of DIABETES.

glycosylation, *n.* the addition of a CARBOHYDRATE to an organic molecule such as a PROTEIN.

gnatho-, *prefix.* denoting the jaws.

Gnathostomata, *n.* the vertebrates that possess true jaws, i.e. all true fish (not lampreys and hagfish), amphibia, reptiles, birds and mammals.

Gnetales, *n.* an order of GYMNOSPERMS whose members have true vessels in the secondary wood and an absence of archegonia in the ovule. In these features they resemble ANGIOSPERMS.

goblet cell or **chalice cell,** *n.* a cell shaped something like a wineglass that is present in the columnar epithelium of the mammalian intestine and secretes MUCIN.

goitre, *n.* an abnormal swelling of the thyroid gland due to a lack of iodine in the diet.

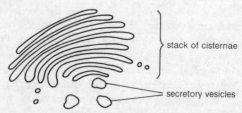

Fig. 173. **Golgi apparatus.** The secretory vesicles carry products from the apparatus to the edge of the cell where they may be released.

Golgi apparatus or **dictyosome,** *n.* a series of cell ORGANELLES consisting of a stack of membrane–lined vesicles called CISTERNAE, first described by Camillo Golgi in 1898 but only clearly defined from studies with the ELECTRON MICROSCOPE. See Fig. 173. The membranes of the Golgi apparatus are often linked temporarily to the ENDOPLASMIC RETICULUM and secretory vesicles are formed by the apparatus which move to the periphery of the cell and may carry out EXOCYTOSIS. The apparatus is thought to have a storage role as well as enabling the assembly of simple molecules into more complex ones. For example, carbohydrates and proteins are packaged into glycoprotein.

gon- or **gono-,** *prefix.* denoting seed.

gonad, *n.* the OVARY or TESTIS of an organism.

gonad hormones, *n.* sex hormones produced by the GONADS.

gonadotrophin or **gonadotrophic hormones,** *n.* hormones secreted by the anterior lobe of the PITUITARY GLAND and in some mammals, once pregnancy is under way by the PLACENTA. FSH and LH are produced by all TETRAPODS though FSH appears to be absent in fish. LTH (Luteotrophin) is present in rats and maintains the CORPUS LUTEUM though it is absent in other mammals. Prolactin is present in some vertebrates and controls milk production (LTH is a form of prolactin). Pituitary output is controlled by the hypothalamus. Gonadotrophins influence other glands connected with reproduction, controlling the activity of the gonads, the onset of sexual maturity, OESTROUS CYCLES, breeding rhythms and LACTATION. See also HCG.

Gondwanaland, *n.* the southern group of continents at the end of the TRIASSIC PERIOD which had split from the northern group (LAURASIA). It consisted of South America, Africa, India, Australia and Antarctica.

gonochorism, *n.* the possession of GONADS of only one sex, either male or female, in an individual.

gonophore, *n.* any structure bearing gonads but particularly the specialized polyp of colonial COELENTERATES which bears the gonads and is shaped like a sessile medusa – a form of GONOZOOID.

gonorrhoea, *n.* a contagious inflammation of the mucus membranes in human reproductive organs, characterized by the discharge of mucus and pus from the urethra or vagina and caused by the gonococcal bacterium *Neisseria gonorrhoeae.* Gonorrhoea is probably the most prevalent communicable bacterial disease in man today and, although a VENEREAL DISEASE, is a distinct condition from SYPHILIS.

gonotheca, *n.* the chitinous cup of perisare surrounding the reproductive structures of colonial COELENTERATES.

gonozooid, *n.* a specialized polyp in colonial coelenterates which bears the gonads (see GONOPHORE).

goodness–of–fit, *n.* an assessment of the similarity between the observed results from an experiment and those expected from theory and supported by a stated NULL HYPOTHESIS. An example of such a statistic is the CHI–SQUARED TEST.

Graafian follicle, *n.* a structure in the ovary of a female mammal, consisting of an OOCYTE surrounded by granular FOLLICLE cells which enclose also a large, fluid filled cavity, the whole structure being encased in a wall of connective tissue. See Fig. 174. The Graafian follicle begins to form deep inside the ovary, stimulated by FSH as the OESTROUS CYCLE develops, gradually enlarging and maturing as it moves to the surface, eventually appearing like a blister on the surface, just prior to release of the egg (ovulation) by rupture of the wall. After ovulation the follicle becomes a CORPUS LUTEUM. Further ovulation is normally prevented by the corpus luteum secreting PROGESTERONE, which in turn inhibits FSH production by the PITUITARY GLAND, so no further follicles develop. The presence of the cavity distinguishes the Graafian follicle (named after Regner de Graaf) from the OVARIAN FOLLICLES of other vertebrates.

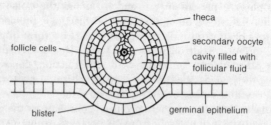

Fig. 174. **Graafian follicle.** General structure.

grade, *n.* a group of organisms having a similar level of organisation.

graded response, *n.* a response that increases with the amount of energy supplied as opposed to the reaction brought about by the ALL-OR-NONE LAW.

graft, *n.* the transfer of a small part of an organism to a relatively larger part where it is transplanted. This may take place from one part of an organism to another part (see AUTOGRAFT), or from one organism to another (see ISOGRAFT, HOMOGRAFT, HETEROGRAFT). Many embryological studies have involved grafting, and from a medical point of view, skin grafting, particularly after serious burns, is commonplace. Organ grafts such as heart transplants are now much more common, though rejection by an animal of the tissues of another is still a serious problem (see IMMUNE RESPONSE). In animals the graft comes from a 'donor' and is transferred to the recipient. In plants, grafts are used often in horticulture, where a scion, the plant to be cultivated, is attached onto a STOCK (1), the rooting portion.

graft-hybrid, see CHIMAERA.

Gram-negative or **Gram-positive,** see GRAM'S STAIN.

Gram's stain, *n.* a stain taken up by the *Gram-positive* bacteria that differentiates these from bacteria which fail to take up the stain (*Gram-negative* types). The initial stain is crystal violet/iodine complex, and

Gram-negative bacteria are decolourized with alcohol whilst *Gram-positive* bacteria retain a blue/purple colour.

grandfather method, *n*. a procedure for estimating the amount of RECOMBINATION between X-linked genes in an individual that depends on the arrangement of genes in the maternal grandfather. For example, if a mother is known to be heterozygous her genes could be in COUPLING or in repulsion. However, this uncertainty can be removed if the phenotype of her father is known since he has only one X-chromosome. Thus if he was *Ab*, his heterozygous daughter must be *Ab/aB*, so ensuring that her children are properly identified as parental or recombinant type (see GENETIC LINKAGE).

granulocyte, *n*. a type of white blood cell (LEUCOCYTE) that possesses granules in the cytoplasm and is formed in the bone marrow. Granulocytes form about 70% of all leucocytes and are of three types: EOSINOPHILS, BASOPHILS and NEUTROPHILS.

granum, see CHLOROPLAST.

graptolite, *n*. any extinct Palaeozoic colonial pelagic animal of the class Graptolithina. Branching forms are found in later strata and up to Carboniferous times.

grass, *n*. any member of the family Gramineae, all of which have jointed tubular stems, sheathing leaves and flowers enclosed in GLUMES. Examples include cereals, reeds, bamboos.

gravity, *n*. the force by which masses are attracted to each other.

green gland, *n*. one of a pair of excretory organs of crustaceans with openings at the antennal base.

greenhouse effect, *n*. **1.** an effect occurring in greenhouses in which the glass transmits short wavelengths but absorbs and re-radiates longer wavelengths, thus heating the interior. **2.** the application of this effect to the earth's atmosphere. Infrared radiation tends to be trapped by carbon dioxide and water vapour in the earth's atmosphere and some of it is re-radiated back to the earth's surface.

green–island effect, *n*. a localized increase of photosynthesis in plant tissues immediately surrounding a lesion caused by a biotrophic pathogen.

green revolution, *n*. the process by which humans have exploited new hybrid varieties of food crops such as barley, rice, maize and wheat to meet their requirements. New varieties have been produced by crossing and selection. The green revolution has had particularly dramatic effects on developing countries. For example, wheat yields in Mexico have increased by about 300% since the introduction of a new dwarf variety in 1960. However, the attempts at plant improvements have not always been of benefit to the peasant farmer.

greeting display, *n.* a form of behaviour in animals, particularly indulged in by some birds, where display arises on the arrival of the mate, say, at a nest site with food.

grey matter, *n.* the material from the CENTRAL NERVOUS SYSTEM of vertebrates containing the nerve cell bodies which give the material a grey colour as distinct from the white colour of the FIBRES (compare WHITE MATTER). Grey matter also contains the DENDRITES of the nerve cells (with no myelin sheath), blood vessels, and glial cells which have a supportive function. Grey matter of the spinal cord is internally arranged in the form of an 'H' shape in cross-section, unlike the brain, where grey matter is external. Coordination occurs in the grey matter of the CNS.

Griffith, Frederick, English medical bacteriologist who, in 1928, published a paper showing that is was possible to convert one type of *Pneumococcus* bacterium into a new type. More specifically, Griffith worked with two strains of bacterium:

Type R	*Type S*
rough outer coat, nonvirulant in mice	smooth outer coat, virulent in mice

His results are summarized in Fig. 175. Griffith proposed that in experiment 4 the heat-killed 'S' type somehow converted the living 'R' type to virulent 'S' types, in a process he called TRANSFORMATION. Later it was shown that the transforming 'principle' was DNA.

Experiment	Type of bacterium injected into mice	Mouse reaction	Type of bacterium recovered from mouse
1	Live Type R	Survive	—
2	Live Type S	Die	—
3	Dead Type S	Survive	—
4	Dead Type S plus Living Type R	Die	Live Type S

Fig. 175. **Griffith, Frederick.** Griffith's results.

gristle, see CARTILAGE.

ground state, *n.* the condition of a molecule when at its lowest energy level.

group, *n.* a collection of closely related taxa (see TAXON), particularly an assemblage of closely related species.

growth, *n.* the process of increase in size which has three distinct components, (a) cell division, (b) assimilation, (c) cell expansion. The basis of growth is CELL DIVISION but in order to increase in size, cells must be able to synthesize new structures that are manufactured from raw materials derived from their immediate environment. This is assimilation and it results in increase in cell size. Growth during development can be continuous but allometric (see ALLOMETRIC GROWTH), as in humans, or can be discontinuous, as in insects where growth occurs at each ECDYSIS.

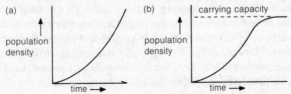

Fig. 176. **Growth curve.** (a) Exponential. (b) Logistic.

growth curve, *n.* the graphic representation of the growth of a population, which could be *exponential* where (theoretically) the density would eventually be increasing at an infinite rate, or could be logistic (see LOGISTIC CURVE) where the density would stabilize near the CARRYING CAPACITY of the population. See Fig. 176. Populations of microorganisms tend to go through a classic four-stage growth curve (see Fig. 177).

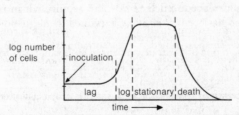

Fig. 177. **Growth curve.** A growth curve in a microbial culture.

The 'lag' phase is a time of adaptation to the new environment where such processes as ENZYME INDUCTION take place and reproduction rate equals death rate. The 'log' phase is a period of exponential growth (reproduction rate much greater than death rate). The 'stationary' phase is a time of equilibrium, representing the response to a limiting factor such as nutrient source, while little or no reproduction occurs during the 'death' phase, so the population declines.

growth, grand period of, *n.* the whole of the time during which a cell, organ or plant part increases in size, slowly at first, then faster and more slowly again until it is mature.

growth habit, *n.* a characteristic manner of growth in plants.

growth hormone or **somatotrophic hormone (STH),** *n.* a hormone secreted by the anterior PITUITARY GLAND mainly in the growth period, where it stimulates the lengthening of the long bones in

TETRAPODS, induces protein synthesis and inhibits INSULIN, thus raising the level of blood sugar. Oversecretion during development produces GIGANTISM but excess later in life gives rise to ACROMEGALY where a massive forehead, nose and lower jaw are produced. Deficiency of the hormone results in DWARFISM.

grub, *n*. an insect larva lacking legs.

guanine, *n*. one of four types of nitrogenous bases found in DNA, having the double-ring structure of a class known as PURINES. Guanine forms part of a DNA unit called a NUCLEOTIDE and always forms complementary pairs with a DNA pyrimidine base called CYTOSINE. Guanine also occurs in RNA molecules. See Fig. 178.

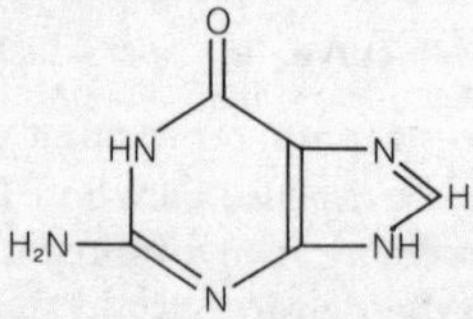

Fig. 178. **Guanine.** Molecular structure.

guano, *n*. the white nitrogenous dried excrement, high in uric acid content, that is produced by birds and reptiles. Guano is a valuable fertilizer and has been extensively harvested from guano bird (cormorant) colonies in South America.

guard cells, *n*. one of a pair of specialized epidermal cells forming a pore (stoma) at the leaf surface. See STOMA for a description of how the stomatal aperture is regulated.

guild, *n*. a group of species that exploits the same class of environmental resources in a similar way.

Gurdon, J.B., (1933–) British experimental embryologist who demonstrated that a nucleus taken from the differentiated gut cell of a clawed toad, injected into an enucleated egg, could produce a normal adult. In other words, he demonstrated that a fully differentiated cell still contains the genetic information to direct the development of the cells in the entire animal (see TOTIPOTENCY).

gustation, *n*. the act of tasting or the faculty of taste. This is brought about by the effect of ions and molecules in solution, stimulating specialized sensory epithelial receptors.

gustatory sensillum, *n*. a taste bud.

gut, see ALIMENTARY CANAL.

guttation, *n*. the emergence of water from the endings of leaf veins, usually occurring at night when TRANSPIRATION is low or absent. Guttation is thus a method of exuding excess water and involves special water pores called HYDATHODES at leaf tips which, unlike normal

STOMATA, are not able to control their aperture size. The process is common in low-growing plants such as grasses. See Fig. 179.

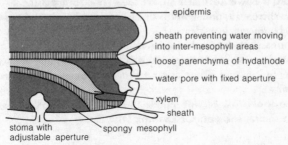

Fig. 179. **Guttation.** Transverse section of a saxifrage leaf.

gymno-, *prefix.* denoting uncovered.

Gymnophiona, see APODA.

gymnosperm, *n.* any member of the class Gymnospermae – the CONIFERS and their allies. Gymnosperms derive their name from the fact that the ovules are borne without a covering on the surface of the megasporophylls, which usually take the form of cones.

gynandromorph, *n.* any animal exhibiting both male and female features, often with the body divided into equal portions of the two sexes. For example, the left-hand side male, the right-hand side female.

gynobasic, *adj.* (of a STYLE) appearing to be inserted at the base of the ovary because of the folding of the ovary wall.

gynodioecious, *adj.* (of female and hermaphrodite flowers) occurring on separate plants, an attribute thought to promote cross-pollination, for example, in thyme.

gynoecium or **gynaecium** or **pistil,** *n.* the female part of a flower composed of one or more CARPELS. When there are several carpels they can be completely fused to form a 'compound' ovary, stigma and style; each carpel can be quite separate, or there can be stages in between the two extremes. See Fig. 180.

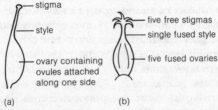

Fig. 180. **Gynoecium.** (a) A simple type, e.g. a bean. (b) A complex type, e.g. a geranium.

When other flower parts arise below the gynoecium the flower is described as *hypogynous* and the ovary is said to be *superior*. Alternatively, when other flower parts arise above the gynoecium and the RECEPTACLE encloses the ovary, the flower is described as *epigynous* and the ovary is said to be *inferior*.

gynogenesis, *n.* the parthenogenetic development of an egg after the egg membranes have been penetrated by a male gamete. There is no nuclear fusion and stimulation of development might be purely mechanical.

gynomonoecious, *adj.* (of female and hermaphrodite flowers) occurring on the same plant, for example, the daisy.

H

habit, *n.* the general appearance and form of branching in plants. For example, dandelions can have an erect or prostrate habit, depending on location. See PHENOTYPIC PLASTICITY.

habitat, *n.* that part of the environment which is occupied by an animal or plant, for example, stream, meadowland, salt marsh, etc.

habitat selection, *n.* the selection by a dispersing organism (see DISPERSAL) of a suitable habitat.

habitable zone, *n.* the area around a star (such as the sun) in which there is sufficient energy present to sustain life.

habituation, *n.* the progressive loss of a behavioural response as a result of continued stimulation.

Haekel's law of Recapitulation or **palingenesis** or **von Baer's law,** *n.* a law stating that ONTOGENY recapitulates PHYLOGENY. This means that an organism will go through embryonic stages similar to those gone through during its evolutionary development. For example, mammals have a fish-like stage in their embryology. The law was formulated by the German biologist Ernst Haekel (1834–1919).

haem- or **haemo-** or **haemato-,** *prefix.* denoting blood.

haem or **heme,** *n.* a ring-like chemical structure containing an atom of ferrous iron that is bound as a PROSTHETIC GROUP to each POLYPEPTIDE CHAIN within HAEMOGLOBIN and MYOGLOBIN, giving blood and muscle their characteristic red colour.

haematin, *n.* an iron-containing pigment that is derived from the breakdown of HAEMAGLOBIN.

haematoblast, *n.* the parent cell of red blood corpuscles.

haematocrit, *n.* the percentage of the blood volume occupied by red blood cells. The percentage is 40.5% in man.

haematocyte, *n.* a blood corpuscle.

haematopoiesis or **haemopoiesis,** *n.* the process leading to red blood cell production.

haemerythrin, *n.* an iron-containing proteinaceous red pigment found in the blood plasma of some ANNELIDS and several minor invertebrate groups. Like HAEMOGLOBIN, the pigment can become oxygenated, but in this case oxygen uptake involves a change in valency from Fe(II) to Fe(III).

haemocoel, *n.* the body cavity within many invertebrates, including arthropods and molluscs, which is an expansion of part of the blood system. In contrast with the COELOM, the haemocoel never opens to the exterior nor does it contain germ cells. See Fig. 181.

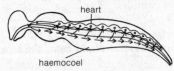

Fig. 181. **Haemocoel.** Transverse section of an insect, showing the positions of the heart, haemocoel and blood flow.

haemocyanin, *n.* a respiratory pigment found in the blood plasma of many molluscs and crustaceans that performs the same function as HAEMOGLOBIN in other animals. The metallic radicle of its PROSTHETIC GROUP, however, is copper rather than iron, giving the molecule a bluish colour.

haemoglobin, *n.* a large PROTEIN molecule with a quaternary structure of four POLYPEPTIDES CHAINS, two alpha and two beta chains, each of which is complexed to a separate HAEM group. About 300 million molecules of haemoglobin occur in each red blood cell of the mammalian circulation, with each molecule binding to a maximum of four oxygen molecules, one per haem group (= oxygenation). See OXYGEN-DISSOCIATION CURVE for an explanation of oxygen carriage.

Haemoglobin is found in all vertebrates and many invertebrates. In mammals the foetal haemoglobin (HbF) has a different polypeptide combination from that in postnatal 'adult' haemoglobin (HbA), consisting of two alpha and two gamma chains, with different oxygen-carrying characteristics (up to 30% more at low oxygen tension). An altered beta chain in HbA produces SICKLE-CELL ANAEMIA.

haemolysis, *n.* the disintegration of red blood cells, with the release of HAEMOGLOBIN. The process can occur (a) when the cells take in excess water by OSMOSIS, (b) when there is an antigen–antibody reaction involving the cells, as in RHESUS HAEMOLYTIC ANAEMIA, (c) as a result of an abnormality such as FAVISM. Addition of glacial acetic acid to a blood sample causes haemolysis of the red blood cells, thus making it easier to observe and count the white blood cells.

haemophilia, *n*. a rare human blood disorder in which BLOOD CLOTTING is deficient, resulting often in severe bleeding internally and externally. The condition is due to a lack of fibrin in the blood and is controlled by two closely linked genes on the X-CHROMOSOME that are responsible for the production of different clotting factors. Haemophilia A individuals lack antihaemophilic globulin (AHG) while haemophilia B individuals lack plasma thromboplastin. Males carrying the mutant ALLELE of either locus or (much more rarely) females homozygous for the recessive mutant alleles of either locus will be affected, although heterozygous females have normal blood. Haemophilia A is by far the most common form of the disease (about 80%) and can be treated by transfusions of AHG.

haemopoiesis, see HAEMATOPOIESIS.

haemorrhage, *n*. an escape of blood from the blood vessels, due to a wound or disease.

haemosporidian, *n*. any protozoan parasite of the order Coccidiomorpha, found in the blood corpuscles of vertebrates, and transmitted by invertebrate mites. See MALARIA PARASITE.

hair, *n*. **1.** (in plants) a filamentous outgrowth from an epidermal cell that may have a secretory function (in glands), an absorbing function (in root hairs), or a function in trapping and preventing air movement of the leaf surface, so lowering the rate of TRANSPIRATION. **2.** (in animals) a filamentous structure of mammalian skin formed of cornified epidermal cells that multiply in the hair follicle.

hair cell, *n*. a sensory epithelial cell of vertebrates that acts as a mechanoreceptor, possessing either one nonmotile cilium (*kinocilium*) or many nonmotile cilium-like projections of the surface (*stereocilia*). Such cells are found in the LATERAL-LINE SYSTEM and the ear.

hair follicle, *n*. a FOLLICLE in the epidermis of mammals through which hair projects into the dermis. The follicle is supplied by a sebaceous gland and erectile muscles for the hair.

half-life, *n*. the time required for half of the mass of a radioactive substance to disintegrate. For example, the half-life of ^{14}C is 5,700 years. See Fig. 182.

hallux, *n*. the first digit on the hind foot of vertebrates, such as the big toe in humans.

halolimnic, *adj*. living in or on salt marshes.

halophyte, *n*. a plant that is tolerant of high concentrations of salt in the

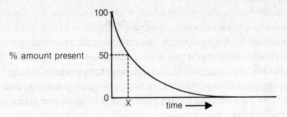

Fig. 182. **Half-life.** X = half-life. Note that the time taken to reach zero amount is *not* 2 × X.

soil and atmosphere and normally found close to the sea. For example, the marsh samphire, *Salicornia europaea*.

haltere or **balancer,** *n.* one of a pair of structures found in DIPTERANS. They are thought to be modified hind-wings, normally being small, clubbed organs very much smaller than the forewing. Halteres are probably associated with the maintenance of equilibrium during flight.

hammer, see MALLEUS.

hamulus, *n.* (*pl.* hamuli) a hook or hooklike process, as in the barbules of interlocking feathers, or on the forewings of bees, linking with the hind-wing.

haplo-, *prefix*, denoting single.

haplochlamydeous, see MONOCHLAMYDEOUS.

haploid, 1. *adj.* (of a cell nucleus) containing one of each type of chromosome. **2.** *n.* a haploid organism in which the main life stage has cell nuclei with one of each type of chromosome, written as 'n'. Such organisms (e.g. fungi, many algae) usually have a brief DIPLOID (2) phase (2n), returning to the haploid state via MEIOSIS. See ALTERNATION OF GENERATIONS.

haplont, *n.* an organism in the haploid stage.

haplotype, see HLA SYSTEM.

hapten, *n.* a substance that is able to combine with specific antibodies but does not produce them unless attached to a large CARRIER MOLECULE.

haptotropism, see THIGMOTROPISM.

Hardy-Weinberg law, *n.* a law stating that in a randomly mating large population the frequency of ALLELES will remain constant from generation to generation. Proposed in 1908 by G.H. Hardy in England and W. Weinberg in Germany, the law explains how the various alleles of a gene could remain constant in a population and yet be inherited by the rules of MENDELIAN GENETICS. The law shows why, for instance, dominant PHENOTYPES are not necessarily in excess of recessive ones in a population. If a gene has two alleles, *A* and *a*, with a frequency of p and q respectively, the genotypic frequencies would have the proportions:

$$AA \qquad Aa \qquad aa$$
$$p^2 \quad + \quad 2pq \quad + \quad q^2 \quad = \quad 1.0$$

This equation is derived from the BINOMIAL EXPANSION $(p+q)^2 = 1.0$. Such GENETIC EQUILIBRIUM only occurs under conditions of: zero selection and mutation, no emigration or immigration, random mating and a large population size. However, selection pressures can produce stability (see GENETIC POLYMORPHISM) and selection and mutation effects might cancel each other out.

Harvey, William (1578–1657) English physician who discovered the mechanism of the human blood circulation, in which blood flows away from the heart in arteries and towards the heart in veins.

Hatch–Slack pathway, *n*. an alternative pathway to the CALVIN CYCLE of photosynthesis, thought to operate in some tropical grasses and some dicotyledons. Phosphoenolpyruvate is carboxylated to form oxalo-acetate as the first product.

Hatscheck's pit, *n*. a mucus gland in the oral hood of AMPHIOXUS.

haustorium, *n*. **1.** a specialized development of the end of a HYPHA in parasitic fungi, penetrating a cell of the host and forming a food-absorbing organ. **2.** (in other parasitic plants) the organ that penetrates the host and acts as a food-absorbing organ.

Haversian canal, *n*. one of many channels formed within bone by the development of OSTEOBLASTS in concentric rings around them, and whose function is to facilitate the linking of the living parts. Each canal may contain an artery, a vein and a nerve, and the canals ramify throughout the bone, communicating with the bone marrow and the PERIOSTEUM. The Haversian systems are mainly longitudinal, but cross-connections are present.

hay fever, *n*. an allergic reaction to atmospheric dust and pollen. Hay fever causes watery eyes, sneezing, etc., due to inflammation of the mucous membranes of the eyes and nose.

HCG (human chorionic gonadotrophin), *n*. a HORMONE secreted by the TROPHOBLAST of the developing foetal tissues. It maintains the CORPUS LUTEUM during the first three to four months in pregnancy, thus ensuring a supply of PROGESTERONE for maintaining the uterus lining and preventing MENSTRUATION. HCG begins to be secreted as soon as the trophoblast implants in the uterine wall and it is at its highest concentration between the 7th and 10th weeks of pregnancy. The urine of women in the first month of pregnancy contains enough HCG to be easily detected by chemical means. After reaching its peak, HCG falls to low concentrations and its only known function is to stimulate the production of TESTOSTERONE by the testes of the male foetus. HCG has almost the same combined properties as LH and LUTEOTROPIC HORMONES.

head, *n.* **1.** the upper or front part of the body in vertebrates, that contains and protects the BRAIN, eyes, mouth, nose and ears when present. There is usually a concentration of sense organs in the head, such as eyes, tentacles, antennae and mouth parts. The head has probably developed because of forward locomotion, as a well-developed head is absent in organisms lacking unidirectional movement (e.g. sea urchin). **2.** the corresponding part of an invertebrate animal. See also CEPHALIZATION.

heart, *n.* the muscular pump of the BLOOD CIRCULATORY SYSTEM. In those invertebrates that possess a heart (e.g. ARTHROPODS, ANNELIDS, MOLLUSCS, ECHINODERMS) the heart is composed of several chambers and lies dorsal to the gut. In vertebrates the heart is made of special CARDIAC MUSCLE and lies in a ventral position surrounded by the PERICARDIUM. The five classes of vertebrates show an increasing complexity of structure, from the simple S-shaped heart with one ATRIUM and one VENTRICLE (2) found in fish, through the amphibians and most reptiles where the heart is divided into two atria but retains a single ventricle, and on to the birds and mammals where the heart shows complex separation into two sides with two atria and two ventricles. The main features of the human heart are:

(a) the right side pumps blood around the pulmonary (lung) circulation for oxygenation, the left side pumping blood around the systemic (body) circulation where it becomes deoxygenated.

(b) blood from the body enters the right atrium via the superior vena cava (upper body) and inferior vena cava (lower body). A coronary sinus also drains into the right atrium bringing blood from the heart itself. The right atrium squeezes blood through the atrioventricular (AV) opening into the muscular right ventricle. Finally, blood is ejected into the single opening of the pulmonary artery which splits to go to the two lungs.

(c) blood enters the left atrium from four pulmonary veins and passes through the left AV opening into the left ventricle. This has a much thicker wall than the right ventricle, reflecting its requirement for greater power. Blood leaves the left ventricle by one great vessel, the AORTA, which supplies all parts of the body, including the heart.

(d) flow of blood through the heart is in one direction only, due to the presence of various valves. Back-flow from ventricles to atria is prevented by AV valves, the *tricuspid* valve on the right side with three flaps, and the BICUSPID (3) *valve* on the left side with two flaps, both valves held in place by cords of connective tissue, the 'chordae tendinae'. Back-flow from arteries to the ventricles is prevented by semilumar valves.

(e) various nerve areas connected with contraction are located in the heart (see HEART, CARDIAC CYCLE): (i) the *sinoatrial node (SAN)* or 'pacemaker' located in the wall of the right atrium near the entry of the venae cavae. (ii) the *atrioventricular node (AVN)* at the junction of all four heart chambers. (iii) a network of *Purkinje tissue* and other fibres spreading out from the AVN across the walls of both ventricles. See Fig. 183.

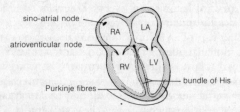

sino-atrial node
atrioventicular node
Purkinje fibres
RA LA
RV LV
bundle of His

Fig. 183. **Heart.** Nerve centres in the mammalian heart.

Other nerve areas are situated in or near the heart: (i) baroreceptors in the walls of the heart, in the aortic arch, the carotid sinus, venae cavae and pulmonary veins where they enter the atria. Such sensory receptors are stimulated by stretching of the structure in which they are found, resulting in a decrease in blood pressure. (ii) chemoreceptors sensitive to blood CO_2 levels are found in the AORTIC BODY and CAROTID BODY.

heart, cardiac cycle, *n.* the muscular contractions that squeeze blood through the chambers of the heart and around the blood circulatory system. The return of blood into the heart differs in ARTHROPODS from the mechanism in all other groups. In arthropods blood is drawn into the heart from the pericardial cavity by a sucking motion. In other animals blood flow into the heart depends on venous pressure being higher than in the relaxed heart, thus pushing blood inwards.

The features of the cardiac cycle in the mammals are as follows:

(a) CARDIAC MUSCLE has a rhythmic beat which arises in the heart itself. MYOGENIC CONTRACTION occurs at about 72 beats per minute in man, although the rate is higher in smaller mammals. The rate is regulated by two nerves connected to the SAN. Sympathetic nerves from the spinal cord accelerate the heartbeat, while the vagus nerve from the brain medulla slows it down.

(b) when the heart is relaxed (the *diastole* phase) pressure from the venae cavae causes blood to enter the atria and, to a lesser extent, the ventricles. Once full of blood, a wave of electrical energy, originating in the SAN, spreads across the atria causing a sharp muscular contraction, or *systole* which forces blood into the relaxed ventricles.

(c) the electrical energy from the atria collects in the AVN and then spreads over the ventricular walls via the bundle of His and Purkinje fibres, causing ventricular systole. Meanwhile, the atria are starting diastole ready to receive further blood from the veins. These changes are illustrated in Fig. 184 which shows a highly simplified heart structure.

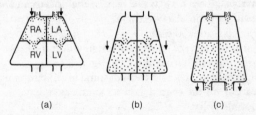

(a) (b) (c)

Fig. 184. **Heart, cardiac cycle.** (a) Atrium and ventricle diastole. (b) Atrium systole and ventricle diastole. (c) Atrium starting diastole, ventricle systole.

heartwood, *n.* the central mass of wood in trees in which there are no living cells and which serves only for mechanical support.

heat exhaustion, *n.* the failure of cooling mechanisms resulting from an excessive rise in body temperature in warm-blooded animals, characterized by cramp and dizziness.

heat gain, *n.* a rise in temperature of a body when obtaining heat by radiation or conduction from an outside source, leading to an increase in METABOLIC RATE in all organs. See TEMPERATURE REGULATION.

heath, *n.* a lowland plant community that is dominated by heathers or ling on a sandy soil or shallow peat.

heat-labile, *adj.* destroyable by heat.

heat-loss, *n.* the removal of heat from a body by radiation, evaporation, conduction and convection during TEMPERATURE REGULATION.

heavy chain, see IMMUNOGLOBULIN.

heavy metal, *n.* any element with an atomic number greater than 20.

heavy-metal pollution, *n.* pollution by HEAVY METALS that is often found in sewage sludge.

heavy isotope, *n.* a stable atom in which there are more neutrons than in the normal isotope of the element, giving it a greater mass. For example, ^{15}N is the heavy isotope, ^{14}N the common form.

hedgerow, *n.* a row of shrubs that forms a habitat similar to a natural woodland edge, planted to form a division between two pieces of land.

Heidelberg man, *n. Homo erectus heidelbergensis*, who lived some 550,000 BP and became extinct some 400,000 BP. Heidelberg man, Java man (*Pithecanthropus*), Peking man (*Sinanthropus*) and *Atlanthropus* are usually all regarded as *Homo erectus*.

HeLa cell, *n.* a cell type used in tissue culture that is grown as a standard in research laboratories all over the world. The culture is derived from a cervical carcinoma obtained from *H*enrietta *La*cks in 1951.

helical, *adj.* of or shaped like a HELIX.

helico-, *prefix.* denoting a spiral.

helicotrema, *n.* an opening at the apex of the COCHLEA connecting the scala tympani and the scala vestibuli.

helio-, *prefix.* denoting the sun.

heliotropism, *n.* a type of PHOTOTROPISM in which the flowers of a plant turn towards the sun during the day.

helix, *n.* (*pl.* helixes or helices) a form of spiral in which any point is at the same distance from the central axis. RNA molecules have a single helix, DNA is a double helix.

helminth, *n.* a PLATYHELMINTH worm.

helminthology, *n.* the study of parasitic worms.

helophyte, *n.* a herb in which the perennating bud is formed in mud.

helper T-cell, see T-CELL.

heme, see HAEM.

hemi-, *prefix.* denoting half.

hemibiotroph, *n.* a pathogen (usually fungal) that has both a BIOTROPHIC and a NECROTROPHIC phase in the course of one life cycle. Examples include apple SCAB and scald of barley (*Rhynchosporium secalis*).

hemicellulose or **hexosan,** *n.* a polysaccharide associated with cellulose and LIGNIN, found in plant cell walls.

hemichordate, *n.* any small wormlike marine animal of the subphylum Hemichordata or Enteropneusta, a division of the PROTOCHORDATES that contains worm-like forms such as *Balanoglossus*.

hemicryptophyte, *n.* a herb in which the perenniating bud is at soil level where it is protected by litter or soil.

Hemimetabola, see EXOPTERYGOTE.

hemiparasite, *n.* an organism that derives part of its sustenance from other organisms.

hemipteran, *n.* any EXOPTERYGOTE of the order Hemiptera, containing the true bugs, including leaf hoppers, cochineal insects, scale insects, bed bugs and aphids. They can cause considerable economic damage, and contain both plant and animal parasites sucking plant sap or blood by means of piercing mouthparts. Normally they possess two pairs of wings, the forewings being at least partially hardened, like the elytra (see COLEOPTERA) of beetles.

hemizygous, *adj.* (of a DIPLOID (2) organism) possessing only one allele of a gene, rather than the normal two. For example, male mammals have only one copy of each gene on the single X-chromosome. See SEX LINKAGE.

Henle, loop of, see LOOP OF HENLE.

Henderson–Hasselback equation, *n.* the formula for calculating the pH of a buffer solution, which states that:

$$pH = pK + \log \frac{(H^+ \text{acceptor})}{(H^+ \text{donor})}$$

where pK is the dissociation constant of an acid. The variable term is really a ratio of concentration values and can be taken as being proportional to the amount of acid and base added in the buffer. For example, when 30 cm^3 of 0.1 M NaOH is added to 100 cm^3 of 0.1 M acetic acid and the pK for acetic acid is 4.73, we can estimate:

$$pH = 4.73 + \log \frac{30}{(100 - 30)}$$
$$= 4.73 + \log 0.429$$
$$= 4.36.$$

heparin, *n.* a mucopolysaccharide molecule produced in the liver that acts as an anticogulant, inhibiting the transformation of prothrombin to thrombin, a vital stage in BLOOD CLOTTING.

hepatic, *adj.* of the liver.

hepatic caecum, see CAECUM.

Hepaticae, *n.* a class of the BRYOPHYTA, a division that contains the liverworts. Liverworts are found in damp and wet conditions and vary in form from species that have a fleshy THALLUS up to 1 cm wide to others (leafy liverworts) which are very MOSS-like. Reproduction in liverworts may be vegetative, by means of GEMMAE, or sexual, ANTHERIDIA and ARCHEGONIA being present. The male gametes are motile and fertilization results in the development of a capsule containing diploid spore-mother cells. Each spore-mother cell carries out meiosis and produces haploid spores that may develop into a PROTONEMA, which in turn gives rise to the new haploid adult gametophyte.

hepatic portal system, *n.* the part of the BLOOD CIRCULATORY SYSTEM of vertebrates in which blood from the alimentary canal is taken through CAPILLARIES in the liver before transport to the heart. Such a system of veins allows some materials absorbed from the gut to be stored or filtered out during passage through the liver.

hepatitis, *n.* a serious disorder of the liver that leads to severe jaundice, liver degeneration and even death. The condition is caused by two viruses: hepatitis A virus, which produces *infective hepatitis* transmitted by the intestinal-oral route, and hepatitis B virus, which produces *serum hepatitis* transmitted via infected blood or its products. Although these two viral types can be distinguished in tests, the acute diseases caused by each may be clinically indistinguishable, the chief difference being that type A usually has a shorter incubation period than type B. No specific therapy is available for hepatitis, although vaccines are being developed.

hepato-, *prefix.* denoting the liver.

hepatocyte, *n.* a liver cell.

herb, *n.* **1.** any seed-bearing nonwoody vascular plant whose aerial parts do not persist above ground at the end of the growing season. **2.** any of various plants, such as rosemary, that are used in cookery and medicine.

herbaceous, *adj.* (of a plant) nonwoody.

herbarium, *n.* a reference collection of preserved plants.

herbicide, *n.* any chemical that kills plants. Herbicides can be highly selective. For example, 2,4-D only kills DICOTYLEDONS (broad-leaved plants), leaving MONOCOTYLEDONS unharmed.

herbivore, *n.* any animal feeding on plant matter.

heredity, *n.* the transmission of characteristics from one generation to another via a mechanism involving GENES and CHROMOSOMES.

heritability (h²), *n.* the proportion of all phenotypic variance in a population that is due to genetic differences and which can be assessed in terms of two main ways, BROAD-SENSE HERITABILITY and NARROW-SENSE HERITABILITY. Heritability is thus a general measure of genetic variation without which SELECTION (both natural and artificial) could not proceed, and is used widely by plant and animal breeders to predict the likely effect of selection. For example, if heritability values for a character are low this indicates high environmental variability, and suggests that the response to selection would not be very rapid.

hermaphrodite, *n.* **1.** any plant possessing stamens and carpels in the same flower. **2.** any animal possessing both male and female sex organs. The condition is usual in many plants and lower animals, but may occur in some unisexual organisms as an abnormality.

heroin, *n.* a white crystalline powder, manufactured from morphine, used as a sedative and narcotic in the hydrochloride form.

herpes simplex, *n.* a common virus of humans that can persist in a quiescent or latent state for long periods and then become active at irregular intervals, producing local blisters in the mucous membranes or in the skin, where local multiplication occurs. A common result of infective activity is 'cold sores' around the mouth and nose, but one form of the virus causes genital herpes, a painful and persistent infection of the genital and anal regions that is transmitted by sexual contact.

Hershey, Alfred (1908–) American scientist who in 1951 with his colleague Martha Chase showed, by means of differential radioactive labelling, that the DNA of T_2 BACTERIOPHAGES (rather than their protein) was necessary for the formation of new phages within the host bacterium, *E.coli.* The experimental results are taken as a demonstration that DNA functions as the genetic material.

hertz (Hz), *n.* the SI UNIT of frequency, measuring cycles per second.

hetero-, *prefix.* denoting dissimilar.

heterocercal fin, *n.* the type of tail fin found on sharks where the ventral lobe is large and the dorsal lobe small.

heterocercal tail, see HETEROCERCAL FIN.

heterochlamydeous, *adj.* (of flowers) having dissimilar sepals and petals in two whorls in the PERIANTH.

heterochromatic, *adj.* (of chromosomal preparations) staining differently from the major chromosome parts.

heterochromatin, *n.* any chromosomal segments or whole chromosomes that appear darkly stained during interphase of the CELL CYCLE (as compared to EUCHROMATIN) due to tight condensation, which may indicate genetic inactivity. Heterochromatin may be condensed at all times (*constitutive*) or only at certain times (*facultative*). See C-BANDING.

heterodont, *adj.* (of an animal) possessing teeth of different kinds, such as incisors or molars.

heteroduplex, *adj.* (of a molecule) possessing double-stranded nucleic acid in which the two POLYNUCLEOTIDE CHAINS come from a different source, either two different DNA strands or a DNA/RNA duplex, held together by COMPLEMENTARY PAIRING.

heteroecious, *adj.* (of rust fungi) having different spore forms on different host plants.

heterogametic sex, *n.* the sex of a particular organism that contains within each nucleus one X-chromosome and one Y-chromosome (or just one X-chromosome), and therefore produces two different types of gamete with respect to the SEX CHROMOSOMES. Usually the male is the heterogametic sex but in some organisms (e.g. birds), it is the female. Since the sex chromosomes usually are responsible for SEX DETERMINATION, it is the heterogametic parent that controls the sex of the offspring.

heterogamy, *n.* the alternation of forms of sexual reproduction. See HETEROGAMETIC SEX.

heterogeneity chi-square test, see CONTINGENCY TABLE.

heterograft or **xenograft,** *n.* a transplant from one species to another.

heterokaryon, *n.* a cell containing nuclei from two different sources, or in fungi a hyphal cell, spore, mycelium or organism that contains genetically different nuclei. See COMPLEMENTATION TEST.

Heterometabola, see EXOPTERYGOTA.

heteromorphic, *adj.* **1.** (of an organism) having morphologically different forms in the life history, for example, where there is ALTERNATION OF GENERATIONS. **2.** (of pairs of homologous chromosomes) differing in size or form.

heteropteran, *n.* any HEMIPTERAN insect of the suborder Heteroptera, having forewings in the form of hemielytra. The suborder includes water boatmen, water scorpions, bed bugs, and most plant bugs.

heterosis or **hybrid vigour,** *n.* the superiority of a HYBRID produced from crossing two different types of parent with one or more inferior characteristics. For example, two different inbred lines of cereals when crossed together can produce hybrids with a far higher yield and general vigour than either parental line.

heterosome, see SEX CHROMOSOME.

heterospory, *n.* the formation of more than one sort of SPORE, usually MICROSPORES and MEGASPORES in ferns and seed plants, giving rise to distinct male and female GAMETOPHYTE generations. The terms mega- and microspore were first used in connection with PTERIDOPHYTES where differences in spore size occur. Later, because of how the structures are formed, these terms were used in reference to seed plants where such size differences are rarer. Here the microspore is the pollen, and the megaspore is the cell which gives rise to the EMBRYO SAC.

heterostyly, *n.* a combination of the morphological and physiological mechanisms that promote cross-pollination in flowering plants. Structurally, there are usually two flower forms (e.g. the 'pin' and 'thrum' forms of the primrose) which ensure that the pollen collected from the stamens of one type is deposited on the stigma of the other (see ENTOMOPHILY). There is also SELF–INCOMPATIBILITY between pollen and stigma from the same flower type, although there is compatibility between flowers of different types. Compare HOMOSTYLY.

heterothallism, *n.* (in algae and fungi) a situation where sexual reproduction occurs only where two self-sterile thalli take part.

heterotherm, see POIKILOTHERM.

heterotrichous, *adj.* (of an algal thallus) prostrate and creeping, and from which branching filaments project.

heterotroph, *n.* an organism dependent on obtaining organic food from the environment because it is unable to synthesize organic material. All animals, fungi, many bacteria and a few flowering plants such as insectivorous plants are heterotrophs, and they obtain almost all their organic material, either directly or indirectly, from the activity of AUTOTROPHS.

heterotroph hypothesis, *n.* a hypothesis suggesting that the first organisms were HETEROTROPHS.

heterozygote, *n.* an individual containing two different allelic forms of the same gene in all DIPLOID(1) cells. Thus if gene *A* has two alleles, *A*1 and *A*2, the heterozygote would contain both alleles (*A*1/*A*2). Compare HOMOZYGOTE.

hex- or **hexa-,** *prefix.* denoting six.

hexacanth, *n.* the six-hooked spherical embryo found in certain tapeworms.

hexapod, see INSECT.

hexosan, see HEMICELLULOSE.

hexose monophosphate shunt, see PENTOSE PHOSPHATE PATHWAY.

hexose sugar, *n.* a sugar containing six carbon atoms (e.g. glucose, $C_6H_{12}O_6$). Hexose sugars are the basic building blocks for more complex CARBOHYDRATES.

Hfr strain, *n.* a strain of *E. coli* bacteria that has the ability to transfer a chromosome to an F^- cell and shows a *high frequency* of recombination. The F FACTOR is integrated into the Hfr cell chromosome.

hibernation, *n.* a state of dormancy entered into by many animals during the winter, particularly those in cooler latitudes. Some temperate and arctic mammals, reptiles and some amphibians hibernate and, in this state, the METABOLISM is slowed down and the body temperature falls. Hibernation is generally triggered by cold weather and whilst body temperatures in mammals normally are maintained at the usual level, they may fall much lower and near to freezing point (e.g. hamster), or the body temperature may fall to that of the surroundings (e.g. bat).

True hibernators tend to be mid-sized animals as these can have sufficient food reserves without too large a surface area for heat loss. Bears are not true hibernators as their body temperature does not drop and they are able to awake from their winter 'sleep' very quickly. In temperate situations, rises in temperature may cause some animals temporarily to come out of hibernation. Compare AESTIVATION(1).

hierarchy, *n.* (in CLASSIFICATION), the system of ranking in a graded order from species to kingdom. See HIGHER CATEGORY.

high-altitude adjustment, *n.* the physiological adjustment to atmospheric pressure that occurs when an animal moves into high altitudes from a lower one, including changes in the respiratory and circulatory system, causing an increase in red blood cells and haemoglobin content of the blood. See ACCLIMATIZATION, and ADAPTATION, PHYSIOLOGICAL.

high-energy bond, *n.* a chemical bond containing several times more stored energy than is usual. See AMP, ATP.

higher category, *n.* (in CLASSIFICATION), any category higher (nearer to kingdom) than the rank of species. See HIERARCHY.

higher taxon, *n.* (in CLASSIFICATION) any TAXON higher (nearer to kingdom) than the one in question.

Hillman and Sartory, see ARTEFACT.

Hill reaction, *n.* a reaction in which oxygen and hydrogen are produced from water when CHLOROPLASTS are illuminated in the presence of a strong oxidizing agent, such as a ferric salt, reported by R. Hill in 1936.

Hill was, however, unable to achieve assimilation of carbon dioxide by chloroplasts (see CALVIN CYCLE). We now know that his experiments were an indication that PHOTOSYNTHESIS is a two-stage process and that he was demonstrating LIGHT REACTIONS.

hilum, *n*. **1.** a scar located down one side of a seed indicating the point of attachment of the OVULE to the ovary. **2.** a notch on the concave side of the mammalian kidney through which the blood vessels and nerves pass.

Himalayan rabbit, *n*. a strain of rabbit in which the coat colour is white but all body extremities are black (nose tip, paws, ears and tail). The condition is controlled by a single autosomal gene with MULTIPLE ALLELISM, one for AGOUTI, one for ALBINISM and the third for Himalayan. The dominance relationship between the three alleles is: agouti > Himalayan > albino. The Himalayan allele is temperature-sensitive, coding for an enzyme that catalyses black pigment formation wherever the skin is cooler, the enzyme becoming inactive at higher temperatures.

hindbrain, *n*. the part of the brain forming the medulla and cerebellum, derived embryologically from the rear third of the brain. See FOREBRAIN, MIDBRAIN.

hip girdle, see PELVIC GIRDLE.

hirsute, *adj*. covered with long hair.

Hirudinea, *n*. a class of the phylum Annelida that includes the leeches, those worms which lack chaetae or parapodia, have an apparent external segmentation much greater than the small number of internal segments and possess suckers for attachment. Most species feed on blood.

His, bundle of, *n*. a group of special cardiac muscle fibres that runs from the artioventricular node down to the septum between the ventricles and, with the Purkinje fibres, transmits electrical energy over the ventricular walls causing ventricular systole. See HEART CARDIAC CYCLE.

hispid, *adj*. (of plant tissues) possessing coarse, stiff hairs.

histamine, *n*. a chemical ($C_5H_9N_3$) produced by LEUCOCYTES and other cells (e.g. MAST CELLS) that causes blood capillaries to become more permeable and so lose fluids into the tissues, producing a local swelling. Histamines are released when foreign ANTIGENS are present. See ANAPHYLAXIS, IMMUNE RESPONSE.

histidine, *n*. one of the 20 AMINO ACIDS common in proteins. It carries an extra basic group making it alkaline and is soluble in water. See Fig. 185. The ISOELECTRIC POINT of histidine is 7.6.

histo-, *prefix*. denoting tissue.

histochemistry, *n*. the study of the distribution of substances in tissues by using staining methods on preparations such as sections or whole mounts.

histocompatibility, *n*. the acceptance by a recipient of tissue trans-

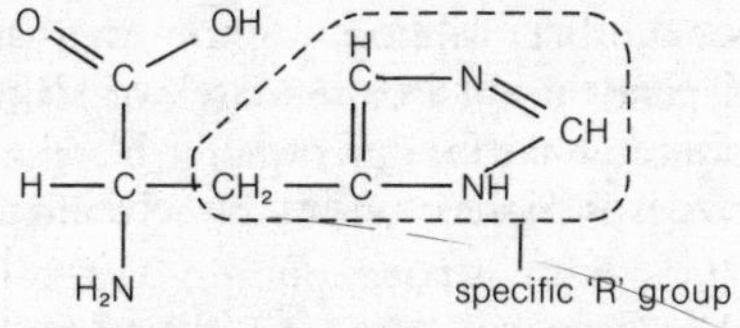

Fig. 185. **Histidine.** Molecular structure.

planted from a donor, a state that is determined by histocompatability ANTIGENS.

histocyte or **clasmatocyte,** *n.* a large phagocyte of irregular shape that occurs in blood, lymph and connective tissue. Histocytes are similar to MONOCYTES but take up stain.

histogen, *n.* a distinct tissue zone found in many plants in the apical meristem, particularly in the root-apical meristem.

histogenesis, *n.* the formation of tissues and organs from undifferentiated cells.

histogram, *n.* a diagram showing a frequency distribution by means of rectangular areas. See FREQUENCY DISTRIBUTION and Fig. 161.

histology, *n.* the study of tissue.

histolysis, *n.* tissue breakdown.

histone, *n.* a type of simple protein that is usually basic and tends to form complexes with nucleic acids (e.g. DNA) forming NUCLEOSOMES. CHROMOSOMES of EUCARYOTES contain large quantities of histones which may regulate DNA functioning in some way. The five major histones are represented as: H1, H2A, H2B, H3 and H4.

HLA (Human Leucocyte A) system, *n.* a major HISTOCOMPATABILITY complex of four genes (perhaps five) called HLA-A, HLA-B, HLA-C and HLA-D, with the D locus nearest to the centromere on the shorter arm of chromosome 6. Each gene has up to 35 alleles, the particular combination of alleles of the four genes on a chromosome being called the *haplotype*. Since humans are diploid organisms we have two haplotypes, these being determined by testing the individual's white blood cells for the specific HLA antigens.

Although these large numbers of alleles would appear to make the system a complex one in terms of possible combinations in each individual, in fact the situation is simpler than it seems. The four loci within the HLA system are so tightly linked together that there is very little CROSSING OVER between them. Thus each haplotype tends to be passed onto an offspring as a block rather than as four individual genes. Given no recombination between the four genes, four rearrangements would be expected to occur in the progeny of a mating. Thus any child of a cross has a one-in-four chance of inheriting precisely the same HLA combination as another sibling. Since the success of organ transplantation depends upon a close match of HLA alleles (even one allele difference can cause the foreign organ to be rejected), the chance of matching HLA types is highest within closely related individuals.

holandric, *adj.* (of genes) carried on the Y-SEX CHROMOSOME and transmitted via the HETEROGAMETIC SEX. In mammals, holandric genes are passed on from father to son.

holdfast, *n.* the basal part of the stripe (stalk) of large sea weeds such as *Fucus, Laminaria*, attaching the plant to the substrate.

holistic, *adj.* viewing a whole as more than the sum of its parts.

holo-, *prefix.* denoting whole.

holoblastic cleavage, see CLEAVAGE.

holocarpic, *adj.* (of the whole adult thallus of a fungus) becoming a reproductive structure.

Holocene epoch or **Recent epoch,** *n.* that part of GEOLOGICAL TIME which has elapsed since the last ice age ended; the last 10,000 years of the QUATERNARY PERIOD.

Holocephali, *n.* a subclass of the CHONDRICHTHYES, including fish having a compressed head and flattened teeth for crushing molluscs; the only present-day representative is CHIMAERA (2).

holocrine, *adj.* **1.** (of a form of cell digestion, particularly in insects) characterized by self-disintegration to produce the digestive fluid. **2.** (of gland secretion) characterized by self-disintegration in releasing its product, as in sebaceous glands.

holoenzyme, *n.* an entire conjugated enzyme consisting of a protein component (an apoenzyme) and a nonprotein component (a coenzyme or an activator).

hologamete, *n.* a gamete formed by a full-sized ordinary individual PROTOZOAN that fuses with another similar individual to form a ZYGOTE.

Holometabola, see ENDOPTERYGOTA.

holophytic, *adj.* (of plants) using the energy of the sun to synthesize organic compounds by means of chlorophyll. See AUTOTROPH.

holothurian, *n.* any ECHINODERM of the class Holothuroidea, including the sea cucumbers.

holotroph, *n.* an organism that is capable of ingesting other whole organisms.

holotype, *n.* the specimen designated by the author of a scientific name as the TYPE specimen at the time of the original publication.

holozoic, *adj.* (of an organism) feeding on solid organic material derived from the bodies of other organisms. This is mainly the method of feeding adopted by animals, though a few specialized plants such as insectivorous plants obtain nutrient matter in this way.

hom- or **homeo-** or **homo-,** *prefix.* denoting the same as.

homeostasis, *n.* the maintenance by an organism of a constant internal environment, such as the regulation of blood sugar levels by insulin. The process involves self-adjusting mechanisms in which the maintenance of a particular level is initiated by the substance to be regulated. See also FEEDBACK MECHANISM.

homeotic mutant, *n.* the mutational change of one structure to another in insects, such as wing to haltere, arista to leg.

homeotherm, see HOMOIOTHERM.

home range, *n.* the area in which an animal normally restricts its movements in search of food or a mate, and in which it cares for its young. Compare TERRITORY.

homing, *n.* (modifier) relating to the ability to return to a place of origin. See NAVIGATION.

hominid, *n.* any member of the family Hominidae, which includes human and human-like fossils from the Pleistocene EPOCH.

homo-, see HOM-.

Homo, *n.* any primate of the hominid genus *Homo.* Only one species exists at the present time, *Homo sapiens*, though several extinct species are recognized, such as *H. erectus, H. habilis.*

homocercal fin, *n.* a type of fish tail fin that is symmetrical in shape, i.e. the upper and lower lobes are of similar size.

homocercal tail, see HOMOCERCAL FIN.

homochlamydeous, *adj.* (of flowers) having the PERIANTH segments which are not distinguishable as separate sepals and petals in two whorls.

homodont, *adj.* (of most nonmammalian vertebrates) having teeth of a similar type.

Homo erectus, see HEIDELBERG MAN.

homogametic sex, *n.* the sex that contains in its nuclei similar sex chromosomes (e.g. XX) and produces gametes all with the same type of SEX CHROMOSOME Compare HETEROGAMETIC SEX.

homogamy, *n.* the state of having anthers and stigmas maturing at the same time.

homogentisic acid, see GARROD, A.

homograft or **allograft,** *n.* a transplant from one individual to another individual that is of the same species but has a different genotype, and is therefore subject to rejection by the recipient as it contains foreign antigens. See HISTOCOMPATABILITY.

homoiotherm or **homeotherm** or **homotherm,** *n.* any warm-blooded animal (mammal or bird). A homoiotherm can maintain its body temperature within a narrow range, usually above that of its surroundings (see CORE TEMPERATURE) despite large variation in environmental temperature. It may be maintained continually or for limited periods only. Compare POIKILOTHERM.

homokaryon, *n.* a cell or mycelium of a fungus, containing identical haploid nuclei.

homologous, *adj.* (of organs, or structures) deriving from the same evolutionary origins. For example, the forelimb of a quadruped, the human arm, the wing of a bird, are said to be homologous (see PENTADACYTL LIMB). Usually similarities are seen best in embryonic development, and are regarded by taxonomists as indications of relationships between present-day organisms.

homologous chromosomes, *n.* those CHROMOSOMES containing identical genetic loci although possibly different allelic forms. In DIPLOID organisms the chromosomes occur in homologous pairs, one member of each pair from a different parent, the pair becoming separated during MEIOSIS and giving rise to genetic SEGREGATION.

homology, *n.* the condition of being HOMOLOGOUS.

homonym, *n.* a specific or generic name that has been used for two or more different organisms. The homonym published first is designated as 'senior', and 'junior' if published last.

homopteran, *n.* any insect of the suborder Homoptera (order HEMIPTERA), in which the forewings are of a uniform nature throughout. Examples include plant bugs, froghoppers, aphids, etc.

homosporous, *adj.* (of an organism) exhibiting HOMOSPORY.

homospory, *n.* the possession of only one form of SPORE. This gives rise to a single GAMETOPHYTE generation carrying both male and female reproduction organs.

homostyly, *n.* the possession of styles of only one length in flowers of a particular species. This is a more common condition than HETEROSTYLY.

homothallism, *n.* the situation in algae and fungi where sexual reproduction occurs in a colony arising from a single spore, each thallus being self-fertile.

homotherm, see HOMOIOTHERM.

homozygote, *n.* an individual containing two identical forms of the same GENE in all DIPLOID(1) cells that is capable of *pure breeding*. Thus if gene A has two alleles, A1 and A2, two homozygous types are possible: A1/A1 and A2/A2. Compare HETEROZYGOTE.

honeybee, *n. Apis melifica*, the species of bee (see HYMENOPTERAN) kept by man to produce honey.

honeycomb, *n*. a structure of waxy hexagonal cells produced by HONEY BEES and in which young are reared and honey stored.

honeydew, *n*. the sugary waste substance passed out by aphids and similar insects.

hookworm, *n*. a NEMATODE parasite of man that gives rise to anaemia and mental and physical retardation. See ANCYLOSTOMIASIS.

hoof, *n*. a horny casing of the toe produced by hardened epidermal cuticle (keratin), found particularly in ungulates.

horizon, *n*. any layer of soil that is distinguishable when soil is examined in vertical sections.

horizontal classification, *n*. the type of CLASSIFICATION that stresses the grouping of species at a similar level of evolution rather than the position on the same phyletic line. Compare VERTICAL CLASSIFICATION.

horizontal resistance, see NONRACE–SPECIFIC RESISTANCE.

hormone, *n*. a chemical that is produced in the body of a plant or animal, sometimes in very small quantities, and which when transported (usually by the blood stream in animals) to another part of the organism, elicits a particular response. Hormones are thus chemical messengers. In animals the ductless glands that secrete hormones are called ENDOCRINE ORGANS. Animal hormones that influence other endocrine glands are called *trophins* (or *tropins*) as in GONADOTROPHINS. See also PLANT HORMONE.

horsetail, see EQUISETUM.

host, *n*. the organism on which a PARASITE lives.

H–substance or **H–antigen,** *n*. the carbohydrate precursor of the A and B antigens found in red blood cells (see ABO BLOOD GROUP) and in various tissue fluids (see SECRETOR STATUS), recognized by an anti-H-reagent made from gorse seeds.

human chorionic gonadotrophin (HCG), *n*. a gonadotrophic hormone secreted by the PLACENTA (1) that has a similar effect to luteinizing hormone (see LH).

humerus, *n*. the bone of the vertebrate forelimb (or arm) nearest to the body, to which it is attached at the shoulder. It is attached distally to the RADIUS and ULNA at the elbow.

humoral, *adj*. of or relating to a body fluid, as in humoral immunity produced by antibodies carried in lymph.

humus, *n*. the organic material that is derived from the breakdown of plant and animal material occurring in the surface layers of the soil. Humus is colloidal in nature and improves the fertility of soil in several ways: (a) in acting as a reservoir of numerous nutrients which it prevents from leaching deeper into the soil layers; (b) in increasing the water holding capacity of the soil and (c) in improving the soil texture (friability).

humus-carbonate soil, see RENDZINA.

Huntingdon's chorea, *n.* an incurable degeneration of the human nervous system characterized by involuntary movements of the head, face and/or limbs, leading to eventual death. The disorder is caused by a single autosomal dominant gene on chromosome 4 (see DOMINANCE) that shows delayed penetrance, fewer than 5% of cases being seen before the age of 25. Thus many individuals carrying the dominant allele have reproduced before their own gene shows itself; their progeny in turn may have to wait until middle age to discover if they have inherited the condition. Named after the American neurologist George Huntingdon (1851–1916).

hyal- or **hyalo-,** *prefix.* denoting glass.

hyaline, *adj.* transparent.

hyaline cartilage, see CARTILAGE.

hyaloid canal, *n.* a canal found in the vertebrate eye that passes through the vitreous humour from the lens to the blind spot.

hyaluronidase, *n.* an enzyme present in snake venom and bacteria that catalyses the hydrolysis of hyaluronic acid, thus making it ineffective in stopping the spread of invading microorganisms and other toxic substances.

hybrid, *n.* an offspring of a cross between two genetically unlike individuals. See also HETEROZYGOTE, HETEROSIS.

hybrid DNA, *n.* a nucleic acid produced by joining together or 'annealing', POLYNUCLEOTIDE CHAINS of DNA from different sources. See HETERODUPLEX.

hybridization, see MOLECULAR HYBRIDIZATION.

hybrid sterility, *n.* the inability of some HYBRIDS to form functional gametes, due to chromosomes mispairing during MEIOSIS. This is a form of REPRODUCTIVE ISOLATION.

hybrid swarm, *n.* a group of genetically variable organisms resulting from the hybridization of two previously distinct and separate populations. Subsequent crossing and backcrossing results in a very variable population as a result of segregation. Often the genetical variability appears as morphological variability, such as different colour forms. Where different colour forms of an animal occur in different parts of the range, the hybrid swarm is characterized by the presence of all or a wide range of these forms.

hybrid vigour, see HETEROSIS.

hydathode, *n.* a gland occurring on the leaf edges of many plants and secreting water. See GUTTATION.

hydatid cyst, *n*. the bladder-like larva of some tapeworms, such as *Echinococcus*, that produces an enormous bladder up to five litres in capacity, which may bud off many smaller cysts. In mammals, cysts occurring in the brain produce symptoms of epilepsy.

hydo- or **hydro-,** *prefix*. denoting water.

Hydra, *n*. a genus of the coelenterate class Hydrozoa. Most species are found in freshwater, and are unusual members of the Hydrozoa in not having an ALTERNATION OF GENERATIONS.

hydranth, *n*. the normal COELENTERATE polyp, having a mouth surrounded by tentacles, found in most colonial forms.

hydration, *n*. the process of chemical combination with water to form a compound referred to as a *hydrate*.

hydrocoel, *n*. the water vascular system of echinoderms, which has branches to the tube feet and is concerned to this extent with locomotion.

hydrogen acceptor, *n*. any substance (e.g. CYTOCHROME OXIDASE) that can become reduced by the addition of hydrogen, thus enabling the transfer and release of energy, as in the ELECTRON TRANSPORT SYSTEM.

hydrogen bond, *n*. the attractive force between the hydrogen atom of one molecule and another molecule forming a noncovalent bond. These weak bonds are relevant to the biological function of particular compounds, being formed in the secondary structure of proteins and between complementary base pairs in NUCLEIC ACIDS.

hydrogen carrier system, see ELECTRON TRANSPORT SYSTEM.

hydrogen ion concentration, see PH.

hydroid, *n*. a member of the COELENTERATE class Hydrozoa. Most are colonial forms which grow in the marine environment on rocks and seaweeds, such as *Obelia*. Usually, there is an ALTERNATION OF GENERATION (not in HYDRA) in which the MEDUSA (free-swimming) stage carries the sex organs. The sedentary, nonsexual stage of this life cycle is itself referred to as the POLYP or (less commonly) *hydroid stage*.

hydrolase, *n*. an enzyme that mediates hydrolytic reactions.

hydrolysis, *n*. a chemical reaction in which large molecules are broken down by the addition of water. For example, fat to fatty acids and glycerol, MALTOSE to glucose, DIPEPTIDE to two amino acids. The reactions are usually enzymically activated. Compare CONDENSATION REACTION.

hydrolytic enzyme, see HYDROLASE.

hydromedusa, *n*. the MEDUSA of a HYDROZOAN.

hydrophilic, *adj*. having an affinity for water.

hydrophobic, *adj*. having an aversion for water.

hydrophyte, *n*. a plant that inhabits wet places and has its perennating buds lying in water. Compare MESOPHYTE, XEROPHYTE.

hydroponics, *n*. the science of growing plants without soil, in which the roots are suspended in aerated water containing known quantities of chemicals that can be adjusted to suit changing conditions. The method is increasingly common in the glasshouse cultivation of, for example, tomatoes and cucumbers.

hydrorhiza, *n*. the root-like base of a colonial COELENTERATE which fixes it to the substrate.

hydrosere, *n*. a plant succession originating in water.

hydrosphere, *n*. the aqeuous envelope of the earth, including underground water and the atmospheric water vapour.

hydrostatic skeleton, *n*. a liquid, usually water, that maintains the shape of an organism by filling internal spaces. See SKELETON.

hydrotheca, *n*. the cup-like part of the perisarc of colonial COELENTERATES which surrounds and protects the polyp.

hydrotropism, *n*. the growth of plant roots in a particular orientation with respect to water, usually towards moisture.

hydrozoan, *n*. any colonial or solitary COELENTERATE of the class Hydrozoa, including most of the species that exhibit an ALTERNATION OF GENERATION between HYDROID and medusoid (see MEDUSA) forms.

hymen, *n*. a fold of mucous membrane stretched across the entrance to the vagina in some mammals. There is normally a small opening in it and the hymen is ruptured during the first copulation.

hymenium, *n*. a layer of spore-producing structures present in the fruiting body of BASIDIOMYCETES and ASCOMYCETES.

hymeno-, *prefix*. denoting a membrane.

Fig. 186. **Hymenoptera.** A winged female ant.

Hymenoptera, *n*. an insect order of the subclass ENDOPTERYGOTA, including some of the most important social insects, such as ants, bees and wasps. They are characterized by the presence of two pairs of membranous wings that are connected together and often there is a marked constriction ('wasp-waist') between the abdomen and thorax. Apart from the social insects, the order also contains many solitary bees and wasps and the ichneumon flies, the larvae of which are often parasitic on other insects. See Fig. 186.

hyo- or **hyoid,** *prefix*. denoting Y-shaped.

hyoid arch, *n*. the visceral skeletal arch that lies directly behind the mandibular arch and supports the floor of the vertebrate mouth.

hyomandibula, *n*. a bone or cartilage present at the end of the hyoid arch that forms part of the jaw structure in fish and the STAPES in higher vertebrates.

hyostylic jaw suspension, *n*. a form of jaw suspension in which the upper jaw is articulated to the skull only by means of ligaments and the hyomandibula, as in the dogfish.

hyper-, *prefix*. denoting greater than.

hypercalcaemia, *n*. having excessive levels of calcium in the plasma.

hypercapnaea, *n*. having excessive levels of carbon dioxide in the plasma.

hyperemia, *n*. an increased blood flow to an organ or tissue.

hyperglycaemia, see GLYCOSURIA.

hypermetropia, see HYPEROPIA.

hyperopia or **hypermetropia** or **long-sightedness,** *n*. an inability to see nearby objects clearly, due to their image being focused behind the retina because the eye lens power is insufficient for the size of eye. Compare MYOPIA.

hyperparasite, *n*. an organism parasitic on another parasite.

hyperplasia, *n*. an increase in tissue mass caused by an increase in cell number.

hyperpnoea, *n*. an increased ventilation of the lungs.

hypersensitivity, *n*. the process of localized plant cell death that occurs immediately after entry of a cell by a plant pathogen. Such a reaction can act as a host-resistance mechanism since the cell death often prevents further growth of biotrophic pathogens, e.g. powdery mildews or rusts.

hyperthyroidism, *n*. a medical condition in which overproduciton of the thyroid hormone causes nervousness, sensitivity to heat and insomnia.

hypertonic, see HYPOTONIC.

hypertrophy, *n*. the excessive growth or development of an organ or tissue.

hyperventilation, *n*. **1.** an increase in air inhalation into the lungs. **2.** the depth or rate of breathing.

hypha, *n*. (*pl*. hyphae) a filament of the plant body of a fungus, the total of which make up the nonreproductive part of the organism, as opposed to the fruiting body. Hyphae may be septate, having internal septae as in some *Phycomycetes*, or nonseptate as in most other groups. However, even in the septate stage, pores are present in the septae so that there is a continuity of cytoplasmic material throughout the hypha. See also HAUSTORIUM.

hypo-, *prefix.* denoting under.

hypocotyl, *n.* the shoot in a young germinating seedling, located below the cotyledon. See GERMINATION and Fig. 166.

hypodermis, *n.* the layer of cells immediately under the epidermis of plant leaves. The layer is sometimes used for water storage or for mechanical strengthening for extra protection.

hypodigm, *n.* all the material available to a taxonomist.

hypogeal, *adj.* (of seed GERMINATION) the COTYLEDONS(2) remaining undergound within the seed coat, with the young shoot and root growing out from the seed. For example, broad bean (*Vicia faba*).

hypoglossal nerve, *n.* the 12th CRANIAL NERVE of higher vertebrates, the motor nerve supplying the floor of the mouth and the tongue.

hypogynous, see GYNOECIUM.

hyponasty, *n.* the more rapid growth of the lower side of an organ, compared with the upper side, that results in curving.

hypopharynx, *n.* a chitinous structure having its origin in the floor of the mouth in insects and normally carrying the salivary apertures. In blood-sucking insects the structure extends into the proboscis.

hypophysectomy, *n.* the removal of the pituitary GLAND.

hypophysis, *n.* **1.** the pouch formed in front of the buccal cavity in the embryo which together with the INFUNDIBULUM forms the pituitary. **2.** a loose term for PITUITARY GLAND.

hypopnea, *n.* the decreased ventilation of the lungs.

hypostasis, *n.* a relationship between two genes whose products act in the same biochemical PATHWAY, where the functional effect of one gene is masked by another. The enzyme coded by the hypostatic gene operates later in the pathway than the enzyme produced by the epistatic gene. See EPISTASIS.

hypostome, *n.* any structure around or below the mouth, such as the oral cone in *Hydra*.

hypothalamus, *n.* the part of the sides and floor of the brain derived from the forebrain. Associated with the control of body temperature, it also partially controls the PITUITARY GLAND and contains centres controlling sleep and wakefulness, feeding, drinking, speech and osmoregulation, the last by secretion of ADH from neurosecretory cells. It is also associated with aspects of reproductive behaviour.

hypothermia, *n.* an abnormally low body temperature that reduces the metabolic rate to a dangerous level and can lead to death, often occurring in elderly people subjected to cold weather. Hypothermia is sometimes induced in medical treatment in order to reduce metabolic activity.

hypotonic, *adj.* (of a fluid) having a less-negative WATER POTENTIAL or a lower solute concentration than another fluid, which is *hypertonic* to it. See Fig. 187. See also PLASMOLYSIS.

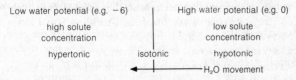

Low water potential (e.g. −6)		High water potential (e.g. 0)
high solute concentration		low solute concentration
hypertonic	isotonic	hypotonic

←——— H$_2$O movement

Fig. 187. **Hypotonic.** The movement of water from a hypotonic fluid to a hypertonic one.

hypoxia, *n.* the reduction of oxygen levels.

hypsodont, *adj.* (of teeth) having open roots that continue to grow as fast as they are worn down. They are common in herbivorous animals such as ungulates and rodents.

H zone, *n.* the lightish area in the middle of a muscle SARCOMERE where MYOSIN and ACTIN filaments do not overlap.

I

IAA (indole-3-acetic acid or **indolacetic acid),** *n.* the most thoroughly investigated type of AUXIN, which has been extracted from various natural sources. See Fig. 188.

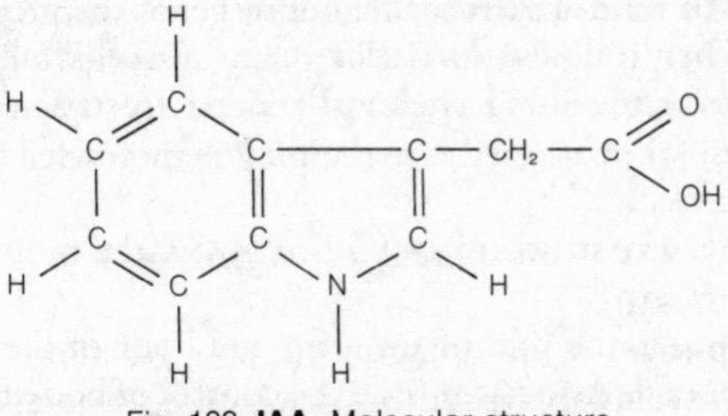

Fig. 188. **IAA.** Molecular structure.

IAN (indole-3-aceto nitrile), *n.* a plant growth-regulating hormone.

I-band or **isotropic band,** *n.* a light band which, in electron micrographs, can be seen traversing skeletal muscle myofibrils that can just be seen on the highest power of the light microscope as fine threads running the length of the muscle fibre. Each myofibril is made up of a series of sarcomeres in the form of a chain (see SARCOMERE and Fig. 265). In electron micrographs the fibrils can be seen to have dark and light bands. Each I-band is traversed by a darker Z-membrane and is made up of thin filaments of ACTIN. See MUSCLE and Fig. 219.

ichthy-, *prefix.* denoting a fish.

ichthyosaur, *n.* a large extinct reptile that existed from the TRIASSIC PERIOD to the CRETACEOUS PERIOD. It had paddle-shaped limbs, a head with long toothed jaws, and a fish-like tail.

identical twins, see MONOZYGOTIC TWINS.

identification, *n.* the determination of the taxonomic identity of an individual. See KEY.

idio-, *prefix.* denoting individual, peculiar.

idioblast, *n.* a cell distinct from others in the same tissue.

idiogram, *n.* a diagrammatic representation of a KARYOTPYE.

ileum, *n.* that part of the intestine lying between the duodenum and the colon where digestion is completed by enzymes that break down carbohydrates, fats and proteins (see SMALL INTESTINE). Absorption of food also occurs here.

ilium, *n.* the dorsal bone of the PELVIC GIRDLE that is joined to the SACRAL VERTEBRAE.

imaginal disc, *n.* a structure found in the pupae of ENDOPTERYGOTES, formed by a thickening of the epidermis and underlying tissue, and which, on metamorphosis, gives rise to an adult structure.

imago, *n.* the sexually mature, adult insect.

imbibition, *n.* the uptake of water, for example, by the dry seed, that causes GERMINATION to start. The process is due, not to SELECTIVE PERMEABILITY through a membrane, but rather to the property of water ADSORPTION by colloidal particles such as cellulose, pectin and cytoplasmic proteins, using chemical and electrostatic attraction.

imbricate, *adj.* (of plant parts) overlapping at their edges like roof tiles, when in bud.

immigration, *n.* the movement of organisms into a specific area. Compare EMIGRATION.

immune response, *n.* an antagonistic host reaction in response to foreign ANTIGENS, involving the formation of ANTIBODIES by B-CELLS (or a cell-mediated response by T-CELLS). When such antibodies are present in the body, the individual is said to possess IMMUNITY against the specific antigen that stimulated the antibody production. The immune response is a vital defence mechanism but creates severe problems with, for example, kidney transplants from donors. In these cases the tissues are 'matched' for *tissue compatibility* and the recipient treated with chemicals that reduce the immune response.

immunisation, see IMMUNIZATION.

immunity, *n.* resistance to foreign ANTIGENS such as a virus. Immunity can be either *active* in which the body produces its own IMMUNE RESPONSE after exposure to a vaccine or *passive* in which ready-made antibodies are supplied in a serum (or obtained naturally from the mother).

immunization, *n.* the administration of an ANTIGEN, in the form of a vaccine, to produce an IMMUNE RESPONSE to that antigen and so protect against future exposure to the antigen. See ATTENUATION.

immunofluorescence, see FLUORESCENT ANTIBODY TECHNIQUE.

immunoglobulin, *n.* a protein (such as gamma globulin) made in B-CELLS that possesses ANTIBODY activity and is made up of four POLYPEPTIDE CHAINS, two identical *light chains* and two identical *heavy chains* joined by disulphide bonds. There are five main classes of human immunoglobins, differentiated principally by their heavy chains:

Type	IgG	IgM	IgA	IgD	IgE
Heavy chain	gamma	mu	alpha	delta	epsilon
M.W. (x 1000)	144	160	144	156	166

Each heavy and light chain consists of a CONSTANT REGION of amino acids that is virtually the same in all antibodies of that class and a VARIABLE REGION containing a sequence of amino acids that makes the antibody specific to a particular ANTIGEN. See Fig. 189.

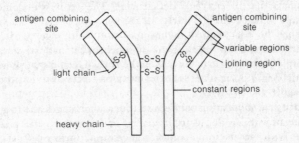

Fig. 189. **Immunoglobulin.** Molecular structure. S = sulphur bonds.

immunological tolerance, *n.* the failure to respond to a potential antigen.

immunotoxin, *n.* an antibody attached to a specific toxic agent that is used to destroy specific cancer cells. Although experimental at present, these so-called 'magic bullets' hold out great hope as a new form of cancer CHEMOTHERAPY.

impedance, *n.* any resistance to the flow of fluids moved by a series of pulses, such as blood flow.

imperfect fungi, see FUNGI IMPERFECTI.

implantation, *n.* the act of attachment of the mammalian embryo to the uterus wall of the mother.

impressed, *adj.* sunk below surface level.

imprinting, *n.* an aspect of learning where there is a rapid development of a response to a particular stimulus at an early stage of development. Young animals 'recognize' the first object they see, such as a mother figure, and they can be 'imprinted' on objects other than their own species. For example, Konrad LORENZ 'imprinted' young greylag geese on himself as a mother figure. Imprinting also occurs in other areas of experience. For example, bird song, where young, inexperienced birds have adult calls 'imprinted' on them.

inactive-X hypothesis or **Lyon hypothesis,** *n.* an explanation for DOSAGE COMPENSATION in the X-chromosomes of mammals including humans, proposed by Mary Lyon. The main features of the hypothesis are that only one X-chromosome is active in a mature mammalian cell, all other X-chromosomes present having been inactivated during embryological development. Different X-chromosomes are thought to be inactivated in different cells, which may produce a MOSAIC phenotype in heterozygous individuals. BARR BODIES are thought to indicate the presence of inactive X-chromosomes.

inborn error of metabolism, *n.* an inherited biochemical abnormality in humans due to enzymic deficiencies. The term was coined by Archibald GARROD in the early 1900s, who theorized that various conditions he had studied in hospital cases were due to defective enzymes, causing the breakdown of biochemical pathways and the build-up of intermediate chemicals in the body. Garrod's work was not recognized as important until the ONE GENE/ONE ENZYME HYPOTHESIS was proposed in the 1940s.

inbreeding, *n.* any mating between relatives, which can lead to a general lowering of viability due to accumulation of deleterious genes passed on through both parents from a common ancestor. Compare OUTBREEDING.

inbreeding depression, *n.* a reduction in viability resulting from increased homozygosity through INBREEDING.

incisor, *n.* a front, chisel-shaped tooth in mammals, normally used for cutting or gnawing purposes. In some species incisors continue to grow throughout life. For example, gnawing rodents and rabbits have 'persistent pulps' where the pulp cavity remains open throughout life, while in other species they are modified as tusks.

inclusion, *n.* a particle or structure contained within a cell or organ.

inclusion body, *n.* a body present in the nuclei or cytoplasm of cells infected by viruses or other intracellular parasites.

incompatibility, see SELF-INCOMPATIBILITY.

incomplete dominance, *n.* a pattern of inheritance in which a cross between two phenotypically different parents produces an offspring different from either parent but containing partial features of both. The classic example is in flower colour where, for example, crossing white and red parents produce a pink offspring. Compare CODOMINANCE. See also DOMINANCE(1).

in coupling, see COUPLING.

incubation, *n.* **1.** the process of brooding or incubating in birds. **2.** the period between infection by a pathogen and appearance of disease symptoms. **3.** the maintenance of microbiological cultures at specific temperatures for a given time.

incurved, *adj.* (of plant structures) gradually bent inwards.

incus or **anvil,** *n.* the anvil-shaped bone of the middle EAR, the central bone of the three ear ossicles, situated between the MALLEUS and STAPES.

indehiscent, *adj.* (of a plant organ) not opening to release spores or seeds.

independent assortment, *n.* the random arrangement and separation of chromosomes during MEIOSIS, giving all possible combinations in equal frequency, unlike the situation with GENETIC LINKAGE. The process is important in understanding MENDELIAN GENETICS and explains the random distribution in the gametes of genes or nonhomologous chromosomes. For example, take two pairs of HOMOLOGOUS CHROMOSOMES in a DIPLOID(1) cell. During anaphase I there are two ways in which the chromosomes can become separated (see Fig. 190). Independent assortment of the chromosomes in the Fig. gives four types of possible gamete $(1 + 3)$, $(2 + 4)$, $(1 + 4)$ and $(2 + 3)$. In fact, the number of combinations is 2 to the power of the number of pairs of chromosome, $2^2 = 4$ in the above example. A human with 23 pairs of chromosomes would produce $2^{23} = 8,388608$ combinations by independent assortment alone.

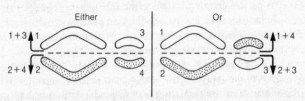

Fig. 190. **Independent assortment.** Separation of chromosomes.

indicator community, *n.* an association of species characteristic of particular conditions, or habitats. For example, a range of CALCICOLE plants indicates a base rich soil.

indicator species, *n.* any species that is indicative of particular conditions or habitats. For example, the leech *Erpobdella testacea* is found only in alkaline or polluted waters, the arrowworm *Sagitta setosa* is characteristic of continental shelf water, whereas *S. elegans* is characteristic of oceanic water.

indigenous, *adj.* native, not introduced by man.

indirect selection, *n.* the selection of mutant organisms (microorganisms) by determining those that do not grow on particular media.

individual variation, *n.* the variation found in a population of the same species.

indolacetic acid, see IAA.

inducer, *n.* a small molecule that inactivates the repressor of an OPERON, usually binding to the repressor and so preventing the repressor binding with the operator of the operon.

inducible enzyme, *n.* an enzyme that is produced in response to the presence of its substrate (the inducer). See OPERON MODEL.

induction, *n.* **1.** (in biochemistry) the synthesis of new enzyme molecules in response to a stimulus, usually another chemical. See OPERON MODEL. **2.** (in embryology) the formation of an alternative cell type during CELL DIFFERENTIATION under the influence of an 'inducer' molecule. **3.** (in microbial genetics) the release of a PROPHAGE for the bacterial chromosome, initiated by, for example, ultraviolet light.

indumentum, *n.* a total hairy covering.

indurated, *adj.* toughened and hardened.

indusium, *n.* the covering of a sporangium or several sporangia.

industrial melanism, *n.* a phenomenon found in several groups, notably moths, in which a heavily pigmented variant (a MORPH) becomes the most frequent type in an area with heavy, man-made atmospheric pollution. Melanism is an excellent example of MICROEVOLUTION, in that rapid evolutionary change has been brought about by strong forces of natural selection acting in favour of a dominant allele for melanism. The agents of selection against moths (such as the peppered moth *Biston betularia*) are birds, which predate more heavily those types resting against an inappropriate background: poorly pigmented forms resting on sooty tree trunks, for example. The proportion of melanics to nonmelanics in an area is correlated with the level of pollution, although neither type appears to be completely absent in any environment, creating a GENETIC POLYMORPHISM. See also KETTLEWELL.

infauna, *n.* the burrowing portion of the BENTHOS.

infection, *n.* the invasion of tissues by microorganisms with or without disease production.

infective hepatitis, see HEPATITIS.

inferior ovary, *n.* an ovary in which the perianth is inserted on top of the ovary, and which appears fused with the receptacle.

inflammation, *n.* a local response to injury or damage, including dilation of blood vessels, and the invasion of blood proteins, blood fluid and LEUCOCYTES into the tissues to combat invading bacteria.

inflexed, *adj.* bent inwards.

inflorescence, *n.* a specialized branching stem bearing flowers, for example, a male CATKIN.

influenza, *n.* an acute respiratory disease affecting the upper respiratory tract, with symptoms of fever, chills and generalized aching, caused by a number of influenza viruses particularly types A and B. Influenza outbreaks occur frequently and can have a worldwide distribution, often being associated with secondary bacterial complications. Over 20 million people died in the PANDEMIC outbreak of 1918–1919, principally young adults who may have been previously exposed to an unusually virulent form of virus.

information theory, *n.* the study of the measurement and properties of codes and messages.

infra-, *prefix.* denoting beneath.

infrared, *n.* the electromagnetic radiation in the region between red light and radio waves. See ELECTROMAGNETIC SPECTRUM.

infraspecific, *adj.* within a species.

infrasubspecific, *adj.* within a subspecies.

infundibulum, *n.* the part of the PITUITARY GLAND that is produced by a downgrowth of the posterior region of the forebrain.

infusion, *n.* the liquid extract of any substance which has been soaked in water.

infusoria, *n.* any organisms found in INFUSIONS of organic material, such as Protozoa, Rotifers, etc. The term is sometimes restricted to the CILIATES.

ingestion, *n.* the act of taking food into the gut system, where it is then subjected to DIGESTION.

inheritance, *n.* **1.** the acquisition of characteristics by the transfer of genetic material from ancestor to descendant. **2.** the total of characters in the fertilized ovum.

inherited abnormality, *n.* any genetically determined malfunction. These fall into two main types:

(a) those controlled by single genes where, for example, a biochemical process is disrupted by faulty enzymes coded by abnormal alleles such as those causing ALBINISM (see also INBORN ERRORS OF METABOLISM). In other cases, such as SICKLE–CELL ANAEMIA, a molecule is produced that does not function in the normal way.

(b) those produced by chromosomal changes (see CHROMOSOMAL MUTATION). Here, loss or addition of parts or of whole chromosomes can have serious effects. Changes to autosomal chromosomes (see DOWN'S SYNDROME) often produce more severe effects than to sex chromosomes (see KLINEFELTER'S SYNDROME).

inhibiting factor, see RELEASING FACTOR.

inhibition, *n.* a state in which an enzyme is unable to catalyse reactions. See COMPETITIVE INHIBITION and NONCOMPETITIVE INHIBITION.

initiation codon, *n.* a triplet of bases in mRNA with the sequence AUG that acts as a 'start' signal for TRANSLATION, specifying the first amino acid at the N-terminus of the POLYPEPTIDE CHAIN. See GENETIC CODE.

initial, *n.* the cell or group of cells that differentiates to give rise to other tissues or organs, such as a MERISTEM.

ink sac, *n.* a sac that opens into the rectum of some CEPHALOPODS, that when stimulated releases dark brown fluid that acts as a 'smoke' screen.

innate behaviour, *n.* any behaviour that is exhibited by animals reared in isolation and which appears to be inherited, and thus instinctive, in some interpretations of the meaning of this word. Innate behaviour develops independently of any experience of the behaviour in other animals. See INSTINCT, IMPRINTING.

innate reflex, *n.* any behavioural response, such as a reflex action (see REFLEX ARC), that is automatic and not learned.

innervate, *vb.* **1.** to supply nerves to (a bodily organ or part). **2.** to stimulate (a bodily organ or part) with nerve impulses.

innominate, *n.* **1.** a short artery arising from the AORTA that gives rise to the subclavian and carotid arteries. **2.** the fusion of ilium, ischium and pubis to form a single bone forming half of the PELVIC GIRDLE.

inoculation, *n.* the introduction of biological material (the *inoculum*) into a medium such as a living organism, synthetic substrate or soil.

inoculum, see INOCULATION.

inotropic, *adj.* affecting or controlling the strength of heart contractions. Chemicals that, for example, increase the force of contraction, have a positive inotropic effect.

input load, *n.* the quantity of inferior alleles in the gene pool resulting from mutation and immigration.

in repulsion, *n.* see REPULSION.

insect, *n.* any small air-breathing arthropod of the class Insecta, containing organisms that normally in the adult have six legs, three distinct regions to the body (head, thorax and abdomen), one pair of antennae and one or two pairs of wings.

Mouthparts are often adapted to the method of feeding, such as biting, piercing and sucking (see Fig. 191).

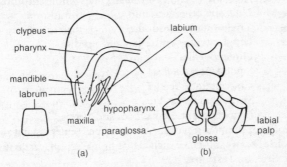

clypeus
pharynx
mandible
labrum
maxilla
hypopharynx
labium
paraglossa
glossa
labial palp

(a)   (b)

Fig. 191. **Insect.** (a) Vertical section, and (b) front view of insect mouthparts.

Abdominal appendages are absent except in the more primitive groups such as springtails. Most insects have a distinct juvenile stage, a nymph (see EXOPTERYGOTA) or a larva (see ENDOPTERYGOTA). These undergo METAMORPHOSIS to form the adult. Insects comprise about five-sixths of all known animal species. The class contains the following groups:

Subclass: APTERYGOTA (wingless)
Orders: Protura
           Collembola – springtails
           Diplura – japygids
           Thysanura – bristletails

Subclass: PTERYGOTA (normally winged)
          EXOPTERYGOTA (HETEROMETABOLA or
          HEMIMETABOLA)
Orders: Odonata – dragonflies
           Ephemeroptera – mayflies
           Orthoptera – grasshoppers
           Dermaptera – earwigs
           Plecoptera – stoneflies
           Isoptera – termites
           Embioptera – embiids
           Mallophaga – biting lice
           Anoplura – sucking lice
           Psocoptera – book lice

Zoraptera
Hemiptera – bugs
Thysanoptera – thrips
ENDOPTERYGOTA (HOLOMETABOLA)
Mecoptera – scorpion flies
Neuroptera – lacewings
Trichoptera – caddisflies
Lepidoptera – butterflies and moths
Diptera – flies
Siphonaptera – fleas
Coleoptera – beetles
Strepsiptera – stylops
Hymenoptera – ants, bees and wasps

The more important insect orders are described under separate headings.

insecticide, *n.* any substance such as DDT or malathion, that is used to kill insects. See also PESTICIDE.

insecticide resistance, *n.* the ability of a member of an insect population to withstand the toxic effects of an insecticide to the point that it now resists control with that chemical. The genes controlling resistance are thought to be present in low frequencies within a generally susceptible population before application of the chemical. After treatment, susceptible members of the population are removed and the survivors thrive, becoming common in the population which is then described as 'resistant'.

insectivore, *n.* any member of the insect-eating mammalian order Insectivora, a group containing shrews, moles and hedgehogs in the British fauna.

insemination, *n.* any means by which male sperm are introduced to the female prior to FERTILIZATION. See ARTIFICIAL INSEMINATION.

insertion, *n.* **1.** a point of attachment of an organ such as a leaf or muscle. **2.** the point of application of force by a muscle.

insertion mutation, *n.* a POINT MUTATION in which one or more bases is inserted into DNA, causing a FRAMESHIFT reading error during TRANSLATION.

insight learning, *n.* the production of a new and adaptive response as a result of 'insight'. For example, presented with a bunch of bananas that is too high to reach, chimpanzees will pile up boxes, or fit two sticks together in order to get them.

inspiration, see BREATHING.

inspiratory centre, see BREATHING.

instantaneous speciation, *n.* the production of an individual (or individuals) that is reproductively isolated from its parent and is capable itself of reproduction and establishing a new population.

instar, *n.* a larval stage of insect development. The first instar hatches from the egg and after the first moult (see ECDYSIS) becomes the second instar. Thus a third instar larva is one which has completed two moults.

instinct, *n.* aspects of behaviour that are not learned, but which appear to be inherited, i. e. INNATE BEHAVIOUR. It is not now used commonly as a scientific term because of the difficulty of distinguishing between some aspects of learning and some aspects of so-called instinctive behaviour.

instinctive behaviour, *n.* any behaviour pattern that is inherited and not clearly learned.

insulin, *n.* the hormone controlling the amount of blood sugar, which is secreted by the beta cells of the ISLETS OF LANGERHANS in the pancreas. Insulin has three targets: the liver, the muscles, and adipose tissue, where its action helps to reduce the blood sugar level in the following ways:

(a) it stimulates the absorption of more glucose from the blood into muscle and adipose cells, by altering cell-wall permeability.

(b) it stimulates the conversion of glucose into GLYCOGEN in the liver and muscles, reducing the supply of free glucose.

(c) it promotes the conversion of glucose into fats in the liver and adipose cells (LIPOGENESIS).

(d) it inhibits GLUCONEOGENESIS.

(e) it promotes GLYCOLYSIS of glucose in all cells.

Underproduction of insulin causes diabetes, resulting in an increase in blood sugar (hyperglycaemia) and sugar appearing in the urine (see GLYCOSURIA). The condition can be fatal if untreated, treatment being by injection of insulin into the blood stream. The hormone cannot be taken orally as, being a protein, it would be digested. Insulin was discovered by BANTING and BEST in 1921. The control of blood sugar, where a change in its level automatically brings about the opposite effect, is a good example of a negative FEEDBACK MECHANISM.

integrated control, *n.* the use of chemical, biological, cultural and legislative methods in a complementary way to control PESTS and PATHOGENS.

integration, *n.* the insertion of DNA from one organism into the recipient chromosome of another.

integument, *n.* **1.** (in flowering plants) the covering of the central tissue (nucellus) of the OVULE that contains the EMBRYO SAC. Most flowering plants possess both an inner and outer integument, which on hardening forms the TESTSA of the seed. **2.** (in insects) the cuticle.

intelligence, *n.* the ability to understand and create abstract ideas. Tests to measure intelligence are rather unreliable since it is not possible to separate completely environmental influences (such as schooling and social background) from innate ability. Nevertheless, such tests are widely used, producing a measure called the Intelligence Quotient (I.Q.) which is:

$$\frac{\text{mental age}}{\text{actual age}} \times 100.$$

Thus, if a person has an average mental age for his age-group he will have an I.Q. score of 100. The HERITABILITY of intelligence is thought to be between 0.5 and 0.7.

inter-, *prefix.* denoting between, amongst.

interbreed, *n.* **1.** to breed within a single family or strain to produce particular characteristics in the offspring. **2.** also called **crossbreed**. to breed using parents of different races or breeds.

intercellular fluid, see INTERSTITIAL FLUID.

intercostal muscles, *n.* the muscles lying between the ribs of vertebrates that, with the ribs, form the walls of the thorax. Contraction of external intercostal muscles brings about rib movement upward and outward, thus expanding the thoracic cavity, reducing the thoracic pressure and drawing air into the lungs. See BREATHING.

interfascicular cambium, *n.* an area of meristematic tissue (see MERISTEM) in plant stems and roots that develops from PARENCHYMA cells between the VASCULAR BUNDLES forming a complete ring of CAMBIUM, and carries out rapid division to produce SECONDARY THICKENING.

interference, see CHROMATID INTERFERENCE.

interferon, *n.* a glycoprotein produced by cells in response to viral attack, whose function seems to be triggering *viral interference* defence mechanisms in uninfected cells of the same species in which it was produced. Since it has been suggested that interferon might prove effective against viral diseases by inhibiting viral multiplication, and even some forms of cancer, strenuous efforts have been made to isolate sufficient quantities with which to run clinical trials. The problem of production has now been solved by GENETIC ENGINEERING but the results of trials are inconclusive, so far.

intergranum, see CHLOROPLAST.

interleukins, *n.* a group of peptides that signal between cells involved in the immune system. See T-CELL.

internal clock, see BIOLOGICAL CLOCK.

international code of zoological nomenclature, *n.* the official regulations dealing with the scientific names and taxonomy of animals.

internal environment, *n.* the medium in which all body cells are bathed and which maintains a constant environment (see HOMEOSTASIS), in terms of pH, osmotic pressure, etc.

interneurone, *n.* a connecting nerve cell between two others.

internode, *n.* that portion of a stem found between two successive NODES. See also BOLTING.

interoceptor, *n.* a receptor that detects stimuli from inside the body.

interpetiolar, *adj.* situated between the petioles.

interphase, *n.* a stage of growth in the CELL CYCLE in which METABOLISM occurs without any visible evidence of nuclear division.

inter-renal bodies, *n.* the endocrine organs that lie between the kidneys in fish and are homologous with the mammalian ADRENAL CORTEX.

interrupted gene, *n.* a EUCARYOTE gene (or one of its viruses) consisting of sequences of INTRON and EXON segments.

intersex, *n.* an individual with characteristics intermediate between those of a male and female.

interspecific behaviour, *n.* any act of behaviour occurring between species.

interspecific competition, *n.* any limited competition between two or more different species populations for a resource such as food. All populations involved are negatively affected by the competition, and may exhibit increased mortality or decreased birthrate.

interstitial cells, *n.* **1.** any cells packing an area between other tissues. **2.** See LEYDIG CELLS.

interstitial cell-stimulating hormone (ICSH), see LH (LUTEINIZING HORMONE).

interstitial fluid or **intercellular fluid,** *n.* a liquid (also called LYMPH when inside lymphatic vessels) bathing all the cells of the body, acting as a connecting link between the blood and the cells. An average human male with 42 litres of total body water would have about 7 litres of interstitial fluid.

interstitium, *n.* the tissue space between cells.

intestine, *n.* the part of the alimentary canal that lies between the stomach and the anus. Usually it is coiled and the internal surface is greatly increased by the presence of folds and projections (VILLI) to enable efficient digestion and absorption. Anteriorly, the intestine is lined with an EPITHELIUM containing ENZYME and MUCUS-secreting glands, and in higher vertebrates it is referred to as the SMALL INTESTINE. The posterior, *large intestine*, dehydrates the faeces which are stored here until voided in DEFECATION.

intine, *n.* the thin, inner coat of a POLLEN GRAIN, composed of cellulose.

intra-, *prefix,* denoting within.

intracellular, *adj.* within a cell.

intracellular tubules, *n.* any tubules that occur within a cell, such as DRAINPIPE CELLS.

intrafusal fibres, *n.* the muscle fibres within a MUSCLE SPINDLE.

intrapetiolar, *adj.* between the stem and the petiole in plants.

intraspecific, *adj.* within the species.

intraspecific competition, *n.* any limited competition within the same species population for a resource such as food. Not all members of the population may be negatively affected by the competition, resulting in differential ability to survive and reproduce (see FITNESS).

intravaginal, *adj.* **1.** (of plants) inside the sheath. **2.** (of animals) within the VAGINA.

introduced species, *n.* one that does not naturally occur in the area and has been brought in accidentally or intentionally by man, for example, rabbits in Australia (introduced originally to the British Isles from Spain).

introgressive hybridization, *n.* the spread of genes of one species into the germ plasma of another species as a result of hybridization.

intron, *n.* a segment of DNA from EUCARYOTES that is transcribed into mRNA but is then excised from the RNA, leaving behind the EXON sequences to be translated into polypeptide.

introrse, *adj.* opening to the middle of the flower.

intussusception, *n.* the growth in the surface area of a cell wall by the inclusion of additional particles in the wall.

inulin, *n.* a complex polymer of FRUCTOSE that is soluble in water and occurs in the cell sap of storage organs such as dahlia TUBERS and dandelion TAP ROOTS.

invagination, *n.* an inpushing of a layer of cells, as in GASTRULATION or in the formation of the PROCTODAEUM.

invasion, *n.* the entry and colonization of a host by an organism.

inversion, *n.* a CHROMOSOMAL MUTATION in which a segment becomes reversed and, although there is no loss or gain of genetic material, there may be a positive or negative POSITION EFFECT on the phenotype.

invertase, see SUCRASE.

invertebrate, *n.* any animal that does not possess a backbone.

in vitro, *adj.* (of biological processes or reactions) made to occur outside the body of an organism in an artificial environment (vitro = glass). For example, in-vitro fertilization of human eggs in the laboratory prior to reimplantation in the mother. Compare IN VIVO.

in vivo, *adj.* (of biological processes or experiments) occurring in the living organism. Compare IN VITRO.

involucre, *n.* **1.** a calyx-like structure formed by bracts below the base of a condensed INFLORESCENCE. **2.** a growth of the tissue of the thallus in liverworts (see HEPATICA) that covers and protects the ARCHEGONIUM.

involuntary muscle or **visceral muscle** or **smooth muscle,** *n.* a type of muscle called 'involuntary' because it is innervated by the AUTONOMIC NERVOUS SYSTEM. It is also called 'visceral' as it is found in the alimentary canal, the blood vessels, CILIARY BODY, respiratory passages and URINOGENITAL SYSTEM. It is also called 'smooth' since the MYOFIBRILS lack striations. Involuntary muscle contracts slowly and takes a long time to fatigue. Compare STRIATED MUSCLE.

involuntary response, *n.* a reaction of INVOLUNTARY MUSCLE brought about by mechanisms not under the control of the will.

involution, *adj.* (of plant organs) having rolled-up margins.

iodopsin, *n.* a photochemical pigment thought to be of three types found in the CONES of the retina of the eye. It is not readily bleached even by high-intensity light. See also RHODOPSIN.

ion, *n.* an atom that carries a charge due to loss or gain of electrons.

ionic bond, *n.* an electrostatic bond.

ionizing radiation, *n.* a beam of short-wavelength electromagnetic energy that can penetrate deeply into tissues, leaving a track of unstable atoms which have lost electrons (IONS). Such radiations, X-rays and gamma rays, are powerful MUTAGENS.

iris, *n.* the pigmented part of the vertebrate eye. It consists of a thin sheet of tissue, attached at its outer edge to the CILIARY BODY, which has radiating muscles which can increase the size of the central pupil and a central ring of muscle around the pupil which, on contraction, causes a decrease in its size. The iris thus regulates the amount of light entering the eye. See ALBINISM for photophobia.

irreversibility, see DOLLO'S LAW OF IRREVERSIBILITY.

irrigation, *n.* moistening or pouring water over a preparation.

irritability, *n.* the responsiveness of organisms to changes in their immediate environment.

ischium, *n.* the ventral bone of the PELVIC GIRDLE in vertebrates, bearing the weight of a sitting human.

islets of Langerhans, *n.* a group of cells found scattered in the pancreas of jawed vertebrates that secretes the hormones INSULIN and GLUCAGON. Named after the German histologist P. Langerhans.

iso-, *prefix.* denoting equal.

iso-alleles, *n.* alleles with so little phenotypic expression that special techniques are necessary to demonstrate their presence.

isobilateral, *adj.* having a similar structure on each side, as in the leaf of a monocotyledon.

isoelectric focusing, *n.* the separation of molecules by differences in their charge, each molecule migrating to the point in a PH gradient where it has no net charge.

isoelectric point, *n.* the PH of an AMPHOTERIC solution when at electrical neutrality, the ionic state of an AMINO ACID at the isoelectric point being a ZWITTERION.

isogamete, *n.* a GAMETE that is usually motile and appears similar to the gamete with which it unites at fertilization. Such sexual reproduction is called *isogamy* and occurs in several primitive plant groups such as many ALGAE.

isogamy, see ISOGAMETE.

isogenic, *adj.* having an identical set of genes.

isograft, *n.* a transplant from one individual to a closely related individual of the same genotype, such as those between MONOZYGOTIC TWINS, or between members of the same inbred line, as in rodents.

isohyet, *n.* a line connecting places with the same rainfall.

isolate, 1. *vb.* to separate a microorganism from fresh material and to establish it in pure culture. **2.** *n.* a single pure culture of a microorganism.

isolation, *n.* **1.** any geographical separation from other populations of the same organism. **2.** a form of genetic isolation, where gene transfer is limited or entirely prevented by geographical separation. Such isolation can be termed behavioural, ecological, seasonal (different breeding seasons) and physiological. The prevention of gene transfer results in new lines of EVOLUTION.

isolating barrier, *n.* any structure that limits the movement of organisms to the extent of preventing gene flow. Examples include oceans, deserts, mountain ranges, salinity, temperature, etc.

isolating mechanism, *n.* any mechanism preventing successful breeding between two populations.

isoleucine, *n.* one of 20 AMINO ACIDS common in proteins, that is nonpolar in structure and insoluble in water. The ISEOELECTRIC POINT of isoleucine is 6.0. See Fig 192.

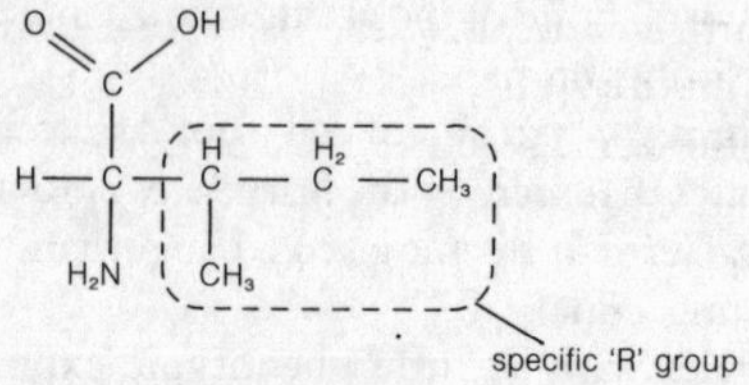

specific 'R' group

Fig. 192. **Isoleucine.** Molecular structure.

isomerase, *n.* a group of enzymes that converts organic compounds from D- to L-, or L- to D- forms.

isomerism, *n.* the phenomenon of two chemical compounds having the same formula but a different arrangement of atoms.

isomerous, *adj.* (of plants) having the same number of parts in different floral whorls, such as a plant possessing five stamens and five carpels.

isomorphic, *adj.* (of organisms, usually plants) having morphologically similar forms in different parts of the life history. For example, where there is ALTERNATION OF GENERATIONS in which the generations are similar.

isophene, *n.* a line connecting parts of equal character expression, i. e. one at right angles to a CLINE.

isopod, *n.* any crustacean of the order Isopoda, containing the woodlice and pill bugs.

Isoptera, *n.* the insect order containing the termites, which live in colonies and have a complex system of castes.

isospores, see ASEXUAL SPORE.

isosmotic, *adj.* (of two solutions) having the same OSMOTIC PRESSURE.

isostatic, *adj.* (of the earth's crust) movements resulting in sea level changes caused by ice-loading from glacier formation during the Ice Age.

isotonic, *adj.* (of a liquid) having a fluid state with the same WATER POTENTIAL as another liquid. See HYPOTONIC and Fig. 187.

isotope, *n.* any of the forms of an element having the same number of protons (atomic number) but a different number of neutrons (atomic mass). Some isotopes of an element may be *radioisotopes* (e.g. ^{12}C is not radioactive while ^{14}C is) and yet can function normally in biological material. Isotopes can thus be 'tagged' (using suitable detection devices such as geiger counters) as biochemical processes occur. See HALF-LIFE, AUTORADIOGRAPH.

isozyme or **isoenzyme,** *n.* one of several forms of an enzyme found in an individual or a population, each form coded by a different allele of a gene. An isozyme is a type of ALLOZYME.

iter (inter a tertio ad quartum ventriculum), *n.* the passage betwen the third and fourth ventricals of the vertebrate brain.

IUD (intrauterine device), *n.* any device such as a coil that is introduced into the female uterus as a means of preventing either fertilization of the egg or implantation of the embryo. See BIRTH CONTROL.

J

Jacob-Monod hypothesis, see OPERON MODEL.

Java man, *n.* a primitive form of human – *Homo erectus*.

jaw, *n.* the mandibles of any animal. The term is usually restricted to the bones surrounding the mouth of vertebrates, the upper jaws being referred to as the maxillae and the lower as the mandible. These bones carry the teeth where present, and are often used for crushing purposes.

jaw articulation, *n.* the joint of the bones forming the skeleton surrounding the mouth of vertebrates, between the upper jaw (maxilla) and lower jaw (mandible).

jejunum, *n.* a part of the SMALL INTESTINE between duodenum and ileum.

jellyfish, *n.* **1.** any large medusa of the class Scyphozoa. **2.** the medusoid stages of any coelenterate.

Jenner, Edward (1741–1823) English physician who discovered that cowpox lymph matter could be safely used as an innoculum (a VACCINE) to prevent smallpox. The first vaccination was carried out in 1796.

Johannsen, Wilhelm (1857–1927) Danish botanist who, in the early 1900s, carried out selection experiments with beans (*Phaseolus vulgaris*) which showed that selection is a passive process that eliminates, but does not produce, variations. Johannsen is, however, chiefly remembered for having coined the terms GENE, GENOTYPE and PHENOTYPE.

Johnston's organ, *n.* an organ situated at the base of the insect antenna consisting of a group of CHORDOTONAL RECEPTORS, probably associated with movement of the antenna.

joint or **articulation,** *n.* a contact between two separate bones.

joule, *n.* an SI unit of energy equal to 10^7 ergs or 0.239 cals. Named after the physicist J.P. Joule (1818–89) who determined the mechanical equivalent of heat and showed that 4.2×10^7 ergs $= 1$ calorie (4.2 joules $= 1$ cal.).

jugular, *adj.* of or relating to the throat or neck, particularly the jugular veins, internal and external, that carry blood from the head to the anterior vena cava and hence to the heart.

junior homonym, *n.* the more recently published of two or more identical names for the same or different taxa.

junior synonym, *n.* the more recently published of two or more synonyms for the same taxon.

Jurassic period, *n.* a period of the MESOZOIC ERA that lasted from 200 to 135 million years ago, during which time the birds evolved. Cycads, ferns and conifers were widespread. Dinosaurs were dominant and the first fossil moth was recorded from this period. *Pangea* began to break up and London was at 30°N.

juvenile hormone, *n.* a hormone present in juvenile insects and secreted by the CORPUS ALLATUM of the brain. So long as it is produced, the cuticle maintains the characteristics of the nymphal or larval form at each moult. Only when it ceases to be present or the level falls below a threshold value does the insect moult to the adult form. Juvenile hormones have been used as an effective means of controlling insects, for example, in preventing the development of adult ants during hospital infestations.

juxta-glomerular complex, *n.* a group of cells lining the efferent glomerular vessels of the kidney, that detect changes in blood volume. The cells secrete the enzyme RENIN when blood volume decreases, with subsequent release of ALDOSTERONE from the adrenal cortex, resulting in increased sodium retention. When the blood osmotic concentration increases, release of ADH occurs, and both these factors reverse the decrease in blood volume, an example of a FEEDBACK MECHANISM.

K

K, a symbol indicating the maximum number of individuals in the LOTKA-Volterra equations.

K selection, *n.* selection proceeding in such a way as to maximize K and reduce R, so optimizing the use of environmental resources without expending excess energy in converting food into offspring.

kappa particles, *n.* the bacteria that inhabit the cytoplasm of the ciliate *Paramecium aurelia.* Kappa-free Paramecia are sensitive to toxins produced by the particles, dying when exposed to them, after CONJUGATION. Kappa carriers (called *Killer stocks*) are immune to the toxin.

kary- or **karyo-,** *prefix.* denoting a nut.

karyokinesis, *n.* the division of the cell nucleus.

karyological, *adj.* of or relating to the structure or number of chromosomes.

karyoplast, *n.* a nucleus surrounded by a thin layer of cytoplasm and membrane.

karyosome, *n.* a chromatin aggregation in the resting nucleus.

karyotype, *n.* the CHROMOSOME complement of a cell or organism, characterized by the number, size and configuration of the chromosomes as seen during metaphase of MITOSIS.

katabolism, see CATABOLISM.

katadromous, *adj.* (of fish) moving from freshwater to the sea in order to breed.

kathepsin, see CATHEPSIN.

keel, see CARINA.

kelp, *n.* a large, coarse seaweed.

Kelvin scale, *n.* an absolute scale of temperature where a range of 1° kelvin (K) is equal to a range of 1°C. $0°K = -273.15°C$ or $-459.67°F$.

keratin, *n.* a hard, fibrous, sulphur-containing protein with an alpha-helix structure, found in the epidermis of vertebrates, mainly in the outermost layers of skin. Keratin can have several forms: in scales, feathers, hooves, horns, claws and nails it is hard, while wool and hair are made up of a soft and flexible form.

ketone, *n.* an organic molecule in which a C=O group is contained within the molecule rather than at one end. Ketones are highly varied in structure, ranging from the simple *acetone* (see Fig. 193) to FRUCTOSE, a KETOSE sugar.

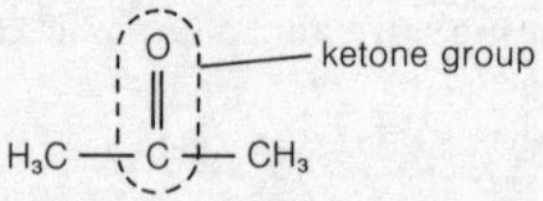

Fig. 193. **Ketone.** The molecular structure of acetone.

ketose, *n.* a MONOSACCHARIDE with a ketonic structure (see KETONE).

Kettlewell, H.B.D, British biologist who in the mid–1950s carried out a series of elegant field and laboratory experiments using light-coloured and melanic forms of the peppered moth *Biston betularia*. He was able to show that the selective forces operating against the moths were due to visual predation by birds, which attacked preferentially those forms resting on a noncryptic background. See also INDUSTRIAL MELANISM.

key, *n.* a character of special significance in an organism. Such characters are used in a system for determining the identity of an organism on the basis of the presence or absence of a succession of characters. Usually such keys are DICHOTOMOUS(2); that is to say, at each step there is a choice between two possibilities. Keys, however, may have more than two characters at each step and there is a gradual elimination of absent characters until identification is made.

kidney, *n.* an organ, found in pairs in a dorsal situation in vertebrates, that serves the dual purposes of EXCRETION and OSMOREGULATION. See Fig. 194. Ultrafiltration takes place in the glomerulus, where the glomerular filtrate contains all the constituents of blood except for blood cells and plasma proteins. Pores of about 0.1μm diameter allow the passage of the filtrate under pressure caused by the efferent blood vessels from the glomerulus being narrower than the afferent vessels. The filtrate which is modified in its flow along the tubule, eventually emerges from the kidney as urine. The LOOP OF HENLE employs the principle of a hairpin COUNTERCURRENT MULTIPLIER. Active transfer of salt (NaCl) takes place from the ascending limb to the descending limb, so raising the concentration in the latter. This results in a region of high salt concentration deep in the medulla of the kidney through which the collecting duct passes. Water is extracted by osmosis from the *distal-convoluted tubule* and collecting duct, so concentrating the urine (see ADH). Over 99% of kidney fluid is thus reabsorbed by the kidney tubules. See NEPHRON.

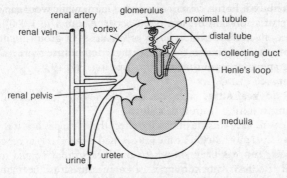

Fig. 194. **Kidney.** General structure.

kinaesthetic, *adj.* capable of detecting movements, as in the sense organs of muscles, tendons and joints.

kinase, *n.* an enzyme that catalyses the transfer of a phosphoryl group from ATP to another compound.

kinesis, *n.* any orientation behaviour in which the organism does not move in a particular direction relative to the stimulus, but instead simply moves at an increasing or decreasing rate until it ends up nearer or further from the stimulus. For example, when a woodlouse finds itself drying out it simply begins to move around a great deal until encountering a moist spot where it may settle. Compare TAXIS.

kinetin, see CYTOKININ.

kineto-, *prefix.* denoting movement.

kinetochore, see CENTROMERE.

kinetodesmata, *n.* the threads connecting KINETOSOMES in ciliates.

kinetosome or **kinetoplast,** *n.* the BASAL BODY of a CILIUM.

kinetoplast, see KINETOSOME.

kingdom, *n.* the highest grouping (taxon) in biological classification. See ANIMAL KINGDOM, PLANT KINGDOM.

kinin, *see* CYTOKININ.

kinomere, *n.* a spindle attachment in the region of a chromosome that attaches it to a spindle.

kin selection, *n.* a form of selection favouring altruistic (self-sacrificing) behaviour towards relatives. Such a process ensures that even if the chances of an individual surviving are reduced, some of his or here genes will survive in the relative.

Klinefelter's syndrome, *n.* a human chromosomal abnormality in which there is an extra X-chromosome making a complement of 44 autosomes plus XXY (47 in all). The affected individuals are male (showing the SEX DETERMINATION role of the Y-chromosomes) but with much reduced fertility, and a number of female secondary sexual characteristics, such as breasts, often occur. About 1 in 1,000 liveborn males are affected by this condition, which arises from NONDISJUNCTION in one or other parent, more probably the female where there is a distinct positive age effect (as in DOWN'S SYNDROME).

klinostat, see CLINOSTAT.

klinotactic response, *n.* a positive orientation movement of an organism, resulting from a stimulus.

knockdown, *n.* a measure of biocide effect that is used when it is difficult to assess actual mortality. The measure is used particularly in biocide tests on insects, and has been shown to produce similar results to actual mortality scores. Knockdown is often measured as the inability to respond to a stimulus such as light or touch. See KNOCKDOWN LINE.

knockdown line, *n.* a graphical representation of the rate of response to a biocide as measured by KNOCKDOWN (y-axis) against time or dose (x-axis). Strains of organisms such as insects are characterized by the slope and position of their knockdown lines relative to other strains.

Koch, Robert (1843–1910) German bacteriologist who first introduced the method of making bacterial smears and fixing them with heat. He worked on bubonic plague and sleeping sickness and discovered how they are transmitted. He also developed means of culturing bacteria and rules for properly identifying the agents of various diseases. He is most famous for establishing a series of criteria necessary to establish whether a specific microorganism causes a specific disease. These criteria are known as *Koch's Postulates*:

(a) the microorganisms must be present in every case of the disease.

(b) the microorganisms must be isolated from the diseased organism and grown in culture.

(c) the disease must be reproduced when a pure culture is reintroduced to a nondiseased, susceptible host.

(d) the microorganism must be recoverable from the experimentally infected host.

Kranz anatomy, *n.* the special structure of leaves in C_4 PLANTS (e.g. maize) where the tissue equivalent to the spongy mesophyll cells is clustered in a ring around the leaf veins, outside the bundle-sheath cells. (The term 'Kranz' means wreath or ring in German). The bundle-sheath cells contain large CHLOROPLASTS whereas the spongy mesophyll cells have few if any chloroplasts, unlike their counterparts in C_3 plants (see MESOPHYLL). See Fig. 195.

Krebs cycle or **tricarboxylic acid cycle (TCA cycle)** or (formerly) **citric-acid cycle,** *n.* a circular series of reactions taking place in the matrix of MITOCHONDRIA, forming part of CELL RESPIRATION in the

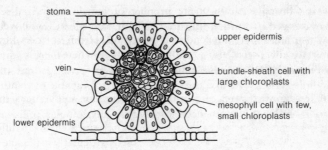

Fig. 195. **Kranz anatomy.** Transverse section through leaf.

presence of oxygen. The steps of the cycle were deduced by Sir Hans Krebs (1900–81), for which he received a Nobel Prize. The reactions leading up to the cycle by which ACETYLCOENZYME A is produced, together with the overall role of the cycle in the breakdown of complex molecules is described fully under AEROBIC RESPIRATION.

Each turn of the cycle releases 2 molecules of carbon dioxide, 8 hydrogen atoms that produce 11 molecules of ATP via the ELECTRON TRANSPORT SYSTEM, and 1 molecule of ATP produced by SUBSTRATE-LEVEL PHOSPHORYLATION. Two turns of the cycle are needed to complete the breakdown of one glucose molecule. See Fig. 196.

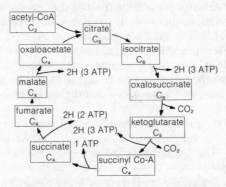

Fig. 196. **Krebs cycle.** The major steps of the Krebs cycle.

krill, see EUPHAUSIID.

Kupffer cells, *n.* the phagocytic cells that form part of the RETICULO–ENDOTHELIAL SYSTEM, destroying old red blood cells and remove foreign particles from blood flowing through the liver. Named after the German anatomist K. Kupffer.

kwashiorkor, *n.* the most widespread and serious human nutritional disease, brought on by acute protein starvation. The condition is characterized by apathy, impaired growth, skin ulcers, swollen hands and feet and an enlarged liver. If untreated, it is fatal. Kwashiorkor typically affects children in the early stages after weaning.

kymograph, *n.* the revolving drum used in investigations of the physiology of nerves and muscles. Usually, it is capable of rotating at different known speeds and carries a piece of graph paper on which a trace is produced by a lever connected to the physiological preparation.

L

label, *n.* any marker, often a radioactive atom, that makes it possible to locate and monitor a particular molecule or organism.

labella, *n.* (*pl.* labellae) an oral lobe present on the distal end of the insect proboscis. They are purely sensory in the mosquito, but are modified for sucking up fluids by means of pseudotrachaea in the blowfly.

labi-, *prefix.* denoting a lip.

labium, *n.* (*pl.* labia) the lower lip of insects, lying immediately behind the maxillae. The labium is formed by the fusion of paired appendages underneath the rear of the head. See Fig. 191.

labrum, *n.* a cuticular plate at the front of the insect head, forming an upper lip. The labrum lies immediately below the clypeus, with which it is hinged, and in front of the MANDIBLES. See Fig. 191.

labyrinth, membranous, see EAR.

labyrinthodont, *n.* any primitive amphibians of the order Labyrinthodontia, which existed from the Devonian to the Triassic. They are related to Crossopterygian fish and some are close to the reptiles.

Labyrinthulales, *n.* a group of aquatic, mainly marine organisms which secrete slimy filaments in the form of a net plasmodium over which cells may glide. The group is a member of the slime moulds – Myxomycophyta.

lacerate, *adj.* having the appearance of being torn.

lacertilian, *n.* any reptile of the suborder Lacertilia, containing the lizards.

lac-operon, *n.* the genetic system of *E. coli* that is required to metabolize lactose and on which the OPERON MODEL is based.

lachrymal gland, *n.* the tear gland that lies below the upper eyelid of mammals and serves to moisten and cleanse the surface of the eye by secreting sterile and antiseptic liquid. Excess fluid is drained away through the lachrymal duct in the corner of the eye which leads eventually into the nasal cavity.

lactase, *n.* an enzyme that splits the disaccharide LACTOSE into galactose and glucose, secreted as part of the intestinal juice by glands in the SMALL INTESTINE wall. Lactase is also produced in ENZYME INDUCTION by bacteria (see OPERON MODEL).

lactation, *n.* the production of milk by the adult female mammal from the mammary gland, and with which it suckles its young. Lactation is induced by LUTEOTROPIC HORMONES from the anterior PITUITARY GLAND.

lacteals, *n.* the central lymph vessels in the villi of the vertebrate intestine into which neutral fat is passed from the columnar epithelium where it has been resynthesized from fatty acids and glycerol. The fat is in the form of a white emulsion of minute globules, giving lymph a milky appearance (hence the name 'lacteal') and is carried to all parts of the body.

lactic acid, *n.* an organic acid ($CH_3CH(OH)COOH$) produced from pyruvic acid as a result of ANAEROBIC RESPIRATION in microorganisms (see LACTOSE) and in the active muscles of animals (see OXYGEN DEBT), the hydrogenation of pyruvate being catalysed by lactic dehydrogenase (LDH). In animals the acid can be oxidized back to pyruvate (using LDH again) when sufficient oxygen is available, most of the conversion taking place in the liver.

lactogenic hormone, see LUTEOTROPHIC HORMONE.

lactogenesis, *n.* the production of milk.

lactose or **milk sugar,** *n.* a DISACCHARIDE carbohydrate found in the milk of mammals. Lactose is produced by CONDENSATION REACTION between galactose and glucose. See Fig. 197.

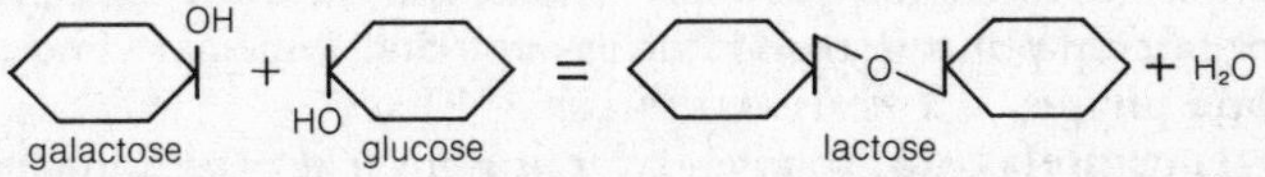

Fig. 197. **Lactose.** The condensation reaction between galactose and glucose.

The sugar can be broken into its component monosaccharides by LACTASE. Souring of milk is due to the conversion of lactose to LACTIC ACID by microorganisms present in the air.

lactose synthetase, *n.* an enzyme that catalyses the synthesis of lactose from glucose.

lacuna, *n.* one of many small spaces between the lamellae of bones that is occupied by individual bone cells. Small canals (canaliculi) radiate from the lacunae and in these are small protoplasmic processes which connect with the osteoblasts in other lacunae. See HAVERSIAN CANAL.

lacunate, *adj.* divided into narrow segments in a deep and irregular manner.

laevorotatory or **levorotatory,** *adj.* (of a crystal, liquid or solution) having the property of rotating a plane of polarized light to the left (e.g. tructose). Compare DEXTROROTATORY.

lag phase, *n.* **1.** the stage of growth of microbial cells where nucleic acids and proteins are synthesized, but there is no cell division. **2.** an adaption stage. See GROWTH CURVE

Lamarck, Jean-Baptiste Pierre Antoine de Monet (1744–1829) French naturalist best known for his theory of the inheritance of acquired characters (see LAMARCKISM). Whilst this theory is now generally discredited, Lamarck deserves recognition for popularizing the word 'biology' and being effectively the founder of modern invertebrate zoology. He was one of the first true scientists to give real consideration to the evolutionary development of life.

Lamarckism, *n.* the theory of inheritance of ACQUIRED CHARACTERS, which suggests that the structures developed during the lifetime of an organism, through use, are passed on as inherited characters to the next generation. Evolutionary change might thus be achieved through the transmission of these acquired characters. This theory, proposed by Jean Baptiste de LAMARCK, is now generally discounted in favour of DARWINISM, where favoured characters of use to a particular organism are maintained by selection, whereas unfavourable characters are selected against. Thus, Lamarck might have claimed that blacksmith's sons were brawny because of their father's profession, whereas Darwin would say that the reason the father was a blacksmith was because he was brawny and brawny men tend to have brawny offspring. LYSENKO attempted unsuccessfully to apply Lamarckian theory to the development of crop plants in the USSR in the 1930s.

lambda phage, *n.* a TEMPERATE PHAGE of *E. coli.*

lamella or **thylakoid,** *n.* a thin layer or plate. The term is used in the plural (*lamellae*) for:

(a) the sheet-like membranes that occur within the CHLOROPLAST, each of which consists of a pair of membranes with a narrow space between. Some 3,000 occur in each chloroplast and their function is to maintain the CHLOROPHYLL molecules within the quantasomes in such a position as to receive the maximum amount of light.

(b) the gills of a basidiomycete fungus that radiate out from the stalk beneath the cap of the fruit and bear the spores.

(c) the layers in which the calcified matrix of bone occurs, each some 5 μm thick.

lamelli-, *prefix.* denoting a thin plate.

lamellibranch, see BIVALVE.

lamina, *n.* a thin, flat structure such as a leaf or petal.

lampbrush chromosomes, *n.* the giant chromosomes in vertebrate oocytes (mainly amphibians) that have hair-like lateral extensions in the form of loops with a DNA core.

lamprey, *n.* a jawless fish of the class AGNATHA.

lamp shell, see BRACHIOPOD.

lanceolate, *adj.* (of plant structures) tapering to a point.

Lansteiner, Karl (1868–1943) Austrian pathologist who spent a lifetime working on human blood groups. In 1901 he found that all humans could be divided into one of four ABO BLOOD GROUPS. In 1928 he and Philip Levine discovered another set of blood antigens which came to be called the MN group and in 1940 Landsteiner and Weiner found the Rhesus factor (see RHESUS BLOOD GROUP), important in haemolytic disease of the newborn (see RHESUS HAEMOLYTIC ANAEMIA).

languets, *n.* small, tongue-like processes.

lanugo, *n.* the hair on the human embryo that is lost before birth.

lapsus calami, *n.* 'a slip of the pen', particularly an error in spelling in nomenclature.

large intestine, see COLON.

larva, *n.* (*pl.* larvae) the preadult form of many animals that is usually morphologically different from the adult, and which in many cases takes up the larger part of the life history. Usually the larva is not sexually mature, but in cases of PAEDOGENESIS, of which the AXOLOTL is an example, breeding may take place at this stage. Often the larva is a dispersal phase, as in many marine invertebrates where larvae occur in the PLANKTON and usually the larva feeds in a different way from the adult and does not compete with it.

larynx, *n.* a dilation of the upper part of the TRACHEA of TETRAPODS (Adam's apple in humans), occurring in the front part of the neck. It is triangular in shape (base uppermost) and is made up of 9 cartilages moved by muscles. It contains the vocal cords which are elastic ligaments embedded in two folds of mucous membrane.

Lassa virus, *n.* a highly contagious virus first isolated in 1969 in Nigeria, that can cause a general malaise leading to severe chest infections and even death.

latent infection, *n.* an infection that does not produce visible signs of a disease, but may be transmitted to another host.

latent period or **reaction time,** *n.* the first period of a simple muscle contraction, being the interval between the stimulus being applied and the contraction occurring, usually around 0.01 seconds. See Fig. 198.

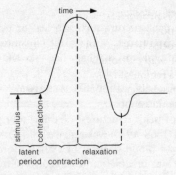

Fig. 198. **Latent period.** A KYMOGRAPH tracing of the 'twitch' of a frog's gastrocnemius muscle.

later- or **latero-,** *prefix.* denoting a side.

lateral, *adj.* at the side of.

lateral inhibition, *n.* the reciprocal suppression of excitation brought about by neighbouring neurons of a sensory network.

lateral-line system or **acoustico-lateralis system,** *n.* a complex system of receptors found in fishes and amphibians, which occurs in a line along the length of the body and in a complex pattern on the head. The receptors occur in pores or canals and are capable of detecting very small changes in pressure in the surrounding medium, such as vibrations. When the animal is moving, impulses from efferent nerve fibres to the receptors prevent the latter from responding, so that they do not react to the animal's own movement. The inner ear of vertebrates is thought to have evolved from a receptor of the lateral-line system.

lateral meristem, see CAMBIUM.

lateral plate, *n.* the mesoderm in the ventrolateral position in the vertebrate embryo.

lateral root or **secondary root,** *n.* the branches that develop from deep inside the plant root (i.e. are ENDOGENOUS) where cells of the PERICYCLE become meristematic (see MERISTEN) and produce a growing point which differentiates into mature tissues. Such branching occurs behind the root hair zone and is initiated by AUXIN.

latex, *n.* a milky plant juice.

latiseptate, *adj.* having a septum across the widest diameter, as on some fruits.

Laurasia, *n.* the northern supercontinent of the MESOZOIC ERA, including North America, Greenland and Eurasia. PANGAEA gave rise to Laurasia and in the south GONDWANALAND. See CONTINENTAL DRIFT.

law of mass action, *n.* a law stating that the rate of progression of a chemical reaction is proportional to the concentration of the reactants.

LD$_{50}$. see LETHAL DOSE.

leaching, *n*. the process by which nutrients are removed from the soil by the percolation of water through it.

leader region, *n*. a region of DNA between the promoter region and the structural genes in a bacterial operon (see OPERON MODEL). The leader sequence is not translated into protein and may contain an ATTENUATOR.

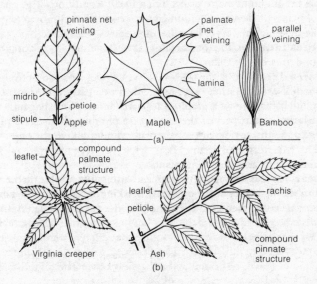

Fig. 199. **Leaf.** (a) Simple leaves. (b) Compound leaves.

leaf, *n*. the principal photosynthetic organ of vascular plants, which typically consists of a flattened lamina joined to the stem by a stalk, or petiole at which junction an axillary bud can be found. There are numerous types of leaves.

(a) *dorsiventral* leaves are held horizontally with the upper half of the lamina containing the majority of photosynthetic cells (see MESOPHYLL for diagram). Typically, the lower epidermis contains the majority of STOMATA through which gas exchange and transpiration take place. Leaf veins are usually arranged in a 'net' venation with principal veins called *midribs*. The lamina can be indented to form a 'compound' leaf or remain as a simple structure. See Fig. 199. Dorsiventral leaves are typical of DICOTYLEDONS.

(b) *isobilateral* leaves grow erectly, with a sword-like shape and are typical of MONOCOTYLEDONS. Both epidermal surfaces contain stomata with palisade mesophyll tissue packed underneath. Isobilateral leaves have a parallel venation and are not divided into compound structures. See also KRANZ ANATOMY.

(c) *centric* leaves are more or less cylindrical with a central region containing the vascular bundles surrounded by mesophyll tissue. Examples are the needles of pine trees, and onion leaves.
Some leaves have special modifications. For example, water storage leaves are found in many plants living in dry conditions (e.g. cacti). In other plants the leaves are modified into tendrils for climbing (e.g. pea) while in some the leaves have become spines, e.g. gorse.

leaf-area index, *n*. the total area of leaves (one side) in relation to the ground area below them.

leaf-area duration, *n*. the LEAF-AREA INDEX over a period of time. Due to the leaf-area duration changing with time, it is proportional to the area under the line on a graph of leaf-area index plotted against time.

leaf blade, *n*. that part of the leaf where photosynthesis takes place.

leaf gap, *n*. the region of the vascular cylinder above the LEAF TRACE where PARENCHYMA occurs. Leaf gaps are characteristic of ferns, gymnosperms and flowering plants.

leaf scar, *n*. a scar left on a stem after leaf ABSCISSION. Visible on the scar is a number of smaller scars, each one marking the position of where a VASCULAR BUNDLE entered the petiole of the LEAF. See Fig. 200.

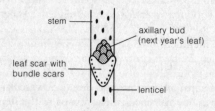

stem
axillary bud
(next year's leaf)
leaf scar with
bundle scars
lenticel

Fig. 200. **Leaf scar.** Position on stem.

leaf sheath, *n*. the part at the bottom of a leaf surrounding a stem, as in grasses.

leaf trace, *n*. a VASCULAR BUNDLE that extends between the main stem and the leaf base.

learning, *n*. an adaptive change in behaviour resulting from past experience. Learned behaviour is distinct from innate BEHAVIOUR, and may begin in the embryo. For example, a chick learns to peck because the heartbeat moves the head forward and causes the bill to open. Learning has been classified by the English behaviourist W.H. Thorpe (b.1902) as follows:

(a) *habituation*, where an animal is subject to repeated stimulation and may cease to respond.

(b) *classical conditioning*, resulting in a CONDITIONED REFLEX action.

(c) *trial-and-error learning*, as in rats learning to follow the correct route through a maze.

(d) *latent learning*, as in rats being allowed to run a maze without a final reward. Once given a reward their performance rapidly reaches that of rats rewarded throughout, thus they must have learnt something (latently) during the nonreward period.

It must be stressed that each of these categories overlap and that other classifications of learning are also accepted.

lecithin, *n.* any of a group of phospholipids, composed of choline, phosphoric acid, fatty acids and glycerol, found in animal and plant tissues.

lectotype, *n.* one of a series of SYNTYPES used by the taxonomist when a organism was named, and which was subsequently designated the TYPE.

leech, *n.* an aquatic annelid of the order Hirudinea.

legionnaires' disease, *n.* a rare human pneumonial condition, caused by the bacterium *Legionella pneumophila*, that can lead to death. The disease was named after an outbreak occurring at an American Legion convention in a Philadelphia hotel in 1976, the bacteria eventually being traced to the air-conditioning plant of the hotel. Since then a number of cases have been reported in connection with recirculated hot water systems such as cooling towers. There is no evidence of person–to–person transmission.

legume, *n.* any member of the pea family (Leguminosae). These plants have fruits which are pods containing one or more seeds. For example, pea, bean, lupin, laburnum, gorse. Most legumes have root nodules and so can grow in soils deficient in nitrogen (see NITROGEN FIXATION).

leishmaniasis, *n.* a human disease caused by infection with the protozoan flagellate *Leishmania*. There are several forms of leishmaniasis. A major type found in tropical and subtropical areas is called *cutaneous*, in which severe pimpling and ulceration can cover face, arms and legs. Another species of *Leishmania* causes a lethal 'visceral' form of the disease in Africa and Asia, in which infection of the spleen, liver and other organs occurs producing symptoms of 'kala-azar' and 'dum-dum fever'.

lemma, *n.* a grass bract that has the appearance of a GLUME and bears a flower in its axil.

lemur, *n.* a primitive primate that is arboreal and nocturnal. Lemurs are Old-World forms with opposable thumbs, a prehensile tail and stereoscopic vision.

lens, *n.* a transparent body at the front of the vertebrate eye, the main function of which is ACCOMMODATION and *not* refraction, though the latter does take place. The CORNEA is the most important refractive structure.

lentic, *adj.* of or relating to standing waters such as ponds, lakes, reservoirs, etc, rather than moving waters such as rivers and streams.

lenticel, *n.* a small pore found on the surface of stems and roots in higher plants. Lenticels usually arise below the STOMATA of the original epidermis, where loose packing tissue becomes waterproofed with SUBERIN, leaving large intercellular spaces through which gas exchange can take place. Lenticels are surrounded by a CORK layer. See Fig. 201.

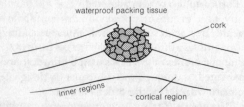

Fig. 201. **Lenticel.** Transverse section of the outer stem.

lenticular, *adj.* having the appearance of a double convex lens, circular in outline, convex on both sides.

lepido-, *prefix.* denoting scaled.

Lepidodendron, *n.* an extinct genus of tree–like giant club moss of the class Lycopodiopsida.

Lepidoptera, *n.* the ENDOPTERYGOTE order of insects containing butterflies and moths, characterized by the presence of scales on the wings and body. Butterflies and moths are no longer classed as separate taxa. The larvae are caterpillars, which feed mainly on plant tissues; the winged adults are usually nectar feeders and are important in pollination. See Fig. 202.

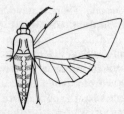

Fig. 202. **Lepidoptera.** General structure.

Lepidosiren, *n.* a genus of South American lungfish that differs from the Australian lungfish by having two lungs, rather than a single one.

leprosy, *n.* a chronic disease characterized by mutilating and disfiguring lesions, with loss of sensation in fingers and toes, caused by infection with the bacterium *Mycobacterium leprae*. About three million people are affected worldwide, the condition being transmitted by contact between an affected area in the donor and skin abrasion in the recipient, although it is not highly contagious. Therapy with sulfone drugs over long periods can produce gradual improvement.

lepto-, *prefix.* denoting small, thin.

leptocephalus, *n.* the transparent oceanic larva of eels of the genus *Anguilla*, that crosses the Atlantic and becomes adult in freshwater on the European continent.

leptosporangiate, *adj.* derived from a single cell and having a wall one cell thick.

leptotene, *n.* the first stage of PROPHASE I of MEIOSIS in which each chromosome resembles a thin thread as it emerges from INTERPHASE. The chromosomes have divided into CHROMATIDS but these are not usually visible at this stage.

lesion, *n.* a localized area of diseased tissue.

lethal allele or **lethal mutation,** *n.* a mutant allele causing premature death in heterozygotes if dominant and in homozygotes if recessive.

lethal dose (LD), *n.* the amount of a treatment (e.g. viral inoculation, insecticide) that induces death in a laboratory animal in a standard time. Because the response to treatment is often nonlinear, it is more conventional to measure the amount of a treatment that causes 50% mortality in a standard time, the so-called LD_{50}. In some tests the treatment dosage is fixed and the duration of the treatment period is varied, producing an estimate of lethal time and an LT_{50}.

lethal gene, *n.* a gene whose effect on the PHENOTYPE is sufficiently drastic to kill the individual. Death from lethal genes may occur at any time from fertilization of the egg to advanced age. Lethal genes can be either dominant (e.g. HUNTINGDON'S CHOREA, a progressive nervous disorder which occurs usually around middle age) or recessive (e.g. STICKLE-CELL ANAEMIA, a disorder of HAEMOGLOBIN causing death in adolescence).

leucine, *n.* one of 20 amino acids common in proteins. It has a NONPOLAR 'R' structure and is relatively insoluble in water. See Fig. 203 on page 320. The ISOELECTRIC POINT of leucine is 6.0.

leuco-, *prefix.* denoting white.

leucocyte or **leukocyte** or **white blood cell,** *n.* any of the large unpigmented cells in the blood of vertebrates. There are several types, formed in both lymph glands and bone marrow. See Fig. 204 (p. 320).

Leucocytes are a primary defence against invading organisms and other foreign material, using two main methods: PHAGOCYTOSIS and the

H
O
H_2N—C—C
OH
CH_2
CH
specific 'R' group
CH_3
CH_3

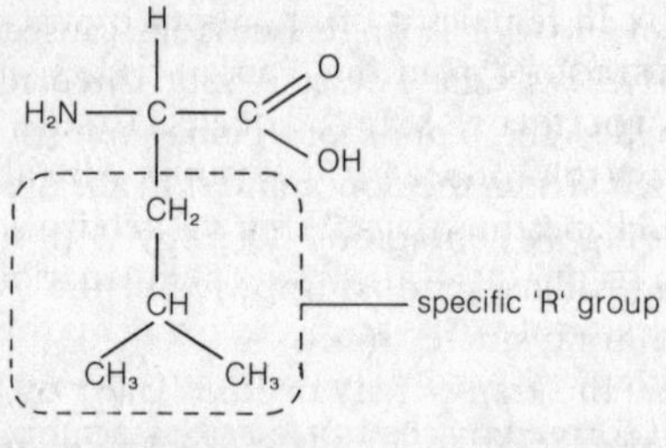

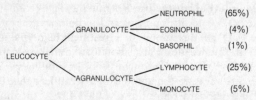

Fig. 203. **Leucine.** Molecular structure.

IMMUNE RESPONSE. The leucocyte count is usually around 10,000 cells per mm³ of blood. This, however, is not the total body count, because leucocytes are found as much in tissues such as thymus, spleen and kidney as in blood. The cells generally have a short life span (2 to 14 days) although the antibody-producing types (B-CELLS) tend to last up to 100 days. Compare ERYTHROCYTE. See also individual entries for the cells in Fig 204.

NEUTROPHIL (65%)
GRANULOCYTE
EOSINOPHIL (4%)
BASOPHIL (1%)
LEUCOCYTE
LYMPHOCYTE (25%)
AGRANULOCYTE
MONOCYTE (5%)

Fig. 204. **Leucocyte.** The types and relative frequencies of leucocytes.

leucocytosis, *n.* the presence of large numbers of leucocytes in the blood, usually resulting from injury to tissue or infection.

leucophyte, *n.* a nonphotosynthetic, colourless alga.

leucoplast, *n.* a colourless PLASTID of plant cells in which starch is often formed (see AMYLOPLAST). Leucoplasts may develop chlorophyll and function as CHLOROPLASTS as, for example, when potato tubers are lifted and exposed to light.

leucosin, *n.* a food reserve (see POLYSACCHARIDE) found in the Chrysophyceae (yellow-brown algae).

leukaemia, *n.* a form of cancer in LEUCOCYTES, resulting in an uncontrolled increase of immature white blood cells in body organs and often in the blood itself.

Leydig cells or **interstitial cells,** *n.* the cells of the testes that secrete TESTOSTERONE when stimulated by LUTEINIZING HORMONE.

LH (luteinizing hormone) or **ICSH (interstitial cell-stimulating hormone),** *n.* a glycoprotein hormone produced by the anterior

PITUITARY GLAND. In females it brings about ovulation on the stimulus of increasing OESTROGEN from the ovarian tissues, and causes a change in the GRAAFIAN FOLLICLE to form a CORPUS LUTEUM. LH also stimulates the corpus luteum to produce PROGESTERONE which in turn inhibits the production of LH and thus the subsequent production of progesterone (an example of a FEEDBACK MECHANISM), menstruation then taking place. In males it causes ANDROGENS to be secreted by the testis.

liana, *n.* a climbing, tropical plant with a rope-like stem.

lichen, *n.* a plant formed by the symbiotic association (see SYMBIOSIS) of algal cells which are surrounded by fungal hyphae. The algae are green (Chlorophyta) or blue-green (Cyanophyta) and the fungus usually an ASCOMYCETE or sometimes a BASIDOMYCETE. The fungus gains oxygen and carbohydrates from the alga and the latter gains water, CO_2 and mineral salts from the fungus. Lichens reproduce vegetatively by means of soredia (hyphal cells enclosing algae) and sexually by means of the fungal apothecia or PERITHECIA. However, where no algae of the usual association occur, the germinating fungal spore dies. Lichens are very common on trees and rocks in unpolluted areas and can be used as an INDICATOR SPECIES.

life cycle, *n.* the serial progression of stages through which an organism passes from FERTILIZATION to the stage producing the GAMETES. In higher animals and plants the life cycle lasts from fertilization until the resulting individual itself produces gametes, but in many organisms there are asexual stages. See ALTERNATION OF GENERATIONS.

life table, *n.* a table giving details of the mortality of a species or organism and all stages of the life history.

ligament, *n.* a band of elastic CONNECTIVE TISSUE that joins bones. Ligaments are composed of tightly packed elastic fibres and they are ideally suited to withstanding sudden stresses applied to the joints.

ligand, *n.* a molecule able to bind to a specific ANTIBODY and used to distinguish closely similar types of antibody.

ligase, *n.* an enzyme that catalyses the joining of short molecules into longer ones, for example, DNA LIGASE.

light, *n.* that part of the ELECTROMAGNETIC SPECTRUM which is visible to the human eye between about 400 nm (blue) and 770 nm (red).

light chain, see IMMUNOGLOBULIN.

light compass reaction, *n.* the ability of some organisms, particularly insects, to orientate in relation to the position of the sun. This involves an internal timing mechanism which is not clearly understood.

light-dependent reaction, *n.* a photosynthetic reaction or sequence of reaction that requires the presence of light. Compare LIGHT-INDEPENDENT REACTION.

light–independent reaction, *n*. a photosynthetic reaction or sequence of reaction, that takes place in the absence of light. Compare LIGHT-DEPENDENT REACTION.

light intensity, *n*. the quantity of light present.

light reactions, *n*. those processes of PHOTOSYNTHESIS requiring light energy in which ATP and NADPH are formed, to be incorporated later into the CALVIN CYCLE. Various constituents are necessary for the light reactions to occur, namely:

(a) CHLOROPLASTS (in higher green plants) within parenchyma cells, particularly the MESOPHYLL tissue of leaves.

(b) various pigments on the inner membranes of the chloroplasts responsible for trapping light energy. Chlorophyll *a* and one or more types of accessory pigment such as chlorphyll *b* and various carotenoids surround a single molecule of specialized chlorophyl *a* (P_{680} and P_{700}), forming a 'photosystem'. *Photosystem I* (PSI) contains P_{700} and *photosystem II* contains P_{680}.

(c) two ELECTRON TRANSPORT SYSTEMS (see Fig. 140).

(d) water.

Two separate processes of PHOTOPHOSPHYLATION occur:

(i) *cyclic photophosphorylation* in which the various pigments in PSI collect light striking the chloroplast, passing the energy on to P_{700} which undergoes PHOTOACTIVATION. An electron with raised energy levels is accepted by ferredoxin and passed on to an ETS where ATP is produced as the energy level falls back to the starting point.

(ii) *noncyclic photophosphorylation*. Here the initial source of electrons is water, which releases electrons by separation of charge (previously thought to be by means of PHOTOLYSIS) and these are passed onto PSII, and then on to plastoquinone at a higher level. As in cyclic photophosphorylation, ATP is produced via the ETS, with the electron dropping down to PSI. Then, however, light causes the energy level of the electron to be raised to a level high enough to be accepted by ferredoxin. At this stage a second ETS is entered, leading to the production of NADPH, with hydrogen coming from the separation of water into ions. Two photosystems are involved in the reduction of NADP because there is insufficient energy in light to energize the electrons from water straight to NADP: it requires two steps.

Thus, the products of these two types of light reactions are ATP, NADPH and oxygen. The first two products enter the DARK REACTIONS of photosynthesis, where they become involved in the CALVIN CYCLE and the synthesis of PGAL and eventually of GLUCOSE. See Fig. 205.

light year, *n*. the distance light travels in one year or 5.9×10^{12} miles.

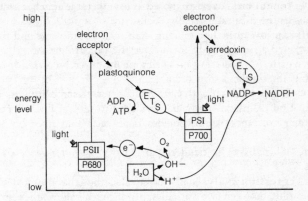

Fig. 205. **Light reactions.** The main chemical pathways in the light reactions of photosynthesis.

lignase, *n.* a group of enzymes that catalyses the condensation of two molecules.

lignin, *n.* a complex, noncarbohydrate polymer found in cell walls, whose function is to provide stiffening to the cell, as in xylem VESSELS, and bark fibres. Such cells are said to be 'lignified' and, since lignin forms an impermeable barrier, the cells are dead.

ligule, *n.* a small, scale-like outgrowth from a leaf, found in club mosses and some grasses.

ligulate, *adj.* strap-shaped.

limb, *n.* **1.** an articulated projection from the body of an animal which is used for locomotion, such as a leg or wing. **2.** a branch of a tree. **3.** the flattened part of a calyx or corolla where the base is tubular.

liming, *n.* the addition of calcium compounds to the soil, producing three main effects: (a) the provision of the major element calcium, (b) the neutralizing of acid soils, (c) the promotion of flocculation (clumping) of clay particles into larger crumbs, so as to encourage aeration and drainage.

limiting factor, *n.* **1.** (in chemical processes) a component that limits the amount of the product that can be formed or its rate of formation, because it is present in small quantities. For example, light intensity can be a limiting factor in PHOTOSYNTHESIS. **3.** (in ecology) a factor that restricts the numbers of a population, such as food or nest sites.

limno-, *prefix.* denoting a marsh.

limnology, *n.* the study of freshwater bodies, such as ponds, lakes, streams and their inhabitants.

limpet, *n.* a marine gastropod MOLLUSC of the genus *Patella* that has a single, conical shell. When exposed at low water it attaches itself very firmly to the substrate, usually rocks.

Limulus, *n.* the horseshoe or king crab – an aquatic arachnid that has maintained a similar body form since the TRIASSIC PERIOD.

lincomysin, *n.* a bacteriostatic antibiotic produced by *Streptomyces* that inhibits the synthesis of protein.

line, *n.* a subpopulation with certain characteristics in common. Thus a population of insects might be divided into several lines, depending on their tolerance to insecticides, and each line may be further inbred.

linear, *adj.* narrow or slender, particularly with reference to plant parts.

linear regression analysis, *n.* a statistical method that aims to define the relationship between two variables, producing a value *b*, the regression coefficient. There are several assumptions that have to be made in carrying out the analysis, particularly (a) that there is an independent variable *x*, e.g. time, which can be measured exactly, and also a dependent variable *y*, e.g. metabolic rate, (b) that for every value of *x* there is a 'true' value of *y*.

Linear regression analysis enables the fitting of a straight line to a scatter graph, using the equation $y = a + bx$, the *a* value being the point at which the regression line crosses the *y*-axis (the intercept).

linear scale, *n.* a scale in which equal intervals represent equal increments in the quantities concerned, such as a temperature scale on a thermometer, as opposed to say a LOGARITHMIC SCALE. See Fig. 206.

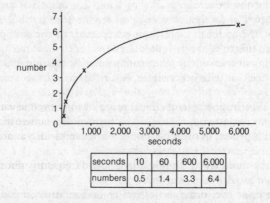

seconds	10	60	600	6,000
numbers	0.5	1.4	3.3	6.4

Fig. 206. **Linear scale.** The lack of observations as the curve flattens renders the shape of the curve at this point uncertain.

line precedence, *n*. the occurrence of a taxonomic name earlier on a page of print than another.

lingual, *adj*. of or relating to the tongue.

lingulate, *adj*. (of plant structures) tongue-shaped.

linkage disequilibrium, *n*. a nonrandom association of alleles at different loci in a population, as is produced when two loci are closely linked and selection operates to keep certain gene combinations together.

linkage, genetic, see GENETIC LINKAGE.

linkage map, see CHROMOSOME MAP.

linkage unit, see MAP UNIT.

Linnaeus, Carl (1707–78) Swedish biologist, author of *Systema Naturae* (1753), in which he outlined the binomial system of classifying organisms (see BINOMIAL NOMENCLATURE). He is thus regarded as the founder of modern systematics. Linnaeus was a highly competent biologist who for a long time supported the idea that species were fixed, an anti-evolutionary view. However, he probably changed his views on this in later life.

linoleic acid or **essential fatty acid,** *n*. an unsaturated fatty acid that cannot be synthesized by man and is therefore described as 'essential'. Deficiency of linoleic acid in the diet results in increased metabolic activity, failure in growth and even death.

lip, *n*. **1.** (of an embryonic BLASTOPORE) the rim of the blastopore. **2.** (of a flower perianth) a group of perianth segments united to a greater or lesser extent, to form a distinct grouping thus divided from the rest of the perianth. **3.** (of vertebrates) one of two fleshy parts (upper and lower) supporting the mouth.

lip- or **lipo-,** *prefix*. denoting fat.

lipase, *n*. the enzyme that breaks down fat into fatty acids and glycerol. The main source is the PANCREATIC JUICE.

lipid, *n*. a biological compound containing carbon, hydrogen and oxygen together with other elements such as nitrogen and phosphorus. Lipids (or fats) are structural components of cell membranes and nervous tissue, and are important sources of energy, being stored in various parts of the body (see ADIPOSE TISSUE). Their large molecular size makes lipids insoluble in water, but soluble in organic solvents such as acetone and ether.

lipin, *n*. a complex fat containing nitrogen and frequently phosphorus or sulphur.

lipogenesis, *n*. the formation of fats from nonfatty sources.

lipoid, *n*. any substance with fat-like properties including true FATS, STEROIDS and LIPINS.

lipophilic, *adj*. having an affinity for LIPIDS.

lipoprotein, *n*. a water-soluble molecule made up of a protein containing a lipid group. It is found, for example, in protoplasm where it is involved in transport of lipids in a soluble form.

liquid feeder, *n*. any organism that feeds exclusively on liquids, such as a mosquito.

lithosere, *n*. a plant succession that originates on a rock surface.

lithotroph, *n*. any organism that makes use of an inorganic electron donor.

litter, *n*. **1.** any material aggregated on the surface of soil from above-ground vegetation. **2.** the offspring produced at any one time by a mammal.

littoral zone, *n*. the shore, or intertidal area of the beach, being the region bounded on its landward side by the extreme high-water mark of spring tides, and on its seaward side by the extreme low-water mark of the same tides.

liver, *n*. the largest and most complex organ of the vertebrate body, with a wide range of functions (see below) several of which are vital for life to continue. In mammals the liver receives a double blood supply, about 70% coming from the HEPATIC PORTAL SYSTEM and 30% from the arterial system. The liver:

(a) removes excess glucose from the blood and stores it as GLYCOGEN.

(b) converts glycogen back to glucose when blood sugar levels are low.

(c) converts food substances to other types, e.g. carbohydrates into fats, amino acids into carbohydrates or fats.

(d) deaminates amino acids, converting the ammonia produced into urea via the ORNITHINE CYCLE, releasing the nitrogenous wastes into the blood.

(e) transaminates amino acids (see TRANSAMINATION) from one type to another via keto acids.

(f) detoxifies many harmful compounds.

(g) manufactures fats, including cholesterol.

(h) manufactures many plasma proteins, including FIBRINOGEN and PROTHROMBIN.

(i) stores several important substances, e.g. iron and fat-soluble vitamins.

(j) excretes bile pigments.

(k) manufactures bile salts.

(l) destroys worn-out red blood cells.

liver fluke, see FLUKE.

liverwort, see HEPATICA.

lizard, *n.* any member of the reptilian order Lacertilia, the majority of which possess a long scaly body and tail, and four obvious limbs.

loam, *n.* a soil formed of sand and clay and having an organic or humus content.

lobed, *adj.* (of leaves) divided but not into separate leaves.

lobule of liver, *n.* one of several lobes or divisions of the liver.

local population, see DEME.

loci, see LOCUS.

lock-and-key mechanism, see ACTIVE SITE.

locomotion, *n.* the progressive movement of an organism.

loculicidal, *adj.* (of an ovary) being split down the middle of each cell.

locus, *n.* (*pl.* loci) the position of a gene along a chromosome.

lodicule, *n.* the small perianth that, in the flowers of grasses, swells and exposes the pistil and stamens, by forcing open the surrounding bracts.

lod score, *n.* a statistical method of estimating linkage between human genes by calculating the odds of a particular frequency of births having been produced by GENETIC LINKAGE as compared with INDEPENDENT ASSORTMENT. The name lod comes from 'log of odds'.

loess, *n.* a fine-grained windblown silt that is characteristic of temperate areas where thick layers of it were distributed during the retreat of the ice after the last glaciation.

logarithmic scale, *n.* a scale in which the values of a variable are expressed as logarithms. Such transformations of data are often employed to simplify the drawing of lines on a graph, where the variable values are spread over a wide range. See Fig. 207. A good example of the use of log scales is seen in Fig. 177 where the LOG PHASE of a microbial GROWTH CURVE is illustrated. Compare LINEAR SCALE.

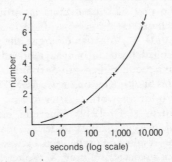

Fig. 207. **Logarithmic scale.** Note the smoothness of the curve as compared to the LINEAR SCALE curve, for which the data are the same.

logistic curve, *n.* an S-shaped curve of numbers against time that represents the growth in numbers of a population of organisms in a limited environment. See GROWTH CURVE.

log phase, *n*. the exponential phase of growth in a bacterial culture. Plotting the log of the number of cells against time gives a straight line. See LOGARITHMIC SCALE and Fig. 207.

log-probit analysis, *n*. a technique for presenting the results of an exposure test on a graph, in which, for example, the dose of a chemical is represented on the *x*-axis as a log scale and the mortality is represented on the *y*-axis as a probability scale. The plotted log-probit line is transformed into a more linear shape as a result of the new scales, making it easier to compare results from different tests.

lomasome, *n*. an INVAGINATION that occurs in the cells of fungal hyphae and sporing structures.

lomentum, *n*. a form of legume seed in which constrictions between the seeds result in their falling singly.

long-day plant, *n*. one that requires a day length exceeding a certain minimum value for the induction of flowering to take place. In fact, the term is somewhat misleading in that such plants (e.g. lettuce, clover) are really sensitive to periods of darkness rather than daylight, requiring a night length of no more than a maximum duration, i.e. these are *short-night plants*. Compare SHORT-DAY PLANT, DAY-NEUTRAL PLANT.

long-sightedness, see HYPEROPIA.

loop of Henle, *n*. the U-shaped, nonconvoluted part of the tubule which leads from a BOWMAN'S CAPSULE to the central cavity of the kidney (the pelvis), and where the urine is concentrated. Together, the capsule and tubule form a NEPHRON. Animals in arid climates have very long loops of Henle and produce small quantities of highly concentrated urine.

lophophore, *n*. the ring of ciliated tentacles around the mouth in polyzoans and brachiopods.

Lorenz, Konrad (1903–) Austrian biologist, the pioneer of modern ETHOLOGY, who founded the school of behavioural studies which regards animal behaviour as a product of adaptive evolution. He is perhaps best known for two of his books, *King Solomon's Ring* (1952) and *Man meets Dog* (1954), and also for his work on IMPRINTING.

Lotka, Alfred James (1880–1949) American biologist, author of *Principles of Physical Biology* (1925), who derived mathematical expressions governing the results of competition between populations and interactions between predator and prey. V. Volterra published similar expressions in Italy, independently, in 1926. See K.

lotic, *adj*. of or relating to an aquatic environment where there is fast moving water.

louse, *n.* any wingless insect of the order Mallophaga (bird lice or biting lice) or the order Anopleura (sucking lice).

lower critical temperature, *n.* the environmental temperature defining the lower limit of the THERMONEUTRAL ZONE of a tachymetabolic animal (see TACHYMETABOLISM). Below the lower critical temperature the animal must undergo REGULATORY HEAT PRODUCTION in addition to its BASAL METABOLIC RATE to maintain homoiothermy (see HOMOIOTHERM).

LSD (lysergic acid diethylamide), *n.* an hallucinogenic drug prepared from lysergic acid.

LTH, see LACTOGENIC HORMONE.

luci-, *prefix.* denoting light.

luciferase, *n.* an oxidizing enzyme that, acting on the bioluminescent substance luciferin, causes luminosity in glow worms and other insects. See LUMINESCENCE.

lugworm, *n.* the common name for the marine annelid *Arenicola*.

lumbar vertebra, *n.* one of the vertebrae present between the thoracic vertebrae and the sacral vertebrae, in the region of the waist in mammals. See VERTEBRAL COLUMN.

lumen, *n.* any cavity enclosed within a cell, or structure, such as the lumen of the gut.

luminescence, *n.* the production of light by living organisms that is brought about by the oxidation of the protein luciferin. The reaction requires ATP and is catalysed by LUCIFERASE. See also BIOLUMINESCENCE.

luminosity, *n.* the amount of light emitted or reflected – brightness.

lumper, *n.* a taxonomist who tends to group organisms together and then form large taxa, as distinct from one who forms many taxa (a SPLITTER).

lung, *n.* the organ that allows an animal to breath air (see AERIAL RESPIRATION). In mammals the paired lungs are each supplied by a bronchus which divides, tree-like, into bronchioles. Each bronchiole ends in an atrium from which arise numerous alveoli which contact a vast capillary system. The surface area of the alveoli in man is about 70 m^2 so that there is a large surface area over which gaseous exchange can take place. See BREATHING.

lung book, *n.* a form of respiratory organ found in spiders and scorpions. The large number of parallel leaflets around which air circulates are contained in the lung book, and have the form of book pages – hence the name.

lungfish, see DIPNOAN.

luteal phase, *n.* the phase of the OESTRUS CYCLE in which the CORPUS LUTEUM forms in the ovary. Secretion of PROGESTERONE takes place followed by the development of uterine glands and a fall off in OESTROGEN secretion.

luteal tissue, *n*. the tissue derived from follicle cells that fills the ruptured Graafian follicle and forms the CORPUS LUTEUM.

luteinizing hormone, see LH.

luteotrophic hormone (LTH) or **lactogenic hormone** or **prolactin,** *n*. a proteinaceous hormone secreted by the anterior PITUITARY GLAND of vertebrates which has a range of functions, including acting as a GONADOTROPHIN. LTH influences the onset of lactation, stimulates PROGESTERONE secretion by the CORPUS LUTEUM in mammals (in tandem with LUTEINIZING HORMONE) and stimulates maternal behaviour in all vertebrates.

luxuriance, *n*. the somatic vigour of hybrids that does not add to their capability to compete.

lyase, *n*. an enzyme that catalyses the addition of (a) groups to double bonds, (b) the removal of groups to form double bonds.

Lycopodiales, *n*. an order of the TRACHEOPHYTA containing the club mosses.

Lycopsida, a subdivision of the TRACHEOPHYTA including the club mosses *Lycopodium* and related species.

Lyell, Charles (1797–1875) British geologist who made extensive studies of PALAEONTOLOGY, but from a biological point of view is best known for arranging for the publication of the views of Darwin and Wallace on the origin of species.

lymph, *n*. the INTERSTITIAL FLUID found in the LYMPHATIC SYSTEM and around the tissues of vertebrates. Although its composition varies with location in the body, lymph is typically a clear, transparent fluid (95% water) which, like blood, will clot when removed from lymph vessels since it contains similar clotting agents to blood (except platelets). Lymph also contains protein, glucose and salts with large numbers of LEUCOCYTES, mainly LYMPHOCYTES.

lymphatic capillary, see LYMPHATIC VESSEL.

lymphatic node, see LYMPH NODE.

lymphatic system, *n*. a system of tubules in vertebrates that drains excess tissue fluid (LYMPH) from the tissue spaces to the blood system. Unlike blood CAPILLARIES, lymph capillaries are blind-ending in the tissue spaces, gradually joining up to larger and larger vessels with two major lymphatic ducts entering the venous system in the upper thoracic cavity. Lymph is not moved along by heart contractions but (as in veins) by the action of skeletal muscles. Lymph vessels contain one-way valves to prevent backflow of fluid to the tissues.

Mammals and some birds possess clumps of lymphatic tissue called

LYMPH NODES that act as filters and are also sites of LYMPHOCYTE formation. Nodes are especially prevalent in the neck, underarm and groin, becoming swollen when adjacent to an infection. Besides returning excess water and protein to the blood system and being active in combating infection, lymph (rather than blood) transports fats from the gut wall, the lymph vessels being called LACTEALS since their contents are milky white.

lymphatic vessel or **lymphatic capillary,** *n.* a vessel that carries LYMPH.

lymph heart, *n.* an enlarged lymphatic vessel that is capable of pumping lymph. It is present in most vertebrates but not birds and mammals.

lymph node or **lymphatic node,** *n.* a mass of lymphoid tissue, important in producing antibodies containing macrophages which remove foreign bodies from the lymph. It is present only in mammals and birds.

lymphocyte, *n.* a type of LEUCOCYTE (white blood cell) of the AGRANULOCYTE group which is formed in the bone marrow from which they differentiate into either B-CELLS or T-CELLS. The typical size is about the same as a red blood cell (about 8 μm in diameter). Lymphocytes are not themselves motile, but are capable of destroying foreign ANTIGENS in the IMMUNE RESPONSE, collecting in the spleen and lymph nodes which may become swollen in a severe attack. The two types of lymphocyte react in different ways, T-cells destroying the antigen themselves (*cell-mediated immunity*) while B-cells produce ANTIBODIES.

lymphoid tissue, *n.* tissue composed largely of LYMPHOCYTES, such as the thymus or lymph nodes.

lymphokine, *n.* a soluble mediator released by lymphocytes on contact with specific antigens.

lymphoma, *n.* a tumor of lymphoid tissue.

lyrate, *adj.* shaped like a lyre.

lysergic acid, see LSD.

Lyon hypothesis, see INACTIVE-X HYPOTHESIS.

Lysenko, Trofim Denisovitch (1898–1976) Russian plant biologist who attempted to apply Lamarckian theory to the development of crop plants. See LAMARCKISM.

lysigenous, *adj.* (of tissues) producing cavities by the breakdown of cells, as in the secretory organs of some plants.

lysin, *n.* a type of ANTIBODY.

lysine, *n.* one of 20 AMINO ACIDS common in proteins, having an extra basic group and being alkaline in solution. See Fig. 208. The ISOELECTRIC POINT of lysine is 10.0.

lysis, *n.* the rupturing of a cell with release of its contents. For example,

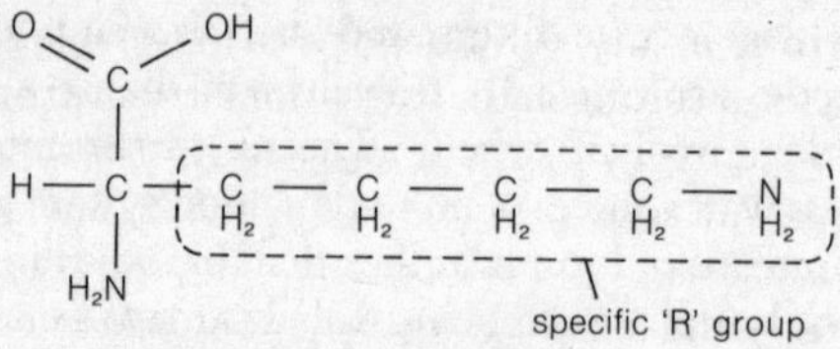

Fig. 208. **Lysine.** Molecular structure.

when a bacterial cell bursts to release BACTERIOPHAGES, or HAEMOLYSIS when a red blood cell bursts (see RHESUS HAEMOLYTIC ANAEMIA).

lysogenic bacterium, *n.* a bacterium that carries a prophage of a temperate virus.

lysogeny, *n.* a state in a living bacterium when it carries a nonvirulent TEMPERATE PHAGE (virus). In this condition the phage DNA is integrated into the bacterial chromosome.

lysosome, *n.* a cytoplasmic organelle of EUCARYOTE cells that contains hydrolytic enzymes and is thought to be produced by the GOLGI APPARATUS. The sac-like structure is surrounded by a single-layered membrane which is impermeable and resistant to the enzymes inside. Lysosomes can act as the digestive system of the cell. When the sac ruptures the enzymes are released into a food vacuole produced by PHAGOCYTOSIS, thus enabling the breakdown of ingested materials.

lysozyme, *n.* an enzyme that breaks down bacterial cell walls and provides protection against bacterial invasion in the skin, mucus membranes and many body fluids.

lytic cycle, *n.* the life cycle of a BACTERIOPHAGE in which many new phages are reproduced and the host bacterial cell undergoes LYSIS, the phages entering new bacterial hosts.

M

McCarty, Maclyn. See AVERY.

MacLeod, Colin. See AVERY.

macro-, *prefix.* denoting great or large.

macrobiotic, *adj.* prolonging life.

macrobiotics, *n.* a dietary system advocating a diet of whole grains and vegetables grown without chemical additives, and in which foods are classified according to the principles of Yin and Yang.

macrocephalic, *adj.* having a large head.

macrocyte, see MONOCYTE.

macroevolution, *n.* any evolutionary change taking place at a level above the species, resulting in the formation of such groups as genera and families.

macrofauna, *n.* **1.** a widely distributed fauna. **2.** any animals visible to the naked eye.

macrogamete, *n.* the female gamete, so designated because of its larger size. See ISOGAMY, ANISOGAMY.

macrolepidoptera, *n.* the larger butterflies and moths. The term has no taxonomic significance.

macromere, *n.* a large cell from the vegetative pole of a developing egg. Such a cell contains yolk and gives rise to the ENDODERM of the embryo.

macromolecule, *n.* a very large molecule, composed of many atoms and having a very large molecular weight. Examples include nucleic acids, proteins.

macronucleus or **meganucleus,** *n.* the larger of two nuclei found in some PROTOZOANS, that is concerned with cell division (see MICRONUCLEUS). It disappears during CONJUGATION and is afterwards reformed from material of micronuclear origin (from the zygote nucleus). It appears to have mainly vegetative functions.

macronutrient, *n.* any element required in large quantities for growth, such as nitrogen or potassium in plants. Compare MICRONUTRIENT.

macrophage, *n.* a type of cell that is widely distributed in large numbers in vertebrate tissues and which is derived from MONOCYTE blood cells. Its function is to remove debris, which it engulfs by PHAGOCYTOSIS, and it is capable of movement by means of PSEUDOPODIA. Macrophages occur in connective tissue, lymph and blood, and one of their important roles is the digestion of old and damaged red blood cells in the spleen. Some macrophages are free-moving, others are located in fixed locations (e.g. KUPFFER CELLS), the whole arrangement forming the RETICULO-ENDOTHELIAL SYSTEM.

macrophagous, *adj.* (of animals) feeding on large (relative to size of animal) food particles.

Macropodidae, *n.* the family of MARSUPIALS containing the kangaroos and wallabies. They are jumping mammals with long hind legs and a balancing tail.

macropterous, *adj.* having large wings, as in some castes of termites and ants.

macrosporangium, see MEGASPORANGIUM.

macrospore, see MEGASPORE.

macrosporophyll, see MEGASPOROPHYLL.

macrurous, *adj.* having a large tail. The term is usually applied to long-bodied CRUSTACEANS, such as the lobster, where the tail projects backwards as opposed to being beneath the abdomen as in the crab.

macula, *n.* an area of acute vision on the retina of many vertebrates which lack a FOVEA.

madreporite, *n.* the porous opening to the hydrocoel in ECHINODERMS.

maggot, *n.* any insect larva lacking appendages and an obvious head, (usually) the larva of a member of the order Diptera.

magnetite, *n.* an iron oxide found in some animals, which may play a part in geomagnetic orientation because of its magnetic properties.

maize, *n.* the Indian corn *Zea mays*, possibly the most intensively investigated higher plant.

major histocompatability complex (MHC), *n.* a cluster of tightly linked genes on chromosome 6 in humans. These genes code for protein molecules that are attached to the surface of body cells and used by the body's immune system to recognize own or foreign material. The proteins of the MHC most easily recognized are found on the surface of leucocytes and are known as the HLA SYSTEM.

malac- or **malaco-,** *prefix.* denoting soft.

Malacostraca, *n.* the crustacean group containing crabs, lobsters, shrimps, crayfish, etc. See also CARIDOID FACIES.

malaria, see MALARIA PARASITE.

malaria parasite, *n.* the PROTOZOAN *Plasmodium* that causes malaria in man. It is transmitted by the *Anopheles* mosquito which takes in *Plasmodium* GAMETOCYTES when feeding on man. The gametocytes develop into male and female gametes in the mosquito stomach, and after fertilization the ZYGOTE bores through the stomach wall, encysts, and gives rise to thousands of sporozoites which then migrate to the salivary glands. The sporozoites are transferred to man at a subsequent blood meal during which anticoagulants are injected with the saliva to prevent blood clotting.

Sporozoites invade the human liver and give rise to merozoites that invade red blood corpuscles. Further production of merozoites occurs in the red blood corpuscle and their release into the blood stream gives rise to bouts of fever ('malaria'). Gametocytes arise from merozoites which invade further red blood corpuscles, grow within them, and are withdrawn by the mosquito during feeding, to complete the life cycle.

male gamete nuclei, *n.* two HAPLOID(1) nuclei of flowering plants that arise from mitotic division of the generative nucleus inside the POLLEN GRAIN, which enter the pollen tube and eventually the EMBRYO SAC, becoming involved in the 'double' fertilization.

malic acid or **malate,** *n.* a four-carbon organic acid which is an intermediate step in the KREBS CYCLE, formed by the HYDROLYSIS of fumaric acid (see Fig. 209). Malic acid is also important in C_4 PLANTS (such as sugar cane), in which carbon is 'captured' from CO_2 to form malic acid and other C_4 acids in MESOPHYLL cells, before being transferred to the CALVIN CYCLE taking place in the CHLOROPLASTS of special cells nearby. See KRANZ ANATOMY.

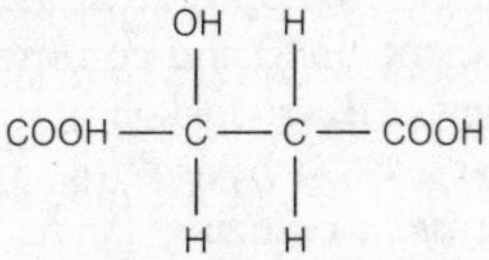

Fig. 209. **Malic acid.** Molecular structure.

malignancy, *n.* a structure (such as a tumour) or condition (such as a fever) the progressive version of which is threatening to life.

malleus or **hammer,** *n.* the hammer-shaped bone that is the outermost of the three ear ossicles.

Mallophaga, see LOUSE.

Malphighian body or **Malphighian corpuscle,** *n.* a structure for filtering blood in the vertebrate kidneys that is formed from the BOWMAN'S CAPSULE with its associated GLOMERULUS.

Malphighian layer, *n.* the layer of EPIDERMIS in the mammalian skin lying immediately above the DERMIS which is continuously dividing by MITOSIS to replace outer epidermal cells that are rapidly worn away by friction.

Malphighain tubules, *n.* the excretory organs of some ARTHROPODS, such as insects and arachnids, that collect wastes from the body fluids and discharge them into the hind gut. They are tubular glands which open into the anterior part of the hind gut, the other end being closed and bathed in the animal's blood in the HAEMOCOEL.

The slender tubules are muscular and capable of movement, taking up soluble potassium urate, water and carbon dioxide from the blood into the tubule cells where they react together to form potassium hydrogen carbonate and uric acid. The $KHCO_3$ and some water are reabsorbed from the tubules into the blood; further water is absorbed by the rectal gland and crystals of uric acid are voided into the gut for excretion.

maltase, *n.* an enzyme that hydrolyses MALTOSE to glucose. In mammals it is produced in the CRYPT OF LIEBERKUHN in the SMALL INTESINE and is present in the SUCCUS ENTERICUS. Maltase is also present in many seeds.

Malthus, Thomas Robert (1766–1834) English political economist whose ideas on population growth influenced the evolutionary theory propogated by Charles DARWIN and Alfred WALLACE. His most interesting observation was that, if unchecked, the human population would grow geometrically whilst the food supply would grow only arithmetically, thus giving rise to mass starvation. He considered that the only checks on population growth were disease, war, famine and abstinence from sex, the latter being the most reliable method of BIRTH CONTROL at the time.

maltose, *n.* a disaccharide CARBOHYDRATE consisting of two GLUCOSE units linked by a GLYCOSIDE bond and common in germinating barley (see Fig. 210). The barley seeds are soaked in water, initiating the release of large amounts of AMYLASE, a hydrolytic enzyme that breaks down stored starch into maltose, a reducing sugar.

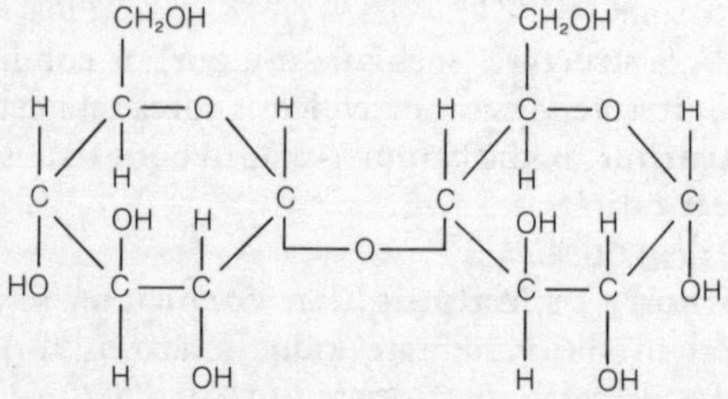

Fig. 210. **Maltose.** Molecular structure.

mammal, *n.* any animal of the class Mammalia, a group often regarded as the most highly evolved animals. There are three living subclasses:

(a) Monotremata − MONOTREMES, primitive egg-laying mammals such as the duck-billed platypus and *Echidna*, the spiny ant eater.

(b) Marsupialia − MARSUPIALS, which transfer their young to pouches for the latter part of their early development.

(c) Eutheria − EUTHERIANS, which have a placenta.

Mammals are characterized by the presence of hair, a DIAPHRAGM used in AERIAL RESPIRATION, milk secretion in the female (LACTATION), presence of the left systemic arch only in the blood circulatory system, three auditory ossicles in the ear and a lower jaw of a single pair of bones. In all classes except the monotremes, the young are born alive.

mammary gland or **milk gland,** *n.* a gland present in female MAMMALS that produces milk used to suckle their young (see LACTATION). It probably evolved from a modified sweat gland and at least two are normally present, though in many mammals more than two are developed, usually concentrated on the underbelly beneath the pelvic girdle. In most mammals the size of the gland is determined by the state of the OESTROUS CYCLE.

mandible, *n.* that part of the mouthparts of an animal which does most of the crushing of food materials. In vertebrates, the term usually denotes the lower jaw. In insects and other arthropods, the mandibles are one of a pair of mouthparts used for crushing food (see Fig. 191).

mantle, *n*. that part of the EPIDERMIS of a mollusc which secretes the shell and covers the dorsal and lateral surfaces.

manubrium, *n*. **1.** any handle-like elongated process. **2.** the tubular mouth of a jellyfish. **3.** the anterior segment of the sternum in mammals. **4.** the first segment of the springing organ in COLLEMBOLA.

manus, *n*. the hand or forefoot of a TETRAPOD.

manuscript name, *n*. a scientific name that has not been published.

map unit or **linkage unit,** *n*. a unit of 'distance' between genes along a chromosome. The number of map units is directly correlated with the amount of RECOMBINATION between loci; 1% recombination is equal to 1 map unit. See CROSS-OVER VALUE, GENETIC LINKAGE.

marker gene, *n*. a gene whose location and characteristics are well-known and which is used to investigate other genes. For example, to determine the location of any gene relative to the marker gene.

marram grass, *n*. *Ammophila arenaria*, a grass found on coastal sand dunes. It is a XEROPHYTE that is capable of rolling its leaves to prevent water loss in dry conditions. It has deeply running stems (RHIZOMES) and roots that bind the sand in the dunes and thus is often planted to stabilize dune systems.

marsh, *n*. an area of wet or periodically wet land where the soils are not peaty.

marrow, *n*. the pulp-like content of the cavities of larger bones that produces red and sometimes white blood corpuscles.

marsupial, *n*. any member of the subclass Marsupialia (also called Didelphia or Metatheria) containing mammals, characterized by the presence of a pouch to which the young, born in an undeveloped state, migrate during early development. The pouch contains the mammary glands, which vary in number between species, and the young complete their development here. The group was at one time widespread, but now is restricted to Australasia and South America. In Australasia, marsupials, free from competition from EUTHERIAN (placental) mammals, have radiated to occupy most niches elsewhere occupied by placental forms.

marsupium, *n*. the pouch of a MARSUPIAL mammal.

mass flow, *n*. an hypothesis to explain the movement of solutes in a plant by means of an hydrostatic pressure gradient. The idea is that in one part of the plant (the *source*) solute particles are actively secreted into sieve cells of PHLOEM, producing a high osmotic concentration which attracts water by OSMOSIS. In another part of the plant (the *sink*) solute particles are leaving the phloem, thus lowering the osmotic concentration which causes water to leave also. There is, therefore, a pressure gradient set up between the source (e.g. leaf) and the sink (e.g. root) and fluids flow from one to the other, down the gradient. See TRANSLOCATION(1).

mass selection, *n.* a method of producing new plant and animal varieties in which many suitable individuals contribute their genes to the new variety, all inferior types in the population having been removed (selected out) by the breeder.

mast cell, *n.* a type of large, amoeboid cell found in the matrix of CONNECTIVE TISSUE, that produces HEPARIN and HISTAMINE and is probably important in quick-acting responses to ANTIGENS, see IMMUNE RESPONSE.

mastication, *n.* the process of chewing and breaking down food particles before swallowing.

mastigo-, *prefix.* denoting whip-like.

mastigoneme, *n.* a projection found in numbers along the length of some flagella that serves to increase the surface area.

Mastigophora or **Flagelleta,** *n.* a group of flagellate PROTOZOANS, sometimes given the status of a class (= Flagellata). The group contains both holozoic and holophyte forms.

mastodon, *n.* a relative of the elephant that existed in MIOCENE and PLIOCENE times. It was, in fact, elephant-like having a short trunk and long tusks.

mastoid process or **mastoid bone,** *n.* the part of the periotic bone that in humans forms a projection behind the ear.

material, *n.* the sample available for any study in TAXONOMY.

maternal effect, *n.* an environmental phenomenon in which the phenotype of an offspring is influenced by the maternal tissue. For example, the adverse influence of maternal smoking on the weight of the unborn baby.

maternal inheritance, *n.* a form of CYTOPLASMIC INHERITANCE in which genes are passed to the offspring from the female only.

mating, *n.* **1.** any reproduction involving two sexes. **2.** (in lower organisms) reproduction between types that differ in physiology but not in physical form. **3.** (in birds and mammals) the behavioural process of pair-formation rather than of copulation leading to sexual reproduction.

mating system, *n.* the social relationship between the sexes during the breeding cycle that ranges from pair formation to forms of polygamy. The pattern of mating can be RANDOM MATING or ASSORTATIVE MATING.

mating types, *n.* the equivalent of sexuality in lower organisms, where physiological differences rather than physical ones distinguish the types, the mating system usually being inherited.

matter, *n.* that which constitutes the substance of physical forms, has mass, occupies space and can be quantified.

matrix, *n.* a ground substance in which other materials or cells are embedded. For example, the matrix of CONNECTIVE TISSUE containing fibres, or blood plasma forming a matrix in which are various blood cells.

matroclinous inheritance, *n.* a system in which all offspring of a cross have the PHENOTYPE of the mother. See CYTOPLASMIC INHERITANCE. Compare PATROCLINOUS INHERITANCE.

maturation of germ cells, *n.* the final stage of development of eggs and sperm in the process of GAMETOGENESIS, producing mature GAMETES capable of FERTILIZATION.

maturation (viral), *n.* the collection of infective VIRIONS produced in the host cell.

maxilla, *n.* **1.** (in vertebrates) the bone of the upper jaw, carrying all teeth except the incisors. The term is occasionally used to denote the whole of the upper jaw. **2.** (in arthropods) that part of the mouthparts lying behind the jaws. See LABRUM and LABIUM.

maxillipede, *n.* any limb that is modified for feeding, normally found immediately in front of the legs, particularly in crustaceans.

maxillule, *n.* one of the first MAXILLAE (2) of a crustacean.

maze, *n.* a labyrinth of winding passages, used in behavioural experiments involving LEARNING.

M.D., see MUSCULAR DYSTROPHY.

mean, see ARITHMETIC MEAN.

mean body temperature, *n.* the CORE TEMPERATURE actually attained by behaviourally thermoregulating bradymetabolic animals (see BRADYMETABOLISM) under natural conditions.

measles, *n.* an acute disease of human childhood in which red circular spots appear all over the body, there is a dry cough, low grade fever and sore throat. The condition is caused by the *Morbilli* virus. It is highly contagious, being spread in respiratory secretion, but once infected the individual is immune for life. Control is by the use of attenuated live virus vaccine given to children.

meatus, *n.* any natural opening or passage, particularly the external auditory meatus leading to the TYMPANUM in the ear.

mechanical tissues, *n.* the tissues that serve to support a plant, such as COLLENCHYMA and SCLERENCHYMA.

mechanoreceptor, *n.* a sensory structure that receives mechanical stimuli such as sound, pressure, movement, etc.

Meckel's cartilage, *n.* one of a pair of cartilages forming the lower jaw of cartilaginous fishes such as sharks and skates. In fish, reptiles and birds it forms the ossified articular bone and in mammals, the MALLEUS.

meconium, *n.* the gut contents of a mammalian foetus, formed of swallowed amniotic fluid and the secretion of glands into the gut.

Mecoptera, *n.* an order of insects containing the scorpion flies.

median, 1. *adj.* (of a structure or character) describing a location along the line of bilateral symmetry. **2.** *n.* (in statistics), the middle value in a frequency distribution, on either side of which lie values with equal total frequencies.

mediastenum, *n.* the space in the chest of mammals containing the HEART, TRACHEA and OESOPHAGUS.

medium, *n.* a substance on or in which microorganisms and other small organisms can be cultured. A medium can be liquid or solid. If solid, it frequently contains agar, a stiffening agent extracted from seaweed. Culture media can contain all necessary nutrients and trace elements for normal growth (a *minimal medium*) but can also be supplemented. For example, ANTIBIOTICS can be added to test for antibiotic resistance in bacteria.

medulla, *n.* **1.** the central part of an animal organ, such as the brain, adrenal gland, or kidney. **2.** see PITH.

medulla oblongata, *n.* the posterior part of the hindbrain of vertebrates that connects with the spinal cord. Ventrally thick-walled and containing groupings of nerve cells called nuclei, the thin-walled posterior CHOROID PLEXUS is situated on its dorsal side. The medulla contains several important areas of nervous control not influenced by conscious action. The respiratory and cardiovascular centres control respiration and heartbeat and organs of the LATERAL-LINE SYSTEM. The ear, taste and touch receptors also have their impulses coordinated in the medulla.

medullary plate, *n.* the neural plate of the vertebrate embryo.

medullary ray, *n.* an area of the stem between the vascular bundles, containing PARENCHYMA cells initially, but which can become converted into meristematic tissue. See INTERFASCICULAR CAMBIUM.

medullated nerve fibre, see MYELIN SHEATH.

medusa, *n.* (*pl.* medusas, medusae) the jellyfish (*medusoid*) stage of the COELENTERATE life cycle, usually free-swimming and propelled by pulsations of the bell. Medusae usually reproduce sexually, giving rise to a POLYP stage from which the medusae are produced asexually. Medusae form the dominant phase of the life history of members of the class Scyphozoa, but are often absent or of lesser importance in other classes of the phylum.

medusoid, see MEDUSA.

mega- or **megalo-**, *prefix*. denoting large.

megakaryocytes, see BLOOD PLATELETS.

megagamete, *n*. the larger of the two GAMETES, usually the female OVUM.

meganucleus, see MACRONUCLEUS.

megaphyll, *n*. the type of leaf found in ferns and seed plants, where the lamina is thought to have developed from a flattened branch and contains VASCULAR BUNDLES.

megapode, *n*. any of the mound birds of Australia and SE Asia that bury their eggs in mounds of earth so that the eggs develop without incubation.

megasporangium or **macrosporangium,** *n*. the organ where MEGASPORES are formed, which in flowering plants is the OVULE.

megaspore or **macrospore,** *n*. the larger of two types of SPORES produced by many ferns and seed plants. The megaspore gives rise to a female GAMETOPHYTE in ferns and in seed plants, and in flowering plants becomes the EMBRYO SAC. Compare MICROSPORE.

megasporophyll or **macrosporophyll,** *n*. **1.** the modified leaf that carries the MEGASPORANGIUM. **2.** the CARPEL of flowering plants.

meiosis, *n*. a type of nuclear division associated with sexual reproduction, producing four HAPLOID(1) cells from a single DIPLOID(1) cell, the process involving two cycles of division. Although meiosis is a continuous process it has been divided into numerous stages, given below. Further details of each stage can be obtained by referring to individual entries.

PROPHASE I: homologous chromosomes pair, split into CHROMATIDS, and carry out CROSSING OVER. The nuclear membrane disintegrates.

METAPHASE I: chromosomes migrate to the spindle equator to which they become attached by their CENTROMERES.

ANAPHASE I: HOMOLOGOUS CHROMOSOMES separate to opposite poles.

TELOPHASE I: new nuclei form, in which there is only one type of each chromosome, although each is divided into two chromatids.

PROPHASE II: nuclear membrane goes.

METAPHASE II: chromosomes attach to spindle.

ANAPHASE II: chromatids separate to poles.

TELOPHASE II: a total of four haploid nuclei is produced, each with one of each types of chromosome.

Meiosis has two major functions: (a) it halves the number of chromosomes to prevent a doubling in each generation, (b) it produces a mixing of genetic material in the daughter cells by the process of INDEPENDENT ASSORTMENT and RECOMBINATION. Note that the second point is only true if variability already is present in the parent cell.

Meiosis occurs at different stages of the life cycle in haploid and diploid organisms. See Fig. 211. See also MITOSIS, GAMETOGENESIS and Figs. 162 and 163.

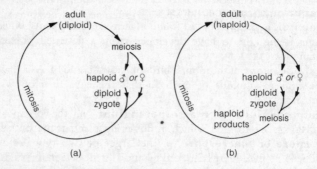

Fig. 211. **Meiosis.** (a) Prezygotic meiosis, e.g. humans. (b) Postzygotic meiosis, e.g. fungus.

meiospore, *n.* a spore formed as the result of MEIOSIS.

meiotic drive, *n.* a mechanical phenomenon during MEIOSIS that results in the two types of gametes produced by a heterozygote occurring in unequal numbers. Meiotic drive can have an important impact when it produces unequal sex ratios, and has been used in attempts at insect control by producing excess males and few females.

Meissner's corpuscle, *n.* a light-pressure receptor in nonhairy mammalian skin.

melanin, *n.* a dark brown or black pigment found in skin or hair. See ALBINISM, DOPA.

melanism, *n.* a condition in which excess dark pigment produces dark colour or blackness in scales, skin or plumage. See INDUSTRIAL MELANISM.

melanoma, *n.* an epithelial cancer in which the cells contain dark pigment.

melanophore or **melanocyte,** *n.* a type of pigment cell (or CHROMATOPHORE) that contains the pigment melanin, usually found in the skin of animals. It has a protective or camouflage function.

melanophore stimulatory hormone, *n.* a hormone in the form of a PEPTIDE produced by the ADENOHYPOPHYSIS. It effects colour changes in fish, amphibians and reptiles by changing melanin distribution within the cells (see MELANOPHORE).

melatonin, *n.* a hormone of the pituitary gland thought to inhibit reproductive activities.

melting, *n.* the denaturation of DNA.

melting point, *n*. the temperature at which a solid liquifies.

membrane, *n*. a thin sheet of tissue.

membrane carrier, *n*. a protein or an enzyme that combines with a substance and transports it across a biological membrane.

membrane potential, *n*. the potential difference between the two sides of a cell membrane.

membranous, *adj*. (of plants) being dry and flexible but not green.

membranous labyrinth, see EAR.

memory, *n*. **1.** the recollection of past events or previously learned skills after the passage of time. **2.** (in computing) the capacity of a computer usually expressed in 'bytes' or Ks, where K = 1024 bytes.

menarche, *n*. the onset of menstruation which occurs at puberty.

Mendel, Gregor (1822–84) Austrian monk and mathematician who,
• working with strains of the garden pea (*Pisum sativum*), managed to deduce several primary facts about how genes are transmitted between generations (see MENDELIAN GENETICS). Mendel's work was published in 1866 but remained unnoticed until 1900, 16 years after his death.

Mendel's laws, see MENDELIAN GENETICS.

Mendelian genetics or **Mendel's laws,** *n*. the basic laws of inheritance, first published by Gregor MENDEL in 1866 to explain the results he obtained from experiments with the garden pea. Working without detailed knowledge of cell structure or nuclear division he suggested that:

(a) each character (e.g. height) is controlled by two *factors*. We would now state this idea as each gene having two alleles, one on each HOMOLOGOUS CHROMOSOME.

(b) each factor segregates in the egg and pollen grains. We would now state that MEIOSIS separates allelic forms of a gene (Mendel's first law – see SEGREGATION).

(c) factors for different characters show INDEPENDENT ASSORTMENT. We would now state that genes are assorted independently during meiosis (Mendel's second law), unless linked on the same chromosome.

(d) factors do not cause blending, but are either dominant (see DOMINANCE(1)) or recessive. We know now that we can explain dominance in terms of enzyme activity, and that sometimes two alleles code for enzymes giving an intermediate phenotype when together (see INCOMPLETE DOMINANCE).

(e) the distribution of factors in the egg cells and pollen grains obey basic statistical laws giving ratios in the progeny which are predictable.

(f) results from crosses are the same whether the dominant form of the character belongs to the female parent or the male parent. We now know that this statement is true only if the gene is located on an AUTOSOME (see SEX LINKAGE).

meninges, *n.* the three membranes (ARACHNOID, DURA MATER, PIA MATER) that cover the CENTRAL NERVOUS SYSTEM of vertebrates. See MENINGITIS.

meningitis, *n.* inflammation of the MENINGES caused by infection.

meniscus, *n.* **1.** the top of a liquid column made either concave or convex by capillarity. **2.** an intervertebral disc of fibro-cartilage.

menopause or **climacteric,** *n.* the time at which women stop ovulating, with the result that the normal MENSTRUAL CYCLE no longer occurs. This is normally at about 45–50 years of age.

menstrual cycle, *n.* the modified OESTROUS CYCLE of most primates, which results in the periodic destruction of the mucosa of the uterine wall at the end of the LUTEAL PHASE. This results in a discharge of blood known as *menstruation*, the shedding of the ENDOMETRIUM, every 28 days in women.

menstruation, see MENSTRUAL CYCLE.

mental prominence, *n.* the projection of the lower jaw – the chin.

mentum, *n.* a structure found on the head of an insect below the labium.

mericarp, *n.* one seeded portion of a dry fruit or SCHIZOCARP split off from a SYNCARPOUS ovary at maturity.

meristele, *n.* a distinct vascular strand surrounded by an endodermis that occurs in some ferns and results in the break-up of the STELE.

meristem or **meristematic tissue,** *n.* a region of a plant in which active cell division (MITOSIS) occurs, the cells of the meristem being undifferentiated into a specialist form. Meristematic tissues occur at the root and shoot tips (see APICAL MERISTEM) giving growth in length, while increase in girth of the plant is produced by the CAMBIUM of VASCULAR BUNDLES and the INTERFASCICULAR CAMBIUM. Note that such specialist dividing zones are not found in animals, where cell division occurs in most tissues.

meristematic tissue, see MERISTEM.

meristic variation, *n.* numerical variation in taxonomic characters such as numbers of bristles, vertebrae, spots etc.

mero-, *prefix.* denoting partial, a portion.

meroblastic cleavage, see CLEAVAGE.

merogamete, *n.* any protozoan GAMETE formed by fission of parent cell.

meromatic, *adj.* (of lakes) stratified so that surface and bottom waters never mix.

-merous, *suffix.* denoting numbers of parts. For example, pentamerous means five-petalled.

mes- or **meso-**, *prefix.* denoting middle.

mescaline, *n.* a psychoactive drug that produces hallucinatory effects in humans probably by interfering with NORADRENALINE at nerve synapses.

Meselson, Matthew, see SEMICONSERVATIVE REPLICATION MODEL.

mesencephalon, see MIDBRAIN.

mesenchyme, *n.* a loose, cellular animal tissue that arises from the embryonic mesoderm, and functions as packing around internal organs. Mesenchyme can be thought of as the animal equivalent of PARENCHYMA in plants.

mesentery, *n.* **1.** the layers of PERITONEUM that attach the gut system and its associated organs (e.g. SPLEEN) to the dorsal surface of the peritoneal cavity in mammals. The blood lymph and nerve supplies to the gut are contained in the mesentery. **2.** one of the filaments dividing the body cavity of sea anemones and corals.

mesoderm, *n.* the layer of embryonic cells lying between ECTODERM and ENDODERM in all higher animals (all METAZOA except COELENTERATES) giving rise to muscle, the blood system, connective tissues, the kidney, the dermis of the skin and the axial skeleton. The mesoderm is often split into two layers with the COELOM in between.

mesodermal pouches, *n.* the segmentally arranged blocks of mesoderm that occur on each side of the NOTOCHORD of the vertebrate embryo and subsequently give rise to SOMITES.

mesocephalic, *adj.* (of humans) having a medium-sized head. See CEPHALIC INDEX.

mesogloea, *n.* the jelly-like material lying between the ectoderm and endoderm of COELENTERATES. In the HYDROZOA it is relatively thin, but in the other two major groups, SCYPHOZOA and ACTINOZOA, it is much thicker. It is acellular and is secreted by the ectoderm and endoderm.

Mesolithic, *n.* the period from 14,000 years ago to 5,500 years ago in Britain, characterized by the use of very small flint implements.

mesonephros or **Wolffian body,** *n.* the middle part of the embryonic kidney which in fish and amphibia is the functional kidney, but which gives rise embryologically to the tubules of the testis in mammals, birds and reptiles. See also WOLFFIAN DUCT.

mesophilic, *adj.* (of microorganisms) having an optimal growth temperature from 20°C to 45°C.

mesophyll, *n.* the internal tissue of a plant leaf, except the VASCULAR BUNDLES. In dorsiventral leaves the mesophyll is differentiated into an upper *palisade mesophyll* and a loosely packed, lower *spongy mesophyll*. In isobilateral leaves there are upper and lower palisade mesophylls and a central spongy area. In C_4 PLANTS there is yet another arrangement called

KRANZ ANATOMY. All mesophyll cells contain CHLOROPLASTS (for PHOTOSYNTHESIS) which lie close to the edge of the cell in order to gain maximum light and gas supply. Mesophyll tissue contains numerous intercellular spaces which communicate with the atmosphere outside the leaf via STOMATA. See Fig. 212.

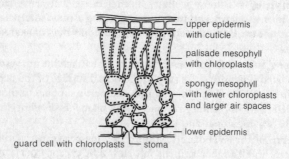

Fig. 212. **Mesophyll.** Transverse section of a dorsiventral leaf.

mesophyte, *n.* a plant growing in soil with a normal water content. Compare HYDROPHYTE, XEROPHYTE.

mesoplankton, *n.* that part of the PLANKTON that occurs below 100 fathoms.

mesosome, *n.* a structure with similar properties to a MITOCHONDRION which is formed in the cell membrane of bacteria by invagination.

mesothorax, *n.* the middle of the three thoracic segments of insects.

mesotrophic, *adj.* (of plants) requiring a moderate level of requirement of soil nutrients.

Mesozoic era, *n.* the geological era that began 225 million years ago and lasted about 155 million years. The era includes the CRETACEOUS PERIOD, JURASSIC PERIOD, and TRIASSIC PERIOD. During this era GYMNOSPERMS became dominant plants, dinosaurs became the dominant animals, reptiles (Pterosaurs) conquered the air, and birds and mammals evolved.

messenger RNA (mRNA), a single-stranded type of POLYNUCLEOTIDE molecule that contains four types of bases: ADENINE, GUANINE, CYTOSINE and URACIL. The function of mRNA is to carry a coded sequence of instructions about protein structure from DNA in the nucleus to RIBOSOMES attached to the ENDOPLASMIC RETICULUM, a process called TRANSCRIPTION.

metabolic intensity, *n.* the metabolic rate of a unit mass of living tissue.

metabolic pathway, *n.* the sequence of enzyme reactions followed in the formation of one substance from another.

metabolic rate, *n.* the rate at which an organism carries out METABO-

LISM, and which is closely linked to temperature. The relationship between metabolic rate and temperature can be expressed in terms of a value called Q_{10}. See also BASAL METABOLIC RATE.

metabolic waste, *n.* any waste substance that is produced during the metabolism of an organism, such as nitrogen in the form of urea.

metabolic water, *n.* water formed by a type of METABOLISM called CATABOLISM in which complex molecules are broken down to release their stored energy, with water as a by-product. In certain insects and desert mammals, which feed primarily on dry seeds, the water conservation mechanisms are so efficient that metabolic water alone is sufficient to replace the normal water loss; 'free' water is not required in the diet.

metabolism, *n.* the sum total of the chemical processes occurring in cells by which energy is stored in molecules (ANABOLISM) or released from molecules (CATABOLISM), life being maintained by a balance between the rates of catabolic and anabolic processes. All metabolic reactions occur in steps, in which compounds are gradually built up or broken down. Each step of the 'metabolic pathway' is catalysed by a different enzyme whose structure is coded by a specific gene, the end product being called a 'metabolite'. A special energy-carrying molecule called ATP is involved in these processes. See BASAL METABOLIC RATE.

metabolic pathway, see METABOLISM.

metabolite, see METABOLISM.

metacarpal, *n.* any one of the series of bones forming the palm of the hand or flat part of the forefoot in vertebrates. See CARPAL.

metacarpus, *n.* a bone of the forelimb of TETRAPODS that occurs between the wrist (CARPALS) and fingers (PHALANGES). In humans the metacarpus is the hand. See PENTADACTYL LIMB.

metacentric chromosome, *n.* a chromosome having the CENTROMERE located about halfway along. Compare ACENTRIC CHROMOSOME.

metachromatic, *adj.* having more than one colour produced by the same stain. For example, methylene blue stains the diphtheria bacterium *Corynebacterium diphtheriae* blue with red inclusions.

metachronal rhythm, *n.* the beat of cilia (see CILIUM) or consecutive limbs (as in the ragworm), where a wave appears to pass forwards through them in much the same way as wind causes the movement of corn in a cornfield. See Fig. 213 on page 348. The result of these actions is that the effective movement of liquid over their surface is backwards and the organism moves forwards.

metameric segmentation, *n.* the division of the body of an organism (e.g. the earthworm) into a series of similar or identical units recurring

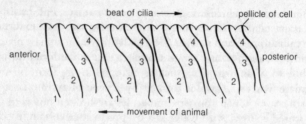

Fig. 213. **Metachronal rhythm.** The 'wave' pattern of metachronal rhythm.

along the length of the body. These units are known as *segments* or *metameres*. In each metamere there is a pattern of muscle blocks, blood vessels, excretory organs, epidermal structures and elements of the nervous system.

metamorphosis, *n.* the change in an organism from larval to adult form, which is often quite rapid, as in tadpole to frog, caterpillar to butterfly. Metamorphosis is said to be 'incomplete' where there is gradual development of a NYMPH to an adult, as in the EXOPTERYGOTA (Hetero- or Hemi–metabola). It is 'complete' where a pupa occurs, as in the ENDOPTERYGOTA (Holometabola).

metanephros, *n.* the functional kidney in adult reptiles, birds and mammals that replaces the MESONEPHROS in the developing embryo.

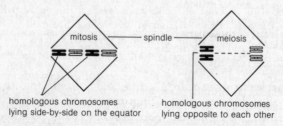

Fig. 214. **Metaphase.** Position of chromosomes in mitosis and meiosis.

metaphase, *n.* a stage of nuclear division in EUCARYOTE cells, occurring once in MITOSIS and twice in MEIOSIS. The main process involved is the organization of CHROMOSOMES at the equator of a spindle, forming a *metaphase plate*. Each chromosome is divided into two CHROMATIDS, joined at a CENTROMERE. The chromosomes are densely staining and highly condensed, properties which makes them suitable for the production of KARYOTYPES. The spindle MICROTUBULES become attached to an area of the centromere called the *kinetochore*, the precise arrangement of chromosomes appearing to be random (this has important genetic implications, see INDEPENDENT ASSORTMENT) and different in mitosis and meiosis (see Fig. 214). The metaphase plate in the second cycle of meiosis is at right angles to the equator in metaphase I.

metaphase plate, see METAPHASE.

metaphloem, *n.* the typical PHLOEM that develops from the procambium next to the *protophloem*, towards the stem centre. It has normal SIEVE TUBES, in contrast with the small simplified sieve tubes of protophloem which lack companion cells.

Metaphyta, *n.* a TAXON, often given the status of kingdom, that includes all multicellular plants.

metaplasia, *n.* the transformation of a tissue to another form.

metaplasm, *n.* the lifeless constituents of protoplasm.

metapneustic, *adj.* (of insects such as mosquito larvae) having SPIRACLES only at the posterior end of the abdomen.

metarhodopsin, *n.* the substance produced when RHODOPSIN is affected by light.

metastasis, *n.* the process of cancerous tissue spreading to various parts of the body.

metatarsal, *n.* any bone of the METATARSUS.

metatarsus, *n.* the skeleton of the foot between the toes and the tarsus of a TETRAPOD consisting of the five bones of the hind limb. In humans this forms the middle and hind part of the foot. See PENTADACTYL LIMB.

Metatheria, see MARSUPIAL.

metathorax, *n.* the posterior of the three thoracic segments in insects.

metaxylem, *n.* those parts of the xylem tissue of plant roots and stems that are produced last by the cambium, tending to have larger cells than the PROTOXYLEM, the first-formed elements.

metazoan, *n.* any animal of the group Metazoa, whose bodies are multicellular, as distinct from the PROTOZOANS where the bodies are unicellular. Sponges (see PARAZOAN) are the only nonprotozoan group excluded from the Metazoa. Most metazoans (except COELENTERATES) have a three-layered body structure with an outer ECTODERM, a middle MESODERM and an inner ENDODERM.

methionine, *n.* one of 20 AMINO ACIDS common in proteins. It has an 'R' group with a nonpolar structure and is relatively insoluble in water. See Fig. 215 on page 350. The ISOELECTRIC POINT of methionine is 5.7.

metazoite, *n.* one of a large number of cells formed by the multiple asexual fission of some Protozoa.

metric character, *n.* a quantitative character, such as human height, in which there is a continuous spread of variability. Such characters are controlled by a system of POLYGENIC INHERITANCE.

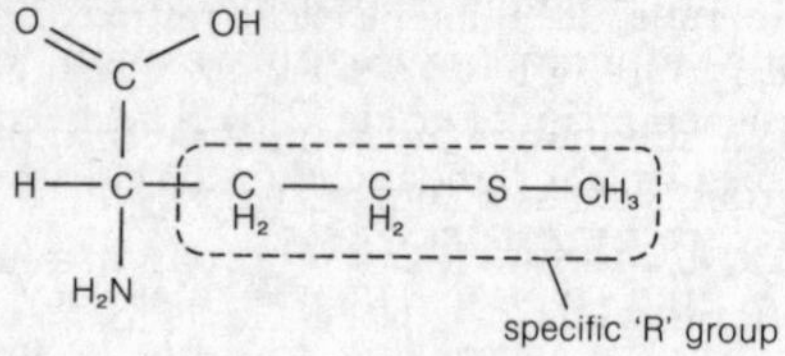

Fig. 215. **Methionine.** Molecular structure.

micelle, *n.* an elongated thread of protein or other material present within the cell protoplasm.

Michaelis–Menten constant K_M, *n.* the concentration in moles/litre of a substrate at half the maximum velocity of an enzymic reaction.

micro-, *prefix.* denoting small. In the metric system of measurement 'micro' means 'one millionth' with a symbol, as in l $(1 \times 10^{-6}$ l). See SI UNITS.

microaerophile, *n.* any organism requiring little oxygen for growth.

microbe, *n.* any microscopic organism. The term is sometimes restricted to pathogenic microorganisms such as BACTERIA and VIRUSES.

microbial degradation, *n.* the beneficial activities of microbes in carrying out biodegradation, e.g. in sewage disposal. COMPARE BIODETERIORATION.

microclimate, *n.* the immediate climate in which an organism lives, such as the climate around a leaf, within the herb layer, or within an adjacent part of the soil.

microdissection, *n.* the dissection of very small structures, often carried out by the use of a micromanipulator, an instrument that significantly reduces hand tremors.

microevolution, *n.* the small-scale changes resulting from genetic adaptation, that are usually expressed as changes within a species rather than the formation of a new species. See INDUSTRIAL MELANISM.

microfauna, *n.* **1.** the flora of a microhabitat. **2.** any microscopic plants not visible to the naked eye.

microfibril, *n.* a group of some two thousand cellulose chains massed together in the CELL WALL. Microfibrils are embedded in an organic matrix giving the cell wall great strength.

microfilament, *n.* a fine, subcellular structure about 5–8 nm in diameter, composed of ACTIN or a similar protein and found in the cytoplasm of EUCARYOTE cells. Microfilaments are able to contract and are also involved in CYTOPLASMIC STREAMING and PHAGOCYTOSIS.

microgamete, *n.* the smaller of two gametes. It is usually motile and called the male.

microgeographic race, *n.* a race that is restricted to a very small geographical area.

microhabitat, *n.* a subdivision of a HABITAT that possesses its own particular properties, such as a MICROCLIMATE.

Microlepidoptera, *n.* a LEPIDOPTERAN subgroup that no longer has any taxonomic significance. The subgroup includes forms only a few mm in length.

micromere, *n.* any of the small BLASTOMERES formed at the animal pole of a developing egg. These eventually give rise to the ECTODERM.

micrometre or (formerly) **micron,** *n.* a unit of length equal to 10^{-6} metre. Symbol $= \mu$m.

micron, see MICROMETRE.

micronucleus, *n.* the smaller of two nuclei found in some protozoans which is concerned with cell division. Compare MACRONUCLEUS.

micronutrient, *n.* any TRACE ELEMENT or compound required in only minute amounts by organisms. Compare MACRONUTRIENT.

microorganism, *n.* any unicellular organism such as a PROTOZOAN, BACTERIUM or a VIRUS. The term is often extended to include fungi.

microphagous, *adj.* (of an aquatic organism) collecting very small particles suspended in water and consuming them as food. See FILTER FEEDER.

microphyll, *n.* a tiny 'leaf' with no vascular tissues that is found in certain mosses.

micropinocytic vesicle, *n.* a flask-like invagination of the cell plasma membrane which eventually cuts off at the neck to form a cellular vacuole. See PINOCYTOSIS.

micropinocytosis, see PINOCYTOSIS.

micropyle, *n.* a small canal in the integument surrounding the ovule of a flowering plant, through which the POLLEN TUBE usually enters the ovule on the way to the EMBRYO SAC. Water enters the seed via the micropyle prior to GERMINATION.

microscope, *n.* an instrument for magnifying the size of an object. The simple microscope is the hand lens, but the instrument commonly used is the *compound microscope* which consists of two sets of lenses: eyepiece and objective. The compound microscope can magnify 600–1000 × using an oil interface between lens and slide (see OIL-IMMERSION OBJECTIVE LENS), and with suitable eyepieces objects can be viewed up to a magnification of 1500 ×, the maximum power of the normal light microscope. Higher magnification can only be obtained using an ELECTRON MICROSCOPE.

microsomal fraction, *n.* a collection of tiny subcellular particles, invisible with the light MICROSCOPE, that are produced during DIFFEREN-TIAL CENTRIFUGATION. Under the ELECTRON MICROSCOPE, these *microsomes* can be seen to consist mainly of membranes and RIBOSOMES from the ENDOPLASMIC reticulum.

microsome, see MICROSOMAL FRACTION.

microsphere, *n.* small bodies of matter formed by the combinations of proteins that are produced in experiments attempting to simulate the creation of life on earth.

microsporangium, *n.* the sporangium from which the microspores are formed, which in higher plants is the pollen sac. See ANTHER.

microspore, *n.* the smaller of the two types of SPORE produced by ferns and higher plants, giving rise to the male GAMETOPHYTE. In TRACHEOPHYTES the microspore is the POLLEN GRAIN. Compare MEGASPORE.

microsporophyll, *n.* the structure bearing the microsporangia – a leaf or modified leaf, and in flowering plants, the STAMEN.

microtome, *n.* an instrument used for cutting very thin sections for microscopic examination.

microtubule or **neurotubule,** *n.* a hollow filament about 20–25 nm in diameter found in EUCARYOTE cells, composed of ACTIN-like protein called *tubulin*. Microtubules are thought to make up the CYTOSKELETON of the cell, the spindle fibres of MEIOSIS and MITOSIS, (see also COLCHICINE) and, in some cells of plants and animals, form the $9 + 2$ structure of the CILIUM and FLAGELLUM.

microvillus, *n.* a minute finger-like process found in numbers projecting from the cell surface where, in epithelium, they form a *brush border*. Microvilli are found, for example, on the villi lining the small intestine.

micturition, *n.* the release of urine from the body. See BLADDER.

midbrain or **mesenchephalon,** *n.* that part of the brain lying between the forebrain and hindbrain. It contains the optic lobe and is particularly concerned with hearing and sight. See FOREBRAIN, HINDBRAIN.

middle ear, see EAR.

middle lamella, *n.* (*pl.* lamellae) the thin membrane separating two

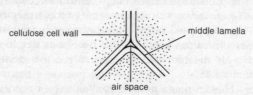

cellulose cell wall — middle lamella

air space

Fig. 216. **Middle lamella.** Transverse section through several cell walls.

adjacent plant cells, whose function is to cement them together. It consists of varying amounts of PECTIN, depending on IAA concentration (see Fig. 216).

midgut, *n.* the middle portion of the ALIMENTARY CANAL.

midparent value, *n.* the average value of a METRIC CHARACTER from two parents. Such values when compared with those of their offspring give a direct measure of NARROW–SENSE HERITABILITY.

midriff, *n.* **1.** the middle part of the human body between chest and waist. **2.** (in anatomy) the DIAPHRAGM.

migration, *n.* any cyclical movements (usually annual) that occur during the life history of an animal at definite intervals, and always including a return trip from where they began. The exact derivation of the word is from the Latin 'migrate' meaning to go from one place to another, but biologically a return journey is part of the accepted definition of the term, the outward journey being termed EMIGRATION and the inward journey IMMIGRATION.

mildew, *n.* **1.** a plant disease caused by a fungus. **2.** any fungus, such as *Erysiphe*, that causes cereal mildew. **3.** a fungus in the mycelial state, growing over a substrate without obvious fruiting bodies.

milk, *n.* **1.** a whitish fluid secreted by the mammary gland in mammals which serves to nourish the young. **2.** any white fluid, such as coconut milk.

milk gland, *n.* **1.** see MAMMARY GLAND. **2.** a nutrient gland in the 'uterus' of viviparous species of insects, such as the tse-tse fly.

milkteeth, see DECIDUOUS TEETH.

Miller, Stanley, see COACERVATE THEORY.

milli-, *prefix.* denoting one thousandth.

millipede or **millepede,** *n.* any member of the subclass Diplopoda, class MYRIAPODA, cylindrical in shape, having two pairs of limbs on each of some 70 segments, herbivorous and terrestrial.

Millon's test, *n.* a test for the presence of protein in a mixture, involving the use of Millon's reagent which contains mercuric nitrate and nitrite. On heating a proteinaceous solution with a few drops of Millon's reagent (using a fume cupboard) a purplish precipitate is formed. The test is less commonly used today because of the danger associated with handling mercuric salts, the BIURET REACTION being employed instead.

mimetic, *adj.* (of an organism) having evolved to resemble another species. See MIMICRY.

mimicry, *n.* the adoption by one species of any of the properties of another, such as colour, habits, structure. Particularly common in insects, two main forms of mimicry are recognized:

(a) *Batesian mimicry*, where two species have the same appearance (often warning colours) but one (the 'model') is distasteful to predators. The mimic gains advantage because predators learn to associate appearance with bad taste and leave both model and mimic uneaten.

(b) *Mullerian mimicry*, where both model and mimic are distasteful to predators and both gain from the other's distastefulness since the predator learns to avoid all similar-looking forms, whichever it eats first.

mineral deficiency, *n*. a lack of minerals in the food intake of an organism.

mineral element, see ESSENTIAL ELEMENT.

mineralization, *n*. the conversion of organic matter to inorganic matter.

mineralocorticoids, see ADRENAL CORTICAL HORMONES.

mineral requirement, *n*. those minerals required to maintain an organism in a normal, healthy state.

mineral salt, *n*. any inorganic homogeneous solid, such as sodium, potassium, phosphorus, chlorine.

minimal medium, *n*. a food source containing the minimum requirement for microbial growth. These are carbon inorganic salts and water. Such media are used in screening for AUXOTROPHS. See ONE GENE/ONE ENZYME HYPOTHESIS.

Miocene epoch, *n*. a division of the TERTIARY PERIOD, lasting from 26 million years ago until 7 million years ago. In this epoch mammals acquired their present form. The European flora became more temperate, grasslands increased and the British Isles more or less reached the present latitude of 54°N.

miracidium, *n*. (*pl*. miracidia) the first larval form of a liver fluke, which develops from the egg as a flat free-swimming, ciliated larva.

miscarriage, *n*. the expulsion of a foetus before it is viable outside the womb.

miscible, *adj*. (of liquids) having the ability to mix.

mismatch of bases, *n*. the noncomplementary pairing of nucleotide bases in a DNA molecule, perhaps following on from a SUBSTITUTION MUTATION event. The error may be corrected at the next replication of the DNA molecule, producing one mutant and one normal molecule.

mis-sense mutation, *n*. a change to the base sequence of a DNA CODON so that it now encodes a different amino acid. For example, a SUBSTITUTION MUTATION in the first base of a triplet might produce the following mis-sense mutation: CCU (pro) ACU (thr). See GENETIC CODE.

Mitchell, P. see ELECTRON TRANSPORT SYSTEM.

mite, *n*. any member of the order Acarina, ARACHNIDS possessing clawed appendages in front of the mouth (chelicarae). They may be free-living (many thousands/m² in soil) or parasitic.

mitochondria, see MITOCHONDRION.

mitochondrial shunt, *n*. a transfer ('shunt') in the presence of oxygen of reduced NAD produced in the cytoplasm by GLYCOLYSIS to a MITOCHONDRION. It enters the ELECTRON TRANSPORT SYSTEM on the mitochondrial inner lining.

mitochondrion, *n*. (*pl*. mitochondria) a subcellular, cylindrical organelle found in EUCARYOTES, of about 0.2–0.5 μm in length. Under the ELECTRON MICROSCOPE, the mitochondrion is seen to consist of a double membrane surrounding a matrix, with the inner membrane folded into projections called CRISTAE. The walls of the cristae are the site of ELECTRON TRANSPORT SYSTEMS producing ATP, while the reactions of the KREBS CYCLE take place within the matrix. Mitochondria have thus been named the 'powerhouses' of the cell and are especially prevalent in cells with a high energy requirement. Mitochondria are selfreplicating and contain DNA by which they can control the synthesis of some of their own proteins.

mitogen, *n*. a growth factor in the form of a small protein found in blood serum that often has a specific target organ in which it induces MITOSIS.

mitogenetic rays, *n*. hypothetical short-wavelength rays emanating from tissues that (in theory) stimulate MITOSIS in other tissues.

mitosis, *n*. a type of nuclear division by which two daughter cells are produced from one parent cell, with no change in chromosome number. Mitosis is associated with asexual growth and repair and, although it is a continuous process, has been divided up into four main stages, given below. Further details of each stage can be obtained by referring to individual entries.

(a) PROPHASE: chromosomes contract and become visible as threads. Each chromosome divides into two CHROMATIDS and the nuclear membrane disintegrates.

(b) METAPHASE: chromosomes migrate to the equator of a spindle and become attached to the spindle microtubes by their CENTROMERES.

(c) ANAPHASE: chromatids separate and go to opposite poles.

(d) TELOPHASE: nuclear membrane reforms, chromosomes lengthen and cannot be distinguished.

See also MEIOSIS.

mitospore, *n*. a SPORE resulting from MITOSIS.

mitotic crossing over, *n*. a recombination between HOMOLOGOUS CHROMOSOMES in a diploid cell undergoing MITOSIS, producing diploid

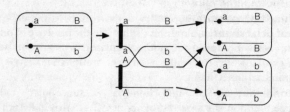

Fig. 217. **Mitotic crossing-over.** Recombination between homologous chromosomes.

daughter cells that have a different combination of alleles than the original parental cell. See Fig. 217.

mitral valve, see BICUSPID(3).

mixed nerve, *n.* a nerve containing both EFFERENT NEURONES and AFFERENT NEURONES.

ml, *abbrev. for* millilitre, represented as cm³.

moa, *n.* a large, extinct, flightless bird of New Zealand, up to 3 m in height, that became extinct in historical times. Eggs are still found preserved in boggy conditions.

mobile genetic element, see TRANSPOSABLE GENETIC ELEMENT.

mobility, electrical, *n.* a measurement of the migration rate of an ion in an electric field.

mobility, mechanical, *n.* a measurement of the rate at which a molecule diffuses in a liquid phase.

mode, *n.* (statistics) the most frequently observed value in a series of observations, i.e. the peak of a FREQUENCY DISTRIBUTION.

modifier gene, *n.* any gene that alters the phenotypic expression of a nonallelic gene. See GENETIC BACKGROUND.

molality, *n.* the number of MOLES of a solute present in a kilogram of pure solvent.

molarity, *n.* the number of moles of a solute present in a litre of solution.

molar teeth, *n.* the crushing teeth that occur at the back of the jaw of mammals. Each molar possesses a complicated pattern of cusps and ridges and several roots. The molars have no preceding milkteeth.

mole, *n.* the basic SI unit of amount of substance. One mole of any substance contains 6.023×10^{23} (Avogadro's number) molecules, atoms or ions of an element or compound equal to the relative molecular mass in grams. Symbol: mol.

molecular biology, *n.* the study of the structures and properties of molecules found in cells.

molecular hybridization or **hybridization** or **annealing,** *n.* the artificial creation of duplex polynucleotide molecules by 'annealing' complementary single strands of RNA and DNA together. Such techniques enable the estimation of complementarity between DNA/DNA, DNA/RNA and RNA/RNA polynucleotide chains. See COMPLEMENTARY BASE PAIRING.

molecule, *n.* the smallest chemical unit of matter that has the characteristics of the substance of which it forms a part.

mollusc, *n.* any member of the phylum Mollusca, including the snails (GASTEROPODS), BIVALVES, the squids and octopuses (CEPHALOPODS), and two minor classes, the Amphineura and Scaphopoda. Most possess one or more shells and have a body structure that is unsegmented, with a COELOM. Some are terrestrial, but the majority are aquatic.

monachasium, *n.* a CYMOSE inflorescence where the branches are alternate or spirally arranged, or where one branch is developed more strongly than the other.

monadelphous, *adj.* (of STAMEN filaments) united into a single bundle.

Monera, *n.* the kingdom that contains the BACTERIA and BLUE-GREEN ALGAE.

mongolism, see DOWN'S SYNDROME.

monkey, *n.* any long-tailed primate excluding the tarsiers and lemurs, comprising the Old World and New World monkeys and marmosets.

mono-, *prefix.* denoting single, one.

monoaminoxidase, *n.* an enzyme that inactivates NORADRENALINE by oxidation in ADRENERGIC nerves.

monocarpic, *adj.* (of plants) flowering once during life.

monochlamydeous or **haplochlamydeous,** *adj.* (of flowers) having only a single whorl of perianth segments.

monoclonal antibody, *n.* an ANTIBODY produced by a CLONE of cells derived from a single, selected cell. Such antibodies are extremely pure and monoclonal techniques have enabled the production of specific antibodies to tumerous cells.

monocotyledon, *n.* any flowering plant of the group Mono-cotyledonae. See DICOTYLEDON for comparative table (Fig. 126).

monocyte or **macrocyte,** *n.* a type of LEUCOCYTE (white blood cell) of the AGRANULOCYTE group that is produced from stem cells in the bone marrow and is 12–15 μm in diameter. Monocytes remain in the blood for a short time and then migrate to other tissues as MACROPHAGES, moving particularly to those areas invaded by bacteria and other foreign materials where they ingest large particles by PHAGOCYTOSIS. See also HISTOCYTE, LYMPHOCYTE.

monoecious, *adj.* (of plants) having both male and female organs occurring on the same plant, in unisexual flowers such as maize and hazel. The term literally means 'one home'. Compare DIOECIOUS.

monogamy, *n.* a state where a single male pairs with a single female in a partnership that may last for several seasons or life. Monogomy is common in birds such as swans, but rare in mammals.

Monogenea, *n.* a group of ectoparasitic trematode FLUKES that have only one host and are parasitic on fish and amphibia. They are distinguished from the DIGENEA which have at least two hosts. See also BILHARZIA.

monogenic, *adj.* of or relating to an inherited characteristic that is controlled by one gene.

monograph, *n.* a publication relating normally to a higher TAXON, which includes an exhaustive treatment of all aspects of the biology of the group as far as they relate to its taxonomy and classification.

monohybrid, *n.* an organism that carries two different ALLELES of one gene. For example character A (controlled by alleles $A1$ and $A2$) would produce a monohybrid GENOTYPE of $A1A2$, i.e. a HETEROZYGOTE.

A monohybrid cross is one in which two parents are crossed to produce monohybrids that are then mated, or a single monohybrid plant is self-fertilized (see SELF–FERTILIZATION). Mendel formulated his law of SEGREGATION from the results of a monohybrid cross involving single characters in the pea plant.

monohybrid inheritance, *n.* a pattern of results from crosses, indicating that a single gene is responsible for the control of a particular character. See MONOHYBRID.

monokaryon or **monocaryon,** *n.* a fungal hypha, characteristic of BASIDIOMYCETES, that contains cells each of which has only one HAPLOID(1) nucleus.

monolayer, see MONOMOLECULAR FILM.

monomer, *n.* any molecule that can exist alone or with other similar molecules to form a polymer.

monomolecular film or **monolayer,** *n.* a single layer of lipid molecules with their hydrocarbon chains at right angles to the surface, as when a lipid is allowed to spread over the surface of pure water. See UNIT MEMBRANE and FLUID–MOSAIC MODEL.

mononuclear phagocyte system, or **reticuloendothelial system,** *n.* a mammalian defence system against foreign bodies, consisting of MACROPHAGE cells located in the lymph nodes (see LYMPHATIC SYSTEM), liver, spleen and bone marrow.

monophyletic, *adj.* (of a group of individuals) descended from a common ancestor.

monoploid, *n.* an organism or cell in which there is one set of chromosomes instead of the normal two.

monopodial, *adj.* (of a stem) having the same apical growing point responsible for consecutive years' growth.

monopodium, *n.* a structure, such as a tree trunk, that has increased in length by apical growth.

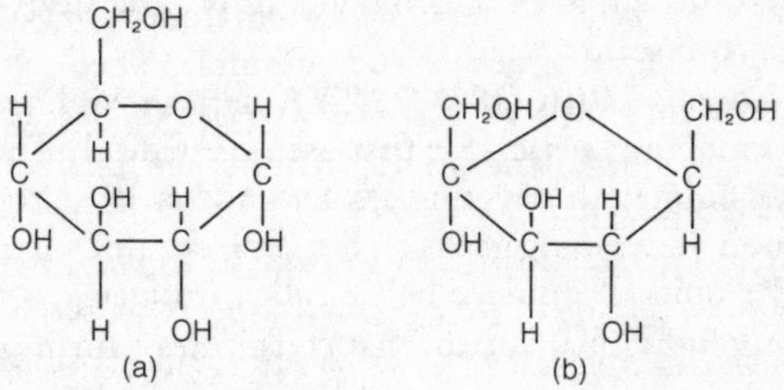

Fig. 218. **Monosaccharide.** Molecular structures of (a) glucose, (b) fructose.

monosaccharide, *n.* a carbohydrate MONOMER, a simple sugar with the formula $(CH_2O)_n$, e.g. $C_6H_{12}O_6$ glucose and fructose. See Fig. 218. Such carbohydrates are generally white, crystalline solids, with a sweet taste, and are usually soluble in water. The carbon chain forming the backbone of such sugars can be of varying lengths. Some monosaccharides contain only three carbons ('triose' types such as glyceraldehyde) others contain five carbons ('pentose' types such as the deoxyribose sugar of DNA), but those with six carbons ('hexose' types such as glucose) are the most important since they can be joined together by CONDENSATION REACTIONS (loss of water) to form DISACCHARIDES and POLYSACCHARIDES.

monosomic, *adj.* having one chromosome of a diploid set of somatic chromosomes missing. For example, TURNER'S SYNDROME. See ANEUPLOIDY.

monotreme, *n.* any mammal of the primitive subclass Monotremata, including the duck-billed platypus and spiny anteater. Monotremes differ from other mammals in laying eggs, and in having a single opening (CLOACA) for the passage of eggs or sperm, faeces and urine. They inhabit Australia and New Guinea.

monotypic, *adj.* (of a TAXON) containing only one subordinate taxon. For example, a genus with only one species, a species with only one subspecies.

monozygotic twins or **identical twins** or **uniovular twins,** *n.*

siblings that develop from one egg, contain identical genetical information, and are (usually) of very similar appearance. Any physical and mental differences detected between identical twins must arise, therefore, from environmental differences, both before and after birth. Such twins are examples of a CLONE and are always of the same sex. Compare DIZYGOTIC TWINS.

moor, *n.* an upland habitat usually dominated by heather growing on peat which is not normally waterlogged.

morbid, *adj.* diseased, as in 'morbid anatomy', the study of structural changes due to disease.

Morgan, Thomas Hunt (1866–1945) American zoologist who was a founder of modern genetics. He first used the fruit fly *Drosophila* as an experimental animal, discovering SEX LINKAGE in 1910, together with a simple method of mapping genes on the same chromosome (see GENETIC LINKAGE), the units of distance being called *centimorgans* originally.

morph, *n.* any individual variant that constitutes a form of a polymorphic species.

morphine, *n.* a white, crystalline, narcotic alkaloid drug obtained from opium. Formula: $C_{17}H_{19}NO_3$.

morphogenesis, *n.* the development of the form or structure of an organism during the life history of the individual.

morphogenetic movement, *n.* the reorientation of masses of cells during embryonic development.

morphology, *n.* the study of the shape, general appearance or form of an organism, as distinct from ANATOMY which involves dissection to discover structure.

morphospecies, *n.* a species recognized solely on the basis of its MORPHOLOGY.

mortality rate or **death rate,** *n.* **1.** the percentage of a population dying in a year. **2.** the ratio of the number of human deaths per 1,000 of population.

morula, *n.* a solid ball of cells resulting from cleavage of a fertilized ovum. It is formed of a mass of BLASTOMERES which subsequently form a BLASTULA.

mosaic, *n.* any organism exhibiting a mixture of cells of different genetic makeup, such as a GYNANDROMORPH. See INACTIVE-X HYPOTHESIS. Plants showing this phenomenon are known as CHIMAERAS(1).

mosaic egg, *n.* any egg that has clearly definable zones which develop into particular tissues and organs at a later stage.

mosaic evolution, *n.* a form of evolution where various characters or structures of a particular phenotype have different rates of change.

mosquito, *n.* a DIPTERAN fly that acts as a vector of numerous tropical diseases, such as MALARIA and YELLOW FEVER.

moss, *n.* any bryophyte of the class Musci. Usually these are small plants (less than 5 cm high) attached to moist or wet substrates by rhizoids; this is the SPOROPHYTE generation. The sexual organs are borne on a GAMETOPHYTE generation and the ANTHERIDIA and ARCHEGONIA are on separate leaf rosettes. The male gametes are motile and after fertilization, a diploid sporophyte is produced, within which haploid spores are developed, each spore giving rise to a protonema from which the new gametophyte develops.

moth, *n.* a LEPIDOPTERAN that is mainly nocturnal, lacks knobbed antenae, and folds its wings flat at rest. The term has no taxonomic status.

motility, *n.* motion.

motivation, *n.* the internal state of an animal prior to a specific behavioural act.

motor, *adj.* relating to the stimulus of an EFFECTOR organ.

motor cortex, *n.* the part of the cerebral cortex of the brain that controls the motor functions.

motor neuron, *n.* a nerve cell, the fibre from which connects with STRIATED MUSCLE under voluntary control and conducts the stimulus from the CENTRAL NERVOUS SYSTEM to the effector which is then stimulated into activity.

mould, see MILDEW(3).

moult, *vb.* **1.** (of birds, mammals, reptiles and arthropods) to shed (feathers, hair, skin and cuticle). See ECDYSIS, MOULTING HORMONE.

moulting hormone or **ecdysone,** *n.* a hormone secreted by the thoracic gland of insects which brings about the moulting of the cuticle and subsequent growth (ECDYSIS). It raises the metabolic rate and increases the build-up of proteins from AMINO ACIDS in growing tissue.

mountain sickness, *n.* fatigue, headache and nausea in humans resulting from the lack of oxygen above about 4000 m.

mouth, *n.* the anterior opening of the ALIMENTARY CANAL of animals through which food is taken into the body. It is often surrounded by mouthparts or tentacles that facilitate feeding. See DIGESTIVE SYSTEM.

mouthbrooder, *n.* any of various African cichlid fishes that retain eggs and young in the mouth, where they are hatched and protected.

mouthpart, *n.* any of the paired appendages in arthropods that surround the mouth and are associated with feeding.

mRNA, see MESENGER RNA.

mucin, *n.* a MUCOPROTEIN that forms MUCUS when in solution.

mucopolysaccharide, *n.* any polysaccharide, such as hyaluronic acid or heparin, that contains either an amino sugar or a derivative.

mucoprotein, *n.* a complex of protein and polysaccharide.

mucosa, *n.* any EPITHELIUM that secretes MUCUS, such as the MUCOUS MEMBRANE lining the alimentary canal.

mucous membrane, *n.* the lining of the gut system and the urinogenital system of animals, consisting largely of moist EPITHELIUM overlying CONNECTIVE TISSUE. It is so-called because of the presence in the epithelium of GOBLET CELLS that secrete MUCUS.

mucronate, *adj.* (of a leaf) having a short narrow point.

mucus, *n.* **1.** any slimy or sticky material secreted by invertebrate animals or plants. **2.** a viscous slimy solution of the protein MUCIN secreted by the MUCOUS MEMBRANES of vertebrates.

muddy shore, *n.* a coastal stretch where deposition of alluvium and detritus has occurred, as opposed to sandy or pebbly shores where beaches are formed. Characterized by the development of SALT MARSH.

mull, *adj.* (of soils, usually in woodlands) having no raw humus layer and usually having a high population of earthworms.

Mullerian duct, *n.* the duct from the embryonic PRONEPHROS which in later development in mammals becomes the oviduct in females and disappears in males. Named after the German anatomist and physiologist Johannes Müller (1801–58).

Mullerian mimicry, see MIMICRY.

multi- or **mult-,** *prefix.* denoting much, many.

multicellular, *adj.* composed of many cells. See ACELLULAR.

multifactorial, *adj.* (of a character) controlled by several genes.

multinucleate, *adj.* having more than one nucleus.

multiple allelism, *n.* the occurrence of more than two ALLELES of a gene, although only two may be present together in an individual at any one time. For example, the ABO BLOOD GROUP gene has at least six alleles. In practice it seems likely that all genes must possess many alternative alleles but these often remain undetected since no noticeable alteration of PHENOTYPE is observed.

multiple factors or **multiple genes,** *n.* genes that collectively produce a particular characteristic, such as height in humans.

multiple fission, *n.* the splitting of the nucleus several times before division of the cytoplasm, as happens in some PROTOZOANS.

multisite activity, *n.* the action of chemicals, e.g. copper, in biological systems where they can disrupt a wide range of biochemical pathways.

multivalent, *n.* (of chromosomes) forming an association during prophase I of MEIOSIS. Compare BIVALENT.

multivariate analysis, *n.* (statistics) an analysis involving several variables simultaneously.

mumps, *n.* an acute, contagious viral disease of humans (particularly of children), characterized by a swelling of the parotid salivary glands lying beneath the ear and caused by a Paramyxovirus virus. The mumps virus is spread by droplets of saliva and respiratory secretions that enter a new host via the repiratory tract. The disease appears most often during the winter and early spring months, and in adults can lead to MENINGITIS. It may cause sterility in males. Once affected, the individual is usually immune for life. Vaccination is possible, using attenuated virus.

murein, *n.* the cross-linked mucopeptides that form the rigid framework of bacterial cell walls.

muricate, *adj.* (of plant structures) having a surface roughened by short, sharp projections.

muscarine, *n.* a poisonous alkaloid occurring in certain mushrooms.

muscarinic, *adj.* (in physiological terminology) having acetylcholine receptors that are sensitive to muscarine but not to nicotine.

muscle, *n.* the fleshy part of any animal that consists of tissue made up of highly contractile cells which serve to move parts of the body relative to each other.

A muscle is composed of many fibres or muscle cells. In STRIATED MUSCLE, each cell contains a bundle of MYOFIBRILS each exhibiting a banding pattern and being made up of a number of SARCOMERES arranged end to end. The sarcomere is the unit of contraction and the banding visible over its surface results from the longitudal filaments which make up the myofibril being of two types, thick (dark) and thin (light). These filaments overlap as shown in Fig. 219. The thick filaments are composed of the protein MYOSIN and the thin filaments of ACTIN. H.E. Huxley and K. Harrison found that on contraction, the light zones (I-BANDS) were comparatively narrow; on relaxation of the muscle the I-bands were broad. Where very strong contraction takes place the H-zone disappears and the thin filaments overlap.

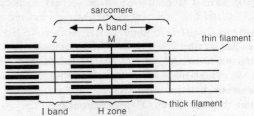

Fig. 219. **Muscle.** A sarcomere.

Huxley and Harrison proposed the *sliding filament hypothesis*, to account for their observations. Bridges occur between thick and thin filaments and in contraction the bridges pull thin filaments past the thick ones using a ratchet mechanism. Some filaments are retained in this 'pulled past' position whilst others detach then reattach and repeat the 'pulling past' action. ACTOMYSIN is formed at the point of contact of bridge and thin filament. For each bridge to go through its cycle of attachment, contraction and reattachment, the splitting of one molecule of ATP is required, the cycles occurring between 50 and 100 times per second. The supply of ATP comes from MITOCHONDRIA between the fibrils. Calcium ions are released from vesicles in the sarcoplasmic reticulum, by the ACTION POTENTIAL passing along the surface of the fibre and these split the ATP. Troponin activated by the calcium displaces tropomyosin, which prevents myosin bridges from binding with actin fibrils. Once binding takes place this activates ATPase and on hydrolysis of ATP the bridge goes through its cycle of movement.

muscle fibre, *n.* a muscle cell consisting of numerous MYOFIBRILS.

muscle spindle, *n.* a PROPRIOCEPTOR found in skeletal muscle in the form of a capsule containing specialized muscle cells and nerve endings. Change in length or tension of muscle cells stimulates the spindle.

muscular dystrophy (M.D.), *n.* a disease characterized by the progressive wasting of muscles and eventual death. One type, called *Duchenne M.D.*, is controlled by a recessive sex-linked gene (see SEX LINKAGE) and, as a result, affects more boys than girls. The disease first shows itself between one and six years, progressing until the patient is confined to a wheelchair by the early teens with death resulting by the late teens in most affected individuals. Other forms of M.D. are controlled by autosomal genes (both dominant and recessive) and they are equally frequent in males and females.

muscularis mucosa, *n.* a thin layer of small muscle cells in the outermost layer of the mucosa of the alimentary canal.

musculoepithelial cell, *n.* a type of cell found particularly in COELENTERATES that is columnar in shape, occurs with ectodem or endoderm cells, and has two contractile processes extending into the mesogloea.

musculoskeletal system, *n.* the basis of locomotion in vertebrates, which is the contraction of MUSCLES against a skeleton.

mushroom, *n.* the common name for the fruiting bodies of the class Basidiomycetes of the family Agaricaceae.

mussel, *n.* the bivalve mollusc *Mytilus edulis*.

mutagen, *n.* an agent that is capable of increasing the MUTATION rate in an organism, for example, X-rays, ultraviolet light, mustard gas.

mutant, *n.* **1.** any gene that has undergone a MUTATION. **2.** an individual showing the effects of a mutation, with a phenotype that is not WILD TYPE.

mutant site, *n.* the area within a gene where a POINT MUTATION has taken place.

mutation, *n.* a change in the genetic material of an organism. If the alteration affects gametic cells the change is a genetic or POINT MUTATION and can be inherited; if body cells (nonsexual) are affected the mutation is called *somatic mutation* and will not normally be inherited. Mutations can be at two levels: (a) alterations to a single gene (point mutation) producing a different ALLELE. (b) alterations to the structure and/or numbers of chromosomes. See also SPONTANEOUS MUTATION, CHROMO-SOMAL MUTATION.

mutation breeding, *n.* a breeding technique in which MUTAGENS are used to produce new genetic forms of useful agricultural species. While novel types can be successfully made in this way, most MUTATIONS are harmful and, being a random occurence, it is difficult to control the changes that are produced, many of which will upset the delicate balance of genes in a particular strain.

mutation frequency, *n.* the proportion of mutants in a population as compared with normal WILD TYPE individuals.

mutation rate, *n.* the number of mutations per gene in a fixed time. In sexual organisms this is often measured as the number of mutations per gamete. Rates vary considerably between genes and between organisms, but a typical rate is 1 mutation per locus per 100,000 gametes.

muton, *n.* the smallest part of a gene that can undergo a MUTATION. It is now known that a muton is the size of a DNA nucleotide base.

mutualism, see SYMBIOSIS.

M.W., *abbrev. for* molecular weight.

mycelium, *n.* the total mass of HYPHAE of a FUNGUS that constitutes the vegetative body (as opposed to a fruiting body).

myceto-, *prefix.* denoting fungi.

Mycetozoa, see MYXOMYCETA.

mycology, *n.* the study of fungi.

Mycophyta, see EUMYCOTA.

mycoplasmas, *n.* a group of minute PROCARYOTES that resemble bacteria but do not possess a cell wall and thus have a variable morphology.

mycorrhiza, *n.* an association between a FUNGUS and the roots of a higher plant. The fungus may be located on or in the root, and in some cases it breaks down PROTEINS or AMINO ACIDS that are soluble and can be absorbed by the higher plant. In most cases, only nitrogen and phosphorus compounds result from fungal activity. Carbohydrates synthesized by the higher plants are absorbed by the fungus, so the relationship is a form of SYMBIOSIS. Some plants which lack chlorophyll, such as the bird's nest orchid, rely on mycorrhizas for carbohydrates in addition to protein.

mycosis, *n.* an animal disease caused by fungal infection.

Mycota, *n.* the plant kingdom containing the fungi and slime moulds.

mycotrophic, *adj.* (of plants) having mycorrhizas associated with the root system.

myel- or **myelo-,** *prefix.* denoting narrow.

myelin, *n.* a white phospholipid. See MYELIN SHEATH.

myelin sheath, or **medullated nerve fibre,** *n.* a sheath of fatty substance associated with protein which surrounds larger nerve fibres of vertebrates. See DENDRITE. The white lipid coating is produced by SCHWANN CELLS. The sheath permits a greater current flow and thus speeds the transmission of nervous impulses. Constrictions in the myelin sheath, known as *nodes of Ranvier*, indicate the division between one Schwann cell and another. Nonmyelinated fibres occur commonly in invertebrates, internally in the spinal cord of vertebrates, and in the AUTOMATIC NERVOUS SYSTEM of vertebrates. Myelinated fibres, white in appearance, occur outside the spinal cord of vertebrates. See Fig. 220.

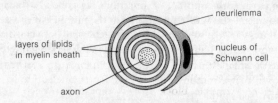

Fig. 220. **Myelin sheath.** T.S. of a myelinated nerve fibre.

myelitis, *n.* inflammation of the MYELIN SHEATH surrounding the spinal cord.

myeloid tissue, *n.* the tissue producing blood cells in vertebrates, normally located in bone marrow.

myeloma, *n.* a cancerous tumour of the bone marrow.

Mymaridae, *n.* an order comprising the fairy flies (HYMENOPTERA), the smallest known insects. They are related to chalcid wasps and are exclusively egg parasites.

myo-, *prefix.* denoting muscle.

myoblast, *n.* a cell that gives rise to a muscle fibre.

myocardium, *n.* the muscular wall of the vertebrate heart.

myocoel, *n.* the part of the coelomic cavity that exists within each muscle block or MYOTOME.

myofibril, *n.* a microscopic contractile unit of STRIATED MUSCLE, made up of a series of SARCOMERES. The myofibril is made up of numerous longitudinal filaments of two forms, thick (MYOSIN) and thin (ACTIN). See MUSCLE, I-BAND. Relatively little is known of the contraction of smooth muscle, where 'contractile elements' are thought to contain both actin and myosin.

myogenic contraction, *n.* contractions that are initiated in the muscle itself rather than by impulses from nerves. For example, in CARDIAC MUSCLE the heart continues to beat under the influence of the PACEMAKER after removal from the body.

myoglobin, *n.* a relatively small globular protein (M.W. = 17,000) consisting of 153 amino acids in a POLYPEPTIDE CHAIN and one HAEM group. The molecule is found in the muscles of vertebrates and some invertebrates (giving the muscles a red colour) and has a high affinity for oxygen.

myoneme, *n.* the contractile fibril that occurs in some PROTOZOANS, such as *Stentor, Vorticella.*

myopia or **short-sightedness,** *n.* an inability to see distant objects clearly, due to their image being focused in front of the retina because the cornea and lens form too powerful an optical system for the size of eye. Compare HYPEROPIA.

myoplasm, *n.* the cytoplasmic SOL content of a muscle cell.

myosin, *n.* a protein found in the thick filaments of the SARCOMERES OF MUSCLE. Myosin combines with ACTIN to produce ACTOMYOSIN during contraction.

myostatic reflex, see STRETCH REFLEX.

myotome, *n.* a muscle block that is repeated in each METAMERIC SEGMENT. Segmentation becomes much modified in the head region, and where paired limbs are developed.

Myriapoda, *n.* the class of ARTHROPODS containing centipedes (Chilopoda) and millipedes (Diplopoda).

myrmecophily, *n.* the use of ants by other insects, usually beetles, as a source of food or a place (ant nest) in which to live.

myxamoeba, *n.* a cell that is produced on the germination of spores. It is found in slime moulds and some simple fungi, and is capable of amoeboid movement.

myxoedema, *n.* the condition occurring in adults due to under-secretion of THYROXINE. There is a decrease of METABOLIC RATE, an increase in subcutaneous fat, a coarsening of the skin and mental and physical sluggishness.

Myxomyceta or **Mycetozoa** or **Myxomycophyta** or **slime mould,** *n.* fungus-like organisms (see PLASMODIUM) having a naked mass of protoplasm and living and feeding on dead, decaying or living plants. They reproduce by spores and are capable of amoeboid movement. The group is now normally classified in the division Myxomyceta of the Kingdom Mycota. See MYXAMOEBA.

Myxomycophyta, see MYXOMYCETA.

Myxophyta, see BLUE-GREEN ALGAE.

Myxosporidia, *n.* a group of SPOROZOANS parasitic on freshwater fish, to which they attach themselves by structures having a superficial resemblance to coelenterate NEMATOCYSTS.

N

nacreous layer, *n.* the pearly, innermost layer of the shell of a mollusc secreted by the mantle epithelium.

NAD (nicotinamide adenine dinucleotide), *n.* a COENZYME that readily accepts or gives up hydrogen.

$$NAD \rightleftarrows NADH$$
oxidized   reduced

NAD acts as an energy carrier in the cell. Reduced NADH transports electrons to the ELECTRON TRANSPORT SYSTEM during AEROBIC RESPIRA-TION. See also FAD.

NADP (nicotinamide adenine dinucleotide phosphate), *n.* a COENZYME that readily accepts hydrogen from an ELECTRON TRANSPORT SYSTEM in PHOTOPHOSPHORYLATION, passing the electrons onto the CALVIN CYCLE of PHOTOSYNTHESIS.

$$NADP \rightleftarrows NADPH$$
oxidized   reduced

naevus, *n.* **1.** any birthmark, usually a sharply defined reddish patch on the skin. **2.** a pigmented mole on the skin.

nail, *n.* a horny, keratinized layer protecting the distal end of each finger and toe in humans and most other primates. In other terrestrial vertebrates the nail is shaped into a claw.

naked, *n.* (of a plant structure) without hairs or scales.

nanometre, *n.* one thousand millionth of a metre (10^{-9}m). Symbol: nm.

narco-, *prefix.* denoting torpor.

narcotic, *n.* any chemical substance that induces a state of stupor or unconsciousness, such as opium.

nares, see NOSTRIL.

narrow-sense heritability, *n.* the proportion of variance among

PHENOTYPES in a population that can be attributed to additive genetic variance, the latter being the sum of the average effects of all the genes carried in the population that affect a particular character. Narrow sense-HERITABILITY is concerned with how much of a parent's phenotype is inherited by its offspring and often is estimated from the results of selection experiments. Compare BROAD-SENSE HERITABILITY.

nasal cavity, *n.* a system of chambers above the hard palate of mammals through which air passes from the nostrils to the pharynx and on to the lungs. During inspiration air is filtered, warmed, moistened and smelled.

nastic movement or **nastic response** or **nasty,** *n.* a growth movement in plants that takes place in response to a diffuse, or nondirectional stimulus. For example, the opening and closing of *Crocus* flowers due to changes in temperature.

nastic response, see NASTIC MOVEMENT.

nasty, see NASTIC MOVEMENT.

national park, *n.* any relatively large area of countryside which is (a) not materially altered by human presence or exploitation, (b) protected by legislation to preserve the characteristics for which it was designated, (c) made accessible to visitors under special conditions for cultural and recreative purposes. This definition is the international meaning adopted by the International Union for the Conservation of Nature and Natural Resources (IUCN) in 1969.

In the UK, a different approach has been adopted based on the bigger problem of integrating conservation of the landscape with the economic and social life of the countryside, i.e. of *not* removing humanity and its effects. National parks in England and Wales, because of their scenery and the opportunities afforded to open-air recreation, and having regard to their character and position in relation to centres of population, are protected by legislation to preserve and enhance their natural beauty. The Countryside Commission has so far designated 10 national parks, for example, the Lake District, Snowdonia.

native, *n.* a species that has colonized an area without human aid.

native protein, *n.* a naturally occurring protein.

natural classification, see CLASSIFICATION.

natural immunity, *n.* immunity due to the possession of suitable genetical characteristics rather than immunity produced in response to a vaccine or serum.

natural order, *n.* an obsolete term for a FAMILY of flowering plants.

natural selection, *n.* the mechanism, proposed by Charles DARWIN, by which gradual evolutionary changes take place. Organisms which are better adapted to the environment in which they live produce more viable young, so increasing their proportion in the population and thus being 'selected'. Such a mechanism depends on the variability of individuals within the population. Such variability arises through MUTATION, the beneficial mutants being preserved by NATURAL SELECTION.

nature and nurture, see PHENOTYPE.

nauplius, *n.* the typical crustacean larva which has a single eye, three pairs of limbs and a rounded, transparent body.

nautiloid, *n.* any mollusc of the subclass Nautiloidea, a group of cephalods related to the fossil ammonites, which produce a large spiral chambered shell, in the last chamber of which the animal lives.

navigation or **orientation,** *n.* a process of direction-finding. Many animals are capable of navigation particularly those performing long MIGRATIONS. Birds are now known to find their way home (see HOMING) by means of the sun and stars, in addition to landmarks. Bees, and other ARTHROPODS, use the pattern of polarized light in the sky (from which they deduce the direction of the sun) and other organisms may use chemical gradients to trace a route. For example, salmon find their specific river for breeding purposes by following a scent in from the sea.

Neanderthal man, *n.* a primitive form of human: *Homo sapiens neanderthalus,* who lived some 100,000 years ago.

necrosis, *n.* the localized death of plant and animal tissue, for example the response of a leaf to invasion of a pathogen. An affected area is described as being 'necrotic'. See DIPHTHERIA.

necrotrophic or **saprophytic,** *adj.* (of microorganisms and plants) feeding on dead tissue (see SAPROPHYTE). Compare BIOTROPHIC.

nectar, *n.* a sugar-containing liquid secretion of the NECTARY of many flowers. It attracts insects, so bringing about pollination.

nectary, *n.* an area of tissue in a flower that secretes sugary nectar, serving as an attraction for visiting insects. Nectaries are frequently deep inside the corolla so that the pollinator both deposits and carries off pollen. See ENTOMOPHILY.

negative feedback, see FEEDBACK MECHANISM.

nekton, *n.* the population of free-swimming animals that inhabits the PELAGIC zone of oceans. Compare PLANKTON.

nemato-, *prefix* denoting thread-like.

nematoblast, see CNIDOBLAST.

nematocyst, *n.* a structure formed of a hollow thread within a bladder in the CNIDOBLAST or thread cell of COELENTERATES. On stimulation, for example by prey, the thread is everted. Several types exist, such as stinging cells that inject poisonous substances into the prey through the thread, and sticky cells which exude a sticky substance which causes the prey to adhere to the coelentrate tentacle. Enormous numbers are present on the tentacles of some forms and nematocysts are responsible for the jellyfish 'sting'.

Such cells can be utilized by other organisms, for example flatworms preying on coelenterates. The nematocysts migrate to the surface layer of a flatworm which has fed on the coelenterate and are used by the flatworms in exactly the same way as coelenterates use them. See Fig. 221.

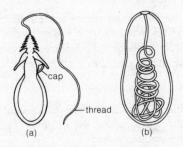

Fig. 221. **Nematocyst.** Nematocysts of *Hydra*. (a) Discharged penetrant. (b) Undischarged glutinant.

nematode, *n.* any member of the phylum Nematoda, containing roundworms such as ASCARIS.

nemertine or **nemertean,** *n.* any member of the phylum Nemertea, containing the ribbon worms, for example *Lineus*, the bootlace worm of rocky shores.

neo-, *prefix.* denoting new or recent.

neoDarwinism, *n.* a view of evolutionary theory that combines MENDELIAN GENETICS with DARWINISM. See CENTRAL DOGMA.

Neogene, *n.* the upper part of the TERTIARY PERIOD including the Pleistocene, Pliocene and Miocene epochs.

Neognathae, *n.* one of the two major groups of living birds, including the flying forms. The group is distinguished by having a small prevomer bone which does not touch the pterygoid bones. Compare PALAEOGNATHAE.

neoLamarckism, *n.* a modern attempt (for which there is scant evidence) to add genetical support to LAMARCKISM, that emphasizes the supposed influence of environmental factors on genetic change.

Neolithic, 1. *n.* the period of the new Stone Age which began approximately 10,000 years ago in the Middle East but not until *c* 5,500 years ago in Britain where it lasted until *c* 4,500 years ago. The period marks the beginning of plant cultivation and domestication of animals by humans. **2.** *adj.* relating to this period.

neonatal, *adj.* (of newborn offspring, particularly human) the first month of independent life.

neontology, *n.* the study of life of recent living organisms. Compare PALAEONTOLOGY.

neopallium, *n.* a nerve mass that forms the main part of the cerebrocortex in man, and the roof of the cerebral hemispheres in other vertebrates. The neopallium is associated with general muscular coordination and intelligence.

neoplasm, *n.* an autonomous growth of tissue in the body which has no apparent physiological function, such as a TUMOUR.

neoplastic disease, *n.* an abnormal condition produced by a NEOPLASM.

Neornithes, *n.* the group containing all birds with the exception of *Archaeopteryx*.

neoteny, see PAEDOGENESIS.

neotype, *n.* a specimen selected as the TYPE subsequent to the publication of the original description where the original types have been lost, destroyed or otherwise invalidated.

nephridiopore, *n.* the external opening of a NEPHRIDIUM.

nephridium, *n.* (pl. nephridia) a primitive excretory organ present in many invertebrates (e.g. the earthworm) in the form of a tube which opens at one end to the exterior. The other end may open into the COELOM or may terminate in a FLAME CELL.

nephromyxium, *n.* a combination of COELOMODUCT and NEPHRIDIUM forming an excretory tubule such as occurs in many ANNELIDS.

nephron, *n.* the MALPIGHIAN BODY and the associated tubule of the vertebrate kidney. In each human kidney there are about one million nephrons, making a total for both kidneys of around fifty miles of tubules.

nephrostome, *n.* the internal opening of a NEPHRIDIUM or NEPHRO-MYXIUM.

neritic, *adj.* of a region of shallow seawater near the coast.

nerve, *n.* **1.** (in animals) a bundle of nerve fibres, usually containing both afferent neurons (to the CENTRAL NERVOUS SYSTEM), and efferent neurones (away from CNS), together with associated connective tissue and blood vessels, situated in a common sheath of CONNECTIVE TISSUE and lying outside the CNS. **2.** (in plants) a characteristic leaf structure consisting of fine strands that are conductive and/or strengthening in function.

nerve cell, see NEURON.

nerve cord, *n.* a strand of nervous tissue that forms part of the CENTRAL NERVOUS SYSTEM of invertebrates.

nerve ending, *n*. the tip of a nerve fibre either in the form of an END ORGAN or small branching structure.

nerve fibre, *n*. the AXON of a NERVE(1) together with its surrounding membranes.

nerve impulse, *n*. the message conducted along the AXON of a NERVE(1) The impulse is a propagated negative charge on the outside of the membrane which results from a wave of DEPOLORIZATION passing along an axon. The RESTING POTENTIAL is reversed and becomes an ACTION POTENTIAL, and this passes down the axon at between 1 and 100 ms, depending on the diameter of the fibre, the presence of a MYELIN SHEATH, temperature, species of animal etc. Once an impulse is initiated it progresses without degeneration, and the strength or nature of the stimulus does not affect it; it either is generated or it is not (ALL-OR-NONE LAW). Varying stimuli produce varying numbers of impulses (see SUMMATION). After each impulse there is a REFRACTORY PERIOD during which a second impulse may not pass.

nerve net, *n*. a system of interconnecting nerve cells, in the form of a net, found in those groups of invertebrates which lack a CENTRAL NERVOUS SYSTEM, such as COELENTERATES and ECHINODERMS. In other groups the nerve net is almost nonexistent. Conduction is slow across the net and is often in all directions, as the synapses are not usually unidirectional.

nerve ring, *n*. the ring of nerves surrounding the mouth of ECHINODERMS.

nerve root, *n*. the points at which a spinal nerve arises from the spinal cord. Each has two roots: a DORSAL ROOT and a VENTRAL ROOT.

nervous system, *n*. the main means by which METAZOAN animals coordinate their activities, through quick, short-lasting reactions (the ENDOCRINE GLAND system also has a coordinating function but this is slower and longer lasting). Receptors receive stimuli and pass them on to effector organs by means of nerve cells. Nerve cells are linked by SYNAPSES, and may connect with other nerve cells, receptors or effectors. Complex connections occur which give rise to complex reactions, and coordination takes place in the CENTRAL NERVOUS SYSTEM (CNS) where synapses are present in profusion. Coordination is brought about by the transmission of NERVE IMPULSES.

nervous tissue, *n*. NERVE cells, together with their AXONS and accessory cells.

net assimilation rate, *n*. a measurement of plant growth (weight increase per unit time) by assessing change in a particular part of the plant (usually leaf area) rather than in the overall plant.

net primary production, see PRIMARY PRODUCTION.

neural, *adj*. relating to nerves or to the nervous system.

neural plate, *n.* that part of the ectoderm which sinks below the surface of the developing vertebrate embryo and forms the spinal cord.

neurenteric canal, *n.* an embryonic structure joining the gut with the neural canal at the time the BLASTOPORE is roofed over by neural folds.

neurilemma, *n.* a thin membrane surrounding the MYELIN SHEATH of myelinated nerve fibres. It is part of the SCHWANN CELL, lying close to the axon.

neurin, *n.* a filamentous protein, similar to muscle ACTIN, that is attached to the inner wall surface of brain cells and possibly has a role in EXOCYTOSIS.

neurocranium, *n.* that part of the vertebrate skull forming the brain case and sense capsules.

neuroglia, see GLIA.

neurohumor, *n.* a hormone secreted by a NERVE ENDING.

neurohypophysis, *n.* see PITUITARY GLAND.

neuromuscular junction, *n.* the area of contact between a MOTOR NEURON and a MUSCLE FIBRE.

neuron or **neurone** or **nerve cell,** *n.* the structural unit of the nervous system, usually consisting of the cell body and cytoplasmic extensions. Long single extensions are known as AXONS and these carry impulses from the cell body to other nerve cells or effectors. Short multiple extensions known as DENDRITES receive impulses. Axons may be myelinated (with MYELIN SHEATHS) or nonmyelinated, but in the former the transmission of impulses is quicker as action potentials jump from one node of Ranvier to another. See Fig. 222.

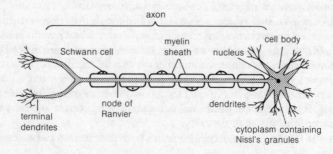

Fig. 222. **Neuron.** A vertebrate motor nerve cell.

neuron theory, *n.* the theory that the nervous system is formed of numerous separate NEURONS that contact only through synapses and not through a continuity of protoplasm.

neuropeptid, see ENDORPHIN.

neurophysiology, *n.* the study of the physiology of NERVES.

neuropil, *n.* the mass of DENDRITES and AXONS surrounding the cell bodies of the central nervous system.

Neuroptera, *n.* the order of ENDOPTERYGOTE insects containing the alderflies and lacewings, possessing two pairs of similar, membranous wings.

neurosecretion, *n.* the process of secretion of hormones by nerve cells.

neurula, *n.* the stage of a vertebrate embryo when gastrulation is largely finished and a neural plate is forming, ending with the formation of the neural tube.

neurotransmitter, see TRANSMITTER SUBSTANCE.

neurotubule, see MICROTUBULE.

neuston, *n.* organisms supported on the water surface.

neuter, *n.* an organism having nonfunctional or absent sexual parts. In plants, a neuter is one without male and female flower parts (e.g. ray florets of sunflower). A neuter in animals is one that is sexually undeveloped or sterile, for example the female worker in a bee colony, or a castrated or spayed animal.

neutral allele, *n.* a form of a gene that when carried in an organism in no way alters the FITNESS of that individual to survive and reproduce.

neutral term, *n.* any term used in taxonomy, such as form or group, that has no reference to any formal category of taxonomic hierarchy.

neutrophil, *n.* the most common type of white blood cell (LEUCOCYTES) formed in the bone marrow and with a normal count in human blood of between 2,500 and 7,500 cells per mm^3. Neutrophils are PHAGOCYTES, and are important in combatting bacterial infections.

niacin, see NICOTINIC ACID.

niche, *n.* the sum of the characters that determine the position of an organism in its ECOSYSTEM. This includes all the chemical, physical, spatial and temporal factors required for the survival of that species and which limit its distribution and growth. A niche is characteristic of a SPECIES, and no two species occupy the same niche in the same environment in each others' presence. However, a different species may occupy the same niche in the absence of the normal occupant from that habitat.

nick, *n.* a single-stranded cut or break in a DNA molecule. Nicking of DNA may form part of a DNA repair mechanism, as occurs after damage caused by ultraviolet light.

nicotinamide, *n.* a crystalline basic amide of the vitamin B complex.

nicotine, *n.* an alkaloid derived from tobacco.

nicotinic acid or **niacin,** *n.* a vitamin of the B COMPLEX found in, for example, meat, yeast and wholewheat. Niacin is converted into nicotinamide in the body, becoming incorporated into NAD, an important coenzyme in cellular respiration. A deficiency of niacin leads to a condition called *pellagra*, with symptoms of dermatitis, muscle weakness and mental disturbance, amongst others.

nictitating membrane, *n.* the third eyelid that occurs in birds, reptiles, and some mammals, sharks and amphibians. It is transparent and flicks over the eyeball from the inner (anterior) corner of the eye to clean and moisten the eyeball without shutting out the light. In some diving birds such as auks, there is a lens-like window in the membrane which adjusts the focus of the eye under water.

nidicolous, *adj.* (of young birds) hatching in a relatively undeveloped stage and remaining in the nest for the fledging period. For example, songbirds.

nidifugous, *adj.* (of young birds) hatching in a condition allowing departure from the nest within a few hours of birth. Such birds spend the fledging period away from the nest. For example, waders.

nipple, *n.* a conical projection at the centre of a mammary gland on which the milk ducts have outlets.

Nissl's granules, *n.* granules that occur in the cell bodies of NEURONS. They are formed of a nucleoprotein combined with iron and are concerned with protein synthesis.

nit, *n.* the egg of the human louse *Pediculus humanus* which is found attached to hair.

Nitella, *n.* a genus of green algae.

nitrification, *n.* the conversion by soil bacteria of organic nitrogen into inorganic nitrates which plants can take up. Decay of animal and plant proteins gives rise to ammonium compounds which the bacteria *Nitrosomanas* and *Nitrococcus* oxidize to nitrate. *Nitrobacter* oxidizes nitrites to nitrates. See NITROGEN CYCLE.

nitrogen cycle, *n.* the circulation of nitrogen in the environment as a result of the activity of living organisms. 80% of the atmosphere is made up of nitrogen and this is maintained by the balancing action of the cycle. See Fig. 223.

nitrogen deficiency, *n.* a supply of nitrogen below an organism's requirements.

nitrogen fixation, *n.* the utilization of atmospheric nitrogen in the synthesis of AMINO ACIDS by some bacteria and blue-green algae. Such PROCARYOTES can be free-living (e.g. *Azotobacter*, an aerobe; *Clostridium*, an obligate anaerobe) while others (e.g. the bacterium *Rhizobium*) live in association with plants, occupying swellings in the root called *root nodules*. The latter relationship is one of SYMBIOSIS, in that the plant gains nutrients and thus can live in nitrogen-poor soils, while the nitrogen-

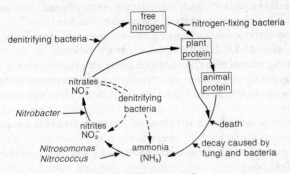

Fig. 223. **Nitrogen cycle.**

fixer obtains a supply of carbohydrates from the plant. The nitrogen is reduced to ammonia in the microbes by action of the enzyme nitrogenase: $N_2 + 3H_2 = 2NH_3$, the ammonia then reacting with keto acids to form amino acids.

nitrogen-fixing bacteria, see NITROGEN FIXATION.

nitrogenous waste, *n.* the waste nitrogen produced during METABO-LISM. Initially this is often in the form of AMINO ACIDS, but few animals can excrete these directly. Many invertebrates and aquatic vertebrates eliminate nitrogen as ammonia, but this would be toxic if excess water were not available. Marine teleost fish excrete TRIMETHYLAMINE OXIDE, which is nontoxic and requires little water for elimination. Mammals, turtles and amphibians excrete UREA, and land reptiles, snails, insects and birds excrete URIC ACID from which water is removed before voiding it.

nm, *abbrev. for* NANOMETRE $(10^{-9}m)$.

Noctuidae, *n.* a family of moths – the owlet moths – whose larvae are known as cutworms and cause damage to crop plant roots.

node, *n.* that part of a plant stem where leaves are attached or may develop from buds. See also INTERNODE.

nodes of Ranvier, see MYELIN SHEATH.

nodule, *n.* **1.** any small spherical swelling. **2.** short for *root nodule*. (see NITROGEN FIXATION).

nomenclature, see BINOMIAL NOMENCLATURE.

nomen conservandum, *n.* a name officially prescribed by the COMMISSION for Zoological Nomenclature.

nomen dubium, *n.* a name given to an organism for which the evidence is insufficient to recognise the species to which it was applied.

nomen oblitum, *n.* a name which has lost its validity.

nominal taxon, *n.* a named taxon objectively defined by its TYPE.

nominate, *adj.* (of a TAXON) bearing the same name as its immediately superior taxon and is its type. For example in the redshank, *Tringa totanus totanus* is the nominate form of the species.

noncompetitive inhibition, *n.* a form of enzyme control in which the enzyme has two kinds of ACTIVE SITE, one for an inhibitor, the other for the enzyme substrate. The inhibitor prevents catalytic activity of the enzyme, perhaps by licking or even changing the shape of the substrate active site. Compare COMPETITIVE INHIBITION. See also ALLOSTERIC ENZYME.

noncyclic photophosphorylation, *n.* one of two main processes occurring in the light stage of PHOTOSYNTHESIS, involving chlorophyll photosystems I and II in the formation of ATP and reduced NADP. See LIGHT REACTIONS.

nondimensional *adj.* (of a species) characterized by the noninterbreeding of two coexisting populations, uncomplicated by considerations of space and time.

nondisjunction, *n.* the failure of chromosomes to go to opposite poles during nuclear division, leading to unequal numbers of chromosomes in the daughter cells (see ANEUPLOIDY). See Fig. 224. Nondisjunction produces abnormal numbers of both AUTOSOMES (e.g. DOWN'S SYNDROME) and SEX CHROMOSOMES (e.g. TURNER'S SYNDROME).

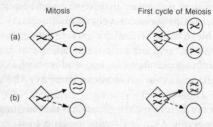

Fig. 224. **Nondisjunction.** (a) Normal disjunction. (b) Nondisjunction.

nonessential amino acid, *n.* any amino acid which is not an ESSENTIAL AMINO ACID.

nonhomologous, *adj.* (of chromosomes or chromosomal segments) containing different genes not pairing during meiosis. Compare HOMOLOGOUS CHROMOSOMES.

nonMendelian genetics, *n.* a mechanism or theory of inheritance that does not obey Mendel's laws about SEGREGATION of genes or their INDEPENDENT ASSORTMENT. For example, GENETIC LINKAGE between genes produces frequencies of progeny different from those predicted from independent assortment.

nonpolar, *adj.* (of a substance) bearing no charge or hydoxyl groups, and thus not combining with water.

nonrace-specific resistance or **horizontal resistance,** *n.* a situation where a large number of genes (POLYGENES) in the host plant confer a general level of resistance to a wide range of attacking organisms (pathogens or parasites). Such a system is usually able to withstand genetic changes in the attacking organism and thus produce stable production of crops. Compare RACE-SPECIFIC RESISTANCE.

nonrandom mating, *n.* any mating in a population where crosses between the various types of individuals do not occur in the frequencies that are expected by chance. Two major types of nonrandom mating can be distinguished: between individuals that are related (see INBREEDING), and between individuals with similar phenotypes (see ASSORTATIVE MATING).

nonreducing sugar, see SUGAR.

nonsense codon or **termination codon** or **stop codon,** *n.* a triplet of MESSENGER RNA bases, such as the OCHRE CODON, that does not code for an AMINO ACID and causes termination of polypeptide synthesis during the translation phase of PROTEIN SYNTHESIS. There are three nonsense codons in the GENETIC CODE.

nonsense mutation, *n.* an alteration in a DNA POLYNUCLEOTIDE CHAIN that results in a NONSENSE CODON when transcribed onto mRNA.

noradrenaline, *n.* a transmitter substance produced at the nerve endings of ADRENERGIC nerves giving effects very similar to those of ADRENALINE. After transmission it is inactivated by monoaminoxidase in order to prevent a build-up. The action of noradrenaline can be inhibited by various drugs, such as mescaline, which produce hallucinatory effects.

normal distribution curve or **Gaussian curve,** *n.* the bell-like shape produced when a 'normal' population is sampled for a continuously varying character (e.g. height in humans) and the FREQUENCY DISTRIBUTION plotted on a graph. The curve is theoretically exactly symmetrical, so that the shape on each side of the mean is the mirror image of the other, with an equal area. See Fig. 225. Thus the MODE, MEDIAN, and MEAN are all at the centre of the distribution.

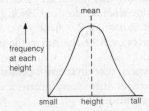

Fig. 225. **Normal distribution curve.** The curve shows the frequency of recorded height in a 'normal' population of humans.

normalizing selection, *n.* a form of STABILIZING SELECTION in which less fit individuals at both extremes of a distribution are eliminated, reducing the genetic variability in the population.

nose, *n.* the projecting part of the head of higher vertebrates that usually carries the nostrils and is associated with the sense of smell.

nostril or **nares,** *n.* an external opening of the nose leading into the nasal cavity.

noto-, *prefix.* denoting back.

notochord, *n.* the longitudinal axial support (skeleton) of the embryos of all chordates, which lies ventral to the nerve cord and dorsal to the alimentary canal. Remnants of the notochord usually remain in the adult between the vertebrae, which come to surround it.

nucellus, *n.* the tissue within the OVULE of a plant, surrounded by INTEGUMENTS(1) and containing the EMBRYO SAC.

nuclear division, *n.* a part of the cell cycle. See MITOSIS, MEIOSIS.

nuclear membrane, *n.* an envelope surrounding the nuclear material of eucaryotic cells that consists of two unit membranes (see UNIT-MEMBRANE MODEL) with a distinct space in between. At intervals the membrane is perforated by large 'pores' allowing the exchange of materials between nucleus and cytoplasm. The nuclear membrane connects up with spaces inside the ENDOPLASMIC RETICULUM.

nuclear pore, *n.* one of many perforations in the NUCLEAR MEMBRANE of cells.

nuclease, *n.* any enzyme that promotes hydrolysis of NUCLEIC ACIDS. For example, DNase, which catalyses the breakdown of DNA into individual DNA NUCLEOTIDES.

nucleic acid, *n.* a molecule that is composed of strings of NUCLEOTIDES forming a POLYNUCLEOTIDE CHAIN. Nucleic acids act as the genetic material of cells and occur as either DNA (two chains) or RNA (one chain).

nucleolar organiser, *n.* an area of nuclear DNA containing genes that code for RIBOSOMAL RNA. The area is associated with the NUCLEOLUS.

nucleolus, *n.* (*pl.* nucleoli) an organelle found within the nuclei of EUCARYOTES, that contains RIBOSOMAL RNA and is associated with that part of the chromosome which codes for rRNA. Each nucleus contains one or more nucleoli, which appear darkly stained in preparations.

nucleoplasm, *n.* the contents of the cell nucleus inside the NUCLEAR MEMBRANE.

nucleoprotein, *n.* a type of protein found especially in the nucleus, such as the various types of HISTONE found complexed with DNA in the nucleosomes of eukaryote cells.

nucleoside, *n.* a chemical compound in which a nitrogenous base is attached to a sugar. Such a structure occurs in DNA nucleotides where a phosphate group is attached to the deoxyribose sugar of the nucleoside.

nucleosome, *n.* the basic structural unit of the eucaryote chromosome, being composed of four pairs of HISTONE proteins (H2A, H2B, H3 and H4) that are combined to form an octomer around which is wrapped about 150 nucleotide pairs of DNA.

nucleotide, *n.* a complex organic molecule forming the basic unit of NUCLEIC ACIDS, with a structure made up of three components: a pentose sugar (ribose, or deoxyribose with one less oxygen atom), an organic base (PURINE type: ADENINE and GUANINE, or PYRIMIDINE type: CYTOSINE, THYMINE and URACIL) and a phosphate group (see Fig. 226). The three elements are linked together by two condensation reactions between the $1'$ sugar carbon and a base forming a NUCLEOSIDE, and the $5'$ sugar carbon and the phosphate (see Fig. 227). The nucleotides are formed into POLYNUCLEOTIDE CHAINS.

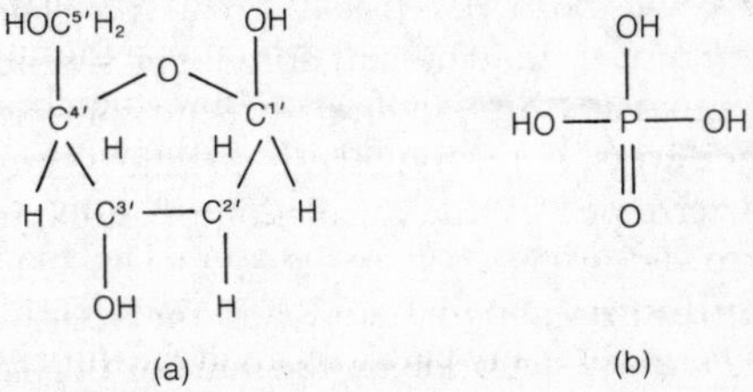

Fig. 226. **Nucleotide.** Basic units of (a) deoxyribose sugar, (b) phosphate. Each carbon atom is numbered (1 prime, 2 prime, etc.).

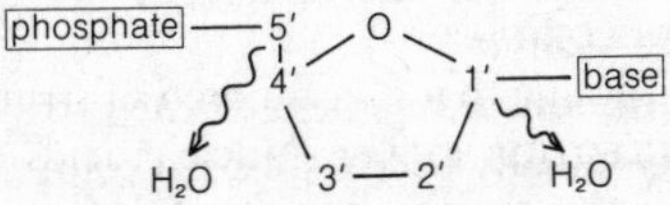

Fig. 227. **Nucleotide.** Linkage of the three nucleotide elements.

nucleotidase, *n.* an enzyme that catalyses the hydrolysis of a nucleotide to a nucleoside and orthophosphate.

nucleotide pair, *n.* two NUCLEOTIDES of opposite strands of DNA, joined by weak hydrogen bonds (see Fig. 228). See COMPLEMENTARY BASE PAIRING.

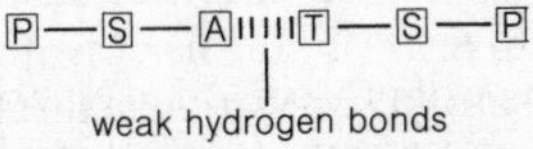

Fig. 228. **Nucleotide pair.** Complementary base pairing. P = phosphate, S = sugar, A = adenine, T = thymine.

nucleus, *n.* **1.** an organelle of eucaryotic cells that is bounded by a

NUCLEAR MEMBRANE and contains the chromosomes whose genes control the structure of proteins within the cell. **2.** (anatomy) the mass of nerve cell bodies, connected by tracts of nerve fibres, which occur in the vertebrate brain.

null allele, *n.* a form of a gene whose product is either absent at the molecular level, or is present but has no measurable phenotypic function. For example, in a heterozygote with INCOMPLETE DOMINANCE one gene form is a 'null allele' and is inactive, while the other allele is unable to compensate for the null allele and so produces an intermediate phenotype.

null hypothesis (N.H.), *n.* a statement that a certain relationship exists, which can be tested with a statistical SIGNIFICANCE test. A typical null hypothesis is the statement that the deviation between observed and expected results is due to chance alone. In biology, a probability of greater than 5% that the N.H. is true ($P > 5\%$) is considered acceptable.

nullisomic, *adj.* (of a mutant condition) the absence of a pair of HOMOLOGOUS CHROMOSOMES, giving a genetic complement of $2n - 2$.

numerical phenetics, *n.* the hypothesis that the relationships between taxa can be determined by the calculation of an overall, unweighted similarity value.

numerical taxonomy, *n.* the study of the relationships between organisms by means of determining the number of characteristics they have in common. See also CHARACTER INDEX.

nunatak, see REFUGIUM.

nut, *n.* a type of fruit with one seed and a hard woody outer layer, the PERICARP, such as the walnut.

nutation, *n.* the continuous movement of the growing part of a plant during growth which usually follows a spiral course, as in the stem of a convolvulus.

nutrient, *n.* any material that organisms take in and assimilate for growth and maintenance.

nutrient cycle, *n.* the passage of a nutrient through an ECOSYSTEM so that it eventually becomes reavailable to the PRIMARY PRODUCERS.

nutrition, *n.* the process of promoting the continued existence and growth of living organisms by taking in materials from the environment. Different organisms have different nutritional requirements. For example, the bacterium *E. coli* is capable of synthesizing all the necessary AMINO ACIDS, COENZYMES, PORPHYRIN structures (e.g. in HAEMOGLOBIN) and NUCLEIC ACIDS, whereas humans need to take in amino acids, VITAMINS, CARBOHYDRATES and FATTY ACIDS, in addition to water and many ESSENTIAL ELEMENTS.

nyct-, *prefix.* denoting night.

nyctinasty, the so-called 'sleep movements' of plants in which the leaves regularly droop at night in a form of DIURNAL RHYTHM. The activity is particularly common in plants with compound leaves, such as *Oxalis acetosella*, the wood sorrel.

nymph, *n.* the immature stages of any EXOPTERYGOTE, such as the mayfly. It has compound eyes and mouthparts like the adult, but usually lacks wings (though traces are sometimes present) and is sexually immature. See METAMORPHOSIS, ENDOPTERYGOTE.

nystatin, *n.* a rod-shaped antibiotic molecule that allows the passage of molecules smaller than 0.4 nm by creating channels through membranes.

O

ob-, *prefix.* denoting against.

obdiplostemous, *adj.* (of plants) having the stamens in two whorls, where the outer ones are opposite the petals and the inner ones opposite the sepals.

objective synonym, *n.* any synonym based on the same TYPE specimen.

obligate, *adj.* (of an organism) being able to live only parasitically. Compare FACULTATIVE.

oblong, *adj.* (of a plant) having one axis longer than the other. **2.** (of a leaf) being elliptical.

obtuse, *adj.* blunt.

Occam's razor, see OCKHAM'S RAZOR.

occipital condyle, *n.* a projection of bone at the back of the vertebrate skull which articulates with the vertebral column. It is absent in most fish.

occipital lobe, *n.* the posterior area of the CEREBRAL HEMISPHERE lying behind the parietal lobe and responsible for the interpretation of visual stimuli from the retina.

occiput or **occipital region,** *n.* **1.** the hard region of the vertebrate cranium. **2.** the dorsilateral plates on the head of an insect.

oceanic, *adj.* occurring in the sea where the depth exceeds 200 m.

ocellus, *n.* a single eye with a single lens found in insects and some other invertebrates.

ochre codon, *n.* one of three NONSENSE CODONS in the GENETIC CODE consisting of the mRNA bases UAA.

ochrea or **ocrea,** *n.* a sheath formed by the fusion of two stipules round a stem to form a tube.

Ockham's razor or **Occam's razor,** *n.* the principle which states that when a selection has to be made from various hypotheses, it is best to start with that which makes fewest assumptions. Named after William of Ockham (d. *c* 1349).

oct-, or **octo-,** *prefix.* denoting light.

octad, *n.* the eight ascospores contained in a fungal ASCUS.

octoploid, *n.* a cell or an individual having eight sets of chromosomes (8n), i.e. four times the DIPLOID(1) number.

ocotopod, *n.* any cephalopod mollusc of the order Octopoda, including eight-armed molluscs such as the octopus.

oculo-, *prefix.* denoting the eye.

oculomotor nerve, *n.* the third cranial nerve of vertebrates, that innervates the muscles which move the eyeball.

Odonata, *n.* the order of insects which includes the dragonflies and damselflies.

odonto-, *prefix.* denoting the teeth.

odontoblasts, *n.* the cells that give rise to dentine in vertebrate teeth and to simple teeth in invertebrates.

odontoid process, *n.* a projection of the second vertebra (the axis), allowing the ATLAS to rotate and thus move the head round.

Oedogonium, *n.* a genus of filamentous algae.

oedema or **edema,** *n.* a swelling of tissues caused by the capilliary blood vessels passing out water into the surrounding tissues, and so increasing the intercellular fluid content.

oesophagus, *n.* the part of the alimentary canal of vertebrates that lies between the PHARYNX and the STOMACH. No digestive juices are produced here, but PERISTALSIS takes place, often moving the bolus upwards from the lowered head to the stomach.

oestradiol, *n.* an OESTROGEN produced by the follicle cells of the vertebrate ovary. It promotes oestrus, the development of the endometrium, and stimulates ICSH secretion.

oestrogen or **estrogen,** *n.* a hormone produced by the OVARY of the female vertebrate which maintains the female SECONDARY SEXUAL CHARACTERS and is involved in the repair of the uterine wall after menstruation (see MENSTRUAL CYCLE). It is also produced by the female PLACENTA and in small quantities by the ADRENAL CORTEX and the male TESTIS.

oestrous cycle or **estrous cycle,** *n.* a reproductive cycle caused by the cyclic production of gonadotrophic hormones by the PITUITARY GLAND. It occurs in adult female mammals and is only complete when pregnancy does not occur, lasting from 5 to 60 days, dependent on species.

Initially, follicle growth occurs in the ovary as a result of FOLLICLE-

STIMULATING HORMONE being secreted by the anterior pituitary. This results in increased OESTROGEN production and a thickening of the uterus lining. The build-up of oestrogen causes the anterior pituitary to produce LUTEINIZING HORMONE which in turn brings about ovulation with the formation of a CORPUS LUTEUM, development of the uterine glands, secretion of PROGESTERONE by the corpus luteum (which inhibits FSH production) and a consequent decrease in oestrogen production. Fertilization may occur at this stage and if it does so, pregnancy results. If not, there is a regression of the corpus luteum and a feedback to the anterior pituitary to produce FSH as progesterone decreases. New follicles are formed, the lining of the uterus thins, menstruation (see MENSTRUAL CYCLE) occurs in human females and some other primates, and there is a cessation of progesterone secretion and a continued decrease of oestrogen. The cycle then begins again as new follicles grow. Only during the initial period of the cycle, during which ovulation takes place, will the female of most mammalian species copulate (the period of 'heat' or oestrus).

official index, *n.* a list of taxonomic names declared invalid by the COMMISSION for Zoological Nomenclature.

oil, *n.* any fat that is liquid at room temperature.

oil-immersion objective lens, *n.* a objective lens used to examine objects at very high optical power (at least 1,000 ×). When in use, the small gap between the lens and the object is filled with oil of a high refractive index, preferably identical with that of the lens itself. In these circumstances no refraction takes place until light leaves the objective lens. See MICROSCOPE.

olecranon process, *n.* a projection of the mammalian ulna forming a process for the attachment of the triceps muscle and other muscles used to straighten the arm.

oleic acid, *n.* an unsaturated, oily liquid found in almost all natural fats, and used in cosmetics, lubricating oils and soaps.

olfactory, *adj.* pertaining to the sense of smell.

olfactory lobe, *n.* a projection of the frontal lobe of a cerebral hemisphere that is well-developed in most vertebrates, but reduced in humans, from which arises the olfactory nerve. It is associated with the sense of smell.

olfaction, *n.* the sense of smell, in which there is chemoreception of molecules suspended in the air.

oligo-, *prefix.* denoting a few or little.

Oligocene epoch, *n.* an epoch of the TERTIARY PERIOD occurring between 38 and 26 million years ago, when modern fauna and flora appeared and the rise of birds and mammals continued. Britain was still moving northwards. See GEOLOGICAL TIME.

Oligochaete, *n.* any ANNELID worm of the class Oligochaeta, including the earthworms. They occur mainly in nonacidic soils and in freshwater, and are distinguished from the other major order of CHAETOPODS, the POLYCHAETES, by a lack of PARAPODIA (projections from the body), few CHAETAE and lack of a well-developed head. Fertilization is internal (external in polychaetes).

oligolectic, *adj.* (of bees) collecting pollen from only a few kind of flowers.

oligomer, *n.* a protein consisting of only a small number of identical POLYPEPTIDE subunits.

oligophagous, *adj.* feeding on only a few species of food plants.

oligosaccharide, *n.* a carbohydrate composed of a small number of monosaccharide residues.

oligotrophic, *adj.* **1.** (of water bodies such as lakes) having a poor nutrient supply and relatively little production of organic material. See EUTROPHIC **2.** (of plants) associated with low levels of nutrition.

omasum, *n.* one of four chambers in the RUMINANT STOMACH, lying between the fermentation section and the true stomach.

ombrogenic or **ombrogenous,** *adj.* (of peat bogs) dependent upon rainfall.

ombrotrophic, *adj.* (of plants or plant communities) associated with a rain-fed substrate which is poor in nutrients.

ommatidium, *n.* (*pl.* ommatidia) any of the numerous facets which make up the compound eye of insects and other anthropods. Each ommatidium has its own lens and is composed of a group of retinal cells surrounded by pigment cells. The light-sensitive part of the ommatidium is the RHABDOM, and on its receiving a stimulus a photochemical reaction takes place which results in impulses being sent to the optic nerve.

omnivore, *n.* an organism feeding on both animals and plants. For example, humans have teeth adapted to chewing both types of material. See also CARNIVORE.

onchosphere, *n.* an embryonic form of a tapeworm (Cestoda), which consists of a spherical chitinous shell containing a hexacanth embryo.

oncogene, *n.* a viral gene causing cancer induction (ONCOGENESIS) in the host.

oncogenesis, *n.* the induction of cancer, sometimes by an ONCOGENE.

one gene/one enzyme hypothesis, *n.* a proposal put forward by George BEADLE and Edward Tatum in 1941 to explain the results of their experiments with the fungus *Neurospora crassa*. By developing mutant strains (AUXOTROPHS) that would not grow on a MINIMAL MEDIUM but would grow on a supplemented medium, and by detecting the accumulation of metabolites, they were able to show that the different enzymes catalysing each step in a particular biochemical pathway were each controlled by a different gene. See also CISTRON.

onomatophore, see TYPE.

ontogeny, *n.* the whole of the development of an organism from fertilization to the completion of the life history. See HAEKEL'S LAW.

onycho-, *prefix.* denoting a nail.

Onychophora, *n.* a small subphylum of primitive arthropods containing the single genus *Peripatus*. They are worm-like, each segment having clawed limbs, and are probably descended from ancestors of the arthropods which diverged at an early and intermediate stage between annelids and arthropods.

oo-, *prefix.* denoting an egg.

oocyte, *n.* an early stage in GAMETOGENESIS of female animals, 'primary' oocytes being DIPLOID(1) and 'secondary' oocytes HAPLOID(1) having undergone the first meiotic division. Human females are born with the primary oocytes already formed, which may increase the risk of NONDISJUNCTION of chromosomes. See also DOWN'S SYNDROME.

oogamy, *n.* a state in which male and female GAMETES of an organism are morphologically different, with the male gamete being motile and the female gamete nonmotile and often larger.

oogenesis or **ovogenesis,** *n.* the process of GAMETE production in female DIPLOID(2) animals. See GAMETOGENESIS.

oogonium, *n.* **1.** An early DIPLOID(1) stage in GAMETOGENESIS of female animals giving rise to the OOCYTES. **2.** (in algae and fungi) the female sex organ containing the OOSPHERES.

oosphere, *n.* the large, nonmotile female GAMETE formed in an OOGONIUM.

oospore, *n.* the thick-walled spore found, for example, in some fungi, and arising from the fertilization of the OOSPHERE derived from an OOGONIUM.

oostegites, *n.* the thoracic limb plates found in some crustaceans that form a brood pouch.

oozooid, *n.* a unit of a colonial organism derived from an egg as opposed to a *blastozooid* derived from a bud. Oozooids occur in TUNICATES.

O.P., see OSMOTIC PRESSURE.

opal codon, *n.* one of the three NONSENSE CODONS in the GENETIC CODE, consisting of the mRNA bases UGA.

open population, *n.* a population into which GENE FLOW is freely possible.

open reading frame, *n.* a sequence of NUCLEOTIDE bases between an INITIATION CODON and a NONSENSE CODON.

operator, see OPERON.

operculum, *n.* **1.** the lid of a MOSS capsule. **2.** a hard bony flap covering the gills of fishes. **3.** the plate of exoskeletal material on the foot of a gastropod mollusc with which it closes off the entrance to the shell.

operon, see OPERON MODEL.

operon model or **Jacob–Monod hypothesis,** *n.* a concept of gene regulation proposed by the French biochemists François Jacob and Jacques Monod in the late 1950s. Using strains of *E. coli* mutant at the 'lac' locus, they demonstrated that production of lactate occurs only in the presence of lactose, an example of ENZYME INDUCTION, and proposed a new organization of related genes. The model has the following components: an *operon* consisting of at least one *structural gene* coding for primary enzyme structure, and two regulatory elements, one called the *operator* and the other called the *promoter* that binds with RNA POLYMERASE. The model proposes the presence of a *regulator gene* producing a repressor substance that binds with the operator, so preventing the TRANSCRIPTION of the structural genes.

When a state of *enzyme repression* is in operation, the operator inactivates the structural genes, preventing the formation of MESSENGER RNA from the structural genes because the RNA polymerase enzyme cannot move past the operator site (see Fig. 229). However, the introduction of a suitable inducer substance (e.g. the substrate of the enzyme coded by the structural gene) causes an inducer–repressor complex to be formed that cannot bind to the operator. Thus the operator site is no longer blocked and RNA polymerase can move from promoter to structural gene(s) which then are transcribed, producing a piece of POLYCISTRONIC MRNA, a process called *enzyme induction* (see Fig. 230).

When the inducer level falls (having been metabolized by the enzyme) the operator becomes blocked again by the repressor so that the structural genes are now repressed, an example of a negative FEEDBACK MECHANISM. The operon model has been extended to encompass a system of enzyme repression where structural genes are active normally, only becoming repressed when too much product is present. Furthermore, although the operon model was developed from bacterial studies, the system has been incorporated into general ideas about CELL DIFFERENTIATION in eucaryotes, in which GENE SWITCHING occurs in an orderly manner throughout development, as may be seen in the sequences of CHROMOSOME PUFFS that occur in some insects.

Ophidia, *n.* **1.** a suborder of the Squamata, comprising the snakes. **2.** (in

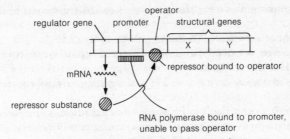

Fig. 229. **Operon model.** Enzyme repression.

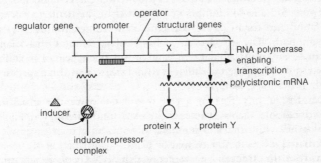

Fig. 230. **Operon model.** Enzyme induction.

some classifications) a separate order of the Reptilia.

Ophiuroidea, *n.* a class of ECHINODERMS, comprising the brittle stars.

ophthalmic, *adj.* of or relating to the eye.

opiate, *n.* a narcotic substance derived from opium.

opposite, *adj.* (of two plant organs, such as leaves) located at the same height on the plant but at opposite sides of the stem.

opsin, *n.* a protein that occurs in rods and cones of the RETINA of the eye, which combines with retinal$_1$ or retinal$_2$ to form visual pigments.

opsonin, *n.* a type of ANTIBODY which binds to ANTIGENS, increasing their susceptibility to phagocytosis by other antibodies.

optic, *adj.* of or relating to the eye or to vision.

optic chiasma, *n.* a point under the hypothalamus of the brain where the two optic nerves meet and cross over, so that stimuli from each eye are interpreted in the OPTIC LOBE of the opposite side of the brain.

optic lobe, *n.* one of a pair of lobes found as swellings on the dorsal side of the midbrain of some lower vertebrates. The lobes integrate sensory information from the eyes and certain auditory reflexes. In mammals, the *corpora quadrigemina* are the equivalent of the optic lobes.

optic vesicle, *n.* an embryological structure formed by the outpushing of the forebrain of vertebrates, which eventually gives rise to the optic cup.

oral, *adj.* of or relating to the mouth or to its uses.

oral groove, *n.* a depression leading to the mouth in some invertebrates.

orbicular, *adj.* rounded.

orbit, *n.* the body cavity or socket in the vertebrate skull containing the eyeball.

order, *n.* a TAXON that occurs at a level of classification between a class (which might contain several orders) and a family (several of which might be found in an order). See CLASSIFICATION.

Ordovician period, *n.* a period of the PALAEOZOIC ERA which occurred between 515 and 445 million years ago. Algae, particularly calcareous, reef-building forms, were common as were graptolites, corals and brachiopods. The first armoured fish evolved in the Ordovician.

organ, *n.* any multicellular structural or functional unit of an animal or plant, often composed of different tissues which perform a specific role, such as the liver, leaf, or eye.

organelle, *n.* any part of a cell that has a particular structural or functional role, such as FLAGELLUM or a MITOCHONDRION. Organelles are analogous with organs in the body of multicellular organisms.

organism, *n.* any living animal or plant.

organization effect, *n.* an interaction that occurs between adjacent loci as a result of some features of chromosome organization.

organizer, *n.* any part of a developing embryo that exerts a morphogenetic stimulus on adjacent parts. See INDUCTION (2).

organizer region, *n.* an area in the animal embryo that appears to control the differentiation of cells in that region. For example, the dorsal lip of the blastopore (see BLASTULA) plays a major part in determining the shape of the early embryo.

organochlorine, *n.* an organic chlorine compound, particularly any of a group of insecticides, e.g. DDT.

organ of Corti, see COCHLEA.

organogenesis, *n.* the period during embryonic development of an animal when the main body organs are formed.

organophosphate, *n.* a phosphate with insecticidal properties. Organophosphates are probably the most commonly used group of insecticides.

organotroph, *n.* an organism that uses organic compounds as electron donors.

orientation, *n.* **1.** The response of an organism in taking up a particular position in relation to a particular stimulus. **2.** see NAVIGATION.

original description, *n.* the statement of the characteristics of an organism in the proposal of a name for a new taxon.

origin of life, *n.* the process by which biomolecules, subcellular structures and living cells have come into existence.

ornithine cycle or **urea cycle,** *n.* a circular biochemical pathway in liver cells with three major steps, in which excess ammonia produced from the breakdown of amino acids and carbon dioxide react together to give UREA which is filtered out by the kidneys and excreted. The reaction is essentially:

$$2NH_3 + CO_2 = CO(NH_2)_2 \text{ (urea)} + H_2O.$$

See Fig. 231.

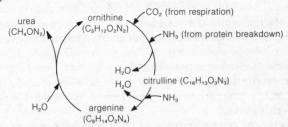

Fig. 231. **Ornithine cycle.** The main steps of the cycle.

Ornithischia, *n.* a large herbivorous order of dinosaurs, most of which were bipedal. They were characterized by a bird-like pelvic girdle, which possessed a postpubic bone stretching on each side below the ischium.

ornitho-, *prefix.* denoting a bird.

Ornithorynchus, *n.* the duck-billed platypus. See MONOTREME.

ortho-, *prefix.* denoting straight.

orthokinesis, *n.* any random movement in response to a stimulus.

orthologous genes, *n.* any genes found in different species that, while nonidentical, can be traced back to a common ancestor.

Orthoptera, *n.* an order of the EXOPTERYGOTA, containing grasshoppers, locusts, stick insects and (in some classifications) cockroaches. Most are good runners or jumpers and some have lost the power of flight. Most forms possess biting and chewing mouthparts, and have hardened forewings which are used in creating sounds by movement against the hind leg (STRIDULATION). See Fig. 232 on page 392.

orthostichy, *n.* **1.** an imaginary vertical line connecting a row of leaves on a stem. **2.** the possession of parallel cartilaginous rays in the fins of fossil shark-like fish, such as *Cladoselache*.

orthotropous, *adj.* (of the angiosperm ovule) having the MICROPYLE directed away from the placenta and erect on the FUNICLE.

Fig. 232. **Orthoptera.** General structure.

os, *n.* **1.** the technical name for a bone. **2.** a mouth or mouth-like part.

osculum, *n.* the large aperture in sponges through which water passes out.

osmic acid, see OSMIUM TETROXIDE.

osmium tetroxide or **osmic acid,** *n.* a substance used in the fixing of cytological preparations that is characterized by the small amount of distortion it causes.

osmometer, *n.* an instrument used for the measurement of OSMOTIC PRESSURE of a solution.

osmoreceptor, *n.* any of the group of structures sited in the HYPOTHALA-MUS that respond to changes in osmotic pressure of the blood by means of neurohypophyseal antidiuretic hormone.

osmoregulation, *n.* the control of OSMOTIC POTENTIAL or WATER POTENTIAL in organisms. Water molecules tend to move from an area of high osmotic or high water potential (low osmotic pressure) to an area of low osmotic potential or low water potential (high osmotic pressure), when separated by a differentially permeable membrane. Where cells are bathed in a solution, water tends to cross the cell membrane in order to equalize the water potential on either side.

In a dry atmosphere or the marine environment, organisms tend to lose water to their surroundings. In a wet atmosphere or in freshwater, organisms may have difficulty in losing water. Various means are thus adopted to maintain a correct water balance. Xerophytic plants (see XEROPHYTE) may reverse the normal stomatal rhythm, develop a waxy cuticle, fold leaves, store water (succulent plants) or survive dry periods as spores or seeds. Excess water may be removed by GUTTATION in some plants, and by CONTRACTILE VACUOLES in some unicellular organisms. In mammals osmoregulation is carried out by the kidney, water passing out of the blood stream into the kidney tubule via the BOWMAN'S CAPSULE and reabsorbed, where necessary, in the tubule itself. Marine fish, reptiles and birds are able to eliminate salt through special excretory cells. Different groups of animals remove NITROGENOUS WASTE in different ways – all associated with osmoregulation. Many marine organisms have blood that is ISOTONIC with sea water, and so do not have an osmoregulatory problem.

osmosis, *n.* the movement of a solvent (usually water in biological systems) through a differentially permeable membrane from a solution with high water concentration and low solute concentration, to one with a low water concentration and high solute concentration.

osmotic potential (O.P. value), *n.* a measure of the tendency of a solution to withdraw water from pure water by OSMOSIS across a differentially permeable membrane. Pure water has an osmotic potential of zero at one atmosphere pressure. The O.P. value of any solution is always negative. *Osmotic pressure* is a measure of the tendency for water to move into a solution by osmosis, and is expressed as a positive value. See Fig. 233.

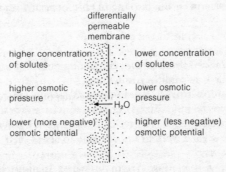

differentially
permeable
membrane

higher concentration of solutes | lower concentration of solutes

higher osmotic pressure | lower osmotic pressure

← H₂O

lower (more negative) osmotic potential | higher (less negative) osmotic potential

Fig. 233. **Osmotic potential.** Osmosis through a differentially permeable membrane.

osmotic pressure, see OSMOTIC POTENTIAL.

ossicle, *n.* any of the bones in the middle ear connecting the eardrum to the oval window. In mammals there are three ossicles (MALLEUS, INCUS and STAPES). In reptiles, birds, and many amphibians, only a single bone (the COLUMELLA AURIS) is present.

Ostariophysi, *n.* a superorder of freshwater TELEOSTS, including the carp and the minnow. The group has several primitive features, including the WEBERIAN OSSICLE.

Osteichthyes, see BONY FISH.

osteo-, *prefix.* denoting a bone.

osteoblast, *n.* a cell that produces the calcified intercellular material of bone. See also PERIOSTEUM.

osteoclast, *n.* **1.** a multinucleate ameoboid cell that breaks down bone during growth and remodelling. **2.** also called **chondrioclast**. a cell of the same kind that breaks down cartilage in the transformation to bone.

Osteolepis, *n.* an extinct genus of Crossopterygian fish that possessed casmoid scales and was closely related to the ancestors of land vertebrates.

Osteostraci, *n.* an order of jawless fish-like vertebrates, possessing a large dorsal head shield, found in the SILURIAN PERIOD and DEVONIAN PERIOD.

ostiole, *n.* a pore through which spores are discharged from the fruiting bodies of some fungi, or from which gametes are passed out from the conceptacles of brown algae.

ostium, *n.* any of numerous openings present in sponges through which water is drawn in.

ostraco-, *prefix.* denoting an oyster shell.

Ostracodermi, *n.* a class of jawless fossil fish which are usually classified, together with the CYCLOSTOMES, in the Agnatha. They existed from the ORDOVICIAN PERIOD through to the DEVONIAN PERIOD and were heavily armoured.

otic, *adj.* of or relating to the ear.

otoconium, see OTOLITH.

otocyst, *n.* an invertebrate organ, containing fluid and OTOLITHS, that is believed to have an auditory function.

otolith or **otoconium,** *n.* a granule of calcareous material, several of which occur in the inner ear of vertebrates, where they are attached to processes associated with sensitive cells, and register gravity. By means of such sense organs, vertebrates are able to assess their position with respect to gravity.

outbreeding, *n.* a mating system in which matings between close relatives do not usually occur. Compare INBREEDING.

oval window, *n.* a membranous wall at the junction of the middle ear and inner ear which connects the stapes with the vestibular canal of the COCHLEA.

ovarian follicle, *n.* the group of cells that surrounds, envelops and probably nourishes the developing oocyte in most animals. See GRAAFIAN FOLLICLE. In vertebrates the ovarian follicle produces OESTROGEN.

ovary, *n.* the organ that produces the female gametes by a process called GAMETOGENESIS. In mammals, it also gives rise to the CORPUS LUTEUM from the GRAAFFIAN FOLLICLE after ovulation, and in vertebrates generally produces the female hormones oestrogen and progesterone. In plants, the ovary occurs at the base of the CARPEL and contains one or more OVULES. See GYNOECIUM for inferior and superior ovaries.

ovate, *adj*. (of a leaf) shaped like the longitudinal section of an egg, with the broader end at the base.

overall similarity, *n*. (in numerical taxonomy) a value of similarity calculated by considering several characters collectively.

overdominance, *n*. a phenotypic condition in which the HETEROZYGOTE expresses a stronger manifestation of the trait than either HOMOZYGOTE. Overdominance produces heterozygous advantage that can maintain a GENETIC POLYMORPHISM in a population.

overfishing, *n*. the removal of fish from a population beyond the point where the population is able to maintain its numbers and goes into decline.

overshoot, *n*. the stage of an ACTION POTENTIAL in which the voltage rises from zero to the positive peak.

overwintering, *n*. the process of passing the winter. For example, some insects overwinter in a pupal stage, some migratory birds overwinter outside the normal breeding area.

oviduct, *n*. a tube that carries the egg to the exterior or, in the case of mammals, to the lower part of the tube, the UTERUS, where it becomes implanted in the wall lining (ENDOMETRIUM) if FERTILIZATION has occurred. Usually oviducts are paired, but only one is present in birds. In higher mammals it is divided into an upper FALLOPIAN TUBE, which is ciliated, and into which the egg passes from the COELOM, a middle uterus, where the embryo develops, and a vagina leading to the exterior. See Fig. 234.

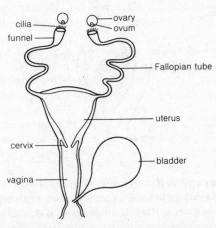

Fig. 234. **Oviduct.** The mammalian oviduct.

oviparous, *adj*. (of an animal) laying eggs in which the embryo develops after egg laying, as in birds and most reptiles. Compare OVOVIVIPAROUS, VIVIPAROUS(1).

ovipositor, *n.* an organ of female insects, usually present at the tip of the abdomen, through which the egg is laid. The ovipositor is sometimes developed to enable the piercing of tissue, particularly where eggs are laid inside other animals and plants.

ovogenesis, see OOGENESIS.

ovoid, *adj.* egg-shaped.

ovotestis, *n.* the organ of some HERMAPHRODITES(2), that functions both as a testis and as an ovary, as in snails.

ovoviviparous, *adj.* (of certain fish, reptiles and many insects) having the embryo developing within the mother, but from which it rarely derives nutrients. The mother and embryo are separated by egg membranes and there is no PLACENTA. Compare OVIPAROUS, VIVIPAROUS(1).

ovulation, *n.* the bursting of the OVARIAN FOLLICLE (see also GRAAFIAN FOLLICLE) on the surface of the ovary with the release of an egg which then normally passes into the OVIDUCT.

ovule, *n.* a structure found in higher plants that contains an egg cell and develops into a seed after FERTILIZATION. In ANGIOSPERMS, the ovule develops within the ovary of a carpel, and consists of an inner EMBRYO SAC containing the egg cell surrounded by the NUCELLUS, which in turn is enclosed within two INTEGUMENTS. The ovule is connected to the ovary wall by the FUNICLE. See Fig. 235.

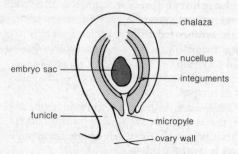

Fig. 235. **Ovule.** General structure.

ovum or **egg** or **egg cell,** *n.* (*pl. ova*) a functional egg cell of an animal, produced by GAMETOGENESIS. Ova are frequently packed with nutritive yolk granules and are usually immobile (see OOGAMY). The human egg is about 0.14 mm in diameter, which is some 50,000 times larger than the human sperm.

oxaloacetic acid, *n.* a dicarboxylic acid that condenses with acetlycoenzyme A, forming citric acid and CoA, thus initiating the KREBS CYCLE.

oxidase, *n.* an enzyme that catalyses a reaction in which oxygen is added to a substance (X) with hydrogen removed and water formed. Thus:

$$O_2 + X - H_2 \xrightarrow{\text{oxidase}} H_2O + X - O$$

Most oxidases are proteins with metallic groups attached. For example, cytochrome oxidase contains iron and oxidizes cytochrome *a* in the ELECTRON TRANSPORT SYSTEM of aerobic respiration:

$$O_2 + \text{reduced cyto. } a \xrightarrow[\text{oxidase}]{\text{cytochrome}} H_2O + \text{oxidized cyto. } a.$$

oxidation, *n.* **1.** the addition of oxygen to a substance to increase the proportion of oxygen in its molecule. Oxidation can be achieved without oxygen by the removal of hydrogen (dehydrogenation). **2.** any reaction involving loss of electrons from an atom. For example,

$$\text{Fe(II)} \rightarrow \text{Fe (III)}.$$

oxidation-reduction reaction, *n.* a reversible chemical process in which electrons are tranferred from one molecule to another.

oxidative deammination, *n.* a deammination that also involves an oxidation, for example, an alpha amino acid to an alpha keto acid.

oxidative decarboxylation, *n.* a decarboxylation (removal of CO_2 from an organic carboxyl group) that also involves oxidation.

oxidative phosphorylation, *n.* a process that takes place in the ELECTRON TRANSPORT SYSTEM of aerobic respiration, in which ATP molecules are synthesized from ADP and inorganic phosphate. The process is the major means by which aerobic organisms obtain their energy from foodstuffs.

oxidoreductase, *n.* one of a group of enzymes that catalyses OXIDATION-REDUCTION REACTIONS.

oxygen, *n.* a colourless, tasteless gas forming about 21% of the earth's atmosphere, that is capable of combining with all other elements except the inert gases. Oxygen is particularly important in physical combustion processes and in AEROBIC RESPIRATION.

oxygen cycle, *n.* the PATHWAY followed by oxygen through the metabolism of organisms which leads back to the regeneration of any given organism.

oxygen debt, *n.* a state that arises in very active muscles when insufficient oxygen is supplied by the lungs, causing the muscle tissue to respire anaerobically with the production of LACTIC ACID. When muscular activities slow down, the rate and depth of breathing remains at a high level until the lactic acid has been oxidized. Oxygen debt can be measured as the difference between the amount of oxygen required after strong muscular activity and the amount required in a resting state. In trained athletes, the debt can be as high as 18 litres of oxygen.

oxygen dissociation curve, *n.* a curve derived from plotting the percentage saturation of blood with oxygen, against the oxygen tension (the *oxygen exchange*). See Fig. 236.

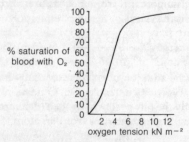

Fig. 236. **Oxygen dissociation curve.** The characteristic S-shaped curve of oxygen dissociation.

The curve is S-shaped and shows that HAEMOGLOBIN has a high affinity for oxygen. Blood can become saturated at relatively low oxygen tensions, but a small drop in oxygen tension brings about a big fall in the saturation of the blood. If tissues use up oxygen, then haemoglobin responds by giving it up.

Where CO_2 is present, haemoglobin has to be under higher oxygen tension in order to become fully saturated. However, under these circumstances it releases O_2 at higher oxygen tensions. CO_2 affects the efficiency of oxygen uptake by haemoglobin but increases its efficiency in releasing it. The oxygen–carrying capacity of blood is also affected by pH (see BOHR EFFECT).

oxygen exchange, see OXYGEN DISSOCIATION CURVE.

oxyhaemoglobin, *n.* HAEMOGLOBIN that is oxygenated.

oxylophyte, see CALCIFUGE.

oxyntic cell or **parietal cell,** *n.* any of the glandular cells lying in pits within the wall of the vertebrate stomach. Their function is to secrete hydrochloric acid, giving the gastric fluids a pH of about 2.0.

oxytocin, *n.* a hormone secreted by the posterior lobe of the PITUITARY GLAND, causing contraction of the uterine muscle. The hormone is important in parturition.

P

P, see PARENTAL GENERATION.

pacemaker, *n.* the sinoatrial node of the right atrium. See HEART.

pachy-, *prefix.* denoting thick.

pachyderm, *n.* any large, thick-skinned, hoofed mammal, such as an elephant, hippopotamus, or rhinoceros.

pachytene, *n.* the stage in prophase I of MEIOSIS where the paired HOMOLOGOUS CHROMOSOMES each split into two CHROMATIDS, the whole structure being called a TETRAD.

pacinian corpuscle, *n.* a heavy-pressure RECEPTOR in mammalian skin.

paedo-, *prefix.* denoting children, immaturity.

paedogenesis, *n.* sexual reproduction in the larval or juvenile stage which may persist either temporarily or permanently, as in an axolotl. Humans resemble the juvenile stages of apes in several features, such as lack of hair, and paedogenesis may well have played an important part in the evolution of *Homo sapiens*. Some structures are said to be *neotonic* if the juvenile appearance persists in adults, as in the downy feathers of an ostrich.

page precedence, *n.* the occurence of a scientific name on an earlier page of the same publication in which a synonym or homonym occurs.

pain, *n.* a conscious sensation produced in the brain and stimulated by pain receptors in, for example, the skin. Pain has a protective function and often produces a reflex action (see REFLEX ARC) in response.

pair bond, *n.* a relationship established between the male and female of higher vertebrates for breeding purposes. In some cases, such as swans, it may last for life.

Palaearctic region, *n.* that part of the Holarctic region comprising the landmass of the Eurasian continent (Europe and Asia) from its northern border to the Sahara and Himalayas. see BIOGEOGRAPHICAL REGION.

palaeo-, *prefix.* denoting old or prehistoric.

palaeobotany, *n.* the study of fossils plants, particularly fossils of pollen grain, which are used in reconstructing past environments.

Palaeocene epoch, *n.* the first epoch in the TERTIARY PERIOD lasting from 65 to 54 million years ago. see GEOLOGICAL TIME. This was the time of the explosive evolution of placental mammals, marine reptiles such as ichthyosaurs, mesosaurs and plesiosaurs and the last of the large land dinosaurs having become extinct at the end of the CRETACEOUS PERIOD.

palaeoecology, *n.* the study of the ecology of past environments through fossil organisms.

Palaeognathae, *n.* the more primitive of the two major groups of living birds, containing the RATITES and the tinamous. In contrast with the NEOGNATHAE, the jaw structure contains a large prevomer bone that contacts the pterygoid bones.

palaeolimnology, *n.* the study of past environments, particularly lake ecosystems, from evidence found in lake muds.

Palaeolithic, *n.* the period of the emergence of primitive man, from *c.* 600,000 to 14,000 years ago. Stone tools were used, and existence was solely by hunting and gathering, with no cultivation.

Palaeoniscus, *n.* a genus of extinct bony fish of the PALAEOZOIC ERA possessing bony head plates and GANOID SCALES. Palaeoniscus probably arose from the same stock as the CHOANICHTHYES from which land animals eventually evolved. Present-day relatives include POLYPTERUS.

palaeontology, *n.* the study of fossil animals and plants. Compare NEONTOLOGY.

Palaeozoic era, *n.* an era of 350 million years duration, from the beginning of the CAMBRIAN PERIOD to the end of the PERMIAN PERIOD, lasting from 590 to 240 million years ago.

palate, *n.* the roof of the mouth in vertebrates formed anteriorly by a bony projection of the upper jaw and posteriorly by the fold of connective tissue (*soft palate*). Mammals possess a *false palate* which has been formed below the original palate, and results in the opening of the nasal cavity at the back of the mouth (in the throat), allowing chewing and breathing at the same time.

palea, *n.* a GLUME-like bract found on the axis of individual flowers in grasses.

palisade mesophyll, see MESOPHYLL.

palindrome, *n.* a region of a DNA POLYNUCLEOTIDE CHAIN where a sequence of bases is followed by a base series so that COMPLEMENTARY BASE PAIRING can occur when the chain is folded back. See Fig. 237.

base sequence: TGAC·GTCA   folded: GTCA
                                       CAGT
                                complementary pairing

Fig. 237. **Palindrome.** Complementary base pairing.

palingenesis, see HAEKEL'S LAW OF RECAPITULATION.

pallium, *n.* **1.** the tissue next to the shell of molluscs. **2.** the roof of the vertebrate CEREBRUM.

palmate, *adj.* (of a leaf) having at least three leaflets joined to the PETIOLE at the same place. See Fig. 238.

Fig. 238. **Palmate.** A palmate leaf.

palp, *n.* a type of appendage associated with the head of invertebrate animals, usually in the mouth region, which has a tactile function. In some crustaceans, palps perform a locomotory function and in others are associated with feeding, as they are in bivalve molluscs where they produce feeding currents. In insects, palps are found on the first and second maxillae, where they are thought to be associated with olfaction.

pancreas, *n.* a gland situated in the mesentary of the DUODENUM of jawed vertebrates that has both an exocrine and an endocrine function. The pancreatic duct carries digestive enzymes (see PANCREATIC JUICES) from the gland into the duodenum, secretion being stimulated by (a) the vagas nerve, (b) the hormone SECRETIN, (c) PANCREOZYMIN.

Groups of cells known as ISLETS OF LANGERHANS secrete two hormones into the blood system: large 'alpha' cells secrete GLUCAGON, while smaller 'beta' cells secrete INSULIN.

pancreatic amylase, see AMYLASE.

pancreatic juice, *n.* the clear alkaline secretion of the PANCREAS that is released into the duodenum and contains the following digestive enzymes:

(a) pancreatic AMYLASE which breaks down STARCH to MALTOSE.

(b) pancreatic LIPASE which breaks down fat into fatty acids and glycerol.

(c) NUCLEASES which break down NUCLEIC ACIDS into nucleotides.

(d) PEPTIDASES which break down POLYPEPTIDE CHAINS to free amino acids.

(e) a group of closely related protein–splitting enzymes such as TRYPSIN in the form of inactive trypsinogen.

pancreatin, *n.* an extract of the pancreas that contains pancreatic enzymes.

pancreozymin, see CHOLECYSTOKININ–PANCREOZYMIN.

pandemic, *adj.* occurring over a wide geographical area, as with a disease such as malaria.

Paneth cells, *n.* the cells that occur at the bases of the CRYPTS OF

LIEBERKUHN. They persist for only about two weeks and then disintegrate. It is thought that they remove ions of heavy metal and secrete AMINO ACIDS and LYSOZYME – an antibacterial enzyme which controls the number of bacteria in the gut. It is also possible that they secrete enzymes such as peptidases and lipases.

Pangaea, see CONTINENTAL DRIFT.

panicle, *n.* **1.** a branched RACEME. **2.** any complex branched inflorescence.

panmixis, see RANDOM MATING.

Pantopoda, a class of ARACHNIDS, including the sea spiders.

pantothenic acid or **vitamin B$_5$,** *n.* a water-soluble organic acid ($C_9H_{17}O_5N$) that is present in all animal tissues, especially the liver and kidney. Pantothenic acid forms part of coenzyme A which, when bonded to acetic acid, forms ACETYLCOENZYME A. The vitamin is present in almost all foods, especially fresh vegetables and meat, eggs and yeast. A deficiency causes nervous disorders with poor motor coordination.

papilla, *n.* (*pl.* papillae) a projection from the surface of a structure such as tongue papillae, which carry the taste buds.

papilloma, *n.* a benign wart.

pappus, *n.* a circle of hairs formed from a modified CALYX found on the seeds of plants of the family Compositae. The pappus assists in the dispersal of the seeds by wind, acting as a kind of parachute, for example dandelion and thistle seeds

parabiosis, *n.* the connecting together of two animals in an experiment so as to allow their body fluids to mix. The technique is used, for example, in the study of the role of insect hormones.

paracentric inversion, *n.* a chromosomal mutation involving INVERSION of a segment of one arm of a chromosome, not involving the CENTROMERE.

parallelism, see CONVERGENCE.

paralogous genes, *n.* genes with a HOMOLOGOUS structure located at different sites on the chromosomes of the same individual, that are probably derived by a process of gene duplication.

Paramecium, *n.* a genus of freshwater PROTOZOAN, having an oval body covered with cilia and a ventral ciliated groove for feeding. See Fig. 239.

paramylum, *n.* a starch-like substance that occurs as a food reserve in flagellate protozoans and algae such as EUGLEMA.

parapatry, *n.* any nonoverlapping geographical contact between separate populations or species, without interbreeding.

parapodium, *n.* (*pl.* parapodia) any of the paired numerous segmentally arranged projections of the body of polychaete worms, containing musculature and often bearing chaetae. It is locomotory in function.

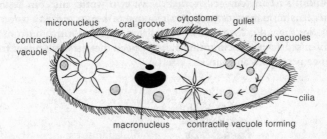

Fig. 239. *Paramecium.* General structure.

parasexual reproduction, *n.* a form of asexual reproduction that involves somatic cell fusion and random loss of individual chromosomes.

parasite, *n.* an organism that lives in association with, and at the expense of, another organism, the host, from which it obtains organic nutrition. Those that live on the outside of the host, such as ticks, are called *ectoparasites*, while others such as tapeworms that live inside the host are called *endoparasites*. Parasites can be FACULATIVE or OBLIGATE and have a range of effects, from inflicting minimum harm to the host which continues to live and reproduce normally (the best-adapted parasites, e.g. TAPEWORMS), to causing the death of the host (e.g. Malaria parasite). COEVOLUTION may occur between host and parasite. See also BIOTROPHIC, NECROTROPHIC.

parasitoid, *n.* any of the alternately parasitic and free-living wasps and flies, such as the ichneumon fly, whose larvae parasitize and often kill members of the host species.

parasympathetic system, see AUTONOMIC NERVOUS SYSTEM.

parathyroid, *n.* an endocrine gland of higher vertebrates that is situated in or near the thyroid gland and controls the calcium level of the blood. When too little $[Ca^+]$ is present in the blood, parathormone is secreted which (a) reduces the OSTEOBLAST activity and thus the amount of bone-matrix formation, (b) increases $[Ca^+]$ uptake from the gut into the blood stream, and (c) acts on the kidney tubules, decreasing calcium and phosphate excretion. Undersecretion of parathormone results in a fall of blood calcium and causes TETANY.

paratype, *n.* any specimen cited by the author of a scientific name, other than the HOLOTYPE.

parazoan, *n.* any multicellular invertebrate of the group Parazoa, comprising the sponges. Compare METAZOAN.

parenchyma, *n.* **1.** a tissue composed of parenchyma cells which are thin-walled 'general purpose' plant cells that often have a packing function. Parenchyma cells remain alive at maturity and can become meristematic, as in INTERFASCICULAR CAMBIUM (see SECONDARY THICKENING). See Fig. 240. **2.** the loose, vacuolated cells that form much of the body tissue of platyhelminths. **3.** any specific organ cells apart from connective tissues and blood vessels.

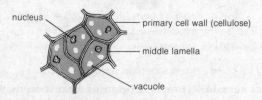

nucleus

primary cell wall (cellulose)

middle lamella

vacuole

Fig. 240. **Parenchyma.** Transverse section of cells.

parental gamete, *n.* a gamete that has a genotype identical to one of the HOMOLOGOUS CHROMOSOMES in the original diploid cell. See GENETIC LINKAGE.

parental generation(P), *n.* the members of a cross between PURE BREEDING LINES, giving rise to offspring called the F_1.

parietal, *adj.* **1.** (of plant organs) joined to a wall, as in parietal ovules attached to the wall of the ovary. **2.** of or relating to the bones of the top of the skull, which have a membranous structure and lie behind the frontal bones. **3.** (of coelomic lining) covering the body wall as distinct from visceral organs (see also OXYNTIC CELL).

pars intercerebralis, *n.* the dorsal part of the insect brain containing neurosecretory cells.

parthenocarpy, *n.* the development of fruit without seeds due to lack of pollination, fertilization or embryo development.

parthenogenesis, *n.* the development of an idividual from an egg without fertilization by a sperm. The process occurs mainly in lower invertebrates, particularly insects. The egg cell can be HAPLOID(1) to produce, for example, male honeybees (drones) or DIPLOID(1) as produced in wingless female aphids which, during the summer months, produce diploid eggs by MITOSIS that develop into female adults, only forming haploid gametes by MEIOSIS in the autumn prior to normal sexual reproduction.

partial dominance, *n.* a form of INCOMPLETE DOMINANCE in which the heterozygote resembles one homozygote more than the other.

partial inflorescence, *n.* a part of a branched inflorescence.

partial pressure, *n.* the total pressure of the mixture of gases within which a gas occurs, multiplied by the percentage of the total volume the gas occupies. Thus if the normal total pressure of the atmospheric gases is

760 mm Hg and there is 21% oxygen in this mixture, the partial pressure of O_2 is: $760 \times 0.21 = 160$ mm Hg.

parturition, *n.* the act of giving birth to an offspring.

passage cell, *n.* any endodermal plant cell which remains unthickened and allows the passage of materials between the cortex and the vascular bundle.

passerine, *n.* any member of the avian order Passeriformes (singing or perching birds), which includes some half of the known species of birds.

passive immunity, see ANTIBODY.

passive transport, see DIFFUSION.

Pasteur, Louis (1822–95) French bacteriologist who was effectively the founder of microbiology as a science. He put forward the theory that disease was caused by microorganisms and established that SPONTANEOUS GENERATION does not occur. He established that inoculation with attenuated forms (see ATTENUATION) of microorganisms provided immunization against virulent forms, and showed that rabies was caused by microscopic agents which could not be seen, so leading to the discovery of viruses. He introduced heat treatment to destroy microorganisms in perishable products such as milk (see PASTEURIZATION).

pasteurization, *n.* a method devised by Louis PASTEUR of partially sterilizing certain foods such as milk (by heating to 62°C for 30 mins) before distribution. Heating destroys many harmful bacteria, including those responsible for tuberculosis.

patagium, *n.* **1.** a membrane of skin stretched on the wing of a bird or bat, or between the fore and hindlimbs of a flying squirrel to form the parachute. **2.** (in some insects) a projection on each side of the prothorax.

Patau syndrome, *n.* a rare human condition characterized by gross developmental malformations and caused by TRISOMY of chromosome 13. The average survival of affected individuals is 6 months.

patella, *n.* **1.** the kneecap bone which is present in most mammals, and in some birds and reptiles, protecting the front of the joint from injury. **2.** the generic name of the LIMPET, *Patella.*

pathogen or **pathogene,** *n.* any organism that causes disease, such as viruses or bacteria.

pathogenic, *adj.* acting as a PATHOGEN.

pathology, *n.* the study of the structural and functional changes caused by disease.

pathotoxin, *n.* a toxin produced by a PATHOGEN, capable of causing symptoms of disease. See BOTULISM.

pathway, *n.* any defined route followed by a series of reactions leading to a specific product, e.g. a biochemical pathway.

patroclinous inheritance, *n.* a form of inheritance in which all offspring have the PHENOTYPE of the father. Compare MATROCLINOUS INHERITANCE.

Pauling, Linus Carl (1901–) American biochemist and Nobel laureate who determined the crystal structure of several molecules by means of X-ray diffraction analysis and, with his coworkers, discovered the structure of abnormal HAEMOGLOBIN, responsible for SICKLE–CELL ANAEMIA, using the technique of ELECTROPHORESIS.

Pavlovian response, see CONDITIONAL REFLEX.

peat, *n.* an accumulation of dead plant material formed in wet conditions in bogs or fens in the absence of oxygen so that decomposition is incomplete. It is usually acidic.

pecking order, *n.* the gradation of behavioural dominance between individuals in active groups of mammals and birds. The leader is dominant, and thus at the top of the 'pecking order' getting the best food and mates, the least dominant animal being at the bottom of the order. Birds often show bald patches where they have been pecked by dominant individuals.

Pecora, *n.* an infra–order of the ARTERIODACTYLA, including ruminants such as cattle, sheep, goats, deer and giraffe, possessing a placenta having villi in small tufts.

pecten, *n.* any comb-like structure or organ. Examples include the oxygenating structure in the eye of a bird, and the organ used by some insects for STRIDULATION.

pectin, *n.* a complex POLYSACCHARIDE often found as calcium pectate in plant cells where it is a component of the MIDDLE LAMELLA of the cell wall. When heated, pectin forms a gel which can 'set', a feature used in the making of jams.

pectinate, *adj.* (of plant structures) with comb-like lobes.

pectoral fin, *n.* one of a pair of propellant/steering organs attached to the PECTORAL GIRDLE of fish.

pectoral girdle or **shoulder girdle,** *n.* the skeletal support for the anterior limbs of vertebrates that transmits the power from the limbs to the body and also serves to protect and support the organs in the thorax. Normally it is a U–shaped structure and is attached to the vertebrae by muscles. See Fig. 241.

pedal, *adj.* of or relating to the foot, particularly those of molluscs.

pedate, *adj.* (of leaves) divided in a PALMATE fashion with further divisions.

pedicel, *n.* the stalk of a flower.

pedigree analysis, *n.* a technique in which patterns of inheritance are traced through several generations.

pedipalp, *n.* either member of the second pair of head appendages of

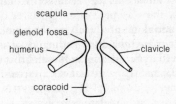

Fig. 241. **Pectoral girdle.** Basic plan.

arachnids, specialized for different functions in different forms. For example, seizing prey in scorpions, locomotion in king crabs, fertilization in male spiders (where the tip becomes a specialized container for sperm transfer), squeezing and chewing food, or for sensory purposes.

peduncle, *n.* the stalk of an inflorescence or a flower.

pelagic, *adj.* **1.** (of organisms) swimming actively in the mass of sea water (NEKTON) rather than living at the bottom, or drifting with the currents in the surface waters (PLANKTON). **2.** (of birds) spending most of their lives at sea.

pelecypod, see BIVALVE.

pellicle, *n.* a thin membrane.

pellagra, see NICOTINIC ACID.

pelleting, *n.* the coating of a seed with material that often includes pesticides, producing a uniform size of seed for sowing.

peltate, *adj.* (of a plant structure) flattened and with the stalk attached to the centre of the lower surface, not the edge.

pelvic fin, *n.* one of a pair of propellant/steering organs attached to the PELVIC GIRDLE of fish.

pelvic girdle or **hip girdle,** *n.* the skeletal support for the posterior limbs of vertebrates that transmits the power in locomotion from the hind limbs to the body. See Fig. 242.

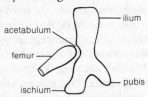

Fig. 242. **Pelvic girdle.** Basic plan.

penetrance, *n.* the percentage of individuals with a particular GENOTYPE that display the genotype in the PHENOTYPE. For example, a dominant gene for baldness is 100% penetrant in males and 0% penetrant in most females, because the gene requires high levels of the male hormone for expression, an example of SEX LIMITATION. Once a gene shows penetrance it may show a range of EXPRESSIVITY of phenotype.

penicillin, *n.* an antibiotic produced by the mould *Penicillium* that is toxic to a number of bacteria, both pathogenic and nonpathogenic. In 1928 it was observed by Sir Alexander FLEMING that the mould inhibited growth of bacteria, and that a substance extracted from it still had this antibiotic property.

penis or **phallus,** *n.* the copulatory organ of a male animal which conveys the sperm to the genital tract of the female.

Pennatulacea, *n.* a COELENTERATE family, including the sea pens.

penta, *prefix.* denoting five.

pentadactyl limb, *n.* a limb bearing five digits (fingers or toes) which is present in amphibia, reptiles, birds and mammals. The limb is thought to have been evolved largely for life on land and consists of three main parts: (a) the humerus (anterior limb) or femur (posterior limb); (b) the radius and ulna (anterior), or tibia and fibula (posterior); (c) a hand or foot made up of carpals (anterior) or tarsals (posterior), metacarpals or metatarsals and phalanges. In many forms the limb is much modified for activities such as flying, running, or swimming.

pentaploid, *n.* a POLYPLOID individual having five sets of chromosomes.

pentosan, *n.* a polysaccharide of pentoses.

pentose, *n.* a monosaccharide with five carbon atoms.

pentose phosphate pathway (PPP) or **pentose phosphate shunt** or **hexose monophosphate shunt** or **phosphogluronate pathway,** *n.* a PATHWAY for the metabolism of glucose. PPP is particularly important in animal cells where it functions as an alternative to GLYCOLYSIS and the KREBS CYCLE, although both mechanisms occur together.

In the PPP, glucose (a hexose sugar) in the form of glucose phosphate molecules, is oxidized by the removal of hydrogen and decarboxylated, producing pentose sugars, carbon dioxide and hydrogen atoms, which are transferred to the coenzyme NADP to form $NADPH_2$. In a second stage the pentose sugars are rearranged to produce hexose again, although in a smaller quantity than at the start. Unlike NADH, NADPH is not involved in producing ATP via an ELECTRON TRANSPORT SYSTEM, but transfers hydrogen atoms and electrons to sites where molecules are being assembled.

The $NADPH_2$ molecules are particularly active in cells carrying out fat metabolism, such as adipose tissue, liver, adrenal cortex, mammary gland tissue.

pentose phosphate shunt, see PENTOSE PHOSPHATE PATHWAY.

pentose sugar, *n.* a MONOSACCHARIDE with a five-carbon ring, particularly sugars such as ribose and deoxyribose found in NUCLEOTIDES.

peppered moth, *n.* the moth *Biston betularia*, which has been extensively studied in many areas of Britain. Its coloration is of two main types: peppered (a mixture of white and grey) and melanic (dark brown), the relative proportion of the two colour types in an area being related to the amount of atmospheric pollution. The colour forms are an example of a GENETIC POLYMORPHISM controlled by a single gene with two ALLELES, the allele for melanism being dominant. See INDUSTRIAL MELANISM.

pepsin, *n.* an enzyme secreted in the inactive form PEPSINOGEN by chief or peptic cells in the gastric pits of the stomach of vertebrates, and which breaks down proteins in acid solution into short polypeptide chains which are subsequently broken down further by PEPTIDASES.

pepsinogen, *n.* a precursor of PEPSIN in the vertebrate stomach which, in the presence of hydrochloric acid (also secreted by the OXYNTIC CELLS in the stomach wall), gives rise to more active pepsin, which itself activates pepsinogen. The reaction is thus autocatalytic.

peptidase, *n.* an enzyme present in PANCREATIC JUICE that releases free AMINO ACIDS from POLYPEPTIDE CHAINS. EXOPEPTIDASES split off terminal amino acids, whereas ENDOPEPTIDASES split links with the chain.

peptide, *n.* any of a group of compounds consisting of two or more amino acids linked by chemical bonding. See PEPTIDE BOND, DIPEPTIDE.

peptide bond, *n.* a chemical compound joining two AMINO ACIDS together, produced by a condensation reaction (see DIPEPTIDE). See Fig. 243.

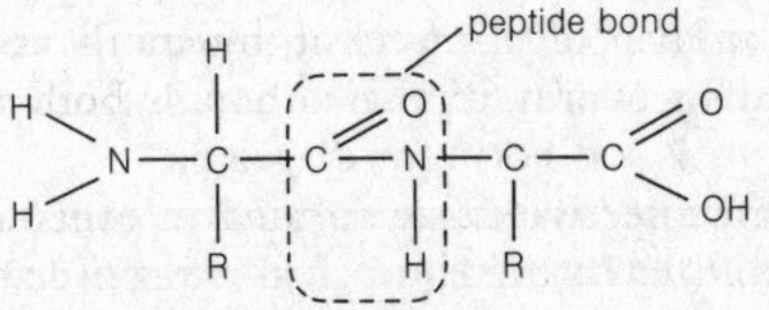

Fig. 243. **Peptide bond.** Molecular structure. R=special group unique to each amino acid.

peptidyl site, *n.* a TRANSFER RNA binding site on the RIBOSOME at which a PEPTIDE BOND occurs.

peptone, *n.* a soluble product resulting from protein hydrolysis. See PEPSIN, PEPTIDASE.

per-, *prefix.* denoting thorough, very, extremely.

perennation, *n.* the survival of an organism from one season to the next, usually with a period of reduced activity between seasons. See PERENNIAL.

perennial, *n.* a plant that lives from year to year. Plants with woody stems are generally perennials (e.g. trees, shrubs and woody vines), but so too are many herbaceous plants (e.g. dandelion, daisy and plantain), in which the aerial parts die back during the winter leaving perennating organs (e.g. tubers, corms and bulbs) underground.

perfusion, *n.* the passage of a liquid through an organ or tissue.

peri-, *prefix.* denoting around or about.

perianth, *n.* the organs of a flower outside the sex organs, consisting of the CALYX and COROLLA; the floral envelope.

perianth segment, *n.* the individual leaves of the PERIANTH.

periblast, *n.* the tissue surrounding the BLASTODERM in meroblastic eggs.

pericardial cavity, *n.* **1.** the part of the coelom surrounding the heart of vertebrates. **2.** the part of the HAEMOCOEL surrounding the heart in arthropods.

pericardial membrane, *n.* the membrane surrounding the heart.

pericarditis, *n.* an inflammation of the PERICARDIUM.

pericardium, *n.* a sac-like envelope surrounding the vertebrate heart and the pericardial cavity in arthropods and molluscs.

pericarp, *n.* the wall of a fruit, developing from the ovary wall after fertilization has occurred. The pericarp is made up of three distinct layers, an outer exocarp, a central mesocarp and an inner endocarp. In dry fruit the pericarp is firm and often rigid (e.g. poppy capsule, strawberry ACHENES); in succulent fruits the pericarp is swollen, often with a high sugar content in the mesocarp (e.g. grape, peach).

periclinal, *adj.* (of planes of cell division in plants) parallel to the surface of the plant.

pericycle, *n.* the layer of plant cells between the ENDODERMIS and the PHLOEM, consisting mainly of PARENCHYMA, from which lateral roots originate. See STEM and ROOT for diagrams.

periderm, *n.* a protective tissue formed in roots and stems that has undergone SECONDARY THICKENING, consisting of an outer cork zone, an underlying phellogen (cork cambium) and with a phelloderm (secondary cortex) beneath that.

perigynous, *adj.* (of flowers) having a flat or concave structure on which the sepals, petals and stamens are borne.

perilymph, *n.* the fluid separating the bony labyrinth from the membranous labyrinth of the inner ear.

perineustic, *adj.* (of insects) having a respiratory system in which spiracles occur laterally on all abdominal segments.

periosteum, *n.* the CONNECTIVE TISSUE which surrounds the bone of vertebrates and to which are connected the MUSCLES and TENDONS. OSTEOBLASTS are present in the periosteum, and are important in bone repair after breakage.

periostracum, *n.* the outermost of the three layers of the shell of molluscs or brachiopods, composed of CONCHIOLIN. On the inner side lie the prismatic layer and the nacreous layer.

peripheral, *adj.* of or relating to the outside or extreme edge of a structure. For example, the peripheral nervous system is the NERVOUS SYSTEM as a whole with the exception of the CENTRAL NERVOUS SYSTEM.

peripheral isolate, *n.* an isolated population at or beyond the normal range of the species.

perisarc, *n.* the horny covering of hydrozoan coelenterates of the orders Gymnoblastea, Calyptoblastea and Hydrocorallinae. In the latter, the perisarc is massive and impregnated with calcium.

perisperm, *n.* a nutritive tissue found in some seeds that is derived from the NUCELLUS of the ovules.

perispore, *n.* a spore membrane.

Perissodactyla, *n.* the odd-toed ungulates, such as horses. Compare ARTIODACTYLA.

peristalsis, *n.* the alternate contraction and relaxation of circular and longitudinal muscle which produces waves that pass along the intestine (and other tubular systems) of animals, moving the tube contents in one direction.

peristome, *n.* any structure that occurs around the mouth or opening of an organ.

perithecium, *n.* a type of fruiting body found in the ascomycete fungi, containing numerous asci (see ASCUS).

peritoneum, *n.* the thin membrane of mesodermal origin that lines the body cavity, covers the heart, and forms mesenteries.

peritrophic membrane, *n.* an inner lining of the gut found in some insects and crustaceans that may serve to protect the gut cells from injury by food particles. There is evidence that in insects harmful substances may be removed from the body by extrusion of the membrane from the anus.

perivisceral cavity, *n.* the body cavity surrounding the viscera.

permafrost, *n.* permanently frozen ground on which only a thin layer thaws in the summer, as in the high Arctic and Antarctic.

permanent teeth, *n.* the second set of teeth in mammals.

permeability of membranes, *n.* a theoretical concept of the permeability of a cell membrane, in which the membrane structure is thought to contain pores through which solutes can sometimes pass down a CONCENTRATION GRADIENT by DIFFUSION (see FLUID–MOSAIC MODEL). Many membranes are, however, differentially permeable to certain molecules. For example, SELECTIVE PERMEABILITY enables OSMOSIS to take place. Others will allow movement against a concentration gradient by

ACTIVE TRANSPORT. For example, the removal of sodium by cells surrounded by a higher sodium concentration.

permease, *n.* a protein thought to act as a carrier in ACTIVE TRANSPORT within membranes.

Permian period, *n.* the last period of the Palaeozoic era which lasted from 280 to 240 million years ago. Amphibians declined in numbers, reptiles increased, and conifers became commoner during this period. Many extinctions occurred, including the trilobites. Mammal-like reptiles appeared and the first beetles, caddis and bugs evolved. London was at 10°N.

pernicious anaemia, *n.* a severe condition in which there is a progressive decrease in the number of red blood cells together with an increase in their size, producing poor colour, weakness and gut disorders. The condition can be fatal but may be treated by dosing with vitamin B_{12}.

peroxidase, *n.* an enzyme such as CATALASE that catalyses the oxidation of various substances by peroxides.

peroxisome, *n.* an organelle of EUCARYOTE cells which is surrounded by a single unit membrane. Peroxisomes contain PEROXIDASES and may be concerned with the production of molecular oxygen from the breakdown of hydrogen peroxide. They are located near MITOCHONDRIA and plant CHLOROPLASTS, and appear to be involved in CELLULAR RESPIRATION.

persistence, *n.* the period of time between a virus being acquired by a VECTOR and its being transferred to a new host.

pest, *n.* any organism that causes nuisance to man, either economically or medically. Our classification of pests is constantly changing and is very dependent on economic circumstances. For example, the fungus causing apple scab is treated as a pest in the UK because it disfigures the fruit. In poorer countries, however, such cosmetic qualities of apples are less important and the fungus is not considered a pest.

pesticide, *n.* any agent that causes the death of a pest. The general definition is usually restricted to chemicals with pesticidal properties, such as herbicides, insecticides, acaricides and fungicides. Pesticide application can produce many problems, for example:

(a) destruction of organisms useful to man ('nontarget' species).

(b) directly harmful effects to man if used incorrectly.

(c) accumulation and concentration in food chains leading to toxicity in animals at a higher TROPHIC LEVEL.

petal, *n.* any of the separate parts of the corolla of a flower. Often brightly coloured, the petal is a modified leaf and is important in flowers pollinated by insects (see ENTOMOPHILY).

petaloid, *adj.* (of plant structures) looking like a petal.

petiole, *n.* the stalk of a leaf, containing vascular tissue which connects with the VASCULAR BUNDLES of the stem. The base of a petiole, where it joins the stem, may have small leaflike structures called STIPULES and axillary buds. See AXIL.

Petri dish, see PLATE.

PGA (phosphoglyceric acid) or **glyceric acid phosphate** or **glycerate 3-phosphate,** *n.* a three-carbon molecule produced from PGAL (see Fig. 244) in GLYCOLYSIS. PGA is a molecule with a similar structure to PGAL, except that a hydroxyl (OH) replaces a hydrogen atom at one point. PGA is also a vital molecule in photosynthesis, being produced in the CALVIN CYCLE from an unstable six-carbon molecule formed when ribulose diphosphate and carbon dioxide combine.

PGAL (phosphoglyceraldehyde) or **triose phosphate** or **glyceraldehyde phosphate,** *n.* a three-carbon molecule produced from fructose diphosphate in GLYCOLYSIS. See Fig. 244. PGAL is also a vital molecule in PHOTOSYNTHESIS, being produced in the CALVIN CYCLE from the phosphorylation and reduction of PGA by ATP and NADPH respectively which are produced in the LIGHT REACTIONS. Via a series of steps, two molecules of PGAL are rearranged to form six-carbon GLUCOSE.

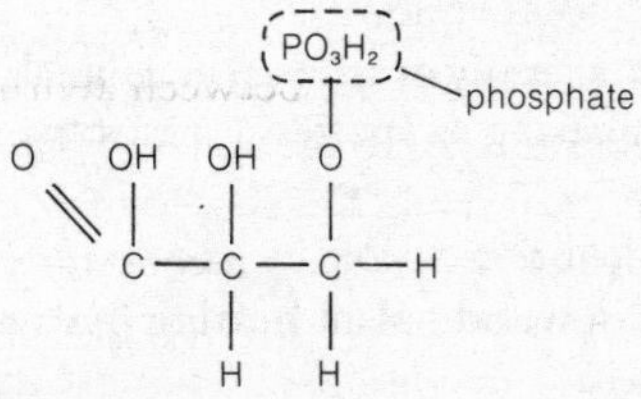

Fig. 244. **PGAL.** Molecular structure.

pH, *n.* a measure of the *hydrogen ion concentration* in an aqueous solution. The formula is:

$$\frac{1}{\log (H^+)}$$

The formula produces a value in which the higher the number of H^+ ions the lower the pH reading. The scale of pH values is from 1.0 (highly acid) to 14 (highly alkaline) with 7.0 as the neutral point. Since the pH is logarithmic, each change of one pH unit means a tenfold change in the number of hydrogen ions. pH can be measured by using indicators which change colour with changing pH, or by electrical means using a pH meter.

Phaeophycae, *n.* a class of algae, comprised of the brown algae. The class is sometimes raised to the status of a division called Phaeophyta. They are mostly marine and their green pigment is masked by fucoxanthin, a yellow pigment. Some members of the group are the largest algae known and may reach several metres in length, for example, *Laminaria*.

Phaeophyta, see PHAEPHYCAE.

phage, see BACTERIOPHAGE.

phago-, *prefix.* denoting eating, absorbing.

phagocyte, *n.* a cell that is capable of flowing round and engulfing material from its surroundings. Such cells are capable of discriminating between different particles. For example, phagocytic white blood cells will engulf only certain BACTERIA. Phagocytes form an important defence mechanism in higher animals, particularly against bacteria which are engulfed and digested. See MACROPHAGE.

phagocytosis, *n.* the ingestion of materials (subcellular particles, cells) from the outside of a cell into its interior, forming a cytoplasmic vacuole.

phalange, *n.* any of the finger or toe bones associated with the PENTADACTYL LIMB. The proximal bone of the phalanges articulates with the METACARPAL or METATARSAL.

Phalangeridae, *n.* a family of arborial marsupials possessing a prehensile tail. The 'flying phalangers' possess a PATAGIUM(1).

phallus, see PENIS.

phanero, *prefix.* denoting visible.

phanerophyte, *n.* a woody plant bearing buds more than 25 cm from the ground.

pharyngo-, *prefix.* denoting the throat.

pharynx, *n.* the canal leading from the mouth to the oesophagus in vertebrates. In humans, the upper pharynx includes the nasal section divided off by the soft palate and the lower pharynx which includes the mouth and throat. In protochordates it is that part of the gut system into which the gill slits open internally.

phasic, *adj.* transient.

Phasmidae, *n.* the family of stick and leaf insects (see ORTHOPTERAN).

phenetic ranking, *n.* an arrangement into categories based on the degree of similarity.

phenocopy, *n.* a disorder or change that appears to be genetic in origin but actually is produced by environmental effects. For example, deafness in an infant that is caused by the GERMAN MEASLES virus rather than by an inherited condition. The distinction can be important since it affects the chances of an affected individual transmitting the condition to the next generation.

phenogram, *n.* a diagram illustrating the degree of similarity between taxa.

phenon, *n.* a sample of specimens of similar appearance. See PHENOTYPE.

phenotype, *n.* the observable features of an individual organism that result from an interaction between the GENOTYPE and the environment in which development occurs. The interaction is that between *nature and nurture*. Variations due to nature are the inherited aspects of the organism, the genotype, while nurture denotes the (usually not inherited) effects of the environment upon the organism.

Sometimes two different genotypes give the same phenotype due to DOMINANCE(1) masking a recessive ALLELE. It is true to say, however, that the closer we look at the effect of an allele the more likely we are to detect a special phenotype unmasked by dominance. For example, an allele may code for a nonfunctional enzyme and thus be hidden in a heterozygote (classifying the allele as recessive) but its effects may be detected by such methods as ELECTROPHORESIS, which can identify different forms of a protein.

phenotypic plasticity or **environmental variation,** *n.* the ability of an organism to alter greatly its PHENOTYPE depending upon environmental conditions. The phenomenon is seen most clearly in plants, perhaps because they are fixed in the ground. For example, a dandelion will produce an erect habit with long flower stalks if in a garden border with other plants. Another dandelion with a similar genotype will produce a PROCUMBENT habit with short flower stalks if grown in a garden lawn.

phenotypic variance, *n.* the amount of variation in a phenotypic trait amongst individuals in a population.

phenylalanine, *n.* One of 20 AMINO ACIDS common in proteins, which has a NONPOLAR 'R' structure and is relatively insoluble in water. See Fig. 245. The ISOELECTRIC POINT of phenylalanine is 5.5. See PHENYLKETONURIA.

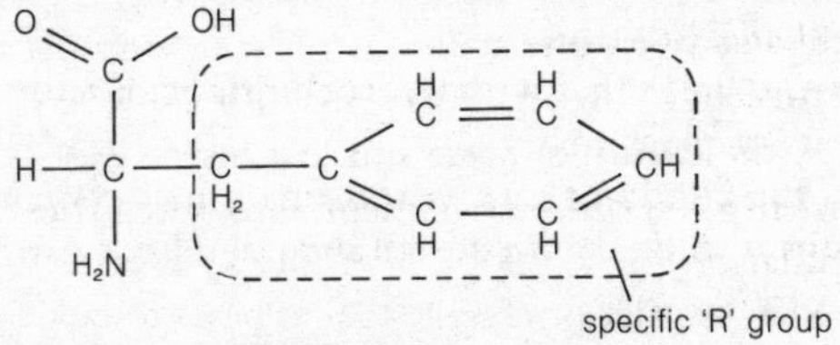

specific 'R' group

Fig. 245. **Phenylalanine.** Molecular structure.

phenylketonuria (PKU), *n.* an INBORN ERROR OF METABOLISM in humans in which there is the inability to convert PHENYLALANINE to tyrosine, due to the absence of a functional phenylalanine hydroxylase enzyme. The condition is controlled by the recessive allele of an autosomal gene, probably on chromosome 1; thus affected individuals are recessive homozygotes having received one recessive allele from each parent, both of whom are usually heterozygous 'carriers'. The effects of PKU are many and extremely serious. Perhaps the principal problem is an abnormal accumulation of phenylalanine in the blood and other tissues which adversely affects the nervous tissue to such an extent that 90% of affected individuals have an IQ of less than 40 and thus are severely mentally retarded. There is microcephaly (reduced brain size) in two thirds of all PKU cases and 75% show 'tailor's posture', in which muscular hypertonicity causes contraction of the leg and arm muscles so that individuals sit cross-legged with their arms drawn into the body.

The condition can be treated if discovered in very early childhood. Treatment consists of a diet low in phenylalanine (some of this amino acid must be present in the diet since it is an ESSENTIAL AMINO ACID). In the UK, a small blood sample is taken from every newborn baby and the sample undergoes a Guthrie test in which high levels of phenylalanine in the blood are detected by inhibition of bacterial growth.

pheromone, *n.* a chemical substance used in communication between organisms of the same species. Pheromones are found mainly in animals, but they occur in some lower plant groups where a chemical is secreted into water by female gametes to attract male gametes. In animals, for example, pheromones are transmitted in the air, as in female emperor and eggar moths, which secrete a chemical that is attractive to males over large distances, or by a dog marking out his territory with urine. Insect pheromones have been used to trap females of serious pests.

philopatry, *n.* the tendency of an animal to remain in its home area, or return to it in the case of migrants.

phloem, *n.* a transport tissue characterized by the presence of sieve tubes and companion cells, found in the VASCULAR BUNDLES of higher plants. Phloem functions in the transport of dissolved organic substances (e.g. sucrose) by TRANSLOCATION.

phloroglucin, *n.* a dye that with hydrochloric acid, stains LIGNIN bright red in sections of plant material.

Phocidae, *n.* the family of carnivorous mammals containing the seals. The hind limbs are united with the tail and point backwards and there are no external ears.

phosphagen, *n.* a type of chemical found in the muscles of all animals, whose function is to pass on high-energy phosphate to ADP to form ATP. Phosphagens thus act as energy-storage molecules and are especially useful when cellular respiration is not providing sufficient ATP molecules, for example when sudden muscular activity takes place. Phosphagens are of two types: *creatine phosphate* found in vertebrates and echinoderms, and *arginine phosphate* found in many other invertebrates.

phosphatase, *n*. an enzyme that catalyses the release of phosphate from a molecule. For example, in the mammalian liver phosphyorylated glucose can be broken down to glucose with a phosphatase enzyme. See GLYCOGEN.

phosphate, *n*. any salt or ester of any PHOSPHORIC ACID.

phosphoargenine, *n*. a phosphagen found in many invertebrates.

phosphocreatine, *n*. a compound used in the production of ATP from ADP in muscles, being converted to creatine with the release of inorganic phosphorus.

phosphoenolpyruvic acid (PEP), *n*. a high-energy compound which when dephosphorylated to pyruvic acid, gives rise to the synthesis of ATP from ADP in GLYCOLYSIS.

phosphogluconate pathway, see PENTOSE PHOSPHATE PATHWAY.

phosphoglyceric acid, see PGA.

phosphoglyceraldehyde, see PGAL.

phospholipid, *n*. a compound fat molecule in which there are two FATTY ACIDS and a phosphate group attached to the glycerol, located mostly in cell membranes. Phospholipids in the blood are responsible for the transport and utilization of fats in the body.

phosphorescence, see BIOLUMINESCENCE.

phosphoric acid, *n*. an important component of NUCLEIC ACIDS, connecting the pentose sugars to form a POLYNUCLEOTIDE CHAIN.

phosphorylation, *n*. the process in which one of more phosphate groups is added to a molecule. Such reactions occur regularly in biological systems. For example, in AEROBIC RESPIRATION glucose is phosphorylated in GLYCOLYSIS; ADP is phosphorylated to ATP at the substrate level and via the ELECTRON TRANSPORT SYSTEM; in plants ATP is produced by PHOTOPHOSPHORYLATION.

photic zone, see EUPHOTIC ZONE.

photo-, *prefix*. denoting light.

photoactivation, *n*. the excitation of atoms by light energy so that electrons become temporarily raised to a higher energy level. The process is the starting point of PHOTOSYNTHESIS, when light strikes a CHLOROPHYLL molecule. See LIGHT REACTIONS.

photoautotroph, *n*. a type of AUTOTROPH that uses light as an energy source to synthesize organic compounds from inorganic materials. Green plants are photoautotrophs.

photolysis, *n*. chemical decomposition using light energy ('splitting by light'). In 1933, Cornelius van Niel (b.1897) proposed that the initial step in PHOTOSYNTHESIS was the photolysis of water, leading to the release of oxygen ions (forming oxygen gas) and hydrogen ions which were involved in the reduction of NADP during the LIGHT REACTIONS. Recently it has become more common to explain the breakdown of water as a 'separation of charge', a fast spontaneous ionic reaction with water splitting into hydrogen (H^+) and hydroxl (OH^-) ions. The end result is the same as photolysis, but light is not thought to be involved in the process.

photon, *n.* a quantum of radiant energy with a wavelength in the visible range of the ELECTROMAGNETIC SPECTRUM.

photo-oxidation, *n.* a chemical reaction occurring as a result of absorption of light in the presence of oxygen.

photoperiod, *n.* the length of daylight as compared with the length of darkness in each 24 hour cycle. See PHOTOPERIODISM.

photoperiodism, *n.* the response observed in an organism to the length and timing of light periods. While all organisms can be affected by PHOTOPERIODS (see also BIOLOGICAL CLOCK) the term is most commonly used in connection with higher plants, particularly their flowering. The regulation of flowering is rather complex and difficult to generalize about, but it is possible to divide plants into three categories: LONG-DAY PLANTS, SHORT-DAY PLANTS and DAY-NEUTRAL PLANTS (the latter being unaffected by photoperiod).

Plants appear to detect the photoperiod with the pigment PHYTO-CHROME which can exist in two forms in the leaves, P^{660} in red light and P^{725} in far-red light. If P^{660} is present in excess in a short-day plant or P^{725} is present in excess in a long-day plant then flowering is induced by production of the hormone florigen, changing a vegetative apical MERISTEM to a flowering one.

Phytochromes are also present in the seed coat, the correct balance of P^{660} and P^{725} stimulating enzyme activation and hence GERMINATION. See also FLORIGEN.

photophase, *n.* the light phase of a dark/light cycle. Compare SCOTOPHASE.

photophosphorylation, *n.* the production of ATP from ADP, with light as the energy source. The process takes place during the 'light reactions' of PHOTOSYNTHESIS which also result in the production of reduced NADP. The details of ATP and NADPH formation are given under LIGHT REACTIONS.

photopigment, *n.* a pigment molecule that can be excited by light, such as CHLOROPHYLL.

photoreceptor, *n.* a structure or pigment sensitive to light. In animals such structures are called eyes, containing sensitive pigments which, when stimulated by light, activate the nervous system. In plants photoreceptors tend to be pigments of various sorts. For example, PHYTOCHROMES sensitive to red light (see PHOTOPERIODISM) and flavoproteins sensitive to blue light trigger AUXIN production, leading to PHOTOTROPISM.

photorespiration, *n.* that part of the respiration of plants that occurs in light and during photosynthesis. Mitochondria are not involved and there is no yield of ATP.

photosynthesis, *n.* the process by which plants convert carbon dioxide and water into organic chemicals using the energy of light, with the release of oxygen. Photosynthesis occurs in green plants which are known as AUTOTROPHS. See LIGHT REACTIONS and CALVIN CYCLE.

photosystems I and II, see LIGHT REACTIONS.

phototaxis, *n.* the movement of a whole organism in response to a directional light stimulus. For example, movement of unicellular algae or fruit flies towards a source of light is *positive* phototaxis (but away from light if the stimulus is too strong is *negative* phototaxis).

phototropism, *n.* a bending growth movement of parts of a plant in response to a light stimulus. The movement produced by unequal growth is due to differences in AUXIN concentration. For example, most seedlings are *positively* phototropic, growing towards a light stimulus, because there is a greater concentration of auxin on the side furthest away from the light, giving greater growth on this side. Roots, on the other hand, are often *negatively* phototropic, growing away from a light source.

phragma, *n.* one of the parts of the insect skeleton projecting inwards in the dorsal region, which serves for the attachment of muscles and other organs.

phragmocone, *n.* the internal shell of a cuttlefish or other cephalopod made up of several chambers.

phycocyanin, *n.* a blue plant pigment with a photosynthetic function found in blue-green and red algae, with an absorption peak of 618 nm (see ABSORPTION SPECTRUM).

phycoerythrin, *n.* a red pigment with a photosynthetic function found in red algae and some blue-green algae, with absorption peaks at 490, 546 and 576 nm (see ABSORPTION SPECTRUM).

phycomycete, *n.* any filamentous fungus of the class Phycomycetes, now included in two separate subdivisions of the Eumycota (see PLANT KINGDOM) which usually lack cross walls (i.e. are nonseptate).

phyletic, *adj.* of or relating to a line of descent.

phyletic correlation, *n*. a correlation of phenotypic characters which reflects a well-integrated ancestral gene complex.

phyletic weighting, *n*. the assessment of taxonomic importance on the basis of PHYLETIC CORRELATION.

phyllo-, *prefix*. denoting a leaf.

phyllode, *n*. an unusual type of leaf in which the lamina is absent but the petiole is so flattened as to appear like a normal leaf, and there is still an axillary bud in the leaf node. Phyllodes are found, for example, in the grass vetch, *Lathyrus nissolia*.

phylloplane, *n*. the surface of the leaf in plants.

phyllopodium, *n*. a broad flattened crustacean limb used for swimming or for creating respiratory currents.

phylloquinone, see VITAMIN K.

phyllotaxy, *n*. the arrangement of leaves on a stem.

phylogenetic tree, *n*. a diagram illustrating lines of descent derived from morphological and palaeontological evidence.

phylogeny, *n*. the whole of the evolutionary history of a species or other taxonomic group of organisms. See Haekel's Law of RECAPITULATION.

phylogram, *n*. a tree-like diagram showing the degree of relationship of different TAXA.

phylum, *n*. a major grouping (TAXON) into which kingdoms (e.g. the animal kingdom), are divided and which is composed of a number of CLASSES.

phys- or **physo-,** *prefix*. denoting a bladder.

physiological drought, *n*. a state in which plants are unable to absorb water even though water may be freely available. The condition can be caused by a high ionic concentration in the water, so OSMOSIS may not occur, or by low water temperature.

physiology, *n*. the study in animals and plants of those internal processes and functions associated with life.

phyt- or **phyto-,** *prefix*. denoting plants.

phytoalexin, *n*. a defensive chemical produced by higher plants that is often fungicidal in action (see PLANT DISEASE). Phytoalexins are specific to particular attacking organisms; about 30 types have been identified in various plants, particularly the Leguminosae.

phytochrome, *n*. a type of plant pigment that occurs in two forms, P_{660} which absorbs red light and P_{725} which absorbs far-red light. P_{725} is biologically active in that it stimulates enzymic reactions, whereas P_{660} is biologically inert. The conversion of one form into the other occurs simultaneously. See Fig. 246.

Daylight contains a mixture of red and far-red light, with an excess of red light. During the day the proportion of P_{725} builds up relative to

Fig. 246. **Phytochrome.** The simultaneous conversion of red light into far-red light.

P_{660}, whereas during the night there is a gradual conversion of P_{725} to P_{660}. Thus phytochrome provides the plant with a method of detecting day and night, with P_{725} predominant when it is day, and P_{660} predominant when it is night. This mechanism is the basis of PHOTOPERIODISM.

phytophagous, *adj.* (of animals) feeding on plants.

phytoplankton, *n.* that part of the plankton made up of plant life. Compare ZOOPLANKTON.

phytosanitary certificate, *n.* a record of plant health before exportation.

pia mater, *n.* a thin, inner membrane or meninge surrounding the brain and spinal cord. See MENINGES.

pigeon's milk, *n.* a crop secretion fed to the young pigeon.

piliferous layer, *n.* a zone of epidermal cells behind the root tip producing root hairs.

piloerection, *n.* the standing up of hairs on the skin to increase thermal insulation.

pilomotor, *adj.* (of autonomic control of smooth muscle) causing erection of body hair.

pilose, *adj.* (of plant structures) bearing hairs.

pilus, *n.* an extension of the cell wall in bacteria that may be converted into a tube for CONJUGATION to take place.

pineal body or **pineal gland,** *n.* an outgrowth of the roof of the forebrain. The posterior part (EPIPHYSIS) has an endocrine function and secretes melatonin which concentrates the melanophores of fishes and amphibians, inhibits the development of the gonads and is involved in the control of CIRCADIAN RHYTHMS. Exposure to light inhibits the production of melatonin. The anterior part forms an eye-like structure (third eye) in some lizards; it is vestigial in the lamprey and absent in other vertebrates.

pineal gland, see PINEAL BODY.

pinna, *n.* **1.** the primary division of a leaf. **2.** a wing or fin. **3.** the projecting part of the external ear of mammals.

pinnate, *adj.* (of compound leaves) made up of leaflets arranged in two rows on opposite sides of the axis, or RACHIS.

pinnatifid, *adj.* (of leaves) having a LAMINA cut into lobes that reach about halfway towards the midrib.

pinnatisect, *adj.* (of leaves) having a lamina that is PINNATIFID in shape but with lobes very deep cut, almost to the midrib.

Pinnipedia, *n.* a suborder of aquatic carnivores, including the seals, sea lions and walruses.

pinnule, *n.* a subdivision of a leaf pinna, often bearing SPORANGIA in ferns.

pinocytosis or **micropinocytosis,** *n.* the active engulfing of very small particles or liquids by cells; a form of ENDOCYTOSIS. The particles become surrounded by the cell membrane on all sides, which eventually forms a channel from which vesicles are pinched off and move within the PROTOPLASM before their contents can be transferred into the cell proper.

pistil, see CARPEL.

pit, *n.* a thin area of the cell wall in higher plants where no secondary wall develops. If two cells have pits in adjacent parts of the wall there are often fine protoplasmic connections called PLASMODESMATA between the cells. Sometimes pits have a more complex structure, as in BORDERED PITS of seed plants.

pitfall trap, *n.* a trap (e.g. a jar sunk into the ground) into which mobile animals may fall and from which they are unable to escape.

pith or **medulla,** *n.* the core of a DICOTYLEDON stem, containing PARENCHYMA cells that have a storage function.

pithec-, *prefix.* denoting an ape.

Pithecanthropus, *n.* an early type of genus of man from the PLIOCENE EPOCH. It possessed a larger cranial capacity than apes and probably walked erect. See HEIDELBERG MAN.

Pithecinae, *n.* a family of New World monkeys lacking a prehensile tail.

pituitary gland or **pituitary body** or **pituitary,** *n.* an ENDOCRINE ORGAN derived from a down-growth of the infundibulum of the brain (pars nervosa) and an upgrowth of the *hypophysis* (pars distalis and pars intermedia). The pars distalis is usually referred to as the anterior pituitary which produces FOLLICLE-STIMULATING HORMONE (FSH), luteinizing hormone (LH), LUTEOTROPHIC HORMONE (LTH), ADRENOCORTICOTROPHIC HORMONE (ACTH), THYROTROPHIC HORMONE and GROWTH HORMONE. The pars intermedia and pars nervosa (*neurohypophysis*) form the posterior pituitary; the former produces 'intermedin' in fish, amphibia and reptiles (absent in birds and mammals) which is associated with change in skin colour; the latter secretes oxytocin and ADH, which are produced in the hypothalamus and stored in the pars nervosa. The pituitary controls the function of other endocrine organs.

placebo, *n.* **1.** any inactive substance given to satisfy a patient's psychological need for medication. **2.** a control in an experiment to test the effect of a drug.

placenta, *n.* **1.** (in animals) the structure formed from the tissues of embryo and mother at the point of attachment of the embryo to the mother, through which the embryo is nourished. In placental mammals the embryonic blood supply may be from the ALLANTOIS (allantoic placenta) or the yolk sac (yolk sac placenta). Small molecules of oxygen and food materials pass to the embryo through the placenta, and urea and CO_2 pass out from the embryo. The placenta itself produces the hormones OESTROGEN, PROGESTERONE and GONADOTROPHIC HORMONE. At birth the foetus is ejected from the uterus first, the umbilical cord broken and the placental afterbirth is discharged last, except in abnormal circumstances. See Fig. 247. **2.** (in plants) a part of the wall of the ovary on which OVULES are borne.

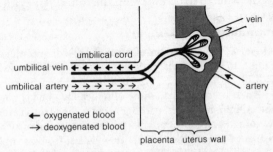

Fig. 247. **Placenta.** Section showing blood circulation.

placentation, *n.* the arrangement of PLACENTAS(2) in the plant ovary, which can be classified into several categories including: (a) apical, at the ovary apex, (b) basal, at the ovary base, (c) parietal, on the ovary wall, (d) free-central, on a column growing up from the ovary base.

placo-, *prefix.* denoting a flat plate.

Placodermi, see APHETOHYOIDEA.

placodont, *n.* a fossil reptile from the Triassic period, which possessed flat crushing teeth and probably fed on molluscs.

placoid or **denticle,** *adj.* the dermal denticles of cartilaginous fish. Placoids are tooth-like structures of enamel, covered with dentine. Compare COSMOID, GANOID.

plankton, *n.* the organisms inhabiting the surface layer of a sea or lake, drifting with the current. Many planktonic organisms are capable of locomotion but are not sufficiently strong swimmers to move against the currents. Members of the plankton vary in size from unicellular organisms to large jellyfish, 1m or more in diameter. Compare NEKTON.

plant, see PLANT KINGDOM.

plant-, *prefix.* denoting the sole of the foot.

plant breeding, *n.* the science that aims to improve the quality and yield of plants by the genetic engineering of varieties. The main tool of the plant breeder is SELECTION, by which it is possible to alter the genetic constitution of a plant type, provided that genetic variability is present in the original strains. Another method of improvement is the crossing together of inbred lines to produce HETEROSIS in the hybrids (see GREEN REVOLUTION). Plant breeding has lead to a great improvement in crop yield, better harvesting (e.g. with dwarf varieties) and improvements in the natural resistance of plants to pathogens. Recently, cell and tissue cultures using EXPLANTS have begun to be used by plant breeders, to produce: (a) multiple copies (CLONES) of sterile types; (b) the development of forms with new chromosome combinations; (c) HAPLOIDS(2) which can be induced to become fertile DIPLOIDS(2) (the latter forming PURE BREEDING LINES).

plant community, see COMMUNITY.

plant disease, *n.* the disruption of normal growth and metabolism in plants leading to reduced viability and even death. Plant diseases may begin with physical attack of the plant by pests such as insects, slugs, mites, birds and other animals, some of which may be VECTORS of plant PATHOGENS. Often, however, pathogens enter the plant via the wounds caused by animal pests. Other pathogens, such as cereal rusts, are able to penetrate plant tissue on their own, often via leaf stomata. Many plants produce special chemicals called PHYTOALEXINS in response to injury, which have an effective defensive function.

plant hormone, *n.* any of several chemical substances produced in very low concentrations in plants, that control growth and development. Plant HORMONES are produced in active MERISTEMS (e.g. root and shoot apices) from where they diffuse to other parts of the plant. A distinct contrast with animal hormonal systems is the fact that plants do not possess specific tissues for hormone production, nor a well-defined distribution system equivalent to the blood system.

Plant hormones are classified generally into several groups, discussed under separate headings: ABSCISSIC ACID, AUXINS, CYTOKININS and GIBBERELLINS.

plantigrade, *adj.* walking on the entire sole of the foot, as in humans and bears.

plant kingdom, *n.* a category of living organisms comprising all members of the kingdom Plantae. Plants lack locomotive movement, possess cell walls and have no obvious nervous or sensory organs. Most plant groups possess members with CHLOROPHYLL. The system in which the plant kingdom is organized is given opposite.

Super-kingdom Procaryonta
  Kingdom Monera
    Division Cyanophyta   Blue-green algae
    Division Bacteria
    (= Schizomycophyta)   Bacteria

Super-kingdom Eucaryonta
  Kingdom Mycota
    Division Myxomyceta   Slime moulds
    Division Eumycota   True fungi
      Subdivision Mastigomycotina
      (= Phycomycetes [Part])
      Subdivision Zygomycotina
      (= Phycomycetes [Part])
      Subdivision Ascomycotina
      (= Ascomycetes [Part])
      Subdivision Basidiomycotina
      (= Basidomycetes)
      Subdivision Deuteromycotina   Fungi imperfecti
                                     (including part of
                                     Ascomycetes)

  Kingdom Plantae (= Phyta)
    Division Chlorophyta   Green algae
    Division Euglenophyta   (included in Flagellata
                                     by zoologists)
    Division Charophyta   Stoneworts and
                                     Brittleworts
    Division Phaeophyta   Brown algae
    Division Chrysophyta   Diatoms, Yellow-green
                                     and golden-brown
                                     algae
    Division Pyrrophyta   Fire algae
    Division Rhodophyta   Red algae
    Division Bryophyta
      Class Hepaticaea   Liverworts
      Class Anthocerotae   Hornworts
      Class Musci   Mosses
    Division Tracheophyta
      Subdivision Psilopsida   Whisk fern, etc.
      Subdivision Lycopsida   Club mosses
      Subdivision Sphenopsida   Horsetails
                                   (continued overleaf)

Subdivision Pteropsida
    Class Filicinae   Ferns
    Class Gymnospermae   Conifers
    Class Angiospermae   Flowering plants
        Subclass Dicotyledonae
        Subclass Monocotyledonae

planula, *n.* the ciliated larva of coelenterates, which lacks a body cavity.

plaque, *n.* a clear area in a 'lawn' of a bacterial colony, in which the bacteria have undergone LYSIS due to BACTERIOPHAGE infection.

plasma, *n.* **1.** the cellular PROTOPLASM inside a plasma membrane. **2.** see BLOOD PLASMA.

plasma gel, *n.* an area of cytoplasm with a GEL structure.

plasma gene, *n.* a selfreplicating particle found in the cellular cytoplasm and showing CYTOPLASMIC INHERITANCE.

plasmalemma, *n.* the CELL MEMBRANE which also lines the connecting PLASMODESMATA between living cells.

plasma membrane, see cell membrane.

plasma proteins, see BLOOD PLASMA.

plasma sol, *n.* an area of cytoplasm with a SOL structure.

plasmid, *n.* a circular piece of DNA found in the cytoplasm of bacterial cells that replicates independently of the host chromosome. Plasmids can be important in public health since some types possess genes for antibiotic resistance, and can be quickly transferred to different types of host cell, thus spreading resistance very rapidly. Plasmids are used extensively in GENETIC ENGINEERING of microorganisms. See also EPISOME.

plasmo-, *prefix.* denoting form.

plasmodesmata, *n.* narrow strands of cytoplasm (see SYMPLAST) that pass through pores in plant cell walls and join the cells to one another. They facilitate movement of material between cells and play an important part in the deposition of cellulose during thickening of the secondary cell wall in plants.

plasmodium, *n.* **1.** any parasitic sporozoan protozoan of the genus *Plasmodium*, esp. *P. vivax*, which causes malaria. **2.** the vegetative phase of slime moulds (see MYXOMYCETE).

plasmolysis, *n.* the shrinkage of plant cell contents due to loss of water, resulting in the CELL MEMBRANE pulling away from the cell wall, leaving a fluid-filled space. Plasmolysis occurs when plant cells are placed in a hypertonic (see HYPOTONIC) medium so that they lose water by OSMOSIS. See Fig. 248.

plastid, *n.* an organelle of plant cells, with a double membrane. Plastids are large (between 3 and 6 μm in diameter) and have either a photosyn-

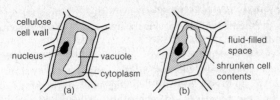

Fig. 248. **Plasmolysis.** (a) Isotonic medium. (b) Hypertonic medium, showing plasmolysis.

thetic function (CHLOROPLAST) or a storage function (AMYLOBLAST).

plate, *n.* **1.** also called **Petri dish.** a flat dish used to culture microorganisms. **2.** *vb.* to spread cells over the surface of a medium in a culture plate.

platelet, see BLOOD PLATELETS.

plate tectonics, *n.* the study of the movement of the large crustal plates that form the surface of the earth on the continental land masses and beneath the seas. See CONTINENTAL DRIFT.

platyhelminth or **flatworm,** *n.* any invertebrate of the phylum Platyhelminthes, containing the TAPEWORMS (Cestoda), FLUKES (Trematoda) and planarians (Turbellaria). The Turbellaria are free-living flatworms which are aquatic and inhabit moist places on land. The two other groups are parasitic. They possess only an anterior opening to the gut system, there is no blood system and the animals show BILATERAL SYMMETRY.

Platyrrhini, *n.* the New World monkeys, distinguished from those of the Old World by their possession of a broad cartilaginous nasal septum and a long, usually prehensile, tail. Compare CATHARRHINI.

Plecoptera, *n.* the insect order containing the stoneflies, which are hemimetabolous and similar to mayflies.

pleiotropism, *n.* a state in which one gene affects two or more aspects of the PHENOTYPE that are apparently unrelated. For example, the 'vestigial wing' mutation of *Drosophila* not only controls the size and shape of the wings but also affects several other features, including reduced FECUNDITY.

Pleistocene epoch, *n.* a division of the Quaternary period lasting from 2 million years ago until 10,000 years ago. The epoch contained four major glaciations and *Homo sapiens* evolved during this time. See GEOLOGICAL TIME.

Pleistocene refuge, *n.* the areas south of the great ice sheets of the PLEISTOCENE EPOCH in which animals and plants survived during periods of glaciation.

pleomorphic, *adj.* (of cells) having several shapes.

plesiomorphic, *adj.* (of characters) primitive or ancestral.

pleura, *n.* the membrane that covers the lung and lines the innermost wall of the thorax.

pleural cavity, *n.* a coelomic cavity surrounding the lungs of mammals that is separated from the rest of the perivisceral coelom by the DIAPHRAGM. The cavity is fluid-filled and small, as the lungs and body wall are in close proximity to each other.

pleuron, *n.* the lateral plate on either side of the arthropod SOMITE.

Pleuronectidae, *n.* a family of TELEOST fish, such as the flounder, that start life normally symmetrical but undergo a development as a result of which the eyes come to be on one side, the head is twisted and the fish, in a resting position, lies on the bottom, on its other side.

plexus, *n.* a network of interlaced blood vessels or nerves.

Pliocene epoch, *n.* the last epoch of the TERTIARY PERIOD, lasting from the end of the Miocene seven million years ago to approximately two million years ago. Early man evolved during this period, some three million years ago. Britain's climate supported a warm temperate flora.

ploidy, *n.* the number of chromosome sets making up the total genome of an organism. Thus the ploidy of normal humans is two and is written 2n. see DIPLOID(1).

plumule, *n.* **1.** (in plants) the embryonic shoot of a germinating seedling, which develops into an EPICOTYL and leaves. **2.** (in animals) a down feather of birds.

pluteus, *n.* a larva of ECHINODERMS that has ciliated arms, sometimes supported by calcareous rods.

pluvial, 1. *adj.* of, relating to, or due to the action of rain. **2.** *n.* any period during which the climate was wetter than now, usually applied to periods during the PLEISTOCENE EPOCH.

pneumaticity, *n.* (of bones), the presence of air spaces which connect with the airsacs and lungs, particularly in birds.

pneumatophore, *n.* the bladder-like polyp of SIPHONOPHORES which enables them to float. It is usually the uppermost polyp of the CORMIDIUM.

pneumonia, *n.* a human lung disease caused by a number of bacterial and viral pathogens, particularly *Streptococcus pneumoniae*.

pneumostome, *n.* a small aperture allowing the passage of air, such as that leading into the respiratory cavity of snails.

pneumotaxic centre, see BREATHING.

pod- or **podo-,** *prefix.* denoting a foot.

podsol or **podzol,** *n.* a forest soil characterized by a surface accumulation of humus, a strongly leached profile, one or more enriched layers in the SUBSOIL, and vegetation consisting predominantly of conifers.

poikilotherm or **ectotherm** or **heterotherm,** *n.* any animal whose body temperature follows that of the surrounding environment. Poikilotherms are described as *cold-blooded*, but the body temperature may be quite high. Aquatic poikilotherms follow the environmental temperatures closely, but terrestrial forms may warm up from the sun's irradiation or cool through evaporation to an extent varying with the temperature of the surrounding air. All animals apart from mammals and birds are poikilothermous. Compare HOMOIOTHERM.

point mutation or **gene mutation,** *n.* a genetic change affecting a single LOCUS, producing an alternative ALLELE of the gene, but not a gross structural change to the chromosome. The genetic change consists of an altered sequence of DNA bases, which can be of three main types: (a) a SUBSTITUTION MUTATION, (b) a DELETION MUTATION and (c) an INSERTION MUTATION, the latter two often causing a major change to the amino acid sequence of the protein structure. Compare CHROMOSOMAL MUTATION.

Poisson distribution, *n.* (statistics) the frequency of sample classes containing a particular number of events (0,1,2,3 . . . n), where the average frequency of the event is small in relation to the total number of times that the event could occur. Thus, if a pool contained 100 small fish then each time a net is dipped into the pool up to 100 fish could be caught and returned to the pool. In reality, however, only none, one or two fish are likely to be caught each time. The Poisson distribution predicts the probability of catching 0,1,2,3 . . . 100 fish each time, producing a FREQUENCY DISTRIBUTION graph that is skewed heavily towards the low number of events.

polar body or **polar nucleus** or **polocyte,** *n.* a small haploid cell, produced during OOGENESIS in female animals, that does not develop into a functional OVUM. During such GAMETOGENESIS three polar bodies are produced, along with one large ovum. A similar situation arises in the development of the egg cell of higher plants, where only one fertile gamete is produced from MEIOSIS. see EMBRYO SAC.

polarity, *n.* the morphological and/or physiological difference between the two ends of an axis, such as root and stem.

polarization, *n.* the act of changing an ordinary light beam consisting of billions of wavetrains each vibrating in a different direction, to a beam in which only those wavetrains vibrating in a particular plane are allowed to continue in the beam which is then less bright.

Light from the sun is scattered by molecules in the upper atmosphere in such a way as to result in light arriving at the earth's surface being partially polarized. The extent of polarization at any point depends on the position of the sun, so there is a pattern of polarization of the sky for any particular position of the sun. Bees, and probably many other arthropods, are capable of navigating by this pattern when the sun is obscured, so long as some blue sky can be seen (see NAVIGATION). Polarization can be brought about by naturally occurring crystals such as calcite, or by Polaroid sheets.

polar nucleus, see POLAR BODY.

polar substance, *n.* any organic substance that combines readily with water. This is due to the presence of side groups of organic molecules that contain the electrical charge of hydroxyl groups.

poliomyelitis, *n.* a paralytic disease in which cells of the CENTRAL NERVOUS SYSTEM become destroyed by the polio virus, leading to crippling, although most infections are not serious and the patient usually recovers full health. The disease can be kept in check by the administration of vaccines, the most popular being a live, attenuated type. In Western countries administration of such vaccines is now routine and the disease is very rare.

pollen, see POLLEN GRAIN.

pollen basket, *n.* that part of the hind leg of a hive or humble bee modified for carrying pollen. In some leaf-cutting bees the pollen basket lies below the abdomen.

pollen grain, *n.* a small structure of higher plants that contains haploid male gamete NUCLEI and is surrounded by a double wall, the EXINE and INTINE. As the pollen grain ripens in the pollen sacs of the ANTHER, the nucleus divides by MITOSIS into two nuclei which have different functions, the 'generative' nucleus and the 'tube' nucleus (see POLLEN TUBE). Pollen grains are transported from the male stamen to the female stigma in a process called POLLINATION.

pollen tube, *n.* a slender structure produced from a POLLEN GRAIN after POLLINATION. The tube protrudes from a pore in the EXINE and grows down through the stigma, style and ovary of the CARPEL, entering the ovule usually by the micropyle. See Fig. 249.

Within the tube, the haploid generative nucleus divides by MITOSIS to give two MALE GAMETE NUCLEI which have an important function in the double FERTILIZATION process of higher plants (see EMBRYO SAC). The

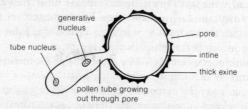

Fig. 249. **Pollen tube.** Section of grain, showing tube.

growth of the tube appears to be controlled by the tube nucleus and there is some evidence that the tube shows negative AEROTROPISM. The stimulus for the original germination of the pollen grain comes from a sugary fluid secreted by the stigma. Sometimes, however, germination is prevented by SELF–INCOMPATIBILITY factors in the stigma cells and pollen grain, which prevent self-pollination in some plants.

pollen tube nucleus, see POLLEN TUBE.

pollination, n. the transfer of pollen from male to female reproductive organs in higher plants, for example, microsporangium to ovule in GYMNOSPERMS, and anther to stigma in ANGIOSPERMS. In flowering plants the flowers show a strong link between structure and the method of pollination (see ANEMOPHILY, ENTOMOPHILY).

Many angiosperms carry out CROSS-POLLINATION. Some of these types, such as brassicas, have SELF–INCOMPATIBILITY systems to prevent self-pollination. Self-pollination, however, is common in certain plants, such as groundsel or chickweed.

pollinium, n. a mass of POLLEN GRAINS stuck together and carried as a whole during POLLINATION.

pollution, n. the contamination of an environment by any substance or energy. Heavy metals, oil, sewage, noise, heat, radiation and pesticides are common pollutants which can affect the environment adversely.

polocyte, see POLAR BODY.

poly-, *prefix.* denoting many.

polyandry, n. a system of POLYGAMY in which the female has several mating partners at one time. Compare POLYGYMY.

polychaete, n. any marine ANNELID worm of the order Polychaeta, including both *errant* (moving) and *tubicolous* (mainly sedentary) forms. They are characterized by the presence of CHAETAE carried on parapodia (projections from the body) and in errant forms the head is well developed. Fertilization is external. The group includes Nereids (bristle worms such as the ragworm), tube worms and fan worms.

polycistronic mRNA, n. a MESSENGER RNA molecule found in many PROCARYOTES, that contains information transcribed from two or more CISTRONS and which is translated into two or more polypeptides.

polyembryony, n. **1.** (in plants) the formation of more than one embryo within the TESTA of a seed. **2.** (in animals) the development of more than one embryo from a single egg as occurs, for example, in certain parasitic HYMENOPTERA which use this phenomenon to increase rapidly numbers of juveniles within the host.

polygamy, n. a mating system of animals in which one sexual partner mates with several of the opposite sex (see POLYANDRY, POLYGYNY). Polygyny is the most common form, and would seem to be present when one parent (usually the male) does not play a large part in rearing the young.

polygenic characters, see POLYGENIC INHERITANCE.

polygenic inheritance or **quantitative inheritance,** *n.* a genetical system in which a number of genes (more than two) is involved in the control of a character, the character being described as 'quantitative' in that it shows a wide range of variability that can be measured (see HERITABILITY). Each 'polygene' in the system has only a small effect on the character and cannot be detected individually. The expression of a polygenic character is strongly affected by the environment in which the individual exists. For example, human height is affected by diet, human intelligence scores are affected by the social background.

polygyny, *n.* a system of POLYGAMY in which the male has several female mating partners at one time. Compare POLYANDRY.

polymer, *n.* a compound of high molecular weight formed of long chains of repeating units (MONOMERS).

polymerase, *n.* an enzyme that catalyses the joining of DNA or RNA nucleotides.

polymorphism, see a GENETIC POLYMORPHISM.

polynomial nomenclature, *n.* a system of nomenclature involving more than two names, such as *Parus major minor*, in which *minor* represents the subspecies.

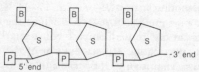

Fig. 250. **Polynucleotide chain.** General structure. P=phosphate, S=ribose or deoxyribose sugar, B=one of four bases.

polynucleotide chain, *n.* a sequence of NUCLEOTIDES joined together to form a string. RNA consists of one polynucleotide chain, while DNA consists of two chains bonded between the bases in a system of complementary pairing: adenine with thymine, guanine with cytosine. The sequence of bases along the chain acts as a GENETIC CODE for the structure of PROTEINS. Polynucleotide chains show polarity based upon the position of the bonds in relation to the sugar component. One end of the chain will end at the 3' position, the other at the 5' position. See Fig. 250. The two chains of DNA show opposite polarities (See Fig. 130).

polyp, *n.* the asexually reproducing, normally sedentary form of COELENTERATES such as the sea anemone (see MEDUSA). In the class Anthozoa the polyp stage is the dominant part of the life history, but in the Scyphozoa (jellyfish) it is often absent.

polypeptide chain, *n.* a sequence of AMINO ACIDS joined together by PEPTIDE BONDS, forming a PROTEIN, the sequence being determined by the order of bases along the POLYNUCLEOTIDE CHAINS of DNA, in the form of a GENETIC CODE. See PROTEIN SYNTHESIS for details.

polypetalous, see GAMOPETALOUS.

polyphenism, *n.* the presence of several phenotypes in a population that are not the result of genetic differences.

polyphyletic, *adj.* (of a group of organisms) having been derived from more than one source, that is, not of a single line of evolution.

polyploid, 1. *adj.* (of cells or organisms) having three or more complete sets of chromosomes. **2.** *n.* an individual or cell of this type. See TRIPLOID, TETRAPLOID.

Polypterus, *n.* a primitive freshwater fish of Central Africa derived, with sturgeons, from Devonian palaeoniscid stock. See PALAEONISCUS.

polyribosome or **polysome,** *n.* two or more RIBOSOMES joined by a molecule of messenger RNA during PROTEIN SYNTHESIS. Such an arrangement ensures that the mRNA is 'read' at the maximum speed.

polysaccharide, *n.* a large carbohydrate molecule with a chain-like or branched structure, made up of many MONOSACCHARIDE units joined together by CONDENSATION REACTIONS. Although most polysaccharides have a terminal monomer present as a REDUCING SUGAR, this forms only a small part of the whole molecule and thus most polysaccharides do not act as reducing sugars. Polysaccharides are insoluble, unsweet and are important as storage molecules (e.g. STARCH, inulin and GLYCOGEN) and as reinforcing materials (e.g. CELLULOSE of plant cell walls, CHITIN of crustacean and insect cuticles).

polysome, see POLYRIBOSOME.

polytene chromosome, see SALIVARY GLAND CHROMOSOME.

polythetic, *adj.* (of a taxon) where each member possesses a majority of a set of characters.

polytocous, *adj.* (of animals) producing many young at birth.

polytopic, *adj.* (of a subspecies) having geographically separate but phenotypically similar populations.

polytypic, *adj.* (of a taxon) containing two or more immediately subordinate taxa, for example a species with several subspecies. A polytypic species is also called a *Rassenkreiss*.

polyuracil, *n.* a synthetic polynucleotide chain formerly used as MESSENGER RNA for the synthesis of phenylalanine. See GENETIC CODE.

polyzoan, see BRYOZOAN.

pome, *n.* a false fruit, such as an apple, where the main part is developed from the RECEPTACLE and not the OVARY.

pooter, *n.* a piece of equipment for collecting insects, similar to a wash-bottle, into which small insects are sucked by an intake of breath.

population, *n.* **1.** the total number of the individuals of a particular species, race or form of animal or plant, inhabiting a particular locality or region. **2.** (in genetics) the total number of BREEDING INDIVIDUALS of a species in a particular location.

population control, see CONTROL (2).

population dynamics, *n.* the processes that contribute to the size and composition of a population.

population genetics, *n.* the study of heredity at the population level, for example, gene frequencies, mating systems.

population growth, *n.* an increase in the numbers of a population as a result of the birth rate exceeding the death rate.

pore, *n.* any small opening in the skin or epidermis or any structure.

Porifera, see SPONGE.

porphyrin, *n.* an organic compound consisting of four pyrrole rings linked by CH bridges with a heavy metal in the centre. Porphyrins form part of several important biological molecules. Examples include the haem group of HAEMOGLOBIN and MYOGLOBIN (see Fig. 251), chlorophyll (with magnesium) and cytochromes (with iron).

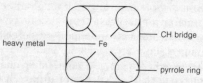

Fig. 251. **Porphyrin.** Compound structure.

portal vein, *n.* any vein that carries blood from one set of capillaries to another, such as renal-portal and hepatic-portal veins.

porrect, *adj.* (of plant structures) extending outward and forward.

position effect, *n.* an alteration to the expression of a gene due to a change in its location within the GENOME.

positive feedback, see FEEDBACK MECHANISM.

post-, *prefix.* denoting after.

postsynaptic membrane, *n.* the excitable membrane of the dendrite next to the axon at a synapse.

potential energy, *n.* any stored energy that can be released to do work.

potometer, *n.* an instrument designed to measure the rate of absorption of water by a cut shoot and hence, indirectly, the rate of TRANSPIRATION by the shoot in a particular set of conditions, which can be varied. See Fig. 252. At the start of the experiment an air bubble is introduced into the tubing adjacent to the graduated scale. Reservoir A supplies water to replace that lost by the shoot, the amount taken up being indicated by the

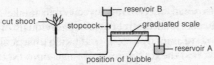

Fig. 252. **Potometer.**

movement of the air bubble along the scale. Opening the stopcock at the end of a time period causes water from reservoir B to force the air bubble along the tube, ready for a fresh start.

PPP, see PENTOSE PHOSPHATE PATHWAY.

preadaptation, *n.* the possession by an organism of characteristics that permit a move into a new NICHE or habitat.

Pre-Cambrian era, *n.* all geological time prior to the beginning of the CAMBRIAN PERIOD 590 million years ago. Very few fossils are known from this time and the fossil record effectively begins with the trilobites and brachiopods of the Cambrian. However, impressions of soft-bodied animals such as medusae, sea pens, annelids and some arthropods have been found in Pre-Cambrian rocks dated between 640 and 580 million years old. Bacteria may have existed for up to 3,800 million years and blue-green algae (Cyanophyta) for 3,500 million years.

precipitin, see ANTIBODY.

predation, *n.* see PREDATOR.

predator, *n.* any animal that lives by preying on other animals (usually) from a lower TROPHIC LEVEL.

predictive values, *n.* the capacity to make predictions on new characters or newly created taxa within a classification.

preferred body temperature, *n.* the CORE TEMPERATURE of behaviourly thermoregulating bradymetabolic (see BRADYMETABOLISM) animals in an artificial temperature gradient.

preformism, *n.* the now discredited theory that an egg or sperm contains a preformed adult.

prehensile, *adj.* adapted for holding or grasping, as in the tails of many New World monkeys.

pre-Linnaean name, *n.* any taxonomic name published prior to 1 January 1758.

premolars, *n.* the cheek teeth in mammals that occur between the molars and the canines. They have a grinding function.

premorse, *adj.* (of plant structures) appearing as if bitten off at the end.

prepuce, *n.* the foreskin or fold of glandular and vascular tissue surrounding the tip of the mammalian penis.

presbyopia, *n.* a loss of ACCOMMODATION (focusing ability) in the eye that occurs frequently in older humans due to a loss of elasticity in the lens. Such affected individuals require two corrective lenses, one for close work, the other for viewing distant objects.

pressure potential (or formerly) **turgor pressure,** *n.* a measure of the tendency of a cell to push water out of the cell, usually a positive value. Compare WALL PRESSURE.

presumptive area, *n.* a part of a developing embryo formed of cells which will become a particular organ or structure, for example presumptive mesoderm.

presynaptic membrane, *n.* the excitable membrane of the axon terminal next to the dendrite at a synapse.

prethroid, *n.* an artificially produced biocide, e.g. permethrin, similar in structure and action to the natural insecticides, PYRETHRUMS.

prey, *n.* any animal species that is fed upon by another (the PREDATOR).

Priestley, Joseph (1733–1804) English chemist who discovered oxygen, and observed that carbon dioxide (fixed air) was produced in FERMENTATION and that oxygen is produced by green plants in sunlight.

prickle, *n.* a hard, pointed plant structure containing no VASCULAR BUNDLE structure.

primary consumer, *n.* any organism that feeds directly on plant material, being the second tier of the PYRAMID OF NUMBERS (biomass). Primary consumers feed directly on primary producers (see PRIMARY PRODUCTION) and are fed upon themselves by SECONDARY CONSUMERS.

primary egg membrane, see VITELLINE MEMBRANE.

primary feather, *n.* the quill feathers on the hand of the bird's wing – the outermost flight feathers.

primary focus, see FOCUS.

primary follicle, *n.* an immature OVARIAN FOLLICLE.

primary growth, *n.* growth from the seedling stage of an ANGIOSPERM to produce the herbaceous structure of stems and roots. Many DICOTYLEDONS and a few MONOCOTYLEDONS carry out SECONDARY GROWTH to produce woody structures.

primary homonym, *n.* either of two specific names proposed in combination with the same generic name at the time of publication.

primary intergradation, *n.* a region of intermediate characteristics between two distinct phenotypic populations that has arisen through SELECTION.

primary oocyte, *n.* an early stage in OOGENESIS.

primary production, *n.* the production by living organisms (AUTOTROPHS) of organic material from inorganic sources. *Gross primary production* is the total amount of organic material synthesized in a given time, whereas *net primary production* is the amount of organic material formed in excess of that used in respiration. Compare SECONDARY PRODUCTION.

primary spermatocyte, *n*. an early stage in SPERMATOGENESIS.

primary structure, *n*. the linear sequence of amino acids in a POLYPEPTIDE CHAIN.

primate, *n*. any member of the mammalian order Primates, including lemurs, tarsiers, monkeys, apes and humans. These mammals have a placenta, possess nails rather than claws, and usually have a thumb and big toe which are opposable to the other digits, allowing objects to be grasped. All possess a relatively large brain and have well developed eyesight, often with BINOCULAR VISION.

primed cell, *n*. a cell stimulated by an ANTIGEN that can produce more cells which respond to the antigen, mediate reactions of cell-mediated immunity, or synthesize immunoglobulin.

primitive streak, *n*. a line of cells in the centre of the embryonic disc of reptiles, birds and mammals that forms the future axis of the embryo.

priority, *n*. the precedence of the earliest of two published taxonomic names.

prismatic layer, *n*. the middle layer of the three layers forming the shell of MOLLUSCS.

pro-, *prefix*. denoting before.

probability, *n*. the likelihood that a given event will occur. Probability is expressed either as values between zero (complete certainty that an event will *not* occur) and 1.0 (complete certainty that an event will occur) or percentage values between 0 and 100. Probability is used widely in SIGNIFICANCE tests.

proband, *n*. an individual considered as the starting point for the study of a family for the inheritance of a particular condition.

proboscis, *n*. **1.** the elongated snout of a tapir or elephant. **2.** the elongated mouthparts of some insects.

procaryote or **prokaryote,** *n*. an organism that is either a bacterium or a blue-green algae, its main characteristic being *procaryotic cells* lacking a membrane-bound nucleus and no MITOSIS or MEIOSIS. Compare EUCARYOTE and see Fig. 150.

procaryotic cell, SEE PROCARYOTE.

process, *n*. any method of operation in the manufacture of a substance or bacteria.

proctodaem, *n*. the inpushing of ectoderm in the embryo where it meets the endoderm to form the anus or cloaca.

procumbent or **prostrate,** *adj*. lying along the ground.

producer, *n*. any AUTOTROPH capable of synthesizing organic material, thus forming the basis of the food web (see FOOD CHAIN). ANGIOSPERMS are the main producers in the terrestrial ecosystem, and PHYTOPLANKTON in the aquatic environment. Compare CONSUMER.

productivity, *n*. the amount of material in terms of BIOMASS or energy generated in a given time in an ECOSYSTEM over and above the STANDING CROP.

proflavin, *n*. a MUTAGEN that can initiate FRAMESHIFTS in the TRANSLATION of an mRNA sequence.

progeny test, *n*. a method by which the genetic capabilities of an organism can be established by examining the performance of its progeny. For example, the 'milk–yield' genes of a bull can be estimated by measuring the milk productivity of its daughters.

progesterone, *n*. a female sex hormone produced by the CORPUS LUTEUM and the PLACENTA which is concerned with preparing the uterus to receive the fertilized OVUM, and in maintaining pregnancy. Progesterone is secreted on the stimulus of the corpus luteum by the pituitary hormones LH and LTH (LUTEOTROPHIC HORMONE).

proglottid, *n*. one of the divisions or apparent segments of a tapeworm, formed by transverse budding in the neck region, which remain connected as they move backwards, increasing in size and maturing. Eventually the proglottid is cut off and is shed from the gut of the host. Each proglottid contains male and female sex organs.

prognathous, *adj*. (of apes or man) having a protruding face or jaws.

prokaryote, see PROCARYOTE.

prolactin, see LUTEOTROPHIC HORMONE.

prolegs, *n*. the false legs present on the anterior part of the abdomen of caterpillars in the form of stumpy, jointless appendages.

proline, *n*. one of 20 AMINO ACIDS common in protein. It has a nonpolar 'R' group structure and is relatively insoluble in water. See Fig. 253. The ISOELECTRIC POINT of proline is 6.3.

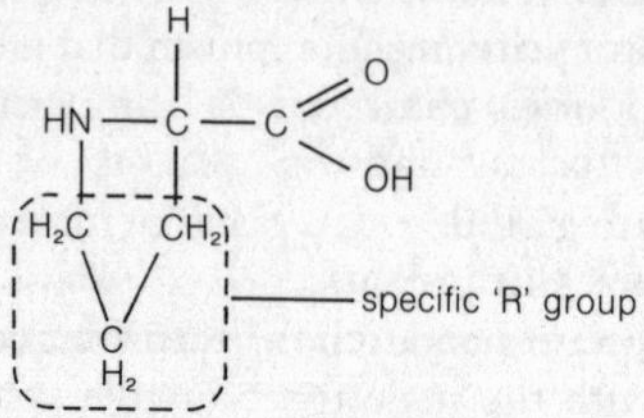

Fig. 253. **Proline.** Molecular structure.

prometamorphosis, *n*. the first stage of METAMORPHOSIS in amphibians.

promoter, see OPERON MODEL.

pronephros, *n*. the first excretory organ in the vertebrate embryo. Its duct (MULLERIAN DUCT) gives rise to the OVIDUCT in the female and degenerates in the male.

pronotum, *n.* the covering of the first thoracic segment in insects.

propagule, *n.* **1.** an infective stage of a plant PATHOGEN such as a fungal spore, by which the organism gains entry into a plant host. **2.** any part of an organism that is liberated from the adult form and which can give rise to a new individual, such as a fertilized egg or spore.

prophage, *n.* a BACTERIOPHAGE 'chromosome' integrated into the DNA of a bacterium.

prophase, *n.* the first stage of nuclear division (MITOSIS and MEIOSIS) in which the chromosomes coil and thicken and become visible with the optical microscope, condensing onto the inner wall of the NUCLEAR MEMBRANE. As the stage proceeds, the NUCLEOLUS disappears from view and the nuclear membrane disintegrates, leaving a clear area at the edge of the nucleus which contains the CENTROSOME.

Meiosis has a much more complicated prophase than mitosis, and can be summarized thus:

(a) meiosis has two prophases, the first one complex (see below) the second rather similar to prophase in mitosis.

(b) prophase 1 of meiosis can be divided into five substages: LEPTOTENE, ZYGOTENE, PACHYTENE, DIPLOTENE and DIAKINESIS. The essential processes occurring are: (i) pairing of homologous chromosomes, (ii) pairing of nonsister CHROMATIDS forming chiasmata with eventual CROSSING OVER.

proprioceptor, *n.* a receptor structure, linked to the nervous system of animals, that detects internal changes, particularly around joints, in tendons and muscles.

prostaglandins, *n.* a group of lipid substances that exert a wide range of stimulatory effects on the body, the most important of which is the enhancement of the effects of CYCLIC AMP. Prostaglandins are derived from many tissues including the prostate gland (or can be made synthetically), and have been used in the induction of labour and abortion.

prostate or **prostate gland,** *n.* **1.** a gland of male animals that produces substances which are added to the semen. ANDROGENS affect the size and secretion of the prostate gland, whose exact function is unknown. **2.** the gland associated with the male reproductive system in ANNELIDS and CEPHALOPODS.

prosthetic group, *n.* a nonprotein group attached to a protein. Such prosthetic groups are usually vital for the functioning of the protein. For example, many enzymes contain metallic ions, as in carboxypeptidase which contains zinc; HAEMOGLOBIN contains HAEM with an iron atom at the centre.

prostrate, see PROCUMBENT.

protagonistic muscles, *n*. muscles that work together to produce a movement.

protandrous, *adj*. **1.** (of male gametes) ripening before the female gametes. Compare PROTOGYNOUS. **2.** (of plant flowers) possessing ANTHERS which ripen before the stigma is able to receive the pollen, thereby preventing SELF-POLLINATION.

protease, *n*. any enzyme that splits proteins, such as PEPSIN, TRYPSIN, EREPSIN or RENNIN.

protective coloration, *n*. any patterning or camouflage that helps an organism blend into its background and thus gives it protection from predators.

protein, *n*. a large complex molecule (M.W. from 10,000 to more than 1 million) built up from AMINO ACIDS joined together by PEPTIDE BONDS. All proteins contain carbon, hydrogen, oxygen and nitrogen and most contain sulphur. Proteins are produced in the cytoplasm at the ribosomes (see PROTEIN SYNTHESIS) and begin as long, unbranched POLYPEPTIDE CHAINS, the 'primary structure'.

All protein molecules undergo a physical rearrangement to give a *secondary structure*. The most common type of shape is *alpha-helix* (right-handed) where the coils are held in place by hydrogen bonds. Some proteins, such as keratin, remain at this stage. An alternative secondary structure is *beta-pleating* where parallel polypeptide chains are cross-linked by hydrogen bonds forming an extremely tough structure, as in silk. Proteins with these relatively simple two-dimensional secondary structures are called *fibrous proteins*.

Some proteins undergo even more complex folding, where the secondary structure is arranged into a three-dimensional *tertiary structure* forming 'globular' proteins held together by forces between side groups. Such molecules are, for example, ENZYMES, ANTIBODIES, most blood proteins, and MYOGLOBIN. Finally, globular proteins can be composed of two or more polypeptide chains loosely bonded together, for example, HAEMOGLOBIN, giving a *quaternary structure*.

proteinase, see ENDOPEPTIDASE.

protein synthesis, *n*. a complex anabolic process occurring in all cells, by which genes control the precise structure of proteins manufactured in the cell. The following summary of events refers to a EUCARYOTE cell:

(a) DNA molecules in the chromosomes of the nucleus carry specific messages about how proteins are to be constructed.

(b) the DNA of each gene carries instructions about one protein chain (see ONE GENE/ONE ENZYME HYPOTHESIS).

(c) the two polynucleotide chains of the DNA molecule separate and messenger RNA nucleotides become attached in complementary pairing, the process being called TRANSCRIPTION.

(d) the mRNA molecule leaves the DNA which reforms a double helix.

(e) mRNA leaves the nucleus via pores in the NUCLEAR MEMBRANE and enters the cytoplasm where it becomes attached to one or more RIBOSOMES (see also POLYRIBOSOMES) near the ENDOPLASMIC RETICULUM.

(f) with the aid of transfer RNA, TRANSLATION of the message in the mRNA takes place, with the formation of a POLYPEPTIDE CHAIN which later becomes coiled (see PROTEIN).

proteolysis, *n*. the splitting of protein molecules by the HYDROLYSIS of their PEPTIDE BONDS.

proteolytic enzyme, *n*. any enzyme concerned with PROTEOLYSIS. See ENDOPEPTIDASE, EXOPEPTIDASE.

prothacic glands, *n*. the glands activated by a hormone from the neurosecretory cells in the brain that secrete the MOULTING HORMONE ecdysone. The glands have a cyclic activity within each INSTAR and break down on METAMORPHOSIS. They induce moulting, which in most insects occurs only in the juvenile stage, and in the absence of JUVENILE HORMONE stimulate metamorphosis.

prothallus, *n*. a small thalloid structure produced by the germination of the spore in ferns and related plants, which bears the ARCHEGONIA and ANTHERIDIA and is the GAMETOPHYTE generation.

prothorax, *n*. the first thoracic segment in insects.

prothrombin, see BLOOD CLOTTING.

Protista, *n*. the kingdom containing all unicellular organisms but now expanded to include many algae, bacteria, fungi, slime moulds and Protozoa.

Protochordata, *n*. a division of the phylum Chordata, including the Hemichordata, Urochorda and Cephalochorda. The group lacks a vertebral column, cranium and organs associated with the region of the head in other chordates.

protogynous, *adj*. (of female gametes) ripening before the male gametes. Compare PROTANDROUS(1).

protonema, *n*. a haploid, branched, filamentous structure, produced by the germination of a moss spore, that produces new plants by budding.

protonephridium, see FLAME CELL.

protophloem, see METAPHLOEM.

protoplasm, *n*. the living contents of a cell; i.e. the cytoplasm and nucleus, the whole cell being called the *protoplast*.

protoplast, see PROTOPLASM.

protopod larva, *n*. a primitive insect larva lacking abdominal segmentation and possessing only rudimentary appendages. Such larvae are usually found as ENDOPARASITES.

prototherian, *n.* any mammal of the subclass Prototheria, including the MONOTREMES.

prototroch, see TROCHOPHORE.

prototroph, *n.* a strain of an organism that does not possess growth requirements in excess of those of the wild type.

protoxylem, *n.* the first XYLEM to differentiate from the primary MERISTEMS, with annular (ring-like) and spiral bands of thickening on the walls. Although dead, the cells are able to elongate as the adjacent cells grow in length. Protoxylem differentiates eventually into METAXYLEM.

protozoan, *n.* any member of the phylum Protozoa (sometimes regarded as a subkingdom of the animal kingdom), including unicellular or acellular organisms. The members of the group feed as HOLOPHYTES, SAPROPHYTES or in a HOLOZOIC manner, move by means of FLAGELLA, CILIA or PSEUDOPODIA, and reproduce by FISSION or CONJUGATION.

proximal, *adj.* a point nearest to the body in any structure or nearest to the centre of the system concerned. For example, the wrist is the proximal part of the hand, being nearest to the body. Compare DISTAL.

proximal tubule, *n.* the convoluted tube between the loop of Henle and Bowman's capsule in the vertebrate kidney.

pruinose, *adj.* (of plant structures) covered with a whitish bloom.

psued– or **psuedo–,** *prefix.* denoting either false or closely resembling.

pseudodominance, *n.* the appearance of a recessive PHENOTYPE due to the DELETION MUTATION of a dominant allele in a HETEROZYGOTE.

pseudogamy, *n.* the parthenogenetic development (see PARTHENOGENESIS) of an egg after penetration of the egg membrane by a male gamete.

pseudoheart, *n.* any of the 10 enlarged blood vessels (5 pairs) in the earthworm, which have valves and act as hearts, pumping blood from the dorsal to the ventral vessel.

pseudopodium, *n.* (*pl.* pseudopodia) a protoplasmic projection (a 'false foot') from the main body of a protozoan – especially Rhizopods such as *Amoeba* – and from white blood cells. The cell moves in the direction from which the pseudopodium is produced and it is also used to engulf particles, both as food (in protozoans) and in PHAGOCYTOSIS. See AMOEBOID MOVEMENT.

pseudopolyploidy, *n.* a set of chromosome numbers in related species of organisms that erroneously suggests that the organisms are POLYPLOIDS(2).

pseudotrachea, *n.* any of the food channels resembling tracheae that cover the oral lobes of some insects (e.g. dipterans), which after converging eventually lead to the mouth.

Psilopsida, *n.* a TRACHEOPHYTE subdivision that includes many extinct forms. Related to the ferns, living forms are placed in the order *Psilotales* and are tropical or subtropical EPIPHYTES.

psittacosis, *n.* a contagious disease of birds such as parrots that can be transmitted to humans where it may cause bronchial pneumonia.

psoriasis, *n.* a noncontagious disease of the skin marked by scaly red patches, due probably to a disorder of the immune system.

psychrophile, *n.* an organism showing optimal growth at temperatures below 20°C and down to below 0°C, e.g. many bacterial species, yeasts and some algae.

pter- or **ptero-,** *prefix.* denoting a wing.

pteridophyte, *n.* any plant of the division Pteridophyta, including ferns, horsetails and club-mosses. The group is now not normally given taxonomic status, and is included in the TRACHEOPHYTA.

pteropod or **sea butterfly,** *n.* a small gastropod mollusc found in the plankton, with the foot modified into wing-like structures used for swimming. Pteropods are known as 'sea butterflies' because of the appearance given to them by these expansions of the foot. See Fig. 254.

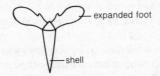

expanded foot

shell

Fig. 254. **Pteropod.** General morphology.

Pteropsida, *n.* a subdivision of the TRACHEOPHYTA, including the ferns, conifers and flowering plants.

Pterygota, *n.* a subclass of the class Insecta into which are placed most insects, including all winged forms and those which have secondarily lost the power of flight, such as lice or fleas. Compare AMETABOLA.

ptyalin, see AMYLASE.

pubescent, *adj.* (of plant structures) covered with fine hairs.

puberty, *n.* the period at the beginning of human adolescence when sexual maturity is attained.

pubis, *n.* the bone forming the anterior part of the PELVIC GIRDLE situated ventrally and projecting forward in most TETRAPODS.

puffs, see CHROMOSOME PUFFS.

pulmo- or **pulmon-,** *prefix.* denoting the lungs.

pulmonary, *adj.* relating to the lungs.

pulp cavity, *n.* the cavity within the tooth of vertebrates, containing nerves, blood vessels, connective tissue and odontoblasts, which opens into the tissues in which the tooth is embedded.

pulse, *n.* the expansion of an artery as the left ventricle contracts (see BLOOD PRESSURE) and which can be detected where the artery is close to the body surface, for example, the radial artery at the wrist of man.

pulvillus, *n.* a cushion or pad between the claws on an insect foot.

pulvinus, *n.* a mass of thin-walled cells at the base of the leaf petiole in certain plants, forming a swollen area surrounding the vascular tissue. The pulvinus is subject to large changes in TURGOR. For example, in the runner bean the pulvinus is turgid during the day and supports the petiole, so that the leaf is held outwards. At night the pulvinal cells loose turgidity and the leaf droops. These diurnal leaf positions are sometimes called *sleep movements*. The sudden change in leaf posture observed in *Mimosa sensitiva* when touched is also controlled by pulvinal cells.

punctate, *adj.* (of plant structures) dotted or pitted.

punctiform, *adj.* (of plant structures) similar to a dot, small and round.

punctuated equilibrium, *n.* the theory that evolutionary change has occurred by short periods of rapid change punctuated by periods of stability. Such a process would be characterized by the absence of an infinite range of intermediate forms, and supporters of the theory point to the vertebrate fossil record as evidence.

pungent, *adj.* (of plant structures) stiffly pointed to the extent that it will prick.

Punnett square, *n.* a table devised by the British geneticist R. Punnett, in which all possible combinations of gametes and progeny are displayed in a grid structure. For example in the cross:

$$♀ \text{ Aa, Bb x aa, Bb } ♂$$

the Punnett square would be as in Fig. 255.

♀ gametes

		AB	Ab	aB	ab
♂ gametes	aB	AaBB	AaBb	aaBB	aaBb
	ab	AaBb	Aabb	aaBb	aabb

Fig. 255. **Punnett square.**

pupa, *n.* a stage found in ENDOPTERYGOTES that occurs as an apparently inactive phase between the larva and adult insect. Whilst locomotion and feeding are absent, extensive developments take place in the formation of adult structures within the pupa which is also referred to as a CHRYSALIS in LEPIDOPTERANS.

pupil, *n.* the central opening in the iris of the vertebrate eye through which the light passes to the lens and retina. It changes in size as a result of muscle contraction and expansion moving the iris.

pure breeding, see HOMOZYGOTE.

pure breeding line or **true breeding line,** *n.* a strain of individuals that are all homozygous for the genes being considered, so that their progeny show no variation in those genes. See BREEDING TRUE.

purine, *n.* one of two types of base found in NUCLEIC ACIDS, having a double ring structure see ADENINE and GUANINE. Purines always pair with PYRIMIDINES in the two strands of DNA, ensuring a parallel-sided molecule.

purinergic, *adj.* (of nerve endings) releasing PURINES as TRANSMITTER SUBSTANCES.

Purkinje cell, *n.* a specialized NEURONE of the CEREBELLUM consisting of an axon, a large cell body and a single dendrite which is much branched, like a tree.

Purkinje tissue, see HEART.

pus, *n.* a yellowish fluid consisting of serum, white blood cells, bacteria and tissue debris formed during the liquiefaction of inflamed tissue (*suppuration*).

pustule, *n.* a blister-like area on a plant structure from which a fungal fruiting body emerges.

putrefaction, *n.* the decomposition of proteins, which gives rise to foul-smelling products.

pycno-, *prefix.* denoting thick or dense.

pycnosis, *n.* the process of shrinking (and becoming more dense) in a cell nucleus, which results in its becoming strongly staining, usually as it dies.

pyloric, *adj.* of or relating to that end of the vertebrate stomach which opens into the intestine. The other end is called the *cardiac* area of the stomach.

pyloric sphincter, *n.* a ring of smooth muscle surrounding the opening of the PYLORIC end of the stomach into the intestine. The sphincter prevents the premature escape of food from the stomach into the intestine before it is properly digested. Constriction of the sphincter (*pyloric stenosis*) is well known in young children, a condition recognised by explosive vomiting. Compare CARDIAC SPHINCTER.

pyogenic, *adj.* pus-producing.

pyramid, biomass, see PYRAMID OF NUMBERS.

pyramid, ecological, see PYRAMID OF NUMBERS.

pyramid, energy, see PYRAMID OF NUMBERS.

pyramid of numbers, *n.* a means of depicting TROPHIC LEVELS to illustrate the numbers of organisms at any one level. The broad base indicates the large numbers of PRODUCERS, the next (narrower) stage the numbers of PRIMARY CONSUMERS, a third stage (narrower still) of SECONDARY CONSUMERS and at the apex the top PREDATOR. It is better expressed in terms of biomass or energy (see ECOLOGICAL PYRAMID), as in some cases, for example, parasites higher in the trophic levels than the host, greater numbers occur higher, and do not produce a pyramid when numbers only are considered. See Fig. 256 and FOODCHAIN.

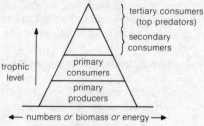

Fig. 256. **Pyramid of numbers.**

pyranose ring, *n.* a monosaccharide with a six-membered ring structure.

pyrenoid, *n.* a region of starch formation found in the CHLOROPLASTS of various algae, for example, *Spirogyra, Chlamydomonas.*

pyrethrum, *n.* an INSECTICIDE prepared from the dried flowers of the chrysanthemum plant, the active ingredient being called *pyrethrin.*

pyridoxine (B_6), *n.* a water-soluble vitamin of the B–COMPLEX found in fresh meat, eggs, liver, fresh vegetables and whole grains. The vitimin acts as a coenzyme in AMINO ACID metabolism from carbohydrates, a deficiency of which causes dermatitis and, sometimes, motor impairment.

pyrimidine, *n.* one of three types of bases found in NUCLEIC ACIDS, with a single ring structure. DNA contains CYTOSINE and THYMINE, RNA contains cytosine and URACIL. Pyrimidines always pair with PURINES in DNA.

pyrogallol or **trihydroxybenzene,** *n.* a solube phenol ($C_6H_4(OH)_3$) which, in alkaline solution, will absorb oxygen and is used to estimate the volume of oxygen in a gaseous sample.

pyrogen, *n.* any substance which alters the body thermostat of HOMOIOTHERMS to a higher setting, giving rise to fever.

pyrrole, *n.* a porphyrin building block that has a five-membered heterocyclic structure and contains nitrogen.

Pyrrophyta, *n.* the fire algae, of which the *Dinophycae* or dinoflagellates form the most important members of the PHYTOPLANKTON. They possess a photosynthetic pigment which is yellow to golden brown in colour (hence 'fire algae'). Most forms have two dissimilar FLAGELLA though others lack them entirely.

pyruvic acid, *n.* an important 3-carbon molecule formed from GLUCOSE and GLYCEROL in glycolysis (see Fig. 257). SEE ALSO ACETYLCOENZYME A. Pyruvic acid is broken down further, the precise reactions depending upon whether oxygen is present or not. See AEROBIC RESPIRATION, ANAEROBIC RESPIRATION.

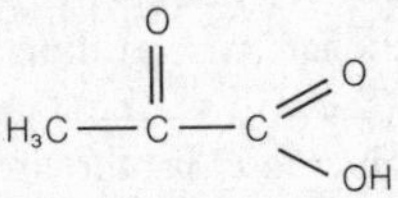

Fig. 257. **Pyruvic acid.** Molecular structure.

Q

Q_{10} or **temperature coefficient,** *n.* a measure of the rate of increase in METABOLIC RATE over a 10-degree range in temperature. Thus if an organism has a metabolic rate at 10°C of T units, a rate of twice T units at 20°C, the $Q_{10} = 2$. A Q_{10} of 2 is the typical exponential increase in rate exhibited by enzymes up to a certain maximum rate, after which DENATURATION occurs faster than increase, causing an overall reduction of the rate.

Qo_2, *n.* the oxygen uptake in microlitres (μl) per milligram dry weight per hour.

Q technique, *n.* a method of analysis used in taxonomic work in which there is an association of pairs of taxa in a data matrix.

quadrat, *n.* an area of ground surface, often 1m², used as a sampling unit in population studies.

quadrate, *n.* a cartilage bone at the back end of the upper jaw in the majority of vertebrates. In mammals it becomes the INCUS of the ear, but in most other vertebrates forms the articulation of the lower jaw.

quadrat sampling, *n.* (in statistics) a method of examining the distribution of organisms by taking samples within squares of known size, normally 1m². Such squares are selected at random and marked usually with wire outlines. A series of quadrats is usually sampled so that comparisons are possible between, say, the distribution of a given plant in two different areas. The mean number in each series of quadrats can be compared. See SIGNIFICANCE.

qualitative inheritance, *n.* a genetical system in which (usually) a small number of genes is involved in the control of a particular CHARACTER. Qualitative characters show discrete forms (rather than continuously varying forms, see POLYGENIC INHERITANCE) and are not normally affected by the environment to any large extent. The major genes in the system can often be detected individually and obey the laws of

447

MENDELIAN GENETICS. The proportions of the various forms of character observed in a progeny can be compared with the expected proportions, using a CHI—SQUARE TEST.

quantasomes, see CHLOROPLAST.

quantitative inheritance, see POLYGENIC INHERITANCE.

Quaternary period, *n.* the last two million years of the Cenozoic Era, from 2 million years BP (the end of the Tertiary) to the present time. During this period the increase in the human population has had significant effect on the populations of both other animals and plants.

quaternary structure of protein, see PROTEIN.

queen substance, see ROYAL JELLY.

quill, *n.* the hollow stalk of the feather of birds. **2.** a whole wing or tail feather. **3.** the spine of a porcupine.

quinone, *n.* any of the various compounds derived from benzene.

R

r, *n.* the intrinsic rate of natural increase of a population.

rabies, *n.* an ACUTE viral disease of the nervous system in many mammals, particularly man, dogs, cattle and foxes, resulting in degeneration of the spinal cord and brain, leading to death. Entry of the virus usually is by a wound or skin abrasion caused by a rabid animal. The incubation period is usually 3–8 weeks, the first signs being increased muscle tone and extreme difficulty in swallowing. The spasmodic contractions of the throat muscles that cause swallowing may become extremely painful and contraction may even be triggered by the sight of water. Thus individuals are described as *hydrophobic*. Despite recent advances, infected individuals must be vaccinated in the first few days after infection for effective treatment.

race, *n.* a population that can be distinguished from other populations of the same species by several genetical characteristics such as frequency of genes or chromosomal arrangements. For example, in humans, different races have been found to have quite different frequencies of alleles for the ABO BLOOD GROUP locus. See FOUNDER EFFECT.

raceme, *n.* an inflorescence in which the main stalk bears the flowers on stalks (pedicels), producing a conical form. For example, the foxglove, in which there is usually no terminal flower.

racemose, *adj.* being or resembling a RACEME.

race-specific resistance or **vertical resistance,** *n.* a situation where a small number of major genes in the host confer high levels of RESISTANCE to a specific attacking organism. Such a system, however, is likely to be ineffective should the attacking organism undergo genetic change. Compare NONRACE-SPECIFIC RESISTANCE.

rachi-, *prefix.* denoting a spine.

rachis or **rhacis,** *n.* any central axis, particularly that of a feather.

radi- or **radio-,** *prefix.* denoting a ray.

radial symmetry, *n.* an animal body structure in which any section through the mouth and down the body length divides the body into identical halves. Coelenterates, ctenophores (sea gooseberries) and echinoderms are all radially symmetrical. Compare BILATERAL SYMMETRY.

radiant energy, see ENERGY.

radiation, *n.* the electromagnetic energy that travels through empty space with the speed of light (2×10^8 ms $^{-1}$). All objects emit radiation, at room temperature mostly in the infrared range, whereas at high temperatures visible radiation is produced. See ELECTROMAGNETIC SPECTRUM, ULTRAVIOLET LIGHT, X-RAY.

radical, *adj.* (of plants) arising from the root or crown.

radicle, *n.* the basal part of the embryo in a seed, developing into the primary root of the seedling.

radioactive isotope, see ISOTOPE.

radioactive label, *n.* any substance that emits an ionizing radiation which is detectable and can be used to locate it.

radioisotope, see ISOTOPE.

Radiolaria, *n.* a group of marine, planktonic protozoans, possessing siliceous skeletons and forming benthic oozes (see BENTHOS).

radius, *n.* the anterior bone in the lower part of the forelimb of vertebrates. See PENTADACTYL LIMB.

radula, *n.* a rasping organ of molluscs situated in a sac on the underside of the BUCCAL cavity. It is used for tearing plant material by rubbing it against the hardened surface of the mouth.

ragworm, *n.* the common name of errant POLYCHAETE worms such as the nereis.

rainfall, *n.* the precipitation of fresh water from the atmosphere, which, through its variability, produces different vegetation in different regions.

Ramapithecus, *n.* a genus of hominid present in Africa, Europe and Asia, between 14 million years ago and 8 million years ago. They largely replaced DRYOPITHECINES in open country.

ramenta, *n.* the brown, flaky epidermis of ferns covering young leaves and stems.

ramet, *n.* a single individual from a CLONE.

ramus, *n.* a branch.

ramus communicans, *n.* the slender trunk that carries nerves from the ventral root of the spinal nerve to a ganglion of the sympathetic nervous system (see AUTONOMIC NERVOUS SYSTEM). See REFLEX ARC.

random fixation, *n.* the accidental loss of one allele resulting in the fixation of another in a population.

random genetic drift (RGD) or **Sewall Wright effect,** *n.* changes in ALLELE frequency in a population from one generation to another due to chance fluctuations. RGD is important in small populations which are subject to sampling error and where an allele can be lost (0% frequency) or fixed (100%). See Fig. 258.

The phenomenon was first described by the American geneticist Sewall Wright.

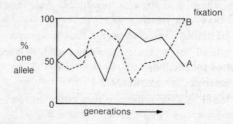

Fig. 258. **Random genetic drift.** The frequency of one allele of a gene over many generations. The continuous line represents population A, the broken line population B.

random mating or **panmixis,** *n.* the selection of mates by chance so that the choice of a partner is not influenced by the genotypes under study. For example, it is normal for human beings to display random mating with respect to the ABO BLOOD GROUP, but ASSORTATIVE MATING with respect to other genetically controlled characteristics, such as body shape.

range, *n.* **1.** (of a species) the extent of its distribution. **2.** (in behaviour) the area within which an individual normally lives.

ranking, *n.* the arranging of a taxon in the appropriate position in the taxonomic hierarchy, for example, selecting the category as an order, rather than a class or family.

raphe, *n.* a longitudinal slit or ridge as found in diatoms, or in some seeds where it marks the position of the FUNICLE(1).

raphide, *n.* a calcium oxalate crystal present in bundles in some plant cells.

raptor, *n.* any bird of prey. The term usually denotes a diurnal bird of prey such as an eagle, hawk or falcon, but can also denote an owl.

Rassenkreiss, see POLYTYPIC.

ratite, *n.* any flightless bird that lacks a keel on the sternum, such as an ostrich, rhea or emu.

Raunkiaer's Life Forms, *n.* a system of classification of plant forms based on the positioning of perennating buds (see PERENNATION) relative to ground level. See PHANEROPHYTE, CHAMAEPHYTE, HELOPHYTE.

Ray, John (1627–1705) English naturalist, regarded as the founder of British natural history, who originated natural botanical classification and the division of flowering plants into monocotyledons and dicotyledons. The Ray Society is named after him.

ray, *n.* a band of PARENCHYMA cells in plant roots and stems leading out from the centre along a radius and separating the vascular tissues, forming xylem rays and phloem rays in primary and secondary growth. Rays enable the transport of water and other materials across the stem or root. See SECONDARY THICKENING.

ray floret, *n.* any of the small flowers radiating from an inflorescence.

reactant, *n.* one of the groups of molecules participating in a reaction, which act upon others to produce new molecules.

reaction, *n.* a chemical process during which one substance is changed to another.

reaction chain, *n.* a chemical or atomic process such as combustion or atomic fission in which the products of the reaction promote the reaction itself.

reaction time, see LATENT PERIOD.

recapitulation, *n.* the apparent repetition in the embryological development of an organism of stages similar to those of the ancestral adult forms which, in terms of evolution, preceded it. The concept is expressed as: ontogeny recapitulates phylogeny. For example, gill slits appear in the development of the embryo of man.

receiving waters, *n.* water bodies into which waste is emptied from treatment plants.

Recent epoch, see HOLOCENE EPOCH.

receptacle, *n.* **1.** also called **torus** in flowering plants, the end of the stalk becoming the flower parts. **2.** in ferns, the mass of tissue that becomes the SPORANGIUM. **3.** in liverworts, the cup containing a GEMMA. **4.** in algae, the swollen tip of a branch carrying the reproductive organs.

receptaculum seminis, *n.* a sac-like structure present in some female or hermaphrodite invertebrate animals in which sperm, previously transferred from a male, is stored.

receptor, *n.* any cell or organ of an animal capable of detecting a change in the external or internal environment, and which subsequently brings about a response in the behaviour of the animal. A receptor, for example the eye, receives and transforms stimuli into sensory nerve impulses. See also CHEMORECEPTOR, REFLEX ARC.

receptor site, *n.* a point or structure in a cell at which combination with a drug or other agent results in a specific change in cell function.

recessive allele or **recessive gene,** *n.* an ALLELE that only shows its effect in the PHENOTYPE when present in a HOMOZYGOTE. When paired with a dominant allele the effect of the recessive allele is hidden. See RECESSIVE CHARACTER.

recessive character, *n.* a character that is controlled by a particular allele of a gene and which will only be displayed when the individual is homozygous for this allele. See DOMINANCE(1).

recessive epistasis, *n.* a form of EPISTASIS in which a pair of recessive alleles of one gene can cause a masking effect on the expression of alleles at another locus. Such an interaction would produce a DIHYBRID(1) ratio of 9:3:4 instead of the more normal 9:3:3:1. Compare DOMINANT EPISTASIS.

reciprocal cross, *n.* a pair of crosses in which the PHENOTYPES of the partners are reversed. Thus, in humans the following are reciprocal crosses:

$$\text{♀ colourblind} \times \frac{\text{normal}}{\text{vision}} \text{♂} \quad \Big| \quad \text{♀} \frac{\text{normal}}{\text{vision}} \times \text{colourblind ♂}$$

Different results from the two crosses indicate SEX LINKAGE of the gene controlling the character.

recombinant DNA, *n.* DNA from PLASMIDS that can be treated to pick up foreign DNA (e.g. from a EUCARYOTE) and carry it into a bacterial cell. Here the new DNA may express, coding for foreign proteins. Such recombinant DNA has been widely used in GENETIC ENGINEERING.

recombinant gamete, *n.* a GAMETE containing a new combination of alleles as compared with those found on the HOMOLOGOUS CHROMOSOMES of the parent, the alteration being produced by RECOMBINATION.

recombination, *n.* a rearrangement of genes during MEIOSIS so that a GAMETE contains a haploid GENOTYPE with a new gene combination. Recombination can occur by INDEPENDENT ASSORTMENT of genes on different chromosomes, but the term is used normally to refer to genes linked on the same chromosome where recombination is achieved by CROSSING OVER. See Fig. 259.

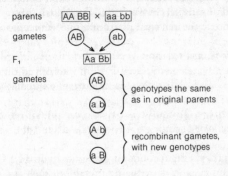

parents $\boxed{\text{AA BB}} \times \boxed{\text{aa bb}}$

gametes (AB) (ab)

F_1 $\boxed{\text{Aa Bb}}$

gametes (AB) (a b) $\Big\}$ genotypes the same as in original parents

(A b) (a B) $\Big\}$ recombinant gametes, with new genotypes

Fig. 259. **Recombination.** The rearrangement of genes during meiosis.

recon, *n.* the smallest genetic unit between two of which RECOMBINATION can take place, effectively a DNA base.

rect-, *prefix.* denoting straight.

rectal gland, *n.* a gland that opens into the rectum in many vertebrate animals, that may produce PHEROMONES.

rectrix, see FEATHER.

rectum, *n.* the terminal part of the intestine of an animal which opens into the ANUS or CLOACA.

rectus, *n.* a straight muscle, such as either of the two muscles that serve, together with the more anterior pair of oblique muscles, to move the eyeball. They are situated behind the eyeball in a posterior position.

recurrent laryngeal nerve, *n.* a branch of the vagus nerve in mammals which loops round the DUCTUS ARTERIOSUS and then goes forward along the trachea. Its peculiar route results from evolutionary lengthening of the neck.

recurved, *adj.* (of plant structures) bent backwards.

red blood cell, see ERYTHROCYTE.

redia, *n.* a larval stage of liver fluke which develops from the sporocyst larva. It possesses a mouth, suctional pharynx and simple gut, and gives rise to secondary rediae or to carcaria larvae.

red nucleus, *n.* a ganglionic centre in the midbrain of mammals, which, if destroyed results in decerebrate rigidity; it therefore plays an important part in the control of movement and posture.

redox potential, *n.* a quantitative measurement of the willingness of an electron carrier to act as a *red*ucing or *ox*idizing agent. Redox potential is measured in volts; the more negative the value, the better the carrier will act as a reducing agent. Thus, in an ELECTRON TRANSPORT SYSTEM the carriers are arranged in order of increasing redox potentials (negative to positive). For example, in Fig. 205 ferredoxin has a lower (more negative) redox potential than plastoquinone which has a higher value (less negative).

redox reaction, *n.* a reaction in which oxidation and reduction occur.

red tide, *n.* sea water discoloured by a bloom of red dinoflagellates in such numbers as to kill other sea creatures, particularly fish, by the production of toxins.

reducing agent, *n.* any substance that is capable of removing oxygen from a molecule or of adding hydrogen, i.e. of contributing electrons to a process.

reducing sugar, *n.* a sugar containing free aldehyde $(H-C=O)$ or keto $(C=O)$ groups capable of reducing metal ions such as Cu^+ and Ag^+. Reducing sugars such as glucose and fructose will reduce solutions in Benedict's test and FEHLING'S TEST.

reduction, *n.* a change in an atom or molecule through losing oxygen, adding hydrogen, or gaining electrons.

reduction division, *n.* nuclear division where the daughter cells have half the genetic material of the parental cell. See MEIOSIS.

reductionism, *n.* an erroneous belief that complex situations may be explained by reducing them to their component parts and explaining these.

redundant character, *n.* taxonomic characters so closely associated with other already used characters (e.g. in a key) that they are of no use in subsequent analysis.

redwood, see SEQUOIA.

reflex arc, *n.* the nervous connections between RECEPTOR and EFFECTOR that result in a simple response being elicited by a specific stimulus (a *reflex action*), such as a knee jerk without the involvement of the brain. The receptor is excited by a stimulus which induces a nerve impulse in the DENDRITE of a sensory (afferent) NEURONE. This passes through the cell body in the dorsal root ganglion and on along with sensory AXON to the SPINAL CORD. There it crosses a SYNAPSE, usually to an intermediate neurone which synapses with the efferent neurone and passes out of the spinal cord via the ventral root along the axon of the motor (efferent) neurone to the effector which produces the stereotyped response. Most reflexes involve more than one synapse, i.e. are *polysynaptic*. See Fig. 260. See RAMUS COMMUNICANS.

reflex action, see REFLEX ARC.

refractory period, *n.* the period of inexcitability, that normally lasts

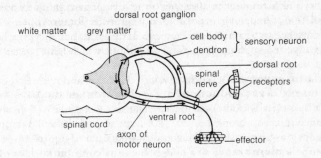

Fig. 260. **Reflex arc.** The arrows indicate the direction of impulse from receptor to effector.

about three milliseconds, during which the AXON recovers after it has transmitted an impulse. During the refractory period it is impossible for the axon to transmit another impulse, because the membrane is being repolarized by ionic movements at this time. During the *absolute* refractory period no NERVE IMPULSE can be transmitted, but during the *relative* refractory period an impulse can be transmitted providing the stimulus is strong.

refugium, *n.* an area that has escaped alteration during glaciation, e.g. a *nunatak* (an area of isolated land such as a mountain peak), and in which relic communities may exist.

regeneration, *n.* the replacement or repair of tissues or organs lost through damage. In animals, the degree of regeneration varies from group to group but is generally more extensive in the lower phyla. In plants, regeneration is common and used extensively in VEGETATIVE PROPOGATION.

regression analysis, *n.* (in statistics) a test in which the size of one variable (the dependent variable) is dependent on another (the independent variable). Often the relationship is a linear one, enabling a line of best fit to be drawn on a SCATTER DIAGRAM.

regressive character, *n.* any character that is reduced during evolutionary change.

regular, *n.* (of plant structures) exhibiting RADIAL SYMMETRY.

regulating factor, *n.* a chemical secretion of the HYPOTHALAMUS, associated with blood sugar levels, that controls the secretion of hormones. The secretion and inhibition of hormones involve negative feedback systems (see FEEDBACK MECHANISM).

regulation, *n.* **1.** (in embryology) the process of determining normal development, even in cases of damage, where a properly formed embryo may result even after the loss of a large part. In many animals regulation after damage is possible only before fertilization but in others it may take place in later development. **2.** the limitation of a population over a period of time by natural factors such as DENSITY DEPENDENT FACTORS.

regulator gene, see OPERON MODEL.

regulatory heat production, *n.* the component of STANDARD META-BOLIC RATE (SMR) additional to BASAL METABOLIC RATE (BMR) produced by an animal exhibiting TACHYMETABOLISM at environmental temperatures below those of thermoneutrality (units: ml O_2/g, ml O_2/g.h).

Reissner's membrane, *n.* a membrane situated in the cochlea of the inner ear, that separates the middle canal from the vestibular canal. Named after the German physiologist E. Reissner.

relationship, *n.* the evolutionary connection between organisms in terms of distance from a common ancestor.

relative growth, *n.* the growth of a structure in relation to the growth of another structure. For example, skull size in developing humans has a smaller relative growth than long bones.

releaser, *n.* any stimulus that gives rise to an instinctive act. For example, the red breast of a male robin causes another male to attack it.

releasing factor, *n.* a special form of REGULATING FACTOR that stimulates the release of hormones into the blood; *inhibiting factors* prevent the release of the hormone.

relic or **relict,** *n.* any organism, community or population surviving from an earlier age.

relic distribution or **relict distribution,** *n.* the much-reduced distribution of organisms resulting from an overall environmental change, such as glaciation.

relict, see RELIC.

remige, see FEATHER.

renal, *adj.* pertaining to the kidney.

renal portal system, *n.* the system of veins in fish and amphibians taking blood from the region of the tail or hind limbs directly to the kidneys. See Fig. 261.

rendzina or **humus-carbonate soil** or **A-C soil,** *n.* a shallow soil formed over limestone that usually shows a dark-coloured organic horizon (A) resting directly on weathering limestone (C).

reniform, *adj.* kidney-shaped.

renin, *n.* a protein-splitting enzyme (sometimes called a hormone) secreted into the blood by the cells lining the efferent glomerular vessels of the kidney. Renin combines with a protein from the liver to form

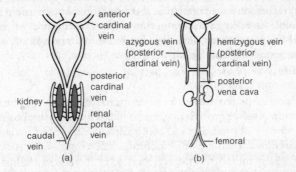

Fig. 261. **Renal portal system.** The venous blood supply to the kidneys in (a) fish (renal portal system present), (b) mammals (renal portal system absent).

ANGIOTENSIN which in turn stimulates the cortex of the ADRENAL GLAND to release ALDOSTERONE.

rennin or **chymase,** *n.* an enzyme present in gastric juice secreted by the gastric glands of the stomach wall that coagulates caseinogen in milk to form casein, which forms an insoluble curd (a calcium–casein compound) which is then attacked by pepsin. It is important particularly in young mammals because it increases retention time in the stomach allowing for a more efficient digestion of the primary food source.

replacement name, *n.* a substitute name proposed in taxonomic work to replace a name given incorrectly. It takes on the original TYPE and TYPE LOCALITY.

replacing bone, see CARTILAGE BONE.

replica plating, *n.* a technique used in the indirect selection of bacterial mutants that involves the transfer of an organism (usually a bacterium or fungus) from one culture to numerous 'plates' using a velvet pile or needles.

replication, *n.* the production of exact copies of complex molecules during the growth of living organisms. See DNA, BASE PAIRING.

replum, *n.* the central part that remains after the valves of a dehiscent fruit have been shed. See PLACENTA (2).

repression, *n.* the state in which a gene is prevented from being transcribed, so that no protein is produced. See OPERON MODEL.

repressor, *n.* a substance, often proteinaceous, that prevents the function of a gene. See OPERON MODEL.

reproduction, *n.* **1.** the production of young. **2.** the mechanisms by which organisms give rise to others of the same kind.

reproductive isolation, *n.* the state of a separated population in which gene flow between the population and other similar populations is either impossible or reduced to a minimum. This can be brought about by several factors, which may be geographical, ecological, seasonal, physiological or behavioural.

reproductive phase, *n.* that part of the life cycle during which reproduction takes place.

reproductive potential, *n.* the population size reached as a result of the maximum possible growth rate, under ideal environmental conditions, in the absence of predators and competition for food and space. Under these circumstances there is no environmental resistance. See FECUNDITY.

reptile, *n.* any member of the vertebrate class Reptilia, including turtles and tortoises (Chelonia), lizards and snakes (Squamata), crocodiles (Crocodilia), the extinct dinosaurs (Ornithischia, Saurischia), pterodactyls (Pterosauria), plesiosaurs (Plesiosauria), and the ancestors of mammals, the Therapsida. All reptiles are egg-laying and present-day forms are POIKILOTHERMS, though some dinosaurs may have been HOMOIOTHERMS.

repulsion, *n.* an arrangement in a double heterozygote where a WILD TYPE allele of one gene is adjacent to a mutant allele of another gene on the same HOMOLOGOUS CHROMOSOME. Such an arrangement of linked genes usually is referred to as *in repulsion*. Compare COUPLING.

reserpine, *n.* an alkaloid extracted from a species of *Rauwolfia*, that is used as a sedative and as an antihypersensitive agent to reduce hypertension.

reserves, *n.* any stored food supplies which may be drawn on in times of food shortage.

residual volume, *n.* the air remaining in the lungs after respiration, *c* 1,200ml.

resilient, *adj.* (of plant structures) springing back into position after being bent.

resistance, *n.* any inherited characteristic of an organism that lessens the effect of an adverse environmental factor such as a pathogen or parasite, a biocide (e.g. herbicide, insecticide, antibiotic) or a natural climatic extreme such as drought or high salinity.

resistance factor, SEE R FACTOR.

resistance transfer factor (RTF), *n.* a gene set that carries genetic information for the transfer of R FACTORS.

resolution, *n.* the minimum distance between two points at which they can be seen as such rather than as a single point. With the light microscope this is approximately half the wavelength of light used in illumination. Only with a shorter wavelength can greater resolution be achieved, as in the ELECTRON MICROSCOPE.

resolving power, *n.* the distance apart of two objects at which they can be distinguished as two separate objects by the microscope.

resorption, *n.* the taking back into an organism of any structure or secretion produced.

resource, *n.* **1.** any potential product of an area. **2.** a food supply.

respiration, *n.* **1.** a process by which gaseous exchange (oxygen and carbon dioxide) takes place between an organism and the surrounding medium. See AERIAL RESPIRATION, AQUATIC RESPIRATION, BREATHING. **2.** a form of CATABOLISM that takes place in all living cells. See ANAEROBIC RESPIRATION, AEROBIC RESPIRATION.

respiratory centre, see BREATHING.

respiratory cycle, *n.* the sequence of events during which an animal inhales and exhales a given volume of air through the respiratory system. In humans some $0.5 \, dm^3$ ($500 \, cm^3$) of air is taken in and expelled during the cycle which lasts about 5 seconds. See BREATHING, VITAL CAPACITY.

respiratory enzyme, *n.* any enzyme that catalyses oxidation reduction reactions.

respiratory gas, *n.* any gas that takes part in the respiratory process. The term usually denotes oxygen and carbon dioxide. See RESPIRATION.

respiratory movement, *n.* any movement by an organism that aids RESPIRATION, such as the movement of air or water through lungs or gills.

respiratory organ, *n.* any organ across the surface of which gases pass to and from the body fluids to the exterior, such as gills or lungs.

respiratory pigment, *n.* a substance present in blood, usually in the corpuscles or the plasma, that combines reversibly with oxygen and acts as a store for it. Oxygen is taken up in a respiratory organ such as the lung, and blood gives up oxygen where it comes into contact with oxygen–deficient tissues. Examples of such pigments include HAEMO-GLOBIN, HAEMOCYANIN, CHLOROCRUORIN, HAEMERYTHRIN. See also OXY-GEN DISSOCIATION CURVE.

respiratory quotient (RQ), *n.* the ratio of volume of CO_2 produced to the volume of O_2 taken up in a given period of time. From RQ estimates it is possible to deduce what type of food is being oxidized. Carbohydrates have an RQ of 1.0, fats of 0.7 and proteins 0.9. An RQ lower than 0.7 indicates that organic acids such as malic acid are being oxidized (see CRASSULACEAN ACID METABOLISM): an RQ greater than 1.0 indicates that ANAEROBIC RESPIRATION is occurring.

respiratory surface, *n.* a special area that is developed in order to satisfy the requirements for gaseous exchange in larger organisms. Examples include external gills, internal gills, lungs, and the insect tracheae. The surfaces are thin and moist with a differential gradient maintained between the air/water and the internal tissues.

respirometer, *n.* any apparatus used to measure the uptake of oxygen by organisms during AEROBIC RESPIRATION. Examples include the Warburg apparatus, Cartesian Diver apparatus, Philippson electrolytic respirometer.

response, see STIMULUS.

resting cell, *n.* any cell that is not undergoing active division (mitosis).

resting egg, *n.* any invertebrate egg that undergoes a period of dormancy during which it is resistant to adverse conditions.

resting potential, *n.* the electrical potential present between the inside and outside of a nerve or muscle fibre when in a resting state. The inside is negatively charged (-60 millivolts) and the outside positively charged. The membrane is polarized and the axon has a 'resting potential'. Compare ACTION POTENTIAL.

restriction analysis, *n.* a genetical technique for fine mapping of a gene using RESTRICTION ENZYMES.

restriction enzyme, *n.* an endonuclease that recognizes specific DNA base sequences and cleaves both strands of DNA at the same time.

resynthesis, *n.* the production of a substance that has previously been broken down.

rete or **rete mirabile,** *n.* a system of arterial and venous capillaries in which COUNTERCURRENT EXCHANGE can occur. Such a network occurs in, for example, the wall of the swim bladder, enabling the wall to be supplied with blood without a large loss of oxygen from the bladder into the blood.

reticular formation, *n.* the part of the CNS that consists of small islands of grey matter separated by fine bundles of nerve fibres running in all directions.

reticulate evolution, *n.* evolution that is dependent on repeated intercrossing between a number of breeding lines. The evolutionary process is thus both convergent and divergent at the same time.

reticulate thickening, *n.* the strengthening of an XYLEM vessel or TRACHEID in the form of a network of lignin deposition.

reticulin fibres, *n.* fibres found intercellularly in many vertebrate tissues which they bind together. They consist largely of COLLAGEN.

reticuloendothelial system, see MONONUCLEAR PHAGOCYTE SYSTEM.

reticulum, *n.* **1.** the second compartment of the ruminant stomach. **2.** a network, particularly of PROTOPLASM.

retina, *n.* the lining of the interior of the vertebrate eye containing a concentration of photoreceptor cells known as rods and cones that are connected to the optic nerve via BIPOLAR CELLS. The retina lies immediately below the vascular choroid layer which nourishes it.

retinaculum, *n.* **1.** a hook-like structure on the forewings of moths that contacts the bristle-like frenulum on the hindwing, thus linking the two. **2.** the hook-like structure that retains the springing organ (furcula) in Collembola.

retinal convergence, *n.* the sharing of a single nerve fibre by several rods in the retina of the vertebrate EYE. The rods share or converge into one nerve fibre. Rods are used particularly in low illumination when the stimulus of light on a single rod may be insufficient to generate an ACTION POTENTIAL in the NEURONE. However, the stimulus of several rods on one neurone will excite it more easily, and six rods stimulated together will fire a neurone. This is an example of SUMMATION. Cones show little or no retinal convergence.

retinene, *n.* the main carotenoid pigment found in the retina of the eye. It turns yellow in light.

retinol, see VITAMIN A.

retuse, *adj.* (of plant parts) blunt or truncate and slightly indented.

reverse mutation, *n.* a genetic change occurring in a MUTANT(2) individual that restores the WILD TYPE phenotype. Such events are far less common than FORWARD MUTATIONS.

reverse transcriptase, *n.* an RNA-dependent DNA polymerase that catalyses the synthesis of DNA from deoxyribonucleoside-5'-triphosphate, using RNA as a template.

revision, *n.* (in taxonomy) the summary and re-evaluation of previous knowledge, including new material and new interpretation if present.

revolute, *adj.* (of parts) rolled down.

R factor or **resistance factor,** *n.* a transferable PLASMID of many enteric bacteria that carries genes for resistance to antibiotics. It is often associated with a transfer factor, which after conjugation results in the transfer of the R factor to another bacterium.

RGD, see RANDOM GENETIC DRIFT.

R genes, *n.* the genes present on the R FACTOR concerned with resistance to antibiotics.

rhacis, see RACHIS.

RHA, see RHESUS HAEMOLYTIC ANAEMIA.

rhabd-, *prefix.* denoting a rod.

rhabdom or **rhabdome,** *n.* a transparent rod that passes down the centre of an insect or crustacean ommatidium. See EYE, COMPOUND.

rhesus blood group, *n.* a form of human blood variation in which about 85% of the UK population possess an *RH factor* (D-ANTIGEN) on the surface of red blood cells. Such people are described as being *Rh-positive*; those without the factor are *Rh-negative*. Unlike the ABO BLOOD

GROUP there is normally no rhesus ANTIBODY, unless an Rh-negative person is sensitized by a rhesus antigen, for example in a blood transfusion or when an Rh-negative mother receives blood cells from an Rh-positive foetus (see RHESUS HAEMOLYTIC ANAEMIA). The blood group factor is controlled by a single autosomal gene on chromosome number 1, with two principal alleles. The allele for the Rh-factor is dominant.

rhesus haemolytic anaemia (RHA), *n.* a serious blood abnormality in newborn children in which LYSIS of red blood cells causes anaemia, the condition arising from a reaction between Rhesus antigens and antibodies (see RHESUS BLOOD GROUP). RHA occurs when a second (and subsequent) Rh-positive child is born to an Rh-negative mother. The mother is sensitized to the Rhesus factor on foetal blood cells transferred from the first baby at PARTURITION and produces rhesus antibodies in response. When she is pregnant again, her Rhesus antibodies pass to the foetus, resulting in RHA when the child is born. The condition can be prevented by the mother being given a large dose of rhesus antibodies immediately after the birth of her first child. These antibodies will destroy the foetal cells in her blood before she has time to become sensitized to them.

RH factor, see RHESUS BLOOD GROUP.

rhin- or **rhino-,** *prefix.* denoting the nose.

rhizo-, *prefix.* denoting a root.

rhizoid, *n.* a hairlike structure that functions as a root in lower organisms, such as certain fungi and mosses. Rhizoids are important in penetrating a substance, giving anchorage and absorbing nutrients.

rhizome, *n.* a horizontal underground stem (with leaves and buds) that serves as a storage organ and a means of VEGETATIVE PROPOGATION. Rhizomes are found in flowering plants such as *Iris*.

rhizomorph, *n.* a densely packed strand of fungal tissue, having the appearance of a root, which is produced by some higher fungi such as *Armillaria*. In some cases HYPHAE can be distinguished in the tissue. Rhizomorphs enable fungi to spread.

rhizoplane, *n.* that part of the RHIZOSPHERE made up of the root surfaces.

rhizopod, *n.* any protozoan of the subclass Rhizopoda or *Sarcodina* including the amoeboid protozoans that feed by means of pseudopodia and lack cilia or flagella.

rhizosphere, *n.* the particular area immediately surrounding the roots of plants in which their exudations affect the surrounding microbial flora. Many exudations are nutrients which the microorganisms utilize and include carbohydrates, vitamins and amino acids. Conversely, the plant roots may take up minerals released by the microorganisms. Bacteria may be present in the rhizosphere in numbers fifty times greater than in surrounding soil. Gram-negative rods predominate whereas gram-positive forms are less common than in the surrounding soil. Fungi are usually present in the rhizosphere in similar quantities to those in the surrounding soil, though some plants encourage particular forms.

Rhodophyta, *n.* members of the Thallophyta division Rhodophyta, in which the chlorophyll is masked by PHYCOERYTHRIN. Rhodophytes vary from filamentous forms to those with an extensive thallus, and have gametes that are nonmotile.

rhodopsin or **scotopsin** or **visual purple,** *n.* a purple pigment found in the rods of the retina of the vertebrate eye. Lack of rhodopsin causes night blindness. When bleached by light, rhodopsin liberates a protein called OPSIN and a yellow pigment called RETINENE. As a result of this reaction energy is released which triggers off an ACTION POTENTIAL. See also IODOPSIN, METARHODOPSIN.

rhomboid, *adj.* (of plant parts) diamond-shaped.

rhyncho-, *prefix.* denoting a snout.

Rhynchocephalia, *n.* a primitive group of DIAPSID reptiles that includes *Sphenodon*, a genus having a pineal eye in the roof of the skull.

Rhynchota, see HEMIPTERAN.

rhythm, *n.* the regular occurrence of strong and weak impulses of a particular phenomenon.

rhytidome, *n.* bark consisting of alternating layers of cork and dead CORTEX or PHLOEM.

rib, *n.* any of the long, curved bones forming the wall of the THORAX in vertebrates, attached dorsally to the VERTEBRAL COLUMN. The anterior ribs are connected ventrally to the STERNUM in higher vertebrates and form a protective ribcage around the contents of the thorax. Ribs are connected by intercostal muscles which cause their movement during breathing in terrestrial vertebrates.

riboflavin or **vitamin B$_2$,** *n.* a member of the B-COMPLEX of water-soluble vitamins. Found in a wide variety of foods, riboflavin is required in the metabolism of all animals (acting as a carrier in the ELECTRON TRANSPORT SYSTEM) and is part of several plant pigments.

ribonuclease, *n.* an enzyme that catalyses the depolymerization of RNA.

ribonucleic acid, see RNA.

ribose, *n.* a pentose sugar. See Fig. 262 on page 464.

ribosomal RNA (rRNA), *n.* a form of RNA transcribed from DNA in the NUCLEOLI of EUCARYOTE cells or clustered ribosomal genes in PRO-CARYOTES, whose function is to join with various proteins to form the RIBOSOME. rRNA is produced in only a few forms of varying length, some of which have been completely analysed, base by base.

ribosome, *n.* a small particle (not an ORGANELLE) found in the cytoplasm

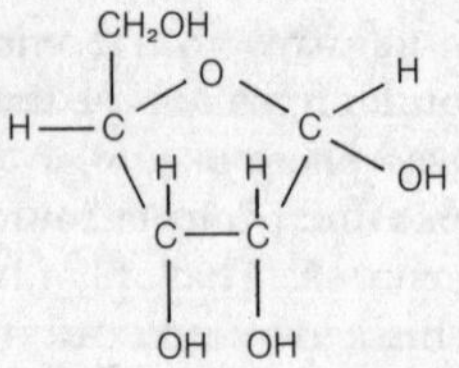

Fig. 262. **Ribose.** Molecular structure.

of all cells, composed of protein and RIBOSOMAL RNA. Each ribosome is composed of two subunits of different sizes which sediment at different rates during centrifugation (see ULTRACENTRIFUGE). PROCARYOTES have ribosome with 70 S size and mass; EUCARYOTES have larger ribosomes with 80 S size and mass. Ribosomes bind to the 5′ end of MESSENGER RNA (see POLYNUCLEOTIDE CHAIN) and travel towards the 3′ end, with TRANSLATION and POLYPEPTIDE synthesis occurring as they go along. Frequently several ribosomes are attached to one piece of mRNA, forming a POLYRIBOSOME.

ribulose, *n.* a pentose sugar, found in syrup, which is important in carbohydrate metabolism. Formula: $C_5H_{10}O_5$.

ribulose biphosphate (RBP) or **ribulose diphosphate,** *n.* a 5-carbon ketose that acts as a receptor of CO_2 in the Calvin cycle.

rickets, *n.* a disease of young children and other mammals where faulty calcification of bones produces bowed limbs. Rickets is due to a diet deficient in VITAMIN D and low in calcium and phosphorus. The condition is now rare in the UK, although it is common in poorer countries.

Ringer's fluids, *n.* saline solutions made up of ions similar to and ISOTONIC with the tissue fluids of the organs to be studied, and in which physiological preparations can be kept alive in vitro. Named after the 19th century physiologist Sidney Ringer.

ringing experiment, *n.* a procedure in which the outer tissues of a stem are removed, leaving ony the xylem and pith intact. Water transport up the stem is unimpeded by such a process, whereas food transport down from the leaves stops at the ring, thus demonstrating the roles of XYLEM and PHLOEM respectively.

RNA (ribonucleic acid), *n.* a NUCLEIC ACID composed of a single POLYNUCLEOTIDE CHAIN of NUCLEOTIDES. RNA is a vital component of PROTEIN SYNTHESIS, and occurs in three forms: (a) MESSENGER RNA involved in TRANSCRIPTION and transport of a coded signal from DNA to RIBOSOMES; (b) RIBOSOMAL RNA forming the site of polypeptide synthesis; (c) TRANSFER RNA carrying amino acids to the ribosomes to be inserted in the correct sequence during translation.

RNA polymerase, *n.* one of several enzymes that catalyses the formation of RNA molecules from DNA templates during TRANSCRIPTION. See OPERON MODEL.

Robertsonian chromosomes, *n.* chromosomes produced as a result of fusion of two separate chromosomes.

rocky shore, *n.* the coastline, usually marine, where rock outcrops at the surface to form at least part of the substrate.

rod or **rod cell,** *n.* a rod-shaped, light-sensitive cell lying in the more peripheral parts of the retina in the vertebrate eye. Rods are particularly associated with vision under conditions of low illumination and they occur in large numbers in nocturnal animals. They are not capable of colour discrimination and their visual acuity is poor (compare CONE CELL). RHODOPSIN (visual purple) is found in rods. There are some 240 million rods in the retinas of a primate.

rodent, *n.* any member of the mammalian order Rodentia, including gnawing mammals which have chisel-like incisors and lack canines. Examples include rats, mice, squirrels.

rogue, *n.* **1.** an unwanted individual in a crop (particularly plants) that usually is removed during a breeding programme. **2.** an animal (particularly one of vicious character) that has separated from the main group and leads a solitary life.

Roentgen or **Röntgen, Wilhelm Konrad** (1845–1923) German physicist who discovered X-RAYS. A *roentgen* or *röntgen* is the quantity of X-rays or gamma radiation used as a unit of radioactivity. Symbol: R or r.

röntgen, see ROENTGEN.

root, *n.* that part of a plant which (usually) grows below ground. The root provides anchorage for the aerial parts, absorbs water and mineral salts from the soil, conducts water and nutrients to other parts of the plant, and often stores food materials over winter. Root structure is variable among higher plants, but generally there is (unlike the young STEM) a central core of conducting tissue (the STELE) which also serves as a strong structural element, assisting the root as it pushes downwards and resisting upward pressures from the aerial parts.

Roots can be classified into three main types: primary, secondary (see LATERAL ROOT), and ADVENTITIOUS(1). The primary root of ANGIOSPERMS develops from the radicle of the seedling. Later, secondary roots emerge from the top of the primary root and then further down. At the apex of the root is a protective ROOT CAP. The typical root structure of a DICOTYLEDON is shown in Fig. 263 on page 466.

In MONOCOTYLEDONS the number of xylem 'arms' is usually much greater than in DICOTYLEDONS, 12 to 20 in monocotyledons, 2 to 5 in dicotyledons. Roots frequently undergo SECONDARY THICKENING in dicotyledons, but not in monocotyledons.

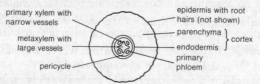

Fig. 263. **Root.** Transverse section of a typical dicotyledon (buttercup) root.

root cap, *n.* a structure found at the apex of all ROOTS (except those of many water plants), produced by the APICAL MERISTEM. The cap consists of a thimble-shaped collection of PARENCHYMA cells which have a protective function. As the root pushes its way through the soil the outer (older) cells of the root cap are sloughed off and replaced by new cells from the MERISTEM.

root crop, *n.* any roots which are used for human or animal foods. Usually they are from plants which use their roots as storage organs, and with few exceptions they have tap roots, for example carrots, parsnips.

root effect or **root shift,** *n.* a change in blood oxygen capacity resulting from a pH change.

root hair, *n.* a hair-like outgrowth from the epidermis of roots. Root hairs occur in large numbers in a zone behind the growing tip, are short-lived, and greatly increase the absorbing area of the root.

root nodules, see NITROGEN FIXATION.

root pressure, *n.* a force exerted within a plant root that pushes water up towards the stem. The phenomenon is produced by the root cells having a solute concentration gradient which increases from outside the root towards the centre of the root. Thus, by OSMOSIS, water passes from the soil, across the root and into the xylem as a result of salt excretion by the ENDODERMIS into the xylem, creating pressure which can be observed in a plant with the aerial parts removed. See Fig. 264.

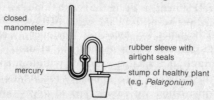

Fig. 264. **Root pressure.** The mercury enables root pressure to be measured.

root shift, see ROOT EFFECT.

root system, *n.* the total below-ground vegetative system of vascular plants. See ROOT.

rotate, *adj.* (of a corolla of a flower) having petals at right-angles to the axis forming a wheel shape.

rotifer, *n.* any minute aquatic multicellular invertebrate of the phylum Rotifera, containing the *wheel animalcules*, the smallest METAZOANS. They move and feed by means of an anterior ring of cilia.

roughage, *n.* **1.** the less-useful part or refuse of a crop. **2.** any indigestible material eaten in food to stimulate the actions of the intestines.

rough endoplasmic reticulum, see ENDOPLASMIC RETICULUM.

round window, see FENESTRA OVALIS AND ROTUNDA.

roundworms, see NEMATODE.

royal jelly or **queen substance,** *n.* a substance produced by the pharyngeal glands in the early adult life of the worker bee that is used for feeding the young brood (for up to four days) and the larvae of queen bees for longer periods to ensure their development into queens. It contains 40% dry matter in the form of lipoproteins, neutral glycerides, free-fatty acids, sugar, amino acids and all the B vitamins. It also has a high content of ACETYLCHOLINE.

RQ, see RESPIRATORY QUOTIENT.

rRNA, see RIBOSOMAL RNA.

r selection, *n.* SELECTION for an opportunist strategy in which traits leading to a high rate of natural increase are selected for.

rubella, see GERMAN MEASLES.

ruderal, *n.* a plant living on waste land in built-up areas.

rugose, *adj.* (of plant parts) wrinkled.

rumen, *n.* a branch of the oesophagus of ruminants in which unchewed food is stored temporarily and from which it is regurgitated to the mouth for chewing (see RUMINANT STOMACH). Some cellulose is digested and absorbed in the rumen and bacterial action results in the synthesis of B vitamins there. Cellulase is produced by bacteria which may number 1 billion per cm^3 in the rumen.

ruminant, *n.* any mammal of the suborder Pecora, (deer, giraffes, antelopes, sheep, goats, cows) of the order Artiodactyla. They usually possess horns in the males, lack incisors in the upper jaw and have a four-compartmented stomach which includes the RUMEN.

ruminant stomach, *n.* a digastric stomach with four chambers of two divisions. The first division is composed of the RUMEN and reticulum, acting as a fermentation vat for unchewed vegetation. The second division is composed of the OMASUM and ABOMASUM, the latter being a 'true' stomach, secreting digestive enzymes.

ruminate, *adj.* (of plant parts) appearing chewed.

rumination, *n.* the chewing of partially digested food, retrieved from the RUMEN by reverse PERISTALSIS.

runcinate, *adj.* (of plant parts, leaves) having pinnate lobes pointing backwards to the leaf base.

runner or **stolon,** *n.* a long, slender stem running along the surface of the ground, arising from the axil of a leaf, whose function is to enable rapid VEGATIVE PROPOGATION in an area. Along the runner are small scale-leaves with buds from which roots emerge and enter the soil. Eventually the runner withers, leaving many new daughter plants (a CLONE). Runners are found in, for example, strawberries and creeping buttercup.

rust, *n.* fungal diseases of plant stems and leaves caused by members of the order Uredinales. Rusts are OBLIGATE parasites, their host plants often being cereals. For example, the fungus *Puccinia graminis tritici* causes black stem rust of wheat.

rut, *n.* the period of maximum testicular activity in male mammals (compare OESTRUS CYCLE), particularly applied to the period of sexual activity in deer.

S

sac, *n.* a bag-like or pouched structure.

saccate, *adj.* bag-like, pouched.

Saccharomyces, *n.* a genus of unicellular fungi, which is included in the order Endomycetales (class Hemascomycetes), most species of which ferment a wide range of sugars. Some forms are used in the manufacture of bread and as 'brewers' yeast' in beer making. They may occur in the form of simple, budding cells or as pseudomycelium (chain of cells). Each ASCUS contains one to four rounded or ovate SPORES.

saccule, *n.* **1.** any small sac. **2.** a membranous labyrinth joined to the UTRICLE and forming part of the vestibular apparatus in the inner ear.

sacral vertebra, *n.* the thick strong vertebrae of the lower backbone, some of which are fused together and attached to the PELVIC GIRDLE to form the SACRUM, giving great strength to the hip region.

sacrum, *n.* the collection of fused sacral vertebrae attached to the PELVIC GIRDLE.

saggitate, *adj.* (of plant structures) shaped like an arrowhead, with two backward-pointing barbs.

sagittal, *adj.* (of a section through an organism) being dorsoventral and longitudinal in the midline, so producing mirror-image halves along the length.

salinity, *n.* the degree of saltiness, as, for example, occurs in enclosed seas, where the salinity is great, or in estuaries where the normal salinity of the sea may be diluted by fresh river water. It is defined as the total amount of dissolved solids in water in parts per thousand by weight and in seawater is approximately 35%.

saliva, *n.* a viscous, transparent liquid containing water, salts, MUCIN and (sometimes) salivary amylase (see PTYALIN). Saliva is secreted by cells of the salivary glands which, in humans, occur in three pairs, one in the cheek and two between the bones of the lower jaw. The quantity of saliva produced depends on the type of food being consumed. Dry foods and acidic foods stimulate a copious volume of nonviscous saliva, while liquids such as milk stimulate small quantities of thick saliva. The amount of amylase can also vary, being high when meat is eaten.

salivary amylase, see AMYLASE.

salivary gland, *n.* any gland that secretes SALIVA.

salivary gland chromosome or **polytene chromosome,** *n.* a giant chromosome present in the salivary glands of certain DIPTERANS, such as midges or fruit flies. The interesting feature of such structures is that, when suitably stained, their CHROMOMERES are readily visible, making it possible to map the physical presence of genes. Studies have shown a clear correlation between the order of genes mapped by genetic RECOMBINATION and visible structures on the polytene chromosomes. At certain times it is possible to see CHROMOSOME PUFFS which are thought to represent areas of gene action.

Salmonella, *n.* a genus of bacteria containing a wide range of species that are pathogenic for man and other animals. Examples include *S. typhi* that causes typhoid fever and *S. typhimurium* that causes severe gastroenteritis, or 'food poisoning'.

salp, *n.* a free-living planktonic TUNICATE with no larval stage. The nervous system is degenerate and the gill clefts lack longitudinal bars.

saltation, *n.* **1.** any changes of an abrupt nature that occur in the thalli of fungi either through mutation or the occurrence of segregation of parts of the thallus with different genetic make-up (HETEROKARYONS and HOMOKARYONS). **2.** the movement of soil particles by wind.

salt gland, *n.* a lead hydathode (see GUTTATION) that secretes a saline solution.

salt marsh, *n.* a community of salt-tolerant plants growing on intertidal mud in brackish conditions in sheltered estuaries and bays.

samara, *n.* a one-seeded indehiscent fruit with a wall flattened into a membrane wing, found, for example, in ash and elm trees.

sample, *n.* any portion of a whole, such as a small part of a population, collected for examination.

sampling, *n.* **1.** the act of taking a fraction of substance to be tested or analysed. **2.** the selection of some parts from a larger whole as in statistical sampling.

sampling error, *n.* (in statistics) a statistical phenomenon in which the variation within a sample is inversely related to the size of the sample. In other words, the smaller the sample the larger the sample error. Thus if 100 people toss 10 coins each, the variation within each sample would be quite high, from perhaps 2 heads: 8 tails, to 8 heads: 2 tails. However, if the same 100 people toss 1,000 coins each, the variation would be expected to be *proportionately* less, from perhaps 350 heads: 650 tails to 650 heads: 350 tails. The variation in a sample can be expressed, for example, as mean ± standard error.

Sanger, Frederick (1918–) English biochemist who carried out extensive research into protein structure and who determined the amino acid sequence of insulin.

sap, *n.* a watery liquid found within the vacuole of a plant cell (cell sap) and within conducting tissues of the VASCULAR BUNDLES.

saprobiont, *n.* any organism that feeds on dead or dying animals or plants. See also SCAVENGER.

saprophyte or **saprotroph** or **necrotroph,** *n.* any plant or microorganism that obtains its nutrition from dead or decaying organic materials in the form of organic substances in solution. Such organisms are of great importance in breaking down dead organic material. See NITROGEN CYCLE.

saprophytic, see NECROTROPHIC.

saprotroph, see SAPROPHYTE.

sapwood, *n.* the ring of living XYLEM tissue in a stem or root which has undergone SECONDARY THICKENING, and which is composed of VESSELS and PARENCHYMA cells. Sapwood lies outside the HEARTWOOD but inside the ring of PHLOEM.

sarco-, *prefix.* denoting flesh.

Sarcodina, see RHIZOPOD.

sarcolemma, *n.* the thin connective tissue sheath that occurs around a muscle fibre and extends and contracts with the contractile material enclosed.

sarcoma, *n.* a cancerous growth derived from muscle, bone, cartilage or connective tissue.

sarcomere, *n.* that part of the muscle fibre which is contained between two Z-MEMBRANES, and is the contractile element of the fibre. See Fig. 265. See MUSCLE, I-BAND.

sarcoplasm, *n.* the cytoplasm of a muscle cell.

sarcoplasmic reticulum, *n.* the cysternae of a single muscle fibre.

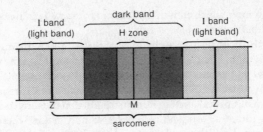

Fig. 265. **Sarcomere.** Muscle fibrils consist of alternating groups of thick and thin filaments (dark and light bands), the distance between two Z-membranes being termed a 'sarcomere.'

saturated fats, see FATTY ACID.

saturated vapour pressure, *n.* the pressure exerted by a vapour when it is completely saturated and in equilibrium with its liquid form. For example, the evaporating surface of a MESOPHYLL cell is saturated with water and has a vapour pressure that is highly affected by the ambient temperature. See Fig. 266.

Temperature °C	Water V.P. (saturated) millibar kPa
10	1.2
15	1.6
20	2.5
25	3.2

Fig. 266. **Saturated vapour pressure.** The effect of ambient temperature on the vapour pressure at the surface of a mesophyll.

saur– or **sauro–,** *prefix.* denoting a lizard.

Saurischia, *n.* an order of dinosaurs lacking a post-pelvic bone and having a normal PELVIC GIRDLE, unlike *Ornithischia*. The group includes the large four-footed dinosaurs such as *Diplodocus*.

Sauropsida, *n.* a nontaxonomic group including all reptiles and birds with the exception of mammal-like reptiles (*Therapsida*).

scab, *n.* a symptom of various plant diseases in which there are local areas of surface roughening, e.g. apple scab caused by the fungus *Venturia inaequalis*, a HEMIBIOTROPH.

scabrous, *adj.* having a rough surface.

scalariform, *adj.* ladder-like.

scalariform thickening, *n.* the internal thickening of XYLEM VESSELS or tracheids in which lignin bands are deposited in a SCALARIFORM pattern.

scala tympani, *n.* the lower chamber of the cochlea in mammals, filled with perilymph which ends in the fenestra rotunda. It connects with the scala vestibuli via the helicotrema.

scala vestibuli, *n*. the upper chamber of the cochlea in mammals, filled with perilymph. It lies between REISSNER'S MEMBRANE and the wall of the bony canal and connects with the SCALA TYMPANI via the helicotrema.

scale, *n*. any plate-like outgrowth of the integument of an organism, each in the form of a flat calcified or horny structure on the surface of the skin. Scales are found in fish, and in reptiles such as snakes where they are derived from both the epidermis and the dermis, and in insects (e.g. *Lepidoptera*), where they are derived from hairs. See PLACOID, COSMOID, GANOID.

scape, *n*. a flower stalk bearing no leaves and arising from a rosette of leaves.

scapho-, *prefix*. denoting a boat.

Scaphopoda, *n*. a class of primitive molluscs, possessing a tube-like shell open at both ends.

scapula, *n*. the shoulder blade. See PECTORAL GIRDLE.

scarious, *adj*. (of plant structures) thin, with a dried-up appearance, especially at the edges and tips.

scarlet fever, *n*. an ACUTE, contagious disease in man, particularly of young children, caused by the bacterium *Streptococcus pyogenes* and characterized by inflammation of the pharynx, nose and mouth and a red skin rash.

scatter diagram, *n*. a graphical figure in which two axes are plotted at right angles to each other, the independent variable on the x (horizontal) axis and the dependent variable on the y (vertical) axis. Each individual or object is then measured for these two variables, for example, age (x) and weight (y), and the position marked on the correct coordinates of the diagram, producing a series of scattered points. These may be given a 'line of best fit' by REGRESSION ANALYSIS.

scavenger, *n*. any organism that feeds on carrion, refuse or material left unconsumed by other organisms. See also SAPROBIONT.

schistosomiasis, see BILHARZIA.

schizo-, *prefix*. denoting a split.

schizocarp, *n*. a dry fruit developing from a SYNCARPOUS ovary that splits into one-seeded portions as, for example, the fruit of fennel.

schizogenous, *adj*. (of secretory organs in plants) derived by spaces being created by separation of cells.

Schwann cell, *n*. a type of cell occurring in the region of the nerve fibres of the peripheral nervous system of vertebrates. It produces the neurilemma which encloses the MYELIN SHEATH (see Fig. 220), and is found in close contact with the nerve axons between each node of Ranvier. See SCHWANN, THEODOR.

Schwann, Theodor (1810–82) German anatomist who first recognized the egg as a cell in 1838. He showed that all living organisms are composed of cells and that fermentation is brought about by living organisms. He discovered PEPSIN in gastric juice and the SCHWANN CELL is named after him.

scientific method, *n.* the way of approaching a problem by means of drawing up a hypothesis based on a series of observations, and then testing the hypothesis by means of experiments designed in such a way as to support or invalidate the hypothesis. On the basis of the experimental evidence a theory is proposed to account for the initial observations. If subsequently the theory is found to be wanting in some respect, new hypotheses are sought and tested experimentally, so the process is a successive refinement which in science never leads to an absolute truth, but to a more reliable knowledge.

scientific name, *n.* the binomial or trinomial designation of an organism in the form of latinized (or Latin or Greek) words.

scion, *n.* the part of a plant, usually a piece of young stem, that is inserted into a rooted STOCK(1) to produce a GRAFT.

sclera or **sclerotic,** *n.* the outer coat of the vertebrate eye to which are attached extrinsic muscles for moving the eyeball. The sclera is lined by a vascular layer, the CHOROID, except for the forward-facing part which is called the CORNEA and is transparent, with no underlying choroid.

sclereid, *n.* a type of SCLERENCHYMA cell of higher plants, which is roughly spherical in shape with a thick wall that can be smooth or spiky and is always heavily impregnated with LIGNIN. Sclereids occur in the flesh of succulent fruits such as pears, and are common in the shells of nuts.

sclerenchyma, *n.* a plant tissue in which the cells have greatly thickened walls impregnated with LIGNIN and no cell contents. The tissue has a mechanical function supporting the plant, and consists of two types of cells: fibres and SCLEREIDS. Fibres are long cells with tapered ends, which are often grouped into bundles.

sclerite, *n.* chitinous plates separated by thinner membranes in the exoskeleton of arthropods.

sclero-, *prefix.* denoting dry, hard.

sclerosis, *n.* **1.** (in animals) a hardening of tissue due to excess growth of fibrous tissue or deposition of fatty plaques or degeneration of the myelin sheath of nerve fibres, as in sclerosis of the liver in humans, which can be lethal. **2.** (in plants), a hardening of the cell wall or tissue, often due to the deposition of LIGNIN.

sclerotic, see SCLERA.

sclerotium, *n.* **1.** a resting stage in many fungi. It takes the form of a ball of HYPHAE varying in size from a pinhead to a football, and usually has a hard, dark-coloured exterior coating. Fruiting bodies may be formed eventually from the sclerotium (either sexual or asexual) or a MYCELIUM may form. Normally the sclerotium does not contain spores. **2.** the firm resting condition of a MYXOMYCETE.

scolex, *n.* the 'head' of a tapeworm, being that part at the anterior end which bears hooks and suckers and is used for attachment to the gut wall of the host. See Fig. 267.

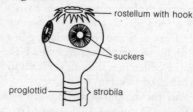

Fig. 267. **Scolex.** The scolex of *Taenia solium*.

scotophase, *n.* the dark phase of a dark/light cycle. Compare PHOTOPHASE.

scotopsin, see RHODOPSIN.

scrapie, *n.* a disease of sheep with no certain cause, in which the animal develops intense itching, increasing weakness and eventually dies.

scrotum or **scrotal sac,** *n.* a sac containing the TESTES of male mammals that is situated outside the posteriorly ventral part of the abdomen. The external location ensures that the testes are cooled to below body temperature (heat can adversely affect the development of sperm). In some organisms, such as bats, testes may lie protected inside the abdomen for much of the year, descending into the scrotum only during the breeding season.

scrub, *n.* a habitat in which shrubs predominate.

scurvy, *n.* a condition in which the blood capillary walls become fragile, leading to excessive bleeding and anaemia, spongy gums with tooth loss, impaired wound healing and eventually death. Scurvy is due to a lack of ASCORBIC ACID in the diet, normal health being restored by administration of fresh fruits, particularly citrus fruits. English sailors on board ocean-going ships of the 18th and 19th centuries were supplied with citrus fruits to prevent scurvy, this habit giving them the nickname 'limeys'.

scutellum, *n.* a plate-like projection from the back of a dorsal plate of the thorax in insects.

scutum, *n.* **1.** the middle of the three dorsal thoracic SCLERITES of an insect. **2.** one of the two lateral plates forming part of the carapace of a barnacle.

scypho-, *prefix.* denoting a bowl or cup-like structure.

scyphozoan, *n.* any marine medusoid COELENTERATE of the class Scyphoza, containing the jellyfish. The medusa is the dominant and sometimes only phase of the life history, the polyp being small (a SCYPHISTOMA) or absent.

scyphistoma, *n.* a larval form of COELENTERATE, arising from a planula larva, which may bud, but in autumn and winter produces EPHYRA larva by STROBILATION. See SCYPHOZOAN.

sea butterfly, see PTEROPOD.

sea mat, see BRYOZA.

seasonal isolation, *n.* any variation in breeding seasons that effectively keeps two populations of organisms from interbreeding.

sea zonation, *n.* the divisions of the sea and sea floor for purposes of reference to specific areas as shown in Fig. 268.

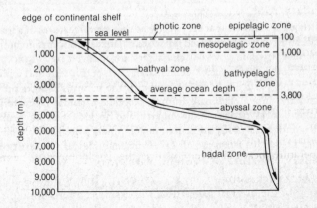

Fig. 268. **Sea zonation.** The zones of the open sea and sea floor.

sebaceous gland, *n.* one of many glands occurring in the skin, that secretes oil or sebum into the hair follicles in mammals. This maintains an oily coating to the hair and contributes to its waterproofing. Secretions from sebaceous glands are the result of cell destruction.

secondary carnivore, *n.* a carnivore that feeds upon other carnivores.

secondary consumer, *n.* a carnivore that feeds upon the herbivores (PRIMARY CONSUMERS) which, in turn, are sustained by PRIMARY PRODUCTION, i.e. PHOTOSYNTHESIS.

secondary growth, see SECONDARY THICKENING.

secondary homonym, *n.* one of the two identical SPECIFIC NAMES that were originally proposed for different genera but which through revision come to be included in the same genus.

secondary intergradation, *n.* a zone of hybridization in which two previously separated populations have come together and interbred.

secondary meristem, *n.* an area of cambial cells between xylem and phloem of roots and stems in DICOTYLEDONS. The MERISTEMS give rise to tissues causing lateral expansion or SECONDARY THICKENING.

secondary metabolite, *n.* a product of microbial cells in culture when growth is slowing down. While having no obvious role in the cellular physiology of the producer, secondary metabolites are sometimes most useful to humans, e.g. as antibiotics.

secondary production, *n.* any production (see PRODUCER, PRODUCTIV-ITY) by organisms which themselves consume organic material. Second-ary production results entirely from the resynthesis of organic matter, as distinct from its synthesis from inorganic materials as in PRIMARY PRODUCTION.

secondary root, see LATERAL ROOT.

secondary sexual character, *n.* any of the several features of a male or female animal that is produced when sexual maturity occurs. In humans, the female secondary sexual characters are induced by OESTROGEN at puberty and include mature genitalia, breasts, bodily hair and feminine body shape, particularly a broadening of the pelvis and fat deposition around the hips. The male human secondary sexual characters are induced by TESTOSTERONE at puberty and include mature genitalia, bodily and facial hair, muscular development, deep voice, narrow pelvis. These characters are all excellent examples of SEX LIMITATION.

secondary structure of protein, see PROTEIN.

secondary succession, *n.* any SUCCESSION of plants that arises after the clearing of the original vegetation, for example, after burning.

secondary thickening or **secondary growth,** *n.* the growth in the girth of stems and roots in DICOTYLEDONS produced by division of the secondary MERISTEM, resulting in woody tissues. The growth begins by the formation of a continuous cambial ring. In stems there is already a fascicular cambium between the xylem and phloem of the vascular bundles which becomes joined up by interfascicular cambium. The xylem tissues become divided into an inner heartwood (dead) and an outer SAPWOOD.

Secondary thickening can sometimes occur in herbaceous ANGIOSPERMS but is usually associated with woody types producing, for example, a stem which tapers in an upward direction.

MONOCOTYLEDONS do not carry out secondary thickening. See Fig. 269.

In roots, a cambial ring is produced from parenchymatous cells between the XYLEM and PHLOEM.

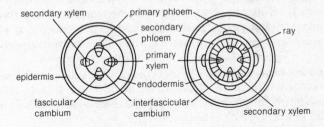

Fig. 269. **Secondary thickening.** Secondary thickening in a typical dicotyledon.

second–division segregation, *n.* the separation of chromatids carrying different gene ALLELES in the second cycle of MEIOSIS. Such an event can be detected by TETRAD ANALYSIS.

second law of thermodynamics, see THERMODYNAMICS.

secretin, *n.* a hormone responsible for the secretion of bile from the liver.

secretion, *n.* **1.** the process by which a useful substance produced in a cell is passed through the plasma membrane to the outside. **2.** the substance itself. Secretions are usually produced by gland cells, but may be the results of cell destruction as in SEBACEOUS GLANDS. Glands of internal secretion (ENDOCRINES) pass their secretions directly into the blood stream whereas glands of external secretion (EXOCRINES) pass their secretions into special ducts.

secretor status, *n.* an individual producing ABO BLOOD GROUP type antigen (including H-SUBSTANCE) in saliva, semen and other body secretions, due to the presence of at least one dominent allele of an autosomal gene. Such fluids can thus be typed, a useful fact in forensic examination.

secund, *adj.* (of plant structures) all turned to one side.

seed, *n.* the structure formed in the fertilized ovule of an ANGIOSPERM, consisting of an embryo surrounded by a food store for nourishment during germination, with an outer hard seed coat, the TESTA. The food store can be located either in a special area called the ENDOSPERM with an outer ALEURONE layer or within the cotyledons, the number of which determines whether a plant is a MONOCOTYLEDON or a DICOTYLEDON. In some plants the so-called seed is really a fruit in which the PERICARP is fused with the testa. See Fig. 270 (b) on page 478.

seed dormancy, see GERMINATION.

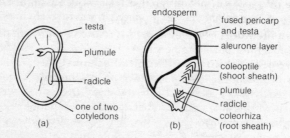

Fig. 270. **Seed.** Longitudinal sections of (a) broad bean seed, (b) a maize fruit 'seed'.

seed germination, see GERMINATION.

seedling, see GERMINATION and Fig. 166.

segmentation, *n*. **1.** the process of cutting off one part of an organism from another, as in CLEAVAGE of an OVUM. **2.** the production of metameres in METAMERIC SEGMENTATION. **3.** the cutting off by a cross wall of a multinucliate portion of a filament or HYPHA.

segregation, *n*. the separation of HOMOLOGOUS CHROMOSOMES during anaphase I of MEIOSIS, producing gametes containing only one allele of each gene. Such an occurrence is the physical mechanism underlying the first law of MENDELIAN GENETICS and is particularly important when the two separated alleles are different.

selection, *n*. the differential rate of reproduction of one phenotype in a population as compared to other phenotypes. Hence an organism that produces more offspring which survive to reproduce than another type is at a 'selective advantage'. The environmental pressures causing selection can be either natural (e.g. competition for food) or man–made (e.g. insecticides, see DDT). See DIRECTIONAL SELECTION, NATURAL SELECTION, STABILIZING SELECTION.

selection coefficient, see SELECTION PRESSURE.

selection pressure, *n*. a measure of the amount of relative reproductive disadvantage of one phenotype over another of the same species living together in the same area. Such pressure is usually represented by the *selection coefficients* which can have a value from zero (no selective disadvantage, maximum FITNESS) to 1.0 (complete selective disadvantage leading to actual death or GENETIC DEATH, minimum fitness).

selective breeding, *n*. the breeding of animals and plants to embrace desirable characters.

selective enrichment, *n*. any technique that encourages the growth of a particular species or group of microorganisms.

selective fishing, *n*. any process by which individuals of a particular size of fish are caught, e.g. with nets of large mesh to allow small fish through.

selective medium, *n*. any medium that restricts the growth of certain microorganisms, but not others.

selective permeability or **differential permeability,** *n*. the capacity of a membrane to allow some particles to pass through but not others. Such 'differentially permeable' membranes (e.g. CELL MEMBRANE, cellular organelle membrane, TONOPLAST) allow water molecules to pass readily through them, whereas solutes dissolved in water can pass less rapidly or not at all. The ability of molecules to travel across the membrane and the velocity with which they do so is dependent on the fat solubility, size, and charge of the molecules. An extreme example of differential permeability is a *semipermeable membrane* which is almost completely impermeable to solute molecules, but is permeable to the solvent. Such membranes, however, are rare. See ACTIVE TRANSPORT.

selective reabsorption, *n*. the re-entry of a specific material into body tissue to avoid a deficiency within the body of that material. This is one of the mammalian kidney's functions.

self-fertilization, *n*. the fusion of male and female GAMETES from the same HERMAPHRODITE individual. Self-fertilization is fairly rare in animals (occurring, for example, in some snails and nematode worms) but is common in some plant groups. See SELF-POLLINATION. Compare CROSS-FERTILIZATION.

self-incompatibility, *n*. a condition in plants where certain types of pollen will not form pollen tubes when deposited on the female stigma, thus preventing fertilization. Self-incompatibility prevents self-fertilization and promotes heterozygosity (the mixing of allelic forms). The system is controlled by an *S* locus with many alleles (see MULTIPLE ALLELISM), and pollen with the same allele as in the stigma will not germinate to form a pollen tube. This is shown in Fig. 271. Self-incompatibility mechanisms are much used to produce HYBRID plants (e.g. various Brassica crops). See also COMPATIBILITY.

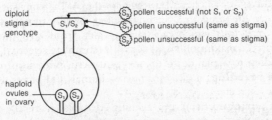

Fig. 271. **Self-incompatibility.** The process depicted results in the progeny being S_1/S_3 and S_2/S_3.

self-pollination, *n.* the transfer of pollen from the stamen of a flower to either the stigma of the same flower or to another flower on the same plants. Since self-pollination leads to self-fertilization which is a form of inbreeding, many plants have developed mechanisms to prevent its occurrence. For example, plants can be DIOECIOUS, show SELF–INCOM-PATIBILITY, and have differential maturation times of male and female floral organs on the same plants (for example, see ARUM LILY). Compare CROSS–POLLINATION.

self-sterility, *n.* the inability of some HERMAPHRODITES to reproduce sexually, that is, to form a viable offspring by self-fertilization.

semen, *n.* the ejaculate from the male reproductive organs which in mammals contains the sperm together with SECRETIONS from the SEMINAL VESICLES and PROSTATE gland which are essential for maintaining the viability of the sperm.

semicircular canals, *n.* the fluid–filled canals present in the inner ear of vertebrates. Usually there are three canals on each side of the head, one in each plane – two vertical and at right angles to each other, and one horizontal and at right angles to the other two. Any movement causes the fluid in the canals to stimulate the receptors in the ampulla at the end of each canal, so that the group of semicircular canals is sensitive to the movements of the head and serves to maintain BALANCE.

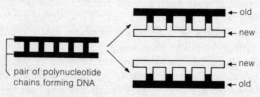

Fig. 272. **Semiconservative replication.** DNA duplication.

semiconservative replication model, *n.* a model of DNA duplication in which each new molecule of DNA contains one POLYNUCLEOTIDE CHAIN from an original molecule and one chain freshly synthesized from individual NUCLEOTIDES. See Fig. 272. That such a model could be tested experimentally was shown by Matthew Meselson and F. Stahl in 1958, using radioactively labelled DNA of *E. coli*. They employed two ISOTOPES of nitrogen of different weight ^{14}N (light) and ^{15}N (heavy), the two types being distinguishable by DIFFERENTIAL CENTRIFUGATION. See Fig. 273. The semiconservative replication model has superseded the CONSERVATIVE REPLICATION MODEL.

semigeographic speciation, *n.* the separation of species along a geographical line of secondary intergradation or along an ecological boundary.

Generation number	DNA grown in	DNA weight		
		light (^{14}N)	medium (^{14}N/^{15}N)	heavy (^{15}N)
1	^{15}N (heavy)	0%	0%	100%
2	^{14}N (light)	0%	100%	0%
3	^{14}N	50%	50%	0%
4	^{14}N	75%	25%	0%

Fig. 273. **Semiconservative replication.** Meselson and Stahl's results.

semilunar valve, *n.* either of the tricuspid heart valves (a) at the orifice of the pulmonary artery, or (b) at the orifice of the aorta.

seminal fluid, *n.* the fluid in which the sperms are bathed.

seminal vesicle or **vesicula seminalis,** *n.* an organ present in male animals that often stores sperm and adds extra secretions to the semen.

seminiferous tubules, *n.* coiled tubules in which SPERMATOGENESIS takes place in the testes of male vertebrates. They are made up of germinal epithelium and hundreds are present in each testis. The germ epithelium proliferates, forms spermatogonia and grows into spermatocytes which give rise to spermatids by GAMETOGENESIS. The spermatids mature to spermatozoa embedded in nutritive SERTOLI CELLS.

semipermeable membrane, see SELECTIVE PERMEABILITY.

semispecies, *n.* a taxonomic group intermediate between a SPECIES and a SUBSPECIES, usually as a result of geographical isolation. See SUPERSPECIES.

senescence, *n.* the process of growing old which occurs in all species and is typified by a gradual slowing down of METABOLISM and breakdown of tissues, often accompanied by endocrinal changes.

senior homonym, *n.* the earliest published of identical names for the same or different taxa.

senior synonym, *n.* the earliest published of synonyms for the same taxon.

sense organ, *n.* any receptor of external or internal stimuli.

sensillum, *n.* (*pl.* sensilla) any small sense organ or receptor (particularly in insects) used for the detection of touch, taste, smell, heat perception, sound or light perception. While most sensilla respond to stimuli from outside the body, for example, taste organs in the feeding of many butterflies and flies, some sensilla act as proprioreceptors, detecting internal changes such as the flexion of joints.

sensitive, *adj.* reacting violently to the effects of a PATHOGEN.

sensory cell, *n.* **1.** see SENSORY NEURON. **2.** a modified cell adapted for the acceptance and transmission of stimuli.

sensory neuron or **sensory cell,** *n.* a neuron that conducts impulses from the periphery of an organ to the CNS.

sepal, *n.* a modified leaf (usually green) forming part of the CALYX of the flower of a DICOTYLEDON.

sepaloid, *adj.* (of plant structures) like a SEPAL.

septi-, *prefix.* denoting a partition.

septicaemia or **blood poisoning,** *n.* an infection of the blood stream by a variety of pathogenic microorganisms, such as *Salmonella* and *Pseudomonas*, usually from a nonintestinal source, leading to fever, lesions in many body organs and even death.

septum, *n.* any dividing wall or partition that occurs between structures or in a cavity.

sequoia, *n.* either of two huge Californian coniferous trees (the redwood *Sequoia sempervivens* or the giant sequoia *Sequoiadendron giganteum*), that can reach heights of up to 100 metres. They are among the oldest living organisms in the world, taking perhaps 2,000 years to reach their mature size. The oldest sequoia, called General Sherman, is thought to be between 3,000 and 4,000 years old, standing 83 metres tall and with a diameter of more than 9 metres.

sere, *n.* a plant SUCCESSION; a progression of plant communities with time.

series, *n.* the sample available for taxonomic study.

serine, *n.* one of 20 AMINO ACIDS common in proteins. It has a polar 'R' group structure and is soluble in water. See Fig. 274. The ISOELECTRIC POINT of serine is 5.7.

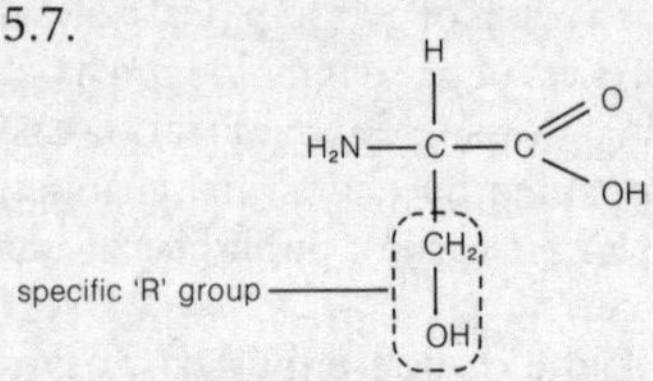

Fig. 274. **Serine.** Molecular structure.

serology, *n.* the branch of biological science that is concerned with the study of SERUMS.

serosa, *n.* **1.** a SEROUS membrane such as the PERITONEUM that secretes a serum. **2.** an epithelial layer formed under the vitelline membrane in the development of the insect egg; it lays down the serosal cuticle of the egg.

serotonin, *n.* a pharmacologically active compound, derived from tryptophan, which acts as a vasodilator, increases capillary permeability, and causes contraction of smooth muscle.

serous, *adj.* of, resembling, or producing, SERUM. See SEROSA.

serrate, *adj.* (of margins) with tooth-like shape.

Sertoli cells, *n.* large nutritive cells occurring in the wall of SEMINIFEROUS TUBULES that serve to nourish the developing sperm. Named after the Italian histologist, E. Sertoli.

serum, *n.* **1.** see BLOOD SERUM. **2.** clear, watery animal fluid, especially that exuded by serous membranes. **3.** also called **antiserum.** antitoxin containing large quantities of ANTIBODIES to a specific ANTIGEN, which confers quick-acting 'passive' IMMUNITY when donated to an individual who may have been exposed to the antigen. For example, tetanus antiserum is injected after the possible entry of tetanus bacterium in an accident. Compare VACCINE.

serum hepatitis, see HEPATITIS.

servomechanism, *n.* a control system that uses negative feedback (see FEEDBACK MECHANISM) to maintain a level of output or performance.

sessile, *adj.* **1.** (of an organism) remaining sedentary. **2.** (of part of an organism) having no stalk.

seta, *n.* **1.** the erect aerial part of the spore-producing structure of mosses or liverworts. **2.** see CHAETA. **3.** a slender, straight prickle.

setaceous, *adj.* bristly.

seventy-five percent rule, *n.* an arbitrary rule stating that a population can be considered subspecifically distinct from another if 75% of the individuals are distinct from all those of the other population.

Sewall Wright effect, see RANDOM GENETIC DRIFT.

sex chromatin, see BARR BODY.

sex chromosome or **heterosome,** *n.* one of a pair of chromosomes that is different in the two sexes and is involved in SEX DETERMINATION. All remaining chromosomes in a KARYOTYPE are called AUTOSOMES. In most organisms, for example mammals and DIOECIOUS plants, the female contains two identical X-chromosomes and is the HOMOGAMETIC SEX while the male contains one X and one Y-chromosome and is the HETEROGAMETIC SEX. In birds, butterflies and moths, some fish and certain plants, the situation is reversed in the two sexes. Genes located on sex chromosomes are described as showing SEX LINKAGE.

sex determination, *n.* the control of maleness and femaleness by genes located on SEX CHROMOSOMES. The actual mechanism differs in various organisms but very often is through a HETEROGAMETIC SEX and a HOMOGAMETIC SEX. In humans the male is heterogametic, sex being determined by the presence of a Y-chromosome rather than the presence of two X-chromosomes (see KLINEFELTERS'S SYNDROME and TURNER'S SYNDROME). In birds, reptiles, some amphibia and lepidopterans, the situation is reversed, with females being the heterogametic sex. In fruit flys, sex is determined by the ratio of the number of X-chromosomes to the number of sets of autosomes (A). When the ratio of X/A = 1.0 the individual is female, when $X/A = 0.5$ a male results. In other DIPTERANS such as the yellow-fever mosquito, sex is determined by a single locus with two alleles, the heterozygote being the male (M/m) and a homozygote the female (mm). Plants too can have XX, XY sex determination where they are DIOECIOUS, but various other mechanisms also can occur. For example, sex in asparagus is controlled by a single pair of alleles, with maleness dominant.

sex factor, see F FACTOR.

sex hormones, *n.* hormones capable of stimulating the development of the reproductive organs and SECONDARY SEXUAL CHARACTERS in mammals. The hormones are ANDROGEN, OESTROGEN and PROGESTERONE. Sex hormones are secreted by the ovaries, testes, adrenal cortex and placenta. See OESTRUS CYCLE.

sex limitation, *n.* the restriction of the expression of a character to one or other of the sexes (e.g. lactation in female mammals, facial hair in male humans), even though both sexes may carry genes for such characters. Sex-limited genes are strongly affected by the level and types of sex hormones present in the body and are mainly carried on AUTOSOMES rather than SEX CHROMOSOMES.

sex linkage, *n.* the location of genes on one or other of the SEX CHROMOSOMES producing an inheritance pattern which is different from that shown by AUTOSOMES. See Fig. 275.

Autosomal Genes	*Sex-linked Genes*
1. RECIPROCAL CROSSES produce the same results.	Reciprocal crosses produce different results.
2. All individuals carry two alleles of each gene.	Males carry only one allele of each gene (= HEMIZYGOUS).
3. Dominance operates in both males and females.	Dominance operates in females only (see 2.).
4. Alleles passed equally to male and female progeny.	'Criss-cross' inheritance pattern produced: father to daughter to grandson, etc.

Fig. 275. **Sex linkage.** The inheritance pattern of autosomal and sex-linked genes.

Most sex-linked genes are located on the X-chromosome (X-linkage), although some conditions are controlled by genes on the Y-chromosome (Y-linkage) that are only transmitted through the males, for example, genes responsible for SEX DETERMINATION in mammals.

sex ratio, *n.* the number of males in a group divided by the number of females, giving a value which is usually about 1.0. The ratio is controlled by the SEGREGATION of SEX CHROMOSOMES in the HETEROGAMETIC SEX during MEIOSIS (see SEX DETERMINATION). For example, in humans the situation is normally as shown in Fig. 276. In fact, in humans there is often a slight excess of males born (100:105 is typical) due either to preferential fertilization of eggs by Y-carrying sperm causing an unequal primary ratio or to preferential survival of males during GESTATION. Whatever the reason, the sex ratio at birth (the secondary ratio) becomes altered to about 1:1 at sexual maturity in natural circumstances, due to a slightly higher mortality of male children.

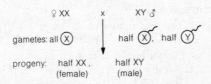

Fig. 276. **Sex ratio.** The ratio of male to female progeny in humans.

sexual behaviour, *n.* any behaviour associated with courtship and reproduction.

sexual cycle, *n.* the series of events in which generations are linked via GAMETES and MEIOSIS takes place. For example, in a DIPLOID(1) organism such as a human, HAPLOID (1) gametes are produced by meiosis and once fertilization has occurred the diploid number is restored in the ZYGOTE. In haploid organisms such as many fungi, meiosis is postzygotic, occurring after fertilization and with the function of restoring the haploid number.

sexual dimorphism, *n.* the presence in a population of two sexes each with a different PHENOTYPE, a classic example of a GENETIC POLYMORPHISM. The underlying mechanism varies in different organisms, but is often controlled by genes on special SEX CHROMOSOMES (see SEX DETERMINATION).

sexual reproduction, *n.* a method of reproduction involving the fusion of haploid GAMETES (fertilization) to form a diploid ZYGOTE. The zygote develops into a new individual organism, in the life history of which MEIOSIS takes place to form the haploid gametes. In the life histories of some organisms sexual reproduction alternates with ASEXUAL REPRODUCTION.

sexual selection, *n.* the selection of a mate by female animals where, for example, the most brightly coloured is favoured, so maintaining brightly coloured males in a population. Some authorities consider that sexual selection explains the existence of SECONDARY SEXUAL CHARACTERS.

shell, *n.* any hard outer covering, such as the carapace of turtles and tortoises, the exoskeleton of crustaceans, the calcareous plates of echinoderms, the outermost membranes of an egg, the skeleton of Foraminifera or the mantle secretions of molluscs.

shipworm, *n.* any marine bivalve mollusc such as *Teredo*, that bores into woodwork by rotary action of the two shell valves and swallows the sawdust, which is then attacked by special enzymes that enable the digestion of cellulose.

shoot, *n.* that part of a vascular plant which is above the ground, consisting of stem and leaves.

short-day plant, *n.* one that requires a daylength of less than a certain minimum values for the induction of flowering to take place. In fact, the term is somewhat misleading in that such plants (e.g. strawberries, chrysanthemums) are really sensitive to periods of darkness rather than daylight, requiring a night length of more than a minimum duration, i.e. these are *long-night plants*. Compare LONG–DAY PLANT, DAY–NEUTRAL PLANT. See PHOTOPERIODISM.

short-sightedness, see MYOPIA.

shoulder girdle, see PECTORAL GIRDLE.

shrub, *n.* a woody plant less than 10 metres high in which there are abundant side branches and no real trunk.

shunt vessel, *n.* a type of bypass channel in the blood circulation. See ARTERIOLE.

sibling or **sib,** *n.* a brother or sister.

sibling species, see SPECIES.

sickle-cell anaemia (SCA), *n.* a human abnormality in which defective HAEMOGLOBIN molecules (Hb^s) cause the red blood cells to have a twisted 'sickle' shape resulting in major circulatory problems and eventually death. The condition is controlled by a single autosomal gene on chromosome 11, with two alleles, S and s. The mutant haemoglobin has one amino acid alteration in its beta chain, number six having changed from GLUTAMIC ACID to VALINE possibly as a result of SUBSTITUTION MUTATION. Sickle-cell anaemics have an s/s genotype, while heterozygotes (S/s) have a condition called sickle-cell 'trait' with a tendency for their blood to sickle at low oxygen tension. Although SELECTION operates strongly against individuals with SCA the mutant allele has remained at fairly high frequency in some populations (e.g. in Central Africa), probably because the heterozygotes are at an advantage relative to the normal (S/s) types in terms of resistance to the MALARIA PARASITE, producing a GENETIC POLYMORPHISM. In North America, however, SCA is a transient polymorphism amongst blacks of African descent since the malarial selection pressure has been removed and its frequency in the population is expected to fall.

sieve cell, see SIEVE TUBE.

sieve plate, see SIEVE TUBE.

sieve tube, *n.* a series of long cells (*sieve cells*) lying end to end forming a tube, found in the PHLOEM of ANGIOSPERMS. Each sieve cell loses its nucleus when mature, but maintains the cytoplasm, which is continuous from one cell to another and crosses the cell boundaries via perforations in the wall called *sieve plates*. Associated with the sieve tube are modified PARENCHYMA cells each with a cytoplasm, nucleus and many MITOCHONDRIA, called *companion cells*. It is considered that the companion cells may regulate TRANSLOCATION of materials within sieve tubes since there are cytoplasmic connections between the sieve cells, and the companion cells and the sieve cells are anucleate.

sigma factor, *n.* a subunit of RNA polymerase that determines the site where transcription begins.

significance, *n.* (statistics) a description of an observed result that shows sufficient deviation from the result expected to be considered different from the expected result. Significance tests such as the CHI-SQUARED TEST can be carried out to produce a value that is converted into the probability that an observed result will match the result expected from a theory. In biology there is a convention that, if there is more than a 5% chance ($P > 5\%$) that the observed result is the same as the expected, it is possible to conclude that any deviations are 'not significant', i.e. have occurred by chance alone. If, however, there is less than a 5% chance ($P < 5\%$) that observed and expected are the same, then it is concluded that the deviations are 'significant', i.e. have *not* occurred by chance alone. For example, tossing a coin 100 times gives 58 heads and 42 tails. The probability that 58:42 is similar to the expected 50:50 is greater than 5%, thus we can conclude that there is no significant deviation between observed and expected results.

silicula, *n.* a type of dry fruit in which the seeds are contained in a short broad structure that splits (dehisces) to allow the seeds to escape, for example, the fruit of Shepherd's Purse.

siliqua, *n.* a type of dry fruit in which the seeds are contained in a long, somewhat cylindrical structure that splits open (dehisces) to allow the seeds to escape, for example, the fruit of the wallflower.

Silurian period, *n.* the period of the PALAEOZOIC ERA lasting from 445 million BP to 415 million BP, during which trilobites declined. The first land plants appeared, though they probably evolved in the Ordovician. Armoured jawless fish and eurypterids were common. See GEOLOGICAL TIME.

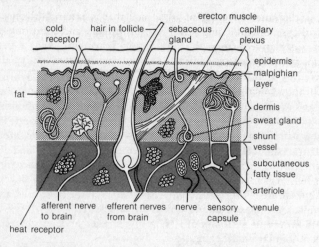

Fig. 277. **Skin.** A section of the human skin showing the dermis.

simple, see COMPOUND.

sinoatrial node (SAN), see HEART.

sinuate, *adj.* (of plant structures) having a wavy outline.

sinus, *n.* **1.** (in plants) a depression or margin between two lobes of a leaf. **2.** (in animals) a cavity within a bone, such as the human facial bone. **3.** (in animals) a large blood-filled cavity such as the HAEMOCOEL of ARTHROPODS or enlarged veins of fish.

sinus venosus, *n.* **1.** (in lower vertebrates) the chamber of the heart to which veins return blood. **2.** (in mammals) a vessel of the embryonic heart, present in the tranverse septum, into which open the vitelline, allantoic and cardinal veins.

Siphonaptera, see APHANIPTERA.

siphonophore, *n.* any marine colonial hydrozoan of the order Siphonophora, including the Portuguese man-of-war.

SI units, *n.* those units of measurement forming the *Système International*, consisting of the metre, kilogramme, second, ampere, kelvin, candela and mole. See Appendix B.

skeletal muscle, see STRIATED MUSCLE.

skeleton, *n.* any structure present in an organism that maintains its shape and supports the structures associated with the body. It can take the form of an internal bony skeleton as in vertebrates, an external calcareous or chitinous exoskeleton as in arthropods, a hydrostatic skeleton as in jellyfish, earthworms etc., or of a subcellular system of support (see CYTOSKELETON).

skin, *n*. the outer covering of an animal that is external to the main musculature, often bearing scales, hair or feathers, and consisting of an EPIDERMIS derived from the embryonic ECTODERM, and a DERMIS originating from a MESODERMIS. The epidermis is often hardened and covered by a cuticle, but may be only one cell thick.

The subcutanerous fat of the dermis acts as insulation and reduces heat loss. Heat is also conserved by the skin in cold conditions by contraction of the superficial blood vessels which diverts blood to lower layers of the skin. In some structures, such as the ear, special shunt vessels occur which dilate in cold conditons and pass blood directly from arterioles to venules, thus bypassing the superficial capillaries. See Fig. 277 on previous page.

skull, *n*. the skeleton of the vertebrate head.

sleeping sickness, see AFRICAN SLEEPING SICKNESS, ENCEPHALITIS.

sliding filament hypothesis, see MUSCLE.

siime moulds, see MYXOMYCETE.

small intestine, *n*. a narrow tube over 7m long, divided into an anterior part, the duodenum (25 cm), a central part, the jejunum (5.6 m) and a posterior part, the ileum (125 cm). CHYME from the stomach stimulates secretion of PANCREATIC JUICE on passing into the duodenum. Bile is added from the liver and flows into the duodenum through the bile duct. The jejunum has a larger diameter than the duodenum and larger VILLI than the rest of the intestine; it is the main absorptive region.

Large amounts of MUCUS are produced by the glandular cells, and a number of enzymes including maltase, peptidases, sucrase, lactase, enterokinase and nucleotidases are also secreted. The enzymes are thought to be largely produced in the CRYPTS OF LIEBERKUHN. The secretions of the small intestine are collectively called the *succus entericus*. The entire intestinal lining in humans is replaced every 36 hours and is the main site of the absorption of food.

smallpox, *n*. a contagious viral disease of humans characterized by pustules on the skin and scar formation. Smallpox was the first disease to be controlled by vaccination (see JENNER) and world-wide immunization programmes organized by the World Health Organization are believed to have virtually wiped out the disease. Unfortunately, with the disappearance of the disease it is likely that the general susceptibility levels will rise in human populations, making wide-scale epidemics a real possibility should the disease return.

smooth muscle, see INVOLUNTARY MUSCLE.

social organization, *n*. the establishment amongst a group or colony of animals of set behaviour patterns.

sodium pump, *n*. the mechanism by which sodium is removed from inside the axon of a neurone so helping to establish the RESTING POTENTIAL. It is also involved in the transfer of ions in the LOOP OF HENLE. The process requires energy from respiration and is an active process; the energy is derived from breaking down ATP. Sodium pumps occur in many cells and should, perhaps, be referred to as *sodium/potassium pumps*, because potassium ions move into the cell as sodium ions move out. However, cell membranes are more permeable to potassium ions than to sodium ions, so the former diffuse out faster than sodium diffuses in and sodium is also pumped out faster than potassium is pumped in. This, together with the mobility of large negative organic ions to move out of the axon, maintains the resting potential.

soil, *n.* the uppermost layer of the earth's crust that supports the majority of terrestrial plant life together with many animals and microorganisms. Soil derives from the erosion of the rock strata and contains minerals and variable amounts of organic material derived from organisms which live or have lived upon it. Soils are affected by climate, living organisms, parent rock, relief, groundwater and age, and usually contain various 'horizons', the succession of which is called the 'soil profile'. MULL soils tend to be alkaline and 'Mor' soils acid. Soils may be heavy, clay, loam (sand and clay), peaty (large amounts of dead plant material), light or sandy.

soil water, *n.* any water held in the soil as a vapour, liquid or solid.

sol, *n.* a suspension of colloidal particles in a liquid. See COLLOID. Compare GEL.

solen- or **soleno-,** *prefix.* denoting a channel.

solenocyte, *n.* a tubular FLAME CELL found in AMPHIOXUS and annelids that is much more elongated than those found in platyhelminths.

solifluction, *n.* a slow movement of soil or rock debris down an incline, usually as a result of lubrication by its water content.

soligenous, *adj.* (of peat bogs) dependent upon an inflow of drainage water.

solubility, *n.* the amount of a substance that will dissolve in a given amount of another substance.

solute, *n.* a substance that is dissolved in a SOLVENT.

solution, *n.* a mixture in which a substance (solid, liquid or gas) is dissolved in another liquid.

solvent, *n.* a liquid in which another substance (a SOLUTE) may be dissolved to form a solution; the solvent is the larger part of the solution.

soma, *n.* the body of an animal excluding the germinal cells that give rise to the gametes.

somatic, *adj.* **1.** of or relating to the SOMA. **2.** of or relating to the human body as distinct from the mind.

somatic cell, *n.* any of the cells of a plant or animal except the reproductive cells.

somatic cell hybridization, *n.* the process of fusion between different SOMATIC CELLS (such as human and rodent cells) to produce hybrid cells in which there is often one fused nucleus. Such a process occurs in cell cultures and the products are useful in establishing the expression and location of particular genes.

somatic mesoderm, *n.* a MESODERM with its embryonic origin on either side of the NOTOCHORD. It develops into (a) a series of blocks of SOMITES from which are derived MYOTOMES, (b) mesenchyme which forms connective tissue and (c) the vertebral column.

somatic mutation, see MUTATION.

somatic nervous system, *n.* that part of the nervous system which supplies the limbs and body wall and which controls the voluntary activities of the body. See AUTONOMIC NERVOUS SYSTEM and NERVE IMPULSE.

somatostatin, *n.* a hormone of the hypothalamus that inhibits the release of growth hormone from the pituitary gland.

somatotrophic hormone (STH), see GROWTH HORMONE.

somite or **metamere,** *n.* a serial segment of the animal body. See METAMERIC SEGMENTATION.

sonogram, *n.* a graphic representation of a sound or sounds made by an animal.

sorus, *n.* a group of sporangia developed on the underside of SPORO-PHYLLS (spore-bearing leaflets) of the SPOROPHYTE generation of a fern. See Fig. 278.

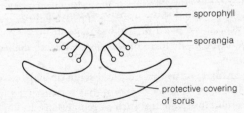

— sporophyll

— sporangia

protective covering of sorus

Fig. 278. **Sorus.** Transverse section.

spadix, *n.* a massive, fleshy spike of flowers protected by a large, sometimes green, bract called a *spathe*. A classic example of a plant with such a structure is the ARUM LILY.

sparging, *n.* the introduction of air into a microbial fermenter under pressure.

spathe, see SPADIX.

spathulate, *adj.* (of plant structures) paddle or spoon-shaped.

special creation, *n.* a theory of evolution that postulates the formation, *de novo*, of species by an all-powerful creator.

speciation, *n.* the process by which new species are formed. Speciation occurs when gene flow has effectively ceased between populations where it previously existed and is brought about by ISOLATING MECHANISMS. See GEOGRAPHICAL ISOLATION.

species, *n.* the lowest (taxonomic) grouping of animals or plants which at least potentially forms an interbreeding array of populations unable to breed freely with other sorts of animals and plants. The species is the only natural unit (taxon) of CLASSIFICATION. It is usually recognized on the basis of morphological characters (a MORPHOSPECIES), but different species can be morphologically identical (*sibling species*), for example, *Drosophila pseudoobscura* and *D. persimilis* exhibit behavioural differences leading to REPRODUCTIVE ISOLATION. See BINOMIAL NOMENCLATURE.

species recognition, *n.* the exchange of specific signals between individual members of a species, for example during courtship.

specificity, *n.* the selective reactivity of an ANTIGEN and its corresponding ANTIBODY.

specific name, *n.* the second of two names in a binominal indicating the name of the particular species in that genus, for example *rubecula* is the specific name of the robin *Erithacus rubecula*.

sperm or **spermatozoon,** *n.* a small, usually motile male GAMETE. See ACROSOME.

spermatheca, *n.* a sac for the storage of sperm, a 'seminal receptacle' as occurs in the female reproductive tract of many lower animals, such as insects and PLATYHELMINTHS.

spermatid, *n.* a haploid stage in male GAMETOGENESIS.

spermatocyte, *n.* a diploid or haploid stage in male GAMETOGENESIS.

spermatogenesis, *n.* the form of MEIOSIS producing male gametes. See GAMETOGENESIS.

spermatogonium, *n.* an early diploid stage of male GAMETOGENESIS.

spermatophore, *n.* a packet of sperm that is transferred from male to female in certain invertebrates such as crustaceans, molluscs, cephalopods, and in a few vertebrates such as aquatic salamanders.

spermatophyte, *n.* any plant of the division Spermatophyta, comprising all forms of seed-bearing plants, ANGIOSPERMS and GYMNOSPERMS.

spermatozoid, see ANTHEROZOID.

spermatozoon, see SPERM.

sperm bank, *n.* a refrigerated source of male gametes (sperm), usually from males of proven breeding value, and used regularly by animal breeders. See ARTIFICIAL INSEMINATION.

sphen- or **spheno-,** *prefix*. denoting a wedge.

Sphenopsida, *n*. a subdivision of the Tracheophyta, including the horsetails (*Equisetum*).

spherule, *n*. a thick-walled structure containing large numbers of fungal spores.

sphincter, *n*. a ring of muscle surrounding a tube or the opening to a tube that controls the size of the aperture it surrounds and, thus, the movement through the tube, for example pyloric and anal sphincters.

spicule, *n*. **1.** a small spiked structure in male nematode worms that assists in copulation. **2.** slender rods of calcium carbonate found in sponges supporting the soft wall.

spike, *n*. a raceme of SESSILE(2) flowers, as found in the plantain and the spotted orchid.

spikelet, *n*. a unit in the inflorescence of a grass consisting of one or more flowers surrounded by other structures, for example one of more sterile GLUMES.

spinal column, see VERTEBRAL COLUMN.

spinal cord, *n*. a more or less uniform tube extending the full length of the body in vertebrates behind the head. It is enclosed in the backbone. Pairs of spinal nerves leave the cord in each segment of the body. The cord contains many NERVE CELLS and bundles of fibres, many associated with simple REFLEX ARCS, others with the brain. Coordination of movement of various parts of the body is brought about in the spinal cord.

spinal nerve, *n*. any of numerous nerves which leave the SPINAL CORD, one pair in each segment, each being connected with the cord by a dorsal and a ventral root. They are 'mixed nerves' (containing both sensory and motor fibres) and carry groups of AXONS to receptors and effectors in different parts of the body. See REFLEX ARC.

spinal reflex, *n*. a reflex action whose pathway goes through the spinal cord, not the brain.

spindle or **spindle fibres,** *n*. a network of fibres or MICROTUBULES formed during late prophase and early metaphase of MITOSIS and MEIOSIS, and which serves as an attachment point for chromosomes before they are pulled to the spindle poles during anaphase. Spindle formation is prevented by COLCHICINE.

spindle fibres, see SPINDLE.

spinneret, *n*. one of a number of tubular appendages, that in spiders and some insects, exudes silk threads. In spiders the silk is used to make webs, and in insects for cocoons. Spinnerets in insects and spiders are not homologous structures.

spiracle, *n*. **1.** (in fish) a gill-like cleft that opens behind the eye and through which water is drawn in for gaseous exchange (as it is through the mouth) by the expansion of the pharyngeal cavity. It is absent in many bony fish. **2.** (in arthropods) the exterior opening of the tracheae, often possessing valves that can close to prevent water loss.

spiral cleavage, *n*. the division of cells in the developing animal egg in a spiral manner (see Fig. 106). Most invertebrate groups exhibit this type of cleavage with the exception of echinoderms, protochordates and a few smaller groups.

spiral thickening, *n*. a deposition of lignin on the inner surface of an XYLEM VESSEL, forming a spiral pattern.

splanchnic, *adj*. of or relating to the viscera.

splanchno-, *prefix*. denoting entrails.

splanchnocoel, *n*. the perivisceral cavity lined on the inside by splanchnopleure and on the outside by somatopleure.

splanchnopleure, *n*. the inner layer of embryonic mesoblast that contributes to the wall of the alimentary canal and parts of the visceral organs.

spleen, *n*. an important part of the MONONUCLEAR PHAGOCYTE SYSTEM made up of lymphoid tissue. It stores excess red blood cells, destroys old cells, and is capable of acting as a reservoir holding 20–30% of all blood cells. It produces LYMPHOCYTES and serves to regulate the volume of blood cells elsewhere in the blood system.

splicing, *n*. the joining together of RNA chains, for example after the excision of INTRONS to link up the AXON segments and produce mature mRNA.

splitter, *n*. a taxonomist who recognizes minute differences and as a result recognizes numerous separate taxa. Compare LUMPER.

sponge, *n*. any member of the phylum Porifera. Sponges are multicellular organisms though many biologists regard them as colonies of single cells. Several types of cells exist in a sponge but they are functionally independent of one another and can exist on their own, or in small isolated groups. Usually they possess an internal skeleton of separate crystalline spicules, irregular organic fibres (bath sponge) or both.

spongin, *n*. a horny substance secreted by cells in the middle layers of the sponge that acts as a cement joining together the spicules in the body wall or forming a fibrous skeleton in other types.

spongy mesophyll, see MESOPHYLL.

spontaneous generation or **abiogenesis,** *n*. a discredited belief that living organisms could arise from nonliving things that was finally shown to be untrue by PASTEUR in his famous swan-neck flask experiments.

spontaneous mutation, *n.* a MUTATION occurring in the absence of a known mutagen. In humans, a typical rate of spontaneous mutation is between 1×110^{-5} and 1×10^{-6} mutations per gamete formed. The rate varies considerably between different organisms, perhaps reflecting different DNA repair systems at work.

sporangiophore, *n.* a fungal HYPHA bearing one or more SPORANGIA.

sporangium, *n.* (*pl.* sporangia) the structure within which asexual SPORES are formed.

spor- or **sporo-,** *prefix.* denoting a seed.

spore, *n.* a reproductive body consisting of one or several cells formed by cell division in the parent organism, which, when detached and dispersed, and if conditions are suitable, germinates into a new individual. Spores occur particularly in fungi and bacteria and also in Protozoa. Some have thick, resistant walls which enable them to overcome unfavourable conditions such as drought. Usually they are produced in very large numbers, and occur as a result of either sexual or asexual reproduction.

spore mother cell, *n.* a DIPLOID(1) cell in which MEIOSIS occurs to give four HAPLOID(1) spores.

sporocyst, *n.* a cyst producing asexual spores.

sporogonium, *n.* (*pl.* sporogonia) the SPOROPHYTE generation of mosses and liverworts, producing asexual SPORES.

sporogony, *n.* the formation of SPOROCYTES in a PROTOZOAN zygote by encystment and division.

sporophore, *n.* any fungal structure that produces SPORES.

sporophyll, *n.* a leaf bearing a SPORANGIUM. In some plants, such as ferns, the sporophyll is identical with ordinary leaves, but in others is considerably modified as in the stamens and carpels of flowering plants.

sporophyte, *n.* the diploid generation of plants which by MEIOSIS gives rise to a HAPLOID(1) generation of spores.

sporozoan, *n.* any parasitic protozoan of the class Sporozoa, such as the MALARIA PARASITE.

sporozoite, *n.* any of the very small mobile spores formed by multiple fission of a protozoan zygote, as in the MALARIA PARASITE.

sport, *n.* an individual that is distinctly different from its parents in a way that was not expected, such as a mutation.

sporulation, *n.* a form of asexual reproduction in which specialized cells become surrounded by a tough, resistant coat and then separate from the parent plant. These SPORES are able to withstand severe environmental conditions and when conditions improve they germinate to produce new plants by repeated MITOSIS. Sporulation is common in fungi.

springtail, see COLLEMBOLA.

spur, *n.* **1.** a short shoot on which flowers are borne. **2.** an extension of a leaf base below its point of attachment to the petiole. **3.** a hollow, conical projection from the base of a petal, as in the larkspur.

Squamata, *n.* an order of reptiles containing the lizards and snakes, which have a skin with horny epidermal scales or shields. The moveable quadrate bone makes it possible to move the upper jaw with respect to the braincase.

squamosal, *n.* either of the paired bones which occur on each side of the skull of vertebrates. In mammals they take part in the zygomatic arch, and the lower jaw articulates with the squamosals.

squamous, *adj.* (of epithelium) flattened and plate-like.

squid, *n.* a 10-armed CEPHALOPOD (cuttlefish) with an internal shell, that moves by means of jet propulsion through a muscular siphon.

S stage, *n.* the synthesis phase of the CELL CYCLE.

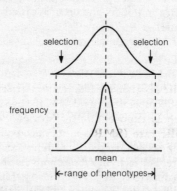

Fig. 279. **Stabilizing selection.** The mean represents the stable average phenotype.

stabilizing selection, *n.* a type of SELECTION in which both extremes of a phenotypic range are selected against, producing a stable average PHENOTYPE with reduced variation in the range. See Fig. 279. A good example of stabilizing selection is human birthweight where, under natural conditions, babies are subject to selection pressure if they are too small at birth (due to immaturity of body system) or too large (due to difficulties in delivery related to the size of the mother's pelvis). Compare DISRUPTIVE SELECTION.

stable polymorphism, see BALANCED POLYMORPHISM.

stable terminal residue, *n.* material that is not broken down by natural means in the soil and remains as the last of the soil degradation products.

stamen, *n.* the male organ in angiosperm flowers, consisting of a stalk, or filament, bearing an anther in which POLLEN grains are produced by MEIOSIS.

staminate, *adj.* (of flowers) having STAMENS but no CARPELS.

staminode, *n.* a rudimentary, infertile stamen that may be reduced in size.

standard deviation (S), *n.* a measure of the variation in a sample, calculated as the square root of the VARIANCE. Mean values are often followed by the standard deviation. See STANDARD ERROR.

standard error (SE), *n.* an estimate of the STANDARD DEVIATION of the means of many samples, calculated as the standard deviation(s) divided by the square root of the number of individuals in a sample (n), i.e. $s/\sqrt{n}$. See Fig 280.

Values in sample (n=7)	48, 27, 36, 52, 35, 41, 33
mean of sample	$\bar{x} = 38.86$
variance of sample	$s^2 = 76.48$
standard deviation	$s = 8.74$
standard error	$SE = 3.31$

Fig. 280. **Standard error.**

The SE can be used to assess if there is any *significance* in the deviation of the mean of a sample from the true mean of the whole population.

standard metabolic level, *n.* the total STANDARD METABOLIC RATE of an animal at a specified temperature divided by the body mass raised to the power of the metabolic exponent for that species or taxonomic group.

standard metabolic rate (SMR), *n.* the metabolic rate of a resting post-absorptive animal in a darkened chamber at a specified environmental temperature. For tachymetabolic animals SMR is BASAL METABOLIC RATE plus REGULATORY HEAT PRODUCTION. Units: ml: O_2/h.

standard temperature and pressure (STP), *n.* the standard conditions of 25°C, 1 atmosphere.

standing biomass, *n.* the BIOMASS present in an ecosystem at a given time.

standing crop, *n.* the mass of a particular organism or organisms present at a particular time within a given environment, i.e. the total BIOMASS of that organism or organisms.

stapes, *n.* the innermost of the three ear osicles which contacts the oval window of the ear. It is stirrup-shaped.

starch, *n.* a polysaccharide carbohydrate consisting of two forms of GLUCOSE units, amylose and amylopectin, which are joined to form a helical molecule. Upon heating, the two components are separated, with amylose giving a purplish/blue colour when iodine is added and amylopectin giving a black colour, this forming the standard test for starch. Starch is the principal storage compound of plants. See DEXTRIN.

starch sheath, *n.* an ENDODERMIS containing starch grains.

starch test, see STARCH.

statoblast, *n.* an internal bud in a BRYOZOAN that develops into a new individual.

statocyst, *n.* an organ of balance in which a STATOLITH(2) rests amongst hairlike projections of sensory cells. A change in position of the animal moves the statolith against a receptor, sending a nervous impulse to the central nervous system.

statolith, *n.* **1.** (in plants) a solid inclusion within a cell, such as a starch granule, by means of which the position of the plant in relation to the gravity is thought to be perceived. **2.** (in animals) a granule that stimulates sensory cells in a STATOCYST as the animal moves. See OTOLITH.

stearic acid, *n.* a saturated fatty acid having 18 carbon atoms, which occurs in animal fat.

stegosaur or *Stegosaurus,* *n.* a quadrupedal herbivorous ornithischian dinosaur of the suborder Stegosauria, a large JURASSIC PERIOD reptile up to 9 m in length. Stegosaurs possessed large neural spines which supported very large, vertical, triangular plates of armour.

stele or **vascular cylinder,** *n.* the cylinder of tissues lying inside the ENDODERMIS of plant stems and roots, containing the vascular tissues.

stellate, *adj.* star-shaped.

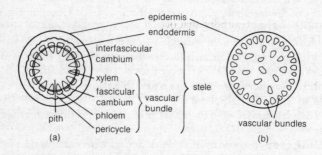

Fig. 281. **Stem.** Transverse sections of herbaceous stems: (a) dicotyledon, (b) monocotyledon.

stem. *n.* the part of the shoot of vascular plants from which are produced leaves at regular intervals (NODES) and reproductive structures. Stems are usually circular in cross section but some are square (for example, members of the family *Labiatae*, such as mint and lavender) while others are ribbed. The internal structure of stems can be herbaceous (non-woody) or show SECONDARY THICKENING. See Fig. 281.

The forms adopted by stems are highly varied, ranging from the oak tree, to climbing plants such as *Clematis* and the pea. Stems are sometimes used as underground storage organs as in RHIZOMES, CORMS, BULBS (underground shoots with food stored in fleshy leaves) and TUBERS, while others are adapted for VEGETATIVE PROPAGATION as in the strawberry RUNNER and suckers of mint.

stereoisomer, *n.* one of two or more isomers that have the same molecular structure but differ in the spatial arrangement of the atoms in the molecule.

stereotropism, see THIGMOTROPISM.

sterile, *adj.* **1.** not capable of producing offspring. **2.** free from living microorganisms.

sterile male release (SRM), *n.* a method of insect BIOLOGICAL CONTROL in which large numbers of sterile males of a pest species are produced and released into the pest population. Since many female insects only mate once, the idea of the technique is that females should mate with sterile rather than fertile males, thus decreasing dramatically the size of the next generation.

sterilization, *n.* **1.** the act of destroying all forms of microbial life on an object, thus making it sterile. **2.** the act of preventing an organism from reproducing by removing the gonads.

stern– or **sterno–,** *prefix.* denoting a breastbone.

sternite, *n.* the skeletal plate(s) on the ventral side of insect and crustacean body segments.

sternum, *n.* the breastbone that occurs in the ventral region of the chest and to which, in flying birds where the bone is keel-shaped, are attached the pectoral muscles associated with flight. Anteriorly it is connected to the shoulder girdle, and the ventral ends of the ribs are attached along its length.

steroid, *n.* an important but unusual type of LIPID, formed of four rings of carbon atoms with various side groups, such as cholesterol, digitoxin (which forms part of the heart-stimulating drug digitalis) and cortisone.

sterol, *n.* a steroid alcohol in which the alcoholic hydroxyl group is attached in position 3, and has an aliphatic side chain of at least 8 carbon atoms at position 17.

stigma, *n.* the upper part of the pistil of a flower on which pollen is deposited. The stigmatic surface secretes a sugary solution which aids germination of the pollen grain, unless a SELF–INCOMPATIBILITY mechanism is operating.

stimulus, *n.* any detectable change in the environment (internal or external) of an organism which is capable of influencing a RESPONSE in the whole organism or parts of it.

sting, *n.* an organ present in many different animal groups that is capable of injecting a poison into other organisms either as a defensive or offensive mechanism. Examples include the modified ovipositor in HYMENOPTERA, cnidoblasts in coelenterata, the tail in scorpion.

stipe, *n.* **1.** the stalk of the thallus of a seaweed. **2.** the fruiting body of a fungus.

stipel, *n.* one of two leaf-like structures at the base of a leaflet in some compound leaves.

stiptate, *adj.* (of plant structures) having a stalk.

stipule, *n.* one of a pair of leaflets found at the base of a leaf where it joins the stem. Stipules are common in many plant families such as Rosaceae.

stock, *n.* **1.** a rooted stem into which a SCION is inserted for grafting (see GRAFT). **2.** a mating group used in a breeding programme.

stochastic model, *n.* any mathematical model of a system that is governed by the laws of probability and contains a randomized element (for example, a computer program that models a population controlled by the mechanisms of MENDELIAN GENETICS).

stolon, see RUNNER.

stoma, *n.* (*pl.* stomata) an opening in the epidermis of leaves (and sometimes stems) that allows gaseous exchange. The size of the stomatal aperture is controlled by two guard cells, whose shape can alter depending upon internal turgidity. See Fig. 282. When flaccid, the thick inner wall causes the guard cells to straighten closing the stoma; when turgid the guard cells become curved, opening the stoma for gaseous exchange.

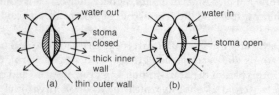

Fig. 282. **Stoma.** (a) closed aperture. (b) open aperture.

The mechanism of guard cell operation is not fully understood. An older theory proposes that high CO_2 values at night cause high acidity which encourages the enzymic conversion of sugar to starch, thus reducing the OSMOTIC PRESSURE in the guard cell sap, so that water is lost and the stoma closes. During daylight, PHOTOSYNTHESIS uses up the available CO_2, the pH rises so favouring the conversion of starch to sugar, thus increasing the flow of water into guard cells by osmosis, causing the stoma to open. A newer theory has been built on the older one, and suggests that the changes in osmotic pressure are caused, not just by differences in levels of CO_2, but by levels of potassium ions (K^+) in the guard cells. According to this theory, K^+ ions are actively pumped into the guard cells during the day so increasing the osmotic pressure and hence the turgidity.

stomach, *n.* that part of the vertebrate gut system which follows the oesophagus, is expanded to form a chamber, and whose walls secrete pepsinogen giving rise to PEPSIN, RENNIN (in young mammals) and hydrochloric acid from the OCYNTIC CELLS. Gastric secretions also include mucin, which lubricates the food mass that is passed, a little at a time, to the SMALL INTESTINE via the PYLORIC SPHINCTER. The mixture of partly digested food and secreted fluids is known as CHYME. See DIGESTION, DIGESTIVE SYSTEM.

stomium, *n.* the part of the wall of a fern SPORANGIUM with thin-walled cells that ruptures during dehiscence, enabling the release of spores.

stone cell, *n.* a cell with thick walls reinforced with lignin, found in the flesh of succulent fruits such as pears.

stop codon, see NONSENSE CODON.

storage material, *n.* any compound that accumulates naturally within a cell, e.g. the starch grains of potato tubers and glycogen in liver cells.

STP, see STANDARD TEMPERATURE AND PRESSURE.

strain, *n.* a group of organisms within a species or variety, distinguished by one or more minor characteristics.

stratification, *n.* a process in which certain seeds are subjected to low temperatures for a period of time, in order for GERMINATION to take place. Natural stratification occurs when seeds are shed in autumn and become covered with soil, leaves, etc. during the winter. The process can be reproduced artificially by alternating layers of seeds with layers of moistened substrate such as sphagnum moss and storing at a low temperature.

stratified epithelium, see EPITHELIUM.

streaming, cytoplasmic, see CYTOPLASMIC STREAMING.

Strepsiptera, *n.* an order of insects containing very small parasitic individuals. The larvae parasitize bees and plant bugs; the females are wingless endoparasites, the males winged (hind wings only). Parasitized insects are said to be *stylopised* (after *Stylops*, one of the parasitizing genera) and cause the host sex organs to degenerate.

strepto-, *prefix.* denoting twisted.

streptomycin, *n.* an ANTIBIOTIC produced by a soil ACTINOMYCETE, which is active against many bacteria, particularly of the gram-negative type (see GRAM'S STAIN). Streptomycin works by causing a misreading of the GENETIC CODE during PROTEIN SYNTHESIS in the bacteria.

stretch receptor, *n*. the receptor for the detection of muscle stretch located in the muscle spindle organ. The receptor is situated centrally and consists of spiral nerve endings (branches of an efferent nerve) surrounding noncontractile muscle fibres. Efferent nerve fibres connect at each end with contractile muscle fibres. Stretching of a muscle causes discharge from spiral nerve endings to the muscle itself thus causing contraction. The more the spindle is stretched the greater the efferent frequency of impulses and subsequently the more the muscle contracts. This is the *stretch reflex*.

Muscle spindles occur in numbers up to 30/g of muscle, shoulder and thigh muscles being spindle-poor and more distal muscles being spindle-rich. The largest numbers occur in the head and upper vertebral column. See Fig. 283.

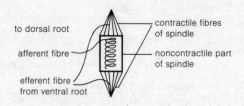

Fig. 283. **Stretch receptor.** The stretch receptor of the muscle spindle.

stretch reflex or **myostatic reflex,** *n*. the contraction of a muscle in response to a sudden longitudinal stretching of the same muscle.

striate, *adj*. (of plant structures) marked with parallel ridges or depressions.

striated muscle or **striped muscle** or **skeletal muscle** or **voluntary muscle,** *n*. contractile tissue which in vertebrates consists of fibrils with marked striations at right angles to the long axis. The muscle fibres are made up of a number of SARCOMERES (see MUSCLE, I-BAND. The muscle is capable of rapid contraction and is concerned with the movement of skeletal parts – hence the alternative name of 'skeletal muscle'. 'Striped' or 'voluntary' muscle are yet further alternative names, the latter because of its stimulation by the voluntary nervous system. Compare INVOLUNTARY MUSCLE.

strict, *adj*. (of plant structures) growing upwards, stiff and rigid.

stridulation, *n*. the sound produced by rubbing together part of the hind legs and the forewing of grasshoppers and locusts (see ORTHOPTERA). The sound produced results in the bringing together of males and females for mating.

striped muscle, see STRIATED MUSCLE.

strobila, *n.* a linear sequence of similar animal structures, such as the segmented body of a tapeworm.

strobilation, *n.* asexual reproduction by division into segments, as in tapeworms and jellyfishes (see EPHYRA, SCYPHISTOMA).

stroma, see CHLOROPLAST.

stromatolite, *n.* a rock-like mound, often found in fossil form, created by blue-green algae. See CYANOPHYTE.

structural gene, see OPERON MODEL.

strychnine, *n.* a poisonous alkaloid obtained from *Strychnus nux-vomica*, used as a stimulus for the CNS.

style, *n.* the stalk of the PISTIL of a flower connecting the STIGMA to the ovary.

sub-, *prefix.* denoting below.

suberin, *n.* a complex of fatty substances present in the wall of cork tissue that waterproofs it and makes it resistant to decay.

suberization, *n.* the deposition of SUBERIN.

subfamily, *n.* a taxon immediately below the level of family and above genus.

subgeneric name, *n.* a little-used taxonomic grouping below the genus; it is usually bracketed when used in a scientific name e.g. *X-us (Y-us) britannicus*.

subjective synonym, *n.* any of two taxonomic synonyms resulting from different TYPES, but considered to be referable to the same taxon by those regarding them as synonyms.

sublittoral, *adj.* growing near the sea, but not on the shore.

subsoil, *n.* the soil below the plough layer.

subspecies, *n.* the division of the population of a SPECIES on the grounds of incomplete REPRODUCTIVE ISOLATION. In many groups of organisms subspecies have been named on minor morphological differences which usually occur in different geographically defined populations. Often, for example in birds, different parts of a CLINE have been given subspecific status; if clines are to be named, only the extremes should have subspecific names. Clearly defined populations such as island forms often merit subspecific names and might be considered as species in the making. See ISOLATING MECHANISMS. ·

substitute name, *n.* a taxonomic name that replaces another pre-occupied name and which therefore takes on the same TYPE specimen and locality.

substitution, *n.* **1.** a reaction in which an atom or group of atoms is removed and replaced by another atom or group. **2.** the replacement of one amino acid by another. **3.** the replacement of one base by another in a nucleotide or nucleic acid.

substitional load, *n.* the cost of replacing one ALLELE by another during evolutionary change in a population.

substitution mutation, *n.* a mutation affecting the base sequence of a DNA molecule, in which one base is substituted for another one, with no loss or gain of base and therefore no risk of a FRAMESHIFT.

A *transition substitution mutation* is the replacement of a PYRIMIDINE base by another pyrimidine (e.g. thymine changing to cytosine) or a PURINE by another purine (e.g. adenine replacing guanine). A *transversion substitution mutation* is the replacement of a pyrimidine by a purine or vice versa.

substrate, *n.* **1.** the medium on which an organism (especially a microorganism) can grow. **2.** the solid object to which a plant is attached, such as a rock forming the substrate for a seaweed STIPE. **3.** any substance on which an enzyme can act.

substrate–enzyme complex, *n.* a chemical combination of a substrate molecule at the ACTIVE SITE of an enzyme.

substrate–level phosphorylation, *n.* the direct transfer of a phosphate group to ADP, thus forming ATP, without the presence of oxygen. The phosphorylation is thus independent of the ELECTRON TRANSPORT SYSTEM used in oxidative phosphorylation. See ANAEROBIC RESPIRATION, GLYCOLYSIS.

subulate, *adj.* (of plant structures) long, narrow and pointed.

succession, *n.* the progression from initial colonization of an area by organisms to the CLIMAX population. The term usually refers to plants and in England the climax vegetation is oak woodland.

succulent fruit, *n.* one in which the PERICARP contains a fleshy mesocarp which is often good to eat. For example, peach, plum, cherry and apricot, which are all DRUPES.

succus entericus, see SMALL INTESTINE.

sucker, *n.* **1.** an axial shoot arising from an adventitious bud of a root and appearing separate from the mother plant. Such structures are common in the family Rosaceae. **2.** a new shoot on an old stem. **3.** a modified root in a parasite enabling it to absorb nutritive materials from the host.

sucrase or **invertase,** *n.* an enzyme that catalyses the hydrolysis of SUCROSE into GLUCOSE and FRUCTOSE. Sucrase is contained in the SUCCUS ENTERICUS.

sucrose, *n.* a DISACCHARIDE nonreducing sugar used in sweetening, being obtained from the juice of the sugar cane and from sugar beet. Sucrose ($C_{12}H_{22}O_{11}$) is formed by a CONDENSATION REACTION between FRUCTOSE and GLUCOSE.

suction pressure, see WATER POTENTIAL.

sudoriferous, *adj.* (of a gland or its duct) producing or conveying sweat.

suffruiticose, *adj.* (of plants) being a dwarf plant, particularly a shrub, in which many upper branches die back after flowering.

sulcus, *n.* a groove or fissure.

sugar, *n.* a simple form of CARBOHYDRATE, formed of MONOSACCHARIDE units. Such molecules can exist in either a straight chain or a ring form. The straight chain contains a $C = O$ group; if this group is terminal the sugar has the properties of an aldehyde (aldose sugar), if nonterminal the sugar acts as a ketone (ketose sugar). Both aldose and ketose sugars can be oxidized and will reduce alkaline copper solutions (see FEHLING'S TEST, BENEDICT'S TEST). They are thus called REDUCING SUGARS. Several disaccharides such as maltose and lactose are also reducing sugars. Sucrose, however, is a *nonreducing sugar* in which the linkage of glucose and fructose masks the potential aldehyde group of glucose and the potential ketone group of fructose, so that no reduction occurs in the Fehling's and Benedict's tests.

The backbone of the sugar can be of varying lengths, containing as little as three carbons (triose sugars) but, more commonly, five carbons (pentose sugar) and six carbons (hexose sugars).

sulphonamide, *n.* a class of organic compounds that are amides of sulphonic acids, some of which are sulpha drugs and act as powerful inhibitors of bacterial activity.

sulphur bridge, *n.* the linkage between two sulphur atoms within molecules that enables complex folding to occur. For example, disulphide bridges often form between cysteine residues in proteins.

summation, *n.* the production of an effect by the repetition of stimuli, any single one of which would be insufficient to produce an effect, as in muscular contraction where summation brings about TETANUS which results from a series of stimuli. See RODS and CONES CELLS for the effect of summation in the eye.

super- or **supra-,** *prefix.* denoting above.

superfamily, *n.* a taxon above a family and below an order.

superficial, *adj.* on or near the surface, e.g. of an artery, vein or ovule.

supergene, *n.* a cluster of genes along a chromosome that, while closely located, may have unrelated functions. Supergenes may act as a single unit during RECOMBINATION in meiosis.

superinfection, *n.* an infection added to one already present.

superior ovary, *n.* a plant ovary in which the PERIANTH is attached to its base leaving the ovary free.

supernatant, *n.* the clear liquid above a precipitate which has settled out.

super-optimal stimuli, *n.* any stimuli to which an organism responds more strongly than to those for which the reaction was originally selected.

superspecies, *n.* a monophyletic group of species which occurs geographically separately, but whose members are too distinct to form a single species. See SEMISPECIES.

supraspecific, *adj.* above the species level.

supressor mutation, *n.* a MUTATION that acts to restore the normal functioning of another gene located elsewhere in the GENOME, or another mutation at a different site in the same gene.

surface area/volume ratio, see BERGMANN'S RULE.

susceptible, *adj.* liable to attack by another organism (pest or pathogen) or a biocide.

suspension, *n.* a system in which denser, microscopically visible, particles are distributed throughout a less dense liquid and maintained there, settlement being hindered or prevented either by the viscosity of the fluid or the molecular impacts of the liquid's molecules on the particles.

suspensor, *n.* **1.** (in higher plants) a mass of cells that pushes the developing embryo down into the nutritive tissues so that it can obtain the maximum nutrition for further development. **2.** (in mycology) a HYPHA that supports a ZYGOSPORE, GAMETE or gametangium.

Sutherland, Earl Wilbur, American pharmacologist and physiologist awarded the Nobel Prize for Physiology and Medicine in 1971 for isolation of CYCLIC AMP. With T.W. Rall, Sutherland discovered that hormones such as ADRENALINE act on their target cells by causing an increase in cyclic AMP.

suture, *n.* **1.** (in surgery) a thread or wire used to join together a wound, in surgery. **2.** the seam found after stitching two parts together. **3.** an immovable joint between the bones of the skull. **4.** (in plants) the line of fusion between two carpels.

Svedberg (S) unit, *n.* a unit of sedimentation velocity used to describe molecules of various sizes. See DIFFERENTIAL CENTRIFUGATION.

Swammerdams's glands, *n.* amphibian glands which secrete calcareous nodules on each side of the vertebral column. Named after the Dutch naturalist J. Swammerdam.

swarm, *n.* **1.** a group of social insects, esp. bees led by a queen, that has left the parent hive in order to start a new colony. **2.** a large mass of small animals, esp. insects.

swarmer cell, *n.* a spherical cell in *Rhizobium* which penetrates the root hairs of legumes in order to establish the symbiotic relationship associated with NITROGEN FIXATION.

sweat gland, *n.* a structure present in the skin of some mammals that secretes sweat containing sodium chloride as part of the cooling system. A typical human has about 2.5 million sweat glands. Dogs, cats and rabbits are amongst the mammals that do not sweat, but use evaporation from the upper respiratory tract to lose excess body heat.

swim bladder, see AIR BLADDER.

swimmeret, *n.* any of the small paired appendages on the abdomen of crustaceans, used for locomotion or reproduction.

switch gene, *n.* a gene that causes the total genetic developmental system to change to a different developmental pathway.

sym- or **syn-,** *prefix.* denoting together, alike, with.

symbiont, *n.* an organism living in a state of SYMBIOSIS with another.

symbiosis or **mutualism,** *n.* a relationship between dissimilar organisms in which both partners benefit. For example, the hermit crab *Pagarus* and the sea anemone *Adamsia palliata* which lives attached to the shell; here the anemone obtains food scraps from the crab and the crab is camouflaged by the anemone and also defended by its stinging cells (NEMATOCYSTS).

sympathetic nervous system, see AUTONOMIC NERVOUS SYSTEM.

sympatric hybridization, *n.* the occasional occurrence of hybrids between well-defined sympatric species.

sympatric, *adj.* (of a population of organisms) occurring together in the same geographical area. The term is used to describe the geographical distribution of organisms which either coincide or overlap. Compare ALLOPATRIC.

sympatric speciation, *n.* the occurrence of species formation without geographical isolation, genetic isolation occurring through other mechanisms.

sympetalous, see GAMETOPETALOUS.

Symphyla, *n.* a group of primitive millipedes possessing features (particularly the mouthparts) that suggest a common ancestry with the insects.

symplast, *n.* that part of a plant consisting of the PLASMALEMMA and the cytoplasm contained within the plasmalemma, outside which is the APOPLAST(2). The symplastic organisation enables *symplastic transport* between live cells, via PLASMODESMATA.

symplesiomorphy, *n.* the sharing of ancestral characters by different species.

sympodial, *adj.* (of a stem) growing in length by development of lateral buds, the terminal bud developing into an inflorescence or dying back each year.

symptom, *n.* any change in normal function or activity associated with a particular disease.

synapomorphy, *n.* the sharing of derived characters by different species.

synapse, *n.* the point at which one nerve cells connects with another and at which transmission of an impulse takes place by chemical means. When an impulse arrives at a synapse it causes a synaptic vesicle to move towards the presynaptic membrane. On contacting the membrane it discharges the contained transmitter substance into the synaptic cleft, across which it diffuses to the postsynaptic membrane which it depolarizes. This causes a positive charge to develop (*excitory postsynaptic potential EPSP*) because of sodium ions flowing into the post synaptic nerve cell. When the positive charge builds up sufficiently it generates an action potential which is usually unidirectional. See ENDPLATE MOTOR and Fig. 143.

Synapsida, *n.* (in the taxonomy of reptiles), a subclass fossil form, having a single inferior temporal FOSSA on each side of the skull. These are the Theromorphs from which mammals probably evolved. The term is sometimes extended to include the mammals.

synapsis, *n.* **1.** the lying together of structures as in, for example, the junction of two nerve cells. **2.** (in genetics) the pairing of HOMOLOGOUS CHROMOSOMES in prophase I of MEIOSIS.

synaptic cleft, *n.* the narrow gap that separates the membrane of an axon terminal of a nerve cell from the membrane of another, across which TRANSMITTER SUBSTANCE diffuses to effect a stimulus of the postsynaptic cell.

synaptic knob, *n.* the expanded distal end of the small terminal branches of a neuron. See END PLATE.

syncarpous, *n.* (of an ovary of a flowering plant) having fused carpels. Compare APOCARPOUS.

synchronic, *adj.* (of species) occurring at the same periods in time.

synchronous muscle, see FLIGHT.

syncytium, *n.* a cellular structure containing many nuclei.

syndactyly, *n.* having two or more digits fused together.

syndrome, *n.* a group of symptoms that occurs together in a disease or abnormality. Syndromes are often named after their discoverer.

Examples include DOWN'S SYNDROME, KLINEFELTER'S SYNDROME, TUR-NER'S SYNDROME.

synecology, *n.* the study of communities, rather than individuals. Compare AUTECOLOGY.

synergid, *n.* either of two small cells present in the embryo sac of seed plants near the egg cell at the micropylar end.

synergism, *n.* a chemical phenomenon in which the combined activity of two or more compounds is greater than the sum of the individual activities. For example, CYTOKININ and AUXIN act synergistically in promoting DNA replication.

syngamy, *n.* the fusion of gametes.

synovial fluid, *n.* a viscous fluid contained within a membrane enclosing moveable joints such as the elbow and knee. The fluid lubricates the cartilages which make the surface contact between the bones at the joint.

synonym, *n.* any (in taxonomy) of a list of different names for the same taxon.

synonymy, *n.* (in taxonomy) a listing, in chronological order, of all the scientific names applied to a given taxon, together with dates and authors.

synopsis, *n.* a summary of current knowledge.

synteny, *n.* the assignment of genes to the same chromosome in SOMATIC CELL HYBRIDS.

synthesis, see ANABOLISM.

synthetic lethal, *n.* a lethal chromosome resulting from RECOMBINA-TION after CROSSING OVER from nonlethal chromosomes.

synthetic medium, *n.* a microbial growth medium in which all components have been chemically identified and quantified.

synthetic theory, *n.* the theory of evolution generally accepted at the present time which has resulted from a synthesis of parts of previous theories. MUTATION and SELECTION are basic components.

syntype or **cotype,** *n.* any specimen of a series, in which no HOLOTYPE was designated, that is used to designate a species. The syntype may become the LECTOTYPE, if so designated.

syphilis, *n.* a human disease of the sexual organs caused by the spirochaete bacterium *Treponema pallidum* that is spread normally by sexual contact. The organism penetrates mucous membranes, giving rise within a few days to primary ulceration of the membranes. This may give way to a secondary stage of low-grade fever and enlargement of the lymph nodes. A final tertiary stage can take place in which serious lesions occur in many organs and blindness is common. The bacterium can be transmitted from an infected mother to her foetus, resulting in congenital syphilis which may be fatal.

Systema Naturae, see BINOMIAL NOMENCLATURE.

systemic, *adj.* distributed throughout the whole of an organism.

systemic arch, *n.* the 4th aortic arch of a vertebrate through which blood flows from the ventral to the dorsal aorta. Birds have only a right arch, and mammals only a left arch.

systemic biocide or **biocide,** *n.* a chemical such as some insecticides or fungicides that is applied to the soil or leaves, and not directly to the target organism. It is absorbed by plant roots to be translocated around the plant, or absorbed from the leaves.

The biocide has its effect when the plant is attacked by a pest organism such as an aphid which sucks up sap containing the insecticide. Compare CONTACT INSECTICIDE.

systemic mutation, *n.* a form of MUTATION postulated by R. Goldschmidt that fundamentally reorganizes the GERM PLASM THEORY. It is suggested that such mutation could cause the origin of completely new types of organisms.

systole, see HEART CARDIAC CYCLE.

syzgy, *n.* the aggregation in a mass of certain protozoans, especially when occurring before sexual reproduction.

T

table, water, *n.* the level at which all fissures and pores are filled with water and which roughly follows the ground surface.

Where the water table comes above ground, standing water, such as a lake, appears.

tachy-, *prefix.* denoting swift or accelerated.

tachycardia, *n.* an increase in heart rate above the normal rate.

tachymetabolism, *n.* the pattern of thermal physiology in which an animal possesses a relatively high BASAL METABOLIC RATE and is able to respond to a fall in CORE TEMPERATURE with an increase in REGULATORY HEAT PRODUCTION to maintain homoiothermy. Most tachymetabolic species are continually both homoiothermic and endothermic. The only extant (living) forms with this physiology are the birds and mammals.

tactile, *adj.* pertaining to touch.

tactor, see TANGORECEPTOR.

tadpole, *n*. the larva of an AMPHIBIAN or of sea squirts (tadpole larva). See TUNICATE.

tagma, *n*. one of the divisions (head, thorax or abdomen) into which the body of an ARTHROPOD is divided.

taiga, *n*. the zone of forest vegetation lying on wet soils between the tundra in the north and steppe hardwood forest in the south, that encircles the northern hemisphere.

tail, *n*. **1.** the rear-most part of an animal. **2.** (in vertebrates) that part behind the anus.

tailor's posture, *n*. a form of muscular hypertonicity commonly observed in PHENYLKETONURIA.

tandem duplication, *n*. a CHROMOSOMAL MUTATION producing idential adjacent segments. For example, A *B C D* E becomes A *B C D B C D* E.

tangoreceptor or **tactor,** *n*. a receptor sensitive to slight pressure differences.

tannin, *n*. a complex organic compound occurring widely in plant sap, particularly in bark, leaves and unripe fruits, that is used in the production of leather and ink.

T-antigen, an antigen found in cells infected by the oncogenic virus. See ONCOGENE.

tapetum, *n*. a nutritive layer surrounding those cells which will become MICROSPORES. The tapetum is found in a wide range of vascular plants, from FERNS to ANGIOSPERMS. In the latter, a tapetal layer is found in the pollen sacs of the ANTHER.

tapeworm, *n*. any parasitic flatworm of the class Cestoda (phylum *Platyhelminthes*). The adults attach themselves inside the gut system of vertebrates by means of hooks and suckers on the SCOLEX, and the long, ribbon-like body made up of a series of proglottides, may reach a length in excess of 10 m in some species. Eggs develop into 6-hooked embryos and pass out with the FAECES of the host, and if eaten by a suitable secondary host, develop into a larval stage called a CYSTICERCUS. When eaten by the primary host, (which may take place in some species only after transfer through a second secondary host) the tapeworm becomes sexually mature.

taproot, *n*. a primary ROOT structure in which one main root forms the major part of the underground system. Taproots are often swollen, serving as storage organs of perennation. Many vegetable crops have taproots, for example, carrot, swede, sugar beet.

Tardigrada, *n*. (in taxonomy, a name that has usually been used to denote the following two different animal taxons) **1.** (formerly) sloths (now Bradypodidae), **2.** a group of small, transparent 8-limbed ARTHROPODS often classified as ARACHNIDS.

target cell, *n.* any cell that responds to specific hormones.

tarsi-, *prefix.* denoting the sole of the foot.

tarsal bone or **tarsus,** *n.* the bone of the hind limb of TETRAPODS that articulates with the TIBIA and FIBULA proximally and with the METATARSALS distally. See PENTADACTYL LIMB.

tarsier, *n.* any arboreal, lemur-like primate of the genus *Tarsius*, of SE Asia, having large eyes and ears.

tarsus, see TARSAL BONE.

tartrazine, *n.* an AZO DYE that produces a yellow colour: widely used as a food additive (E102).

taste bud, *n.* a receptor organ of taste that contains groups of slender modified epithelial cells with hair-like microvilli which are clustered in a small external pore. Usually taste buds are associated with the mouth of most vertebrates, but in fish they also occur over the body surface and on the fins, and on the tarsi of some insects (e.g. DIPTERANS). Humans distinguish between sweet, salt, sour and bitter tastes, using taste buds in different locations on the tongue, sweet on the front, sour on the sides, bitter on the back and salt all over. Tastes depend on the relative activity of neurones that innervate the taste buds and not on any structural differences on the buds. See Fig. 284.

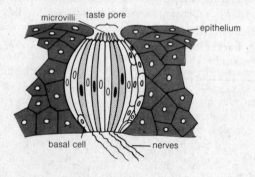

Fig. 284. **Taste bud.** Cross section of the human tongue.

Tatum, Edward, see BEADLE.

tautomeric shift, see BASE ANALOGUE.

taxis, *n.* a behavioural response in which an animal orientates itself toward or away from a given stimulus. For example, GEOTAXIS, PHOTOTAXIS. Compare KINESIS.

taxon, *n.* (pl. taxa) any grouping within the CLASSIFICATION of organisms, such as SPECIES, GENUS, ORDER, etc.

taxonomic category, *n.* the ranking of an organism within an hierarchical CLASSIFICATION.

taxonomic character, *n.* any characteristic used to separate members of different TAXONS.

taxonomy, *n.* the study of the CLASSIFICATION or organisms. Classical taxonomy involves the use of morphological features, Cytotaxonomy the use of somatic chromosomes, experimental taxonomy involves the determining of genetical inter-relationships, and numerical taxonomy involves quantitative assessments of similarities and differences in an attempt to make objective assessments.

Tay-Sachs disease, *n.* a lethal human condition in which children who are apparently normal at birth show signs within six months of marked deterioration of brain and spinal cord. By the age of one year the child can only lie helplessly, becoming mentally retarded, increasingly blind and paralysed. Death occurs between three and four years, with no known survivors and no cure. The condition is controlled by the recessive alleles of a gene located on chromosome 15, double recessives producing a deficient amount of the enzyme hexosaminidase A which leads to the accumulation of complex fatty substances in the CENTRAL NERVOUS SYSTEM.

T$_2$ bacteriophage, *n.* a particularly virulent BACTERIOPHAGE that infests *Escherichia coli* and was used by HERSHEY and Chase to demonstrate that the genetic material of phages is DNA.

T$_4$ and T$_6$ bacteriophage, *n.* fever phages of *Escherichia coli*. See T$_2$ BACTERIOPHAGE.

TCA cycle (Tricarboxylic Acid Cycle), see KREBS CYCLE.

T-cell or **helper T-cell,** *n.* a type of LYMPHOCYTE produced in bone marrow and differentiating in thymus tissue, that counteracts the presence of foreign ANTIGENS by a process of cell-mediated immunity in which the antigen is slowly destroyed by the T-cell or by a toxin released from the T-cells. The 'priming' of the T-cell to a specific antigen and the long-term growth of activated T-cells is thought to be influenced by INTERLEUKINS produced by MACROPHAGES.

tect- or **tecti,** *prefix,* denoting a roof.

tectorial membrane, *n.* a membrane, present in the COCHLEA of the inner ear, that runs parallel with the BASILAR MEMBRANE. Between the tectorial and basilar membranes, sensory cells span the gap and these cells are connected with nerve fibres which join the auditory nerve. The tectorial membrane forms part of the organ of Corti.

tel- or **tele-** or **telo-,** *prefix.* denoting far, distant.

teleo-, *prefix.* denoting complete.

teleost, *n.* any bony fish of the subclass Teleostei, containing all the present–day bony fishes, except lungfish, sturgeon and garpike. They are characterized by the presence of an AIR BLADDER.

telluric, *adj.* of, relating to, or originating on or in the earth or soil.

telocentric chromosome, *n.* a chromosome in which the CENTROMERE is terminal (at one end), so that there is effectively only one arm.

telolecithal, *adj.* (of eggs) having the yolk aggregated at the vegetative pole.

telomere, *n.* the rounded tip of a chromosome.

telophase, *n.* a stage of NUCLEAR DIVISION in EUCARYOTE cells, occurring once in MITOSIS and twice in MEIOSIS. During telophase, the two sets of chromosomes aggregate at the spindle poles, each set having the same number of chromosomes as the parental cell in mitosis, but half the parental complement in meiosis. See Fig. 285.

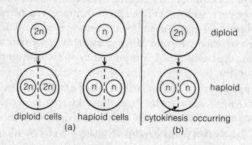

Fig. 285. **Telophase.** (a) Mitosis. (b) Meiosis.

The chromosomes uncoil, become extended and cannot be identified individually. A NUCLEAR MEMBRANE forms around each set of chromosomes and the NUCLEOLI are reformed. The process of cell division (CYTOKINESIS) is completed during this stage.

telson, *n.* the terminal appendage of the last abdominal segment of some arthropods, such as the tail fan in many crustaceans or the sting in the scorpion.

temperate phage, *n.* a BACTERIOPHAGE which may become a PROPHAGE by becoming integrated with the host DNA or by replicating outside the host chromosome and causing LYSIS. Compare VIRULENT PHAGE.

temperature, *n.* the degree of hotness or coldness, usually related to a zero at the melting point of ice (Celsius scale) or absolute zero (Kelvin scale).

temperature coefficient, see Q_{10}.

temperature regulation, *n.* the maintenance of body temperature at a steady level. This occurs to some extent in all vertebrates and many invertebrates, but applies particularly to HOMOIOTHERMS. In humans, a constant body temperature of 36.9°C is maintained, and this is the optimum temperature for normal metabolic reactions involving enzymes.

Various mechanisms bring about temperature regulation. In mammals and birds, hair and feathers trap air which acts as insulation. Sweating acts as a cooling mechanism, and animals in cooler climates have a smaller surface area/volume ratio (see BERGMANN'S RULE). Subcutaneous fat also acts as an insulation; superficial blood vessels constrict in response to cold and dilate in warm conditions, so taking blood away from the skin surface when it is cold and to the skin surface when warm. The centre for controlling body temperature lies in the HYPOTHALAMUS.

Plants also control their temperature and keep cool by transpiration, losing the latent heat of EVAPORATION. On the loss of too much water they wilt, but this results in their leaves being moved (by drooping) from the direct rays of the sun so that in effect they remain cooled until sunset.

template, *n.* the molecule that forms the mould for the synthesis of another.

tendon, *n.* a bunch of parallel COLLAGEN fibres making up a band of CONNECTIVE TISSUE which serves to attach a muscle to a bone.

tendril, *n.* a modified plant structure used for attachment to a support. *Clematis* has tendrils which are modified PETIOLES (leaf stalks), the vine uses a modified stem, and in some species of pea (e.g. *Lathyrus aphaca*) the whole leaf is specialized to form a tendril, with normal leaf functions performed by greatly enlarged STIPULES.

tensor tympanni, *n.* the muscle connecting the MALLEUS to the wall of the tympanic chamber in the mammalian ear.

tentacle, *n.* any long slender organ of touch or attachment.

tentaculocyst, *n.* any of the numerous sensory tentacles surrounding the rim of coelenterate medusae. Tentaculocysts are associated with sense organs such as STATOCYSTS and light-sensitive pigment spots.

tepal, *n.* a single member of the perianth of a flower, especially where sepal and petals are not clearly differentiated, as in the tulip.

terat- or **terato-,** *prefix.* denoting a monster or an abnormality.

teratogenic, *adj.* producing malformation in a foetus.

teratology, *n.* the study of abnormalities in structures, such as malformations.

terete, *adj.* (of plant structures) lacking ridges, grooves or angles.

tergum, *n.* the plate of thickened, often chitinous material on the dorsal side of an insect segment, forming part of the EXOSKELETON.

terminal, *adj.* (of plant structures) carried at the end of a stem.

terminalization, *n.* the process of apparent movement of CHIASMATA to the ends of the chromosomes in late prophase of MEIOSIS.

termination codon, see NONSENSE CODON.

termite, see ISOPTERA.

terpene, *n.* an unsaturated hydrocarbon of plant resins and oils.

terrestrial, *adj.* of the land as opposed to water or air.

territorial behaviour, see TERRITORY.

territory, *n.* any area defended by an animal against members of the same species. During the breeding season breeding territories may be defended, but territoriality may occur outside the breeding season, for example in defending feeding territories. In breeding territories, the male usually defends the area against other males, and territoriality may last all, or only a part of, the breeding season. The term is not to be confused with HOME RANGE.

tertiary consumer, see TOP CARNIVORE.

Tertiary period, *n.* a period of the Cenozoic era lasting from 65 million years ago until the beginning of the Quaternary, 2 million years ago. This is the period of the emergence of mammals, following the extinction of the DINOSAURS.

tertiary structure of protein, see PROTEIN.

testa, *n.* a protective coat around the seed, formed from the INTEGUMENTS of the OVULE. At one area on the testa is the hilum, at one end of which is often found the MICROPYLE. Sometimes the testa is responsible for seed dormancy, a state which is broken when the coat ruptures (see GERMINATION).

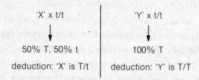

Fig. 286. **Testcross.** A testcross to find the genotype of the unknown parent.

testcross, *n.* a cross in which one partner is homozygous (see HOMOZYGOTE) for the recessive alleles of the genes being investigated. The testcross progeny will indicate the GENOTYPE of the unknown parent. See Fig. 286. Test crosses are used extensively in experiments to establish GENETIC LINKAGE. See also BACKCROSS.

testis, *n.* (pl. testes) the organ of the male animal, producing the male gametes, or sperms. It also produces ANDROGENS – male sex hormones. See SCROTUM, SEMINIFEROUS TUBULES, GERMINAL EPITHELIUM, SERTOLI CELLS.

testosterone, *n.* a steroid ANDROGEN produced by the LEYDIG CELLS between the SEMINIFEROUS TUBULES of the TESTIS. It causes the development and maintenance of accessory sex organs, the genitalia and the SECONDARY SEXUAL CHARACTERS.

tetanus, *n.* **1.** a sustained contraction of muscle due to the fusion of many small contractions (twitches) that follow one another in very rapid succession. **2.** a disease produced by toxins from the bacterium *Clostridium tetani* which usually enters the body through a wound, producing spasm of the voluntary muscle, especially of the jaw (*lockjaw*). Bacterial tetanus can be treated by administering antitetanus serum containing ready-made ANTIBODIES, or by antitetanus vaccine which induces the formation of antibodies by the recipient.

tetany, *n.* an abnormal increase in nerve and muscle excitability resulting in spasms of the arms and legs, caused by a deficiency of PARATHYROID secretion.

tetra-, *prefix.* denoting four.

tetracycline, *n.* an ANTIBIOTIC of the *broad spectrum* type, so-called because of its effect on a wide variety of bacterial types. Since they can adversely effect host cells as well as bacteria, tetracyclines are prescribed medicinally with extreme caution.

tetrad, *n.* **1.** the four homologous CHROMATIDS which associate during prophase (the PACHYTEN stage) and metaphase of MEIOSIS and are involved in CROSSING OVER. **2.** the four haploid cells produced by one complete meiotic division.

tetrad analysis, *n.* a genetic analysis of events at MEIOSIS as shown by the arrangement of cells in a TETRAD, the four products of a single meiotic cell. The technique usually involves the determination of the precise order of the eight ascospores within a fungal ASCUS. See FIRST-DIVISION SEGREGATION.

tetradynamous, *adj.* (of plant stamens), being six in number, two shorter than the other four.

tetraploid, 1. *adj.* having four times the HAPLOID(1) number of chromosomes in the nucleus. **2.** *n.* an individual with four sets of chromosomes per cell. See TRIPLOID.

tetrapod, *n.* any vertebrate with two pairs of PENTADACTYL LIMBS.

tetrapterous, *adj.* having four wings.

tetraspore, *n.* the product of MEIOSIS in many red and brown algae, being produced in fours in testrasporangia and forming the first cells of the GAMETOPHYTE generation.

thalamus, *n.* the major sensory coordinating part of the vertebrate forebrain.

thalassaemia or **Cooley's disease,** *n.* a type of human anaemia in which there is a deficiency of either alpha or beta HAEMOGLOBIN chains. Various causes have been found for the condition, including a recessive mutant allele for beta chain deficiency that is present in high frequencies in certain areas of the Mediterranean, often associated with high incidence of mosquito activity. Named after the American paediatrician Thomas B. Cooley (1871–1945).

thallophyte, *n.* any plant of the group (or former division) Thallophyta, containing the most primative types of plants characterized by the possession of a THALLUS. They vary in size from unicellular types to giant seaweeds up to 75 m in length. The group includes Algae, Bacteria, Fungi, slime fungi (Myxomycetes) and lichens. The Thallophyta clearly have several different evolutionary origins and the term is not used in modern systematics.

thallus, *n.* the vegetative part of simple plants, ranging from unicellular structures to large seaweeds which shows no differentiation into root, stem and leaves. See THALLOPHYTE.

thec- or **theco-,** *prefix,* denoting a sheath or case.

theca, *n.* **1.** a sheath or case. **2.** the protective sheath surrounding the GRAAFIAN FOLLICLE. **3.** an obsolete term for ASCUS.

thecadont, *adj.* possessing teeth in sockets as in mammals and some reptiles.

thelytoky, *n.* a form of PARTHENOGNESIS in which females give rise to more females without fertilization.

therapsid, *n.* any extinct reptile of the order Therapsida, from which mammals are thought to have evolved.

theri- or **thero-,** *prefix.* denoting a wild animal.

thermocline, *n.* a boundary layer, found in lakes or enclosed seas, between warm upper water and cooler lower water, that is usually maintained only under calm conditions in summer.

thermodynamics, *n.* the science concerned with the relationships between heat and mechanical work. The laws of thermodynamics are:

(1st) when one form of energy is converted to another there is no loss or gain.

(2nd) when one form of energy is converted to another a proportion is turned into heat.

thermonasty, *n.* a response of plants to a general temperature stimulus. See NASTIC MOVEMENT.

thermoneutral zone, *n.* the range of environmental temperature across which a tachymetabolic organism (see TACHYMETABOLISM) is able to regulate CORE TEMPERATURE solely by modulating its thermal conductance. Within the thermoneutral zone the STANDARD METABOLIC RATE (SMR) of an animal is at a constant minimum level – the basal metabolic rate (BMR).

thermophilic, *adj.* of or relating to heat-preferring fungi or other microorganisms whose optimum temperature for growth is above 45°C.

thermoregulation, *n.* the control of body heat in HOMOIOTHERMS.

thermoreceptor, *n.* a sensory nerve ending which is responsive to temperature change.

therophyte, *n.* a herb which survives the winter, or other unfavourable conditions, as seeds – one of RAUNKIAER'S LIFE FORMS.

thiamine or **vitamin B₁,** *n.* a water-soluble organic compound found in cereals (e.g. rice) and yeast, whose main function is to act as a COENZYME in sugar breakdown by forming part of the NAD molecule. Thiamine is part of the B-COMPLEX of VITAMINS and deficiency results in the disease BERI-BERI.

thigmotropism or **haptotropism** or **stereotropism,** *n.* a form of plant growth (TROPISM) in which the plant responds to contact. The response is usually positive, in that there is movement towards the contacted surface, and is best seen in climbing plants which utilize tendrils.

thoracic cavity, *n.* the space within the THORAX.

thoracic duct, *n.* the main collecting duct of the LYMPHATIC SYSTEM of mammals that leads into the left anterior vena cava.

thoracic gland, see PROTHACIC GLAND.

thorax, *n.* **1.** (in vertebrates) that part of the body which contains the lungs and heart, and which in mammals is divided from the abdomen by the DIAPHRAGM. **2.** (in ARTHROPODS) that part of the body directly behind the head and in front of the abdomen. **3.** (in insects) three segments bearing the legs and wings.

thorn, *n.* a sharply pointed woody plant structure formed from a modified branch.

thread cell, see CNIDOBLAST.

threat display, *n.* a type of animal behaviour in which aggressive postures are adopted or intimidating markings revealed. Such displays are common in fish and birds involved in defence of TERRITORY, and often result in the retreat of the threatened animals, so avoiding direct conflict. See PECKING ORDER.

threonine, *n.* one of 20 AMINO ACIDS common in proteins. It has a polar 'R' group structure and is soluble in water. See Fig. 287. The ISOELECTRIC POINT of threonine is 5.6.

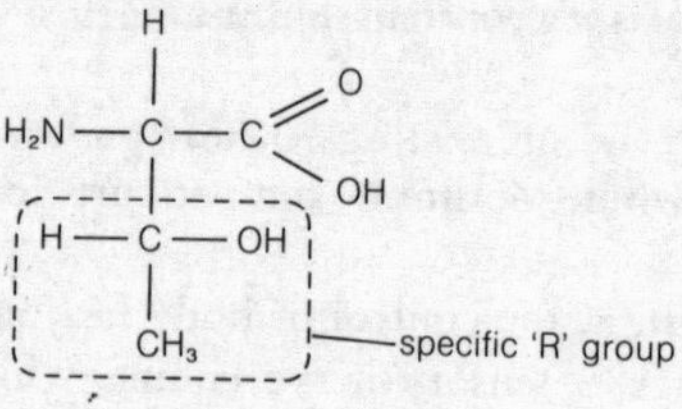

Fig. 287. **Threonine.** Molecular structure.

threshold, *n.* the level at which a STIMULUS results in a response and below which there is no response despite the application of a stimulus.

thrombin, see BLOOD CLOTTING.

thrombocyte, see BLOOD PLATELETS.

thromboplastin, see BLOOD CLOTTING.

thrombosis, *n.* the formation of a clot of blood in a blood vessel.

thrush, *n.* **1.** an acute or chronic condition produced by the fungus *Candida albicans* in which lesions occur in the mucous membranes of mouth, vagina and respiratory tissues. Thrush can occur also in skin areas that are subjected to long periods of immersion in water. **2.** a member of the PASSERINE genus *Turdus*.

thylakoid, see LAMELLA.

thymine, *n.* one of four types of nitrogenous bases found in DNA, having the single-ring structure of a class known as PYRIMIDINES. Thymine forms part of a DNA unit called a NUCLEOTIDE and always forms complementary pairs with a DNA purine base called ADENINE. See Fig. 288. Thymine is replaced by URACIL in RNA molecules.

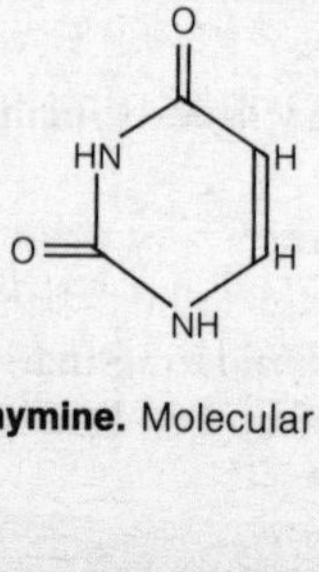

Fig. 288. **Thymine.** Molecular structure.

thymonucleic acid, *n.* a NUCLEIC ACID which was first extracted from the THYMUS gland.

thymus, *n.* an endocrine gland situated in the neck region of most vertebrates, but close to the heart in mammals. It produces LYMPHOCYTES which then move to lymph nodes. The thymus produces a hormone called *thymosin* which causes the lymphocytes to form ANTIBODY-producing plasma cells immediately after birth, but regresses in adult animals.

thyroglobulin, *n.* a protein containing and storing THYROXINE and tri-iodo-thyroxine in the THYROID GLAND.

thyroid gland, *n.* an endocrine organ situated in the region of the neck which produces THYROXINE, a hormone that contains iodine obtained from the diet. Thyroxine controls BASAL METABOLIC RATE (BMR), and also controls metamorphosis in amphibians. Undersecretion during development causes *hypothyroidism* (reduced physical and mental development) leading to CRETINISM in subadults. In the adult stage MYXOEDEMA results, and this gives rise to general slowness of thought and action, coarse skin and fatness. Overproduction, (*hyperthyroidism*) causes *exophthalmic goitre*, where the thyroid, and thus the neck, swells and the eyeballs protrude. Those so affected become restless and overactive, lose weight and have an accelerated heart rate. Deficiency or excess of iodine in the diet causes malfunction of the thyroid gland. A deficiency occurs naturally, for example, in parts of Derbyshire (*Derbyshire Neck*).

thyrotrophic hormone or **thyroid stimulating hormone (TSH),** *n.* a hormone secreted by the anterior lobe of the PITUITARY GLAND which triggers the shedding of THYROXINE from the colloidal follicles of the THYROID GLAND into the bloodstream. Excess thyroxine inhibits the production of TSH – an example of a negative FEEDBACK MECHANISM.

thyrotrophic releasing hormone, *n.* a hormone secreted by the neurosecretory cells of the HYPOTHALAMUS which releases THYROTROPHIC HORMONE from the anterior lobe of the PITUITARY GLAND.

thyroxine, *n.* a complex organic compound containing iodine which is the main hormone produced by the THYROID GLAND.

Thysanoptera, *n.* the insect order containing the thrips, major pests of cereals and fruits.

Thysanura, *n.* the insect order containing the bristletails, primarily wingless insects inhabiting soil and leaf-litter.

tibia, *n.* the anterior of the two long bones articulating in the hind limb, proximally with the femur, and distally with the tarsal bones. See

PENTADACTYL LIMB. In humans, the tibia is the shinbone. In insects, it is the fourth segment from the base of the leg.

tibial, *adj.* of or in the region of the TIBIA.

ticks, *n.* arthropods of two families of the ACARINA, that are mainly ectoparastic blood suckers, having an anitcoagulant in the saliva and are important in carrying disease.

tidal volume, *n.* the amount of air taken in to the lungs by an animal breathing normally, at rest, during each respiratory cycle (each breath). In humans this is about 0.5 dm³ (500 cm³).

tiller, *n.* a branch that forms from the base of a plant or the axils of the lower leaves. Such 'tillering' is common in cereal plants.

Tinbergen, Niko (1907–) British zoologist, born in Holland, who made animal behaviour his speciality and, with Konrad LORENZ, can be considered to be the cofounder of the science of ETHOLOGY.

tissue, *n.* any large group of cells of similar structure in animals or plants, that performs a specific function, such as MUSCLE, PHLOEM.

tissue compatibility, see IMMUNE RESPONSE.

tissue culture, *n.* a technique in which individual cells grow and divide in a bath of sterile, nutritive fluid. The method is used extensively in biological laboratories, for example in cancer research (see HELA CELLS), plant breeding and routine analysis of chromosome KARYOTYPES. See AMNIOCENTESIS.

tissue fluid, *n.* the intercellular fluid forming the internal environment immediately surrounding cells. It is formed from blood by a process of ULTRAFILTRATION.

tissue respiration, see AEROBIC RESPIRATION, ANAEROBIC RESPIRATION.

titre, *n.* the concentration of a substance in a solution, for example the amount of specific antibody in a serum. The concentration is usually expressed as the reciprocal dilution. For example, when 1:200 gives a positive test and greater dilutions are negative, the titre is 200.

toadstool, *n.* the common name for the fruiting bodies of Basidiomycete fungi.

tobacco mosaic virus, *n.* a rod-shaped viral particle consisting of an RNA helix surrounded by a protein coat. The virus was the first to be purified and crystallized, and causes yellowing and mottling in the leaves of tobacco and other crops (e.g. tomato) which affects adversely the quality and quantity of crop produced.

tocopherol, see VITAMIN E.

tolerance, *n.* **1.** the ability of an organism to withstand harsh environmental pressures such as drought or extreme temperatures. **2.** the ability of an organism to withstand the build up of an adverse factor such as pesticides or endoparasites within itself without showing serious symptoms of attack.

tomentum, *n.* the dense covering of short cottony hairs on some plant structures.

tongue, *n.* a muscular organ on the floor of the mouth in most higher vertebrates that carries taste buds and manipulates food. It may act as a tactile or prehensile organ in some species.

tonoplast, *n.* the CELL MEMBRANE surrounding the vacuole in the cytoplasm of a plant cell controlling movements of ions into and out of the vacuole. See SELECTIVE PERMEABILITY.

tonsil, *n.* one of the two large outgrowths at the back of the human oral cavity which have a lymphatic function. The tissue can become infected, resulting in considerable soreness and general ill-health. The removal of children's tonsils by surgery (tonsillectomy) used to be common, but is less popular at the present time.

tonus, *n.* the state of moderate contraction of muscle as a result of continuous low-level stimulation which maintains the posture of an animal.

Tonus occurs because some cells in every muscle are in TETANUS(1) and the muscle feels firm. The condition continues without fatigue because some cells are relaxed whilst others are contracted.

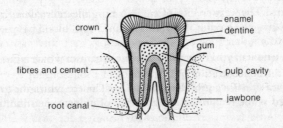

Fig. 289. **Tooth.** Vertical section of a human molar.

tooth, *n.* **1.** one of a series of structures found in the mouth of vertebrates associated with the biting, tearing and crushing of food. Each tooth is a hard structure consisting of a very hard external enamel layer of minerals bound by KERATIN. Underneath this is dentine which has a similar structure to the bone, but is harder, again due to mineral content. The dentine is perforated by fine channels containing processes of the odontoblasts (tooth cells). Centrally the pulp cavity contains blood capillaries and nerve endings. The root is covered by cement and embedded in the jaw bone. See Fig. 289. See INCISORS, CANINES,

PREMOLARS and MOLARS. **2.** any structure with the general appearance of a tooth, such as a dogfish tooth (denticle) which is a modified scale.

tooth bud, *n.* the germinal material from which teeth develop.

top carnivore or **tertiary consumer,** *n.* an organism found at the top of the PYRAMID OF NUMBERS that preys on other organisms but which is not itself preyed on.

topogenous, *adj.* (of peat bogs) dependent upon landscape (topography) for its origins.

topotype, *n.* any specimen of the same species collected in the type-locality of that species.

topsoil, *n.* the surface layer of soil to plough depth.

torpor, *n.* a state of inactivity into which some HOMOIOTHERMS enter in order to conserve energy. There is usually a reduction in both temperature and general body metabolism.

tormogen, *n.* a cell that in insects, secretes the socket for a bristle.

tornaria larva, *n.* a HEMICHORDATE free-swimming larva possessing two bands of cilia.

torsion, *n.* a phenomenon occurring in embryonic gastropods in which the visceral hump rotates through 180°.

torus, *n.* **1.** a central thickened area in the BORDERED PITS of GYMNOSPERMS. **2.** see RECEPTACLE.

totipotency, *n.* the ability of a cell or tissue to give rise to adult structures. The capacity is often lost in adult cells (particularly in animals) which, having differentiated into one specific type, cannot change to another type of cell. See GURDON, CELL DIFFERENTIATION.

toxin, *n.* a nonenzymic metabolite of one organism which is injurious to another organism. See TETANUS, BOTULISM.

toxoid, *n.* a substance produced from a toxin but in which the toxicity is destroyed while remaining the property of inducing immunity to the toxin.

trabeculae, *n.* **1.** a pair of cartilage bars in the vertebrate embryo which form the front part of the floor of the cranium, **2.** the rod-shaped supporting structures of various plant organs.

trace element, *n.* any element that is necessary for the proper working of biological systems in concentrations less than 10^{-5}M. Absence can cause disease and death. For example, boron deficiency causes 'heart rot' in sugar beet, and cobalt deficiency causes 'coast disease' in Australian sheep and cattle. See THYROID GLAND for iodine deficiency, and ESSENTIAL ELEMENTS.

tracer, *n.* any rare ISOTOPE, for example radioactive forms such as ^{14}C, which is administered in some way to organisms so that its fate may be subsequently followed within the organism or in the products of its METABOLISM.

trach– or **trache–** or **tracho–,** *prefix*. denoting the trachea.

trachea, *n*. **1.** (in vertebrates) also called **windpipe**. the main tube leading from the glottis in the neck to the point where the bronchia branch to the lungs. **2.** (in insects) a series of TRACHEOLES by which air is conducted into the body from externally openings called spiracles. See also TRACHEAL GILL. **3.** (in plants) see XYLEM VESSEL.

tracheal gill, *n*. the TRACHEA(2) of aquatic insects formed by slender outgrowths of the body into which TRACHEOLES project.

tracheid, *n*. any of several structures found in XYLEM, consisting of a cell which is long, slender and tapered with heavily lignified walls surrounding an empty lumen, the protoplasm having died.

The wall thickening can be continuous with BORDERED PITS, or else arranged in a variety of patterns, annular, helical and ladder-like. See Fig. 290. Tracheids form the water-conducting tissues of GYMNOSPERMS, but are found also amongst the xylem vessels of ANGIOSPERMS.

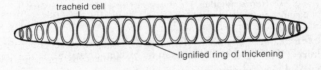

tracheid cell

lignified ring of thickening

Fig. 290. **Tracheid.** General structure.

tracheole, *n*. the small branching tubes from the TRACHEAE(2) of insects.

tracheophyte, *n*. any plant of the division Tracheophyta, comprising all forms of vascular plants, PTERIDOPHYTES and SPERMATOPHYTES. The division includes the Subdivisions PSILOPSIDA, LYCOPSIDA, SPHENOPSIDA and PTEROPSIDA.

trachosphere, see TROCHOPHORE.

Trachylina, *n*. an order of hydrozoan COELENTERATES in which the hydroid stage is suppressed.

trachymedusa, *n*. a medusa of the hydrozoan COELENTERATE order Trachylina.

tract, *n*. a bundle of nerve fibres that may run within the CENTRAL NERVOUS SYSTEM or to the peripheral nervous system.

trans–, *prefix*. denoting across or over.

transaminase, *n*. an enzyme that transfers amino (NH_2) groups. See TRANSAMINATION.

transamination, *n.* the process by which amino groups are transferred from one AMINO ACID to form another, using a keto acid as an intermediary. The mechanism takes place in the liver, and is important in the breakdown of excess amino acids to form keto acids and in the formation of new amino acids (perhaps not available in the diet) from keto acids. An example is shown in Fig. 291. ESSENTIAL AMINO ACIDS cannot be produced by transamination.

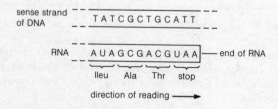

Fig. 291. **Transamination.** The transfer of an amino group to form a different amino acid.

transcription, *n.* the formation of RNA from a DNA template during PROTEIN SYNTHESIS. The process can be broken down into a number of steps (see Fig. 292):

```
sense strand   -- --
of DNA         -- --  T A T C G C T G C A T T  -- --
                      _____
RNA            -- --  A U A G C G A C G U A A  ——— end of RNA
                      ‿‿‿ ‿‿‿ ‿‿‿ ‿‿‿
                      Ileu  Ala  Thr  stop

                      direction of reading ——→
```

Fig. 292. **Transcription.** Uracil replaces thymine in the RNA stage. See GENETIC CODE for details of triplets.

(a) the DNA double helix unwinds.

(b) only one DNA POLYNUCLEOTIDE CHAIN serves as a template (the 'sense strand') in any one region, although both chains will act as templates in different locations.

(c) the enzyme RNA polymerase catalyses the synthesis of an RNA molecule from RNA nucleotides with bases complementary to the DNA base sequence, starting at the 'promoter' region of DNA. Growth of the RNA chain is from 5′ to 3′, the same as in DNA synthesis.

(d) transcription carries on until a DNA stop signal is reached (see GENETIC CODE).

(e) the new RNA molecule leaves the DNA and moves to the cytoplasm, where it will function as TRANSFER RNA, MESSENGER RNA or RIBOSOMAL RNA.

(f) the DNA molecule 'rezips'.

transduction, *n.* the transport of DNA from one bacterium to another, using a VIRUS as a vector.

transect, *n.* a line across a habitat or habitats along which organisms are sampled in order to study changes that may occur along that line. Transects are most frequently used in studying changes in vegetation (often a particular species) across a physically changing habitat such as blanket bog or sand dunes.

transferase, *n.* any enzyme that catalyses the transfer of a chemical group such as amino, methyl or alkyl from one SUBSTRATE to another substrate.

transfer RNA (tRNA), *n.* a form of RNA molecule with about 80 NUCLEOTIDES and a 'cloverleaf' shape, whose function is to carry specific AMINO ACIDS to the ribosomes during TRANSLATION. At one end (3′ end) there is attachment of the amino acid producing AMINOACYL–tRNA, with an unpaired region about half way along called the *anticodon* in which three bases are complementary to the codon on MESSENGER RNA. See Fig. 293.

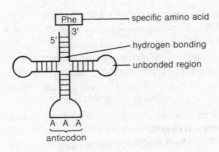

Fig. 293. **Transfer RNA.**

transformation, *n.* the incorporation of a piece of foreign DNA into the GENOME of a bacterial cell. The process is important historically since, following transformation experiments by Frederick GRIFFITH, on *Pneumococcus* bacterium, DNA was shown to be the genetical material of cells by AVERY, MacLeod and McCarthy.

transfusion tissue, *n.* tissue lying on either side of the vascular bundles in leaves of GYMNOSPERMS that probably represents an extension of the vascular system. It is composed of empty parenchymatous cells with pitted and thickened walls.

transient polymorphism, *n.* a temporary form of GENETIC POLYMOR-PHISM existing at a time when an ALLELE is being replaced by a superior one.

transition substitution mutation, see SUBSTITUTION MUTATION.

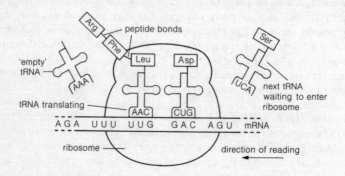

Fig. 294. **Translation.** A polypeptide chain, formed by amino acids joined by peptide bonds.

translation, *n.* the formation of a POLYPEPTIDE CHAIN in a RIBOSOME during PROTEIN SYNTHESIS, using a sequence that is contained in MESSENGER RNA.

The process can be broken down into a number of steps:

(a) mRNA produced from the DNA during TRANSCRIPTION becomes attached to one or more ribosomes (see POLYRIBOSOME).

(b) the RNA passes through the ribosome, beginning with a start signal at the 5' end of RNA and ending with a stop signal at the 3' end (see POLYNUCLEOTIDE CHAIN).

(c) surrounding the ribosomes are various TRANSFER RNA molecules each attached to its specific AMINO ACID. As a triplet of RNA bases moves through the ribosome it is 'read' by the correct tRNA molecule which 'leaves behind' the correct amino acids.

(d) the amino acids become joined by PEPTIDE BONDS, forming a polypeptide chain. See Fig. 294.

See GENETIC CODE for details of polypeptide sequence.

translocation, *n.* **1.** the transport of organic substances in the PHLOEM of higher plants, the mechanism for which is not fully understood. A popular theory is by MASS FLOW but there is evidence that there is active uptake of solutes into the sieve tubes of phloem involving the use of ATP. There may also be movement of solutes along the protein filaments which stretch from one sieve element to the next via the sieve plates, using a form of CYTOPLASMIC STREAMING. **2.** a type of CHROMOSOMAL MUTATION in which non–HOMOLOGOUS CHROMOSOMES break and exchange pieces. During the process, pieces of chromosomes can be lost or gained, causing serious problems, particularly with gamete viability. See TRANSLOCATION HETEROZYGOTE.

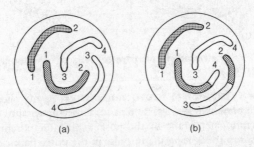

Fig. 295. **Translocation heterozygote.** (a) Normal cell. (b) Reciprocal translocation heterozygote.

translocation heterozygote, *n.* an individual carrying cells with one normal set of chromosomes and one set in which a reciprocal TRANSLOCATION has occurred between two non–HOMOLOGOUS CHROMOSOMES. See Fig. 295. Such an event causes pairing problems in prophase I of MEIOSIS when the chromatids undergo SYNAPSIS(2). These problems are solved by producing a cross-shaped configuration where all chromosomal segments are correctly paired.

When the chromatids separate they can do so in two ways, 'adjacent' or 'alternate' disjunction. See Fig. 296 on page 530. *Adjacent disjunction* produces gametes that have duplicated and deficient chromosomal segments, and therefore produces gametes that can cause death in the offspring. *Alternate disjunction*, produced by a twisting of the cross, results in a mixture of completely normal gametes and ones containing a balanced translocation. These types are both viable and the progeny will all survive. The overall result of a translocation heterozygote is that about half of the progeny survive, a fact that has been used in experiments to control insect populations.

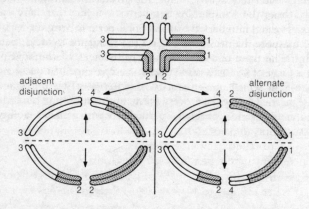

Fig. 296. **Translocation heterozygote.** Adjacent and alternate disjunction.

transmitter substance or **neurotransmitter,** *n.* any chemical substance that is present in SYNAPSES and passes across the synaptic cleft, from the presynaptic membrane to the postsynaptic membrane which it depolarizes, so propagating the impulse in the postsynaptic axom. See ACETYLCHOLINE, ADRENERGIC, CHOLINERGIC.

transovarial, *adj.* passing infection from one generation to another through the egg.

transpiration, *n.* the loss of water vapour from the inside of a leaf to the outside atmosphere, via STOMATA and LENTICELS. Transpiration exerts a considerable upward pressure in the stem and is thought to be part of the explanation of how water ascends from roots to leaves.

The rate at which transpiration proceeds depends upon several physical factors:

(a) the Water Vapour Pressure at the MESOPHYLL cell surface inside the leaf. The evaporation surface is saturated and will have a VAPOUR PRESSURE (WVP_{satn}) that is highly affected by ambient temperature. For example, at 20° the $WVP_{satn} = 2.5$ kPa, at 10°C the $WVP_{satn} = 2.5$ kPa.

(b) the Water Vapour Pressure in the outside air (WVP_{air}), the maximum value being equal to WVP_{satn} at that temperature. If there is the same temperature inside and outside the leaf, the rate of flow of water vapour between the surface of the mesophyll cell and the outside is the difference between WVP_{satn} and WVP_{air}; i.e:

$$WVP_{diff} = WVP_{satn} - WVP_{air}$$

Thus the greater the WVP_{diff} value, the greater the diffusion gradient and the higher the transpiration rate.

(c) the size and number of stomatal pores per unit area of a leaf. The smaller the pore diameter, the greater the resistance to water vapour diffusion. The presence of 'vapour shells' over each stoma creates a boundary layer of high Water Vapour Pressure in still air, which will slow down the transpiration rate since it increases the WVP_{air} value. This effect is most important when the pore diameter is large. In moving air, the vapour shells cannot form and thus transpiration rates increase.

(d) the stomatal and leaf structure, which are modified in some XEROPHYTES to reduce transpiration rates.

(e) a constant supply of water from the roots.

transpiration stream, *n.* the flow of water through a live plant as a result of water loss due to TRANSPIRATION through the leaves.

transplantation, *n.* the transference of an organ or tissue from a donor to a recipient in need of a healthy organ or tissue. In recent years kidney, lung, heart and liver transplants have taken place. For successful transplantation to occur similar tissues types must be involved (see HLA SYSTEM) and genetical similarity is one of the best ways of ensuring this. Drugs which inhibit the normal IMMUNE RESPONSES are used, but these also inhibit the body's defence against microorganisms. Rejection of foreign tissue is part of the normal response of the body, and the development of drugs that will prevent rejection but which will not affect the normal response to microorganisms is actively being researched.

transport, *n.* the movement of materials through a system, as in an ELECTRON TRANSPORT SYSTEM.

transposable genetic element or **mobile genetic element,** *n.* any genetic unit that can become inserted into a chromosome but has the ability to become relocated.

In recent years phage, DNA, PLASMIDS and certain regions of bacterial chromosomes have been shown to be transposable, but the phenomenon was first reported in the 1950s by Barbara McClintock working on maize. She received a Nobel Prize in 1983 for her discoveries in this area. A transposable element in a bacterium carrying bacterial genes is called a *transposon*.

transposon, see TRANSPOSABLE GENETIC ELEMENT.

transudate, *n.* any substance that passes through a membrane, especially through the wall of a capillary.

transude, *vb.* to ooze or pass through interstices, pores or small holes.

transverse process, *n.* lateral projections from each node of the neural arch of tetrapod vertebrates to which ribs are attached.

transversion substitution mutation, see SUBSTITUTION MUTATION.

trematode, *n.* any parasitic flatworm of the class Trematoda, including the FLUKES.

triangular, *adj.* (of plant structures) having the shape of a triangle.

Triassic period, *n.* the first period of the MESOZOIC ERA, immediately following the PERMIAN PERIOD. It lasted from some 240 million years BP to 200 million year BP and was the age of reptiles in which the DINOSAURS emerged. Horsetails, cycads, *Ginko* and conifers were common, ammonites reached their peak and the first primitive mammals and flies appeared. London was at 20°N and Britain was moving northwards.

tribe, *n.* a taxonomic category between genus and subfamily, used mainly in the CLASSIFICATION of plants.

tricarboxylic acid cycle, see KREBS CYCLE.

tricho-, *prefix.* denoting a hair.

trichocyst, *n.* a thread–like structure of uncertain function found in some PROTOZOANS, which is ejected from a pit on stimulation.

trichogyne, *n.* a projection from the female sex organ of some ASCOMYCETES, red algae and lichens, that receives the male gamete before fertilization.

Trichomonas, *n.* a genus of parasitic protozoan occurring in the digestive and reproductive systems of humans and many other animals.

trichome, *n.* an epidermal outgrowth on a plant, forming a hair or spine. For example, the hairs growing out from the epidermis of cotton plant seeds are unicellular trichomes with thick, CELLULOSE walls.

Trichonympha, *n.* a flagellate PROTOZOAN that is capable of producing CELLULASE (a very rare enzyme in animals), and which lives in the gut of termites, so enabling them to digest wood.

Trichoptera, *n.* the insect order containing the caddis flies. The larvae are aquatic, and the adults have reduced mouthparts and only feed rarely.

trichotomous, *adj.* (of a plant) branching into three divisions or shoots.

trichromatic theory, *n.* the theory that all colours can be produced by the mixing of blue, green and red. See colour vision.

tricuspid, 1. *n.* a tooth having three cusps. **2.** *adj.* (of a valve). See HEART (section d).

triecious, see TRIOECIOUS.

trifid, *adj.* split into three.

trigeminal nerve, *n*. the fifth cranial nerve of vertebrates, usually possessing opthalmic, mandibular and maxillary branches.

triglyceride, see FAT.

trigonous, *adj*. having three prominent angles, e.g. a plant stem or ovary.

trihydroxybenzene, see PYROGALLOL.

tri-iodothyroxine, *n*. a minor hormone of the thyroid gland that has the same function as THYROXINE, though it is present in smaller amounts. Formula: $C_{15}H_{12}I_3NO_4$.

trilobite, *n*. a fossil arthropod of the class Trilobita from the CAMBRIAN and SILURIAN PERIODS.

trimethylamine oxide, *n*. the compound found in marine TELEOST fish that is the end product of nitrogen metabolism. The compound has evolved because it requires relatively little water for its elimination as compared with ammonia, which is the nitrogenous end product in freshwater fish.

trimorphism, *n*. the property exhibited by certain species of having or occurring in three distinct forms.

trioecious or **triecious,** *adj*. (of a plant species) having male, female, and hermaphrodite flowers occurring on three different plants. Compare MONOECIOUS, DIOECIOUS.

trinomial nomenclature, *n*. the addition of a subspecific or trivial name to BINOMIAL NOMENCLATURE.

triose, *n*. a monosaccharide that has three carbon atoms.

triose phosphate, see PGAL.

triplet codon, see CODON.

triploblastic, *adj*. having three primary embryonic cell layers, ECTODERM, MESODERM and ENDODERM. All animals except PROTOZOANS, SPONGES and COELENTERATES are triploblastic.

triploid, 1. *adj*. having three times the HAPLOID(1) number of chromosomes. **2.** *n*. an individual of this type. See TETRAPLOID.

trisaccharide, *n*. an OLIGOSACCHARIDE whose molecules have three linked MONOSACCHARIDE molecules.

trisected, *adj*. cut into three nearly separate parts.

trisomy, *n*. a state in which a DIPLOID(2) organism has three chromosomes of one type in a cell or in all cells. For example, DOWN'S SYNDROME individuals are trisomic for chromosome number 21, but have two of every other type of chromosome. See also EDWARDS' SYNDROME and PATAU'S SYNDROME.

tritium, *n*. a radioactive isotope of hydrogen with an atomic mass of three (H^3).

trivial name, *n.* the subspecific name in CLASSIFICATION.

tRNA, see TRANSFER RNA.

troch-, *prefix.* denoting a wheel.

trochanter, *n.* **1.** (in insects) the second segment from the base of the leg. **2.** (in vertebrates) one of several large processes near the head of the femur to which muscles are attached.

trochlear nerve, *n.* the fourth cranial nerve on vertebrates, which innervates the superior oblique muscles of the eye.

trochophore or **trachosphere,** *n.* the characteristic larva of polychaet worms, molluscs and some other invertebrates. It is more or less spherical, with a pre-oral ring of cilia called the *prototroch*.

trophallaxis, *n.* the exchange of regurgitated food that occurs between adults and larvae in colonies of social insects.

trophic, *adj.* of or relating to nutrition. See FOOD CHAIN. See TROPHIC LEVEL.

trophic level, *n.* any of the feeding levels that energy passes through as it proceeds through the ecosystem. For example, primary producers (see PRIMARY PRODUCTION) form the first level in most ecosystems, followed by PRIMARY CONSUMERS (herbivores) up to the top predator level. See PYRAMID OF NUMBERS, BIOMASS.

trophic substance, *n.* a hypothetical substance that is thought to be released from the terminals of neurones and influence the postsynaptic cell.

trophoblast, *n.* the outermost layer of cells surrounding the BLASTOCYST, consisting of embryonic epithelium, which subsequently encloses all the embryonic structures of the developing mammal and forms the outer layer of the CHORION and the embryonic side of the placenta.

trophozoite, *n.* the form of a protozoan of the class Sporozoa in the feeding stage, containing SPOROZOITES. In the MALARIA PARASITE this stage occurs in the human red blood cell.

tropic, see TROPISM.

tropism, *n.* a bending growth movement in a plant either away from or towards a directional stimulus. Tropic movements are brought about by unequal growth on the two sides of an organ (such as the stem) brought about by unequal distributions of AUXIN. See PHOTOTROPISM, GEOTROPISM, THIGMOTROPISM.

tropo-, *prefix.* denoting a change or a turning, as in a TROPOPHYTE.

tropomyosin, *n.* a long protein molecule found in the grooves of muscle ACTIN filaments. It inhibits muscle contraction by blocking the interaction between the filaments and MYOSIN.

troponin, *n.* a complex of calcium-binding proteins that allows

TROPOMYOSIN to move out of the myosin–blocking situation on muscle actin filaments.

tropophyte, *n.* a plant that is unable to adapt to seasonal changes in temperature, rainfall, etc.

tropotaxis, *n.* the movement or orientation of an organism in relation to a source of stimulation (esp. light).

true breeding line, see PURE BREEDING LINE.

truffle, *n.* any of various edible saprophytic ascomycete subterranean fungi of the order Tuberales. Truffles are an expensive table delicacy.

truncate, *adj.* (of plant structures) terminating abruptly.

trunk, *n.* **1.** the main stem of a tree. **2.** the body excluding the head, neck and limbs, i.e. the torso. **3.** the thorax of an insect. **4.** the elongated prehensile proboscis of an elephant. **5.** the main stem of a nerve, blood vessel, etc.

trypanosome, *n.* any parasitic protozoan of the genus *Trypanosoma*, which is parasitic in the gut of tsetse flies and in the blood of vertebrates. They cause important diseases in horses and cattle and AFRICAN SLEEPING SICKNESS in man.

trypsin, *n.* an enzyme secreted as part of the PANCREATIC JUICE, which breaks down PROTEIN into POLYPEPTIDES. It is secreted as an inactive precursor *trypsinogen* which is converted to trypsin by ENTEROKINASE secreted in the SMALL INTESTINE.

trypsinogen, see TRYPSIN.

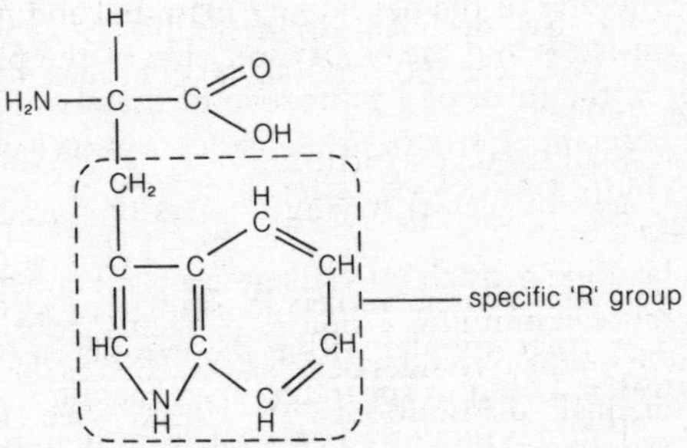

Fig. 297. **Tryptophan.** Molecular structure.

tryptophan, *n.* one of 20 AMINO ACIDS common in protein. It has a nonpolar 'R' group structure and is relatively insoluble in water. See Fig. 297. The ISOELECTRIC POINT of tryptophan is 5.9.

tsetse fly, *n.* any of various bloodsucking DIPTERANS of the genus *Glossina* which transmit AFRICAN SLEEPING SICKNESS. See TRYPANSOME.

TSH, see THYROTROPHIC HORMONE.

t-test, *n.* (statistics) a test of SIGNIFICANCE that enables the means of two small samples (less than 30) to be compared, in order to assess if they came from the same population or from populations with different means.

tube feet, *n.* the locomotory organs of ECHINODERMS that are protruded from the body by, and are retracted by, fluid pressure from the water-vascular system.

tube nucleus, *n.* one of two haploid nuclei within a developing POLLEN TUBE, whose function may be control of tube growth.

tuber, *n.* an enlarged undergound root or stem containing PARENCHYMA cells packed with STARCH for overwintering. The best known example is the stem tuber of potato which has buds in the axils of tiny leaves (forming the potato 'eyes'). Root tubers often have a finger-like branched structure, as in the lesser celandine, *Ranunculus ficaria.*

tubercle, *n.* a spherical or ovoid swelling.

tuberculate, *adj.* (of plant parts) warty, having small blunt projections.

tuberculosis, *n.* a contagious human disease (the *consumption* of Victorian times) affecting particularly the lungs, that is caused by the bacterium *Mycobacterium tuberculosis.* Response to infection is varied amongst individuals, some showing no signs while a few will die of the effects, these variations in host resistance being under genetic control. Tuberculosis is endemic in many parts of the world but, since the introduction of drugs and immunization with vaccines such as BCG, the world death rate has declined dramatically. A typical European mortality rate in 1900 was 190 per 100,000. This has now dropped to around 10 per 100,000.

tubule, *n.* any small tubular structure.

tubuliflorous, *adj.* (of plants) having flowers or florets with tubular corollas.

tubulin, *n.* a globular protein molecule, similar to ACTIN, that is the building block of microtubules within the cytoplasm of cells.

Tullgren funnel, *n.* a piece of apparatus for extracting animals from soil or litter. The soil litter is supported on a gauze, heated from above by a light bulb, and organisms are collected in a container as they fall through the funnel as they move down the heat/desiccation gradient.

tumour, *n.* a swelling, usually a morbid growth.

tundra, *n.* a vegetation type typical of arctic and antarctic zones, consisting of lichens, grasses, sedges and dwarf woody plants.

tunic, *n.* the thin dry, papery covering of a BULB or CORM.

tunica, *n.* any layer of tissue or membrane that encloses a structure or organ in either an animal or plant.

tunicate or **urochordate,** *n.* any minute marine chordate of the subphylum Tunicata, containing the sea squirts. They feed by ciliary action and lack a NOTOCHORD, though one is present in the tail of the tadpole larva.

turbellarian, *n.* any aquatic free-swimming PLATYHELMINTH of the class Turbellaria, the body of which is covered with cilia.

turbinate, *n.* (of plant parts) shaped like an inverted cone.

turgor, *n.* the cell state in which it has taken in a maximum amount of water, causing distension of the protoplast. The term is used mainly in connection with plant cells, which have a maximum size when turgid that is governed by how much the cellulose cell wall will stretch. See WALL PRESSURE, PRESSURE POTENTIAL, TRANSPIRATION.

turgor pressure, see PRESSURE POTENTIAL.

turion, *n.* a swollen bud of many water plants that contains stored food. It becomes detached from the parent plant, enabling it to survive the winter.

Turner's syndrome, *n.* a human chromosomal abnormality in which the individual has 45 chromosomes, 44 AUTOSOMES and one X-CHROMOSOME.

The main features of the syndrome are (a) a female phenotype but with little or no SECONDARY SEXUAL CHARACTERS, (b) a broad shield-like chest, (c) slight mental retardation. The condition is important in understanding SEX DETERMINATION in humans since the fact that a single X-chromosome produces a female (rather than a male) suggests that normal maleness (with XY) is not due to the possession of a single X but rather to the presence of a Y-chromosome. See also KLINEFELTER'S SYNDROME.

turnover rate, *n.* an assessment of the ability of an enzyme to catalyse a reaction, as measured by the number of molecules of substrate which react per second at one ACTIVE SITE when the enzyme is saturated with substrate. The turnover rate varies widely between different enzymes. The fastest known enzyme is CATALASE with a rate of 10 million molecules of hydrogen peroxide reacting per second at one active site.

twins, *n.* two offspring born at the same time of the same mother. See DIZYGOTIC TWINS, MONOZYGOTIC TWINS.

tylose or **tylosis,** *n.* an outgrowth from xylem parenchyma in the XYLEM VESSELS, produced in response to either severe water shortage or, more usually, to the presence of pathogenic organisms. The growth of tyloses often causes the vessels to become blocked, so preventing water transport. See Fig. 298 on page 538.

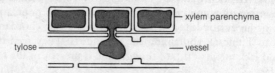

Fig. 298. **Tylose.** An outgrowth into the xylem vessel.

tympanic bone, *n.* a bone that is homologous with the angular bone of reptiles, found only in mammals, forming the base of the auditory capsule and the external auditory meatus. See EAR.

tympanic cavity, see EAR.

tympanic membrane or **tympanum** or **eardrum,** *n.* a membrane situated at the end of the external auditory meatus at the junction of the outer and middle ear. Vibrations are set up in the membrane by sound waves and these are transmitted by the ear ossicles to the oval window.

tympanum, see TYMPANIC MEMBRANE.

type or **type specimen** or **onomatophore,** *n.* the specimen that serves for the original description of a TAXON. 'Onomatophore' means 'bearer of the name'.

type genus, *n.* the genus from which the name of the family or subfamily is derived.

type locality, *n.* the locality in which a HOLOTYPE, LECTOTYPE or NEOTYPE was collected.

type method, *n.* the process of associating the name of a TAXON with a single specimen of that taxon.

type species, *n.* the species from which the name of the genus is derived.

type specimen, see TYPE.

typhlosole, *n.* a dorsal unfolding of the earthworm's intestine, along its whole length, that increases its surface area for digestive purposes.

typhoid fever, *n.* an acute intestinal disease caused by infection by the bacterium *Salmonella typhi*, that is characterized by high fever, skin spots and abdominal pain. Treatment can be effected by administration of the drug chloramphenicol, although resistant strains have appeared in which other drugs must be used.

typhus, *n.* an acute, infectious disease caused by intracellular rickettsial bacteria, that is characterized by high fever, skin rash and severe headache. The bacteria are transmitted from human to human via the body louse *Pediculus humanus*, and control of the insect vector is a powerful means of preventing the spread of typhus fever.

typical, *adj.* having most of the characteristics of a particular taxonomic group.

typological thinking, *n.* a CLASSIFICATION concept that disregards individual variation and considers all members of a population to be replicas of the TYPE.

7–8 nm thick
← protein (dark)
← lipid (light)
← protein (dark)

Fig. 299. **Tyrosine.** Molecular structure.

tyrosine, *n.* one of the 20 AMINO ACIDS common in PROTEINS. It has a polar 'R' group structure and is soluble in water. See Fig. 299. The ISOELECTRIC POINT of tyrosine is 5.7.

U

ulna, *n.* the posterior of the two bones of the forearm of TETRAPODS which articulates proximally with the HUMERUS and distally with the CARPALS. See PENTADACTYL LIMB.

ultracentrifuge, *n.* a machine capable of spinning a rotor at speeds of up to 50,000 revolutions per minutes, producing up to 500,000 g forces. The high speeds enable the separation of tiny particles, which are identified by the rate at which they move down the centrifuge tube. The units of rate are called Svedbergs (S), after the inventor of the ultracentrifuge. Thus RIBOSOMES are found to consist of two subunits after ultracentrifugation, called 30S and 50S. See also DENSITY–GRADIENT CENTRIFUGATION, DIFFERENTIAL CENTRIFUGATION, MICROSOMAL FRACTION.

ultrafiltration, *n.* the process by which small molecules and ions are separated from larger molecules in blood to form the INTERSTITIAL FLUID.

ultraviolet light (UV), *n.* a type of electromagnetic radiation beyond the wavelength of visible violet light, ranging from 18,000 to 33,000 nm (see ELECTROMAGNETIC SPECTRUM). UV light is not an ionizing radiation like X-RAYS and can only penetrate a few cells. However, it is used as a powerful MUTAGEN of microorganisms to cause the formation of thymine DIMERS in DNA, and can be harmful to the human RETINA. Some organisms can detect UV light (see ENTOMOPHILY).

umbel, *n.* an umbrella-shaped inflorescence that forms a RACEME in which the central axis has not elongated. Examples include the flowers of cow parsley, parsley, lovage.

umbilical cord, *n.* the cord that joins the embryo of placental mammals to the PLACENTA(1) consisting of blood vessels supported by connective tissue. The cord is usually severed at birth.

umbo, *n.* the pointed centre of a clam shell, being the oldest part around which are many concentric lines of growth.

unarmed, *adj.* (of plant structures) without prickles of thorns.

uncinate process, *n.* a hook-like projection on each of the first four thoracic ribs of birds that overlaps the rib behind and serves to strengthen the thorax during flight.

uncini, *n.* the hooked chaetae of tubulous polychaetes with which they cling to the sides of their tubes.

undulate, *adj.* (of plant leaves) with up-and-down wavy edges.

undulent fever, *n.* a type of persistent human BRUCELLOSIS caused by *Brucella melitensis* in which vague gastrointestinal and nervous symptoms are common, together with enlarged lymph nodes, spleen and liver. Most cases occur in humans such as meat packers or vets, who may be exposed to infected cattle.

unequal crossing over, *n.* a CROSSING OVER event between HOMOLO-GOUS CHROMOSOMES that are not perfectly aligned, producing duplicated segments on one homologue and deleted segments on the other.

ungul-, *prefix.* denoting a hoof.

ungulate, *n.* one of the numerous species of herbivorous mammals possessing hooves. See ARTIODACTYLA, PERISSODACTYLA.

unguligrade, *adj.* walking on the tips of the digits sheathed in cornified hooves, as in horses and cattle.

uni-, *prefix.* denoting one.

unicellular, see ACELLULAR.

unilocular, *adj.* (of plant structures) made up of one cavity.

uninomial nomenclature, *n.* the system of reference to a taxon by a single word − as in all taxa above the rank of species.

uniovular twins, see MONOZYGOTIC TWINS.

unisexual, *adj.* having separate male and female organisms in a species. See DIOECIOUS.

unit-membrane model, *n.* a model for the CELL MEMBRANE in which the membrane consists of a central lipid layer sandwiched between two layers of protein. See Fig. 300. The unit-membrane model was proposed in the 1950s by the American physician David Robertson, who found his

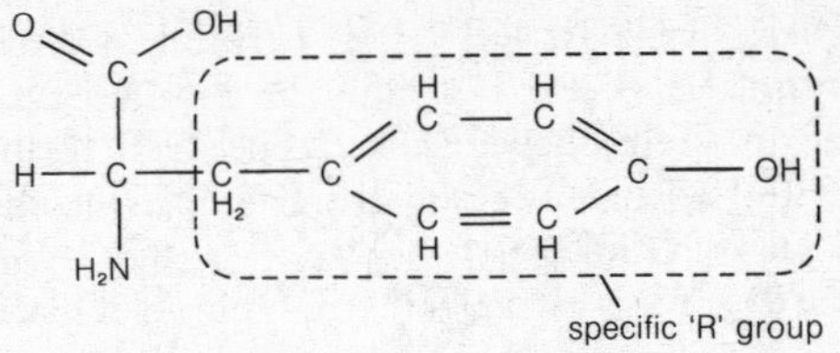

Fig. 300. **Unit-membrane model.**

ELECTRON MICROSCOPE studies tallied with deductions reported by H. Davson and James Frederick Danielli in the 1930s. More recently the FLUID–MOSAIC MODEL has come to be more generally accepted.

univalent chromosome, *n.* a chromosome that is unpaired during prophase I of MEIOSIS and fails to carry out CROSSING OVER with another HOMOLOGOUS CHROMOSOME. Univalent chromosomes occur often in POLYPLOIDS with odd numbers of chromosome sets.

univariate analysis, *n.* (statistics) an analysis of a single character.

universal donor/recipient, *n.* the ABO BLOOD GROUP compatibility between a donors and a recipient. Individuals with Group O have no A or B ANTIGENS and are therefore called *universal donors* because their blood type will not react with the recipient's and thus can be given to *any* other ABO type. Individuals who are Group AB, with no anti-A or anti-B ANTIBODIES, are *universal recipients* since they have no antibodies to react with any donated antigens. The terms do not apply to any other type of BLOOD GROUPING and, since these other groups are important in blood transfusion, the terms are rather misleading.

unsaturated fat, see FATTY ACID.

upper critical temperature, *n.* the environmental temperature defining the upper limit of the THERMONEUTRAL ZONE of a tachymetabolic animal (see TACHYMETABOLISM). Above this temperature, STANDARD METABOLIC RATE rises as a consequence of an increase in CORE TEMPERATURE.

upwelling, *n.* an upward movement of water masses that results in nutrients being brought to the surface. Regions of upwelling are regions of high ocean productivity, occurring, e.g. off the Peruvian coast where there are major stocks of anchovies.

uracil, *n.* one of four types of nitrogenous bases found in RNA, having the single-ring structure of a class known as PYRIMIDINES. Formula: $C_4H_4N_2O_2$. See Fig. 301 on page 542. Uracil always forms complementary pairs with a base called adenine in DNA (during TRANSCRIPTION) or RNA (during TRANSLATION).

urceolate, *adj.* (of flowers) urn-shaped.

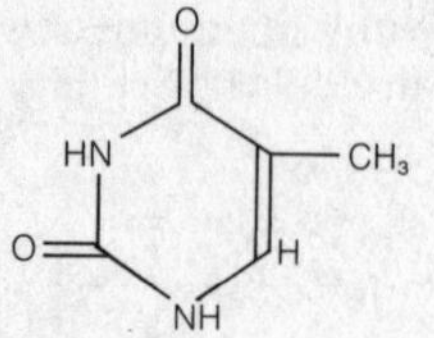

Fig. 301. **Uracil.** Molecular structure.

urea, *n.* an organic molecule that forms the major end product of protein metabolism in mammals, being derived mainly from the ammonia released by the deamination of AMINO ACIDS. Formula: $CO(NH_2)_2$. See ORNITHINE CYCLE.

urea cycle, see ORNITHINE CYCLE.

urease, *n.* an enzyme that catalyses the hydrolysis of UREA into carbon dioxide and ammonia. See Fig. 302.

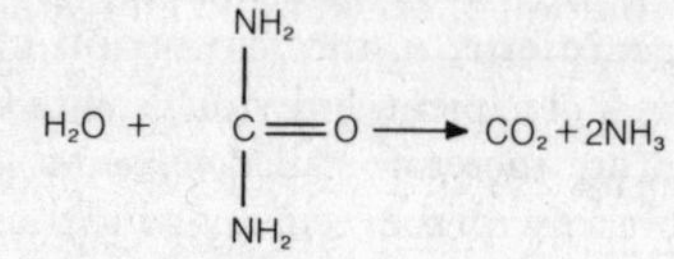

Fig. 302. **Urease.** The hydrolysis of urea.

Uredinales, *n.* an order of basidiomycete fungi containing the rust fungi, important plant pathogens.

ureide, *n.* any compound formed between urea and organic acids.

ureotelic, *adj.* (of animals) excreting UREA as the end product of the breakdown of amino acids. Urea is produced by the ORNITHINE CYCLE, and is characteristic of embryonic development in which soluble urea can easily diffuse away.

ureter, *n.* the duct that carries URINE from the kidney to the CLOACA or urinary BLADDER.

urethra, *n.* the duct by means of which URINE is discharged from the BLADDER in mammals. In males the distal part of the duct also has a reproductive function, carrying semen from the VAS DEFERENS and voiding to the exterior via the penis.

Urey, Harold, see COACERVATE THEORY.

uric acid, *n.* an organic molecule which is the primary end product of protein metabolism in various animals, particularly those well-adapted for life on land, such as insects, reptiles and birds. Formula: $C_5H_4N_4O_3$. Uric acid is quite insoluble in water and is thus nontoxic when released during embryonic development within the egg. Uric acid also permits the removal of nitrogen with a minimum of water loss and is eliminated as a thick paste or even dry pellets. See Fig. 303.

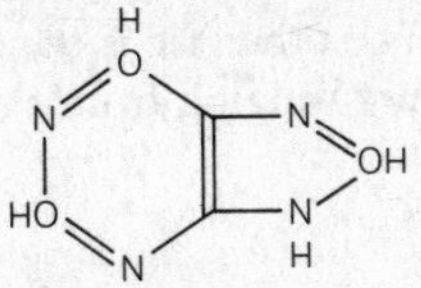

Fig. 303. **Uric acid.** Molecular structure.

uricotelic, *adj.* (of animals) excreting uric acid as the end product of the breakdown of amino acids. Such excretion is characteristic of embryonic development in a shell, where waste products must be stored.

uridine, *n.* a molecule formed from a combination of a ribose sugar and the RNA base URACIL. Further addition of a phosphate group produces the NUCLEOTIDE.

urine, *n.* an aqueous solution of organic and inorganic substances, that is the waste product of METABOLISM. In mammals, elasmobranch fishes, amphibia, tortoises and turtles, nitrogen is excreted in the form of UREA which in humans forms 2% of the urine on average.

uriniferous tubule, *n.* the coiled tube that leads from the BOWMAN'S CAPSULE of the vertebrate kidney to the collecting tubes which convey urine to the ureter.

urinogenital system, *n.* the excretory and reproductive organs and functions of animals. The two are often grouped together because they are closely associated, particularly in male animals where, for example, in mammals the URETHRA is a common duct for urine and sperm.

uro-, *prefix.* denoting a tail.

urochordate, see TUNICATE.

urodaeum, *n.* the region of the CLOACA into which the genital ducts and URETERS open in birds and some reptiles.

Urodela, *n.* the amphibian order containing newts and salamanders. They usually have a well-developed tail, four limbs and have a larva with external gills.

uropygeal gland, *n.* a large gland opening on the dorsal side of the UROPYGIUM in birds, from which an oily fluid, used in preening, is secreted.

uropygium, *n.* the fleshy and bony swelling at the rear of a bird that supports the tail feathers.

urostyle, *n.* a slender, pointed bone at the base of the backbone in amphibians, which is formed from fused vertebrae.

uterus, *n.* the enlarged posterior portion of the OVIDUCT in which the embryo implants and develops in VIVIPAROUS(1) species. It is also called the **womb** in female humans.

utricle, *n.* one of two connecting sacs from which the semicircular canals of the inner ear arise. The other sac is the SACCULE(2).

uvula, *n.* a fleshy structure hanging from the back of the soft palate in man.

V

vaccine, *n.* a preparation containing inactivated (see ATTENUATION) or dead microorganisms, that is used to stimulate an IMMUNE RESPONSE in the recipient, which gains IMMUNITY. Vaccines are thus not quick-acting, but rely on the recipient to build up a supply of ANTIBODIES gradually. For example, the Salk polio vaccine contains attenuated viruses. Compare SERUM(3).

vacuolar membrane, *n.* the membrane surrounding the VACUOLE.

vacuolation, *n.* the formation of a VACUOLE.

vacuole, *n.* a space within the cytoplasm of a plant cell, containing cell sap. The vacuole is surrounded by the TONOPLAST.

vacuum activity, *n.* an animal behaviour pattern performed without apparent need or stimuli. See DISPLACEMENT ACTIVITY.

vagina, *n.* the portion of the female reproductive tract (OVIDUCT) of mammals into which the penis is introduced during copulation.

vagus nerve, *n.* the 10th cranial nerve of vertebrates that arises on the side and floor of the brain MEDULLA and supplies the pharynx, vocal cords, lungs, heart, oesophagus, stomach and intestine. In the vocal cords and lungs it has a sensory function, but elsewhere in the parasympathetic nervous system (see AUTONOMIC NERVOUS SYSTEM) it has a motor function including inhibition of the heartbeat.

valid name, *n.* any name available to a taxonomist that has not been previously used.

valine, *n.* one of 20 AMINO ACIDS common in proteins. It has a nonpolar 'R' group structure and is relatively insoluble in water. See Fig. 304. The ISOELECTRIC POINT of valine is 6.0.

valvate, *adj.* (of PERIANTH segments) having the edges in contact in the bud, but not overlapping.

valve, *n.* **1.** a piece of tissue that enables the movement of a liquid (e.g. blood), in one direction only. **2.** the lid-like part of the shell of brachiopods and barnacles. **3.** either shell of a bivalve molusc. **4.** the lid of some ANTHERS.

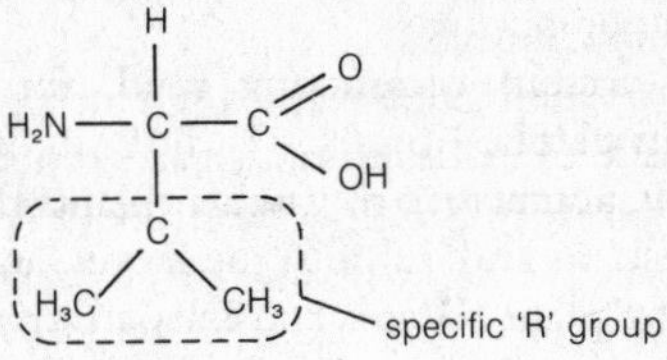

Fig. 304. **Valine.** Molecular structure.

van der Waals interactions, *n.* the weak bonds formed between electrically neutral molecules or parts of molecules when they lie close together. Such interactions are common in the secondary and tertiary structure of protein.

vane, *n.* the part of a bird's feather formed by the barbs and excluding the RACHIS.

vapour pressure, *n.* the pressure exerted by a vapour (gas) that is in equilibrium with its solid or liquid form. The term is particularly important in the diffusion of water vapour. See TRANSPIRATION.

variable (V) region, *n.* an area of an IMMUNOGLOBIN molecule that is specific to that particular molecule. Compare CONSTANT REGION.

variance (s^2), *n.* (in statistics) the variation around the ARITHMETIC MEAN. It is calculated as the average squared deviation of all observations from their mean value. The square root of variance is the STANDARD DEVIATION.

variant, *n.* a particular form of a species or variety that has not been given a special name.

variation, *n.* **1.** ecophenotypic variation (see ECOPHENOTYPE) caused by local factors, as opposed to genetic factors, in an organism. **2.** any differences (both genotypic and phenotypic) between individuals in a population or between parents and their offspring. See GENETIC VARIABILITY.

varicosites, *n.* swellings that occur on a blood vessel or nerve fibre.

variegation, *n.* colour variation in different parts of the leaves or flowers of a plant. It can occur genetically, for example by a somatic mutation affecting PLASTIDS, or through disease, particularly virus infection.

variety, *n.* a group of organisms that differs in some way from other groups of the same species. Botanically this is often below the level of subspecies, but in zoology the term is often synonymous with subspecies or race. However, it is also often used for a morphological variant, for example forms displaying MELANISM.

vasa recta, *n.* one of a series of long loops of thin-walled blood vessels (efferent arterioles) that dip down alongside the loop of Henle in the vertebrate kidney. They pass blood into the interlumber veins and then into the renal vein.

vascular, *adj.* (of vessels) conducting fluid, for example blood in mammals, water in plants.

vascular bundle, *n.* a structure of vascular plants that runs up through the roots, into the stems and out into the leaves, and whose function is transport within the plant. Water and ions are transported mainly by XYLEM, and dissolved organic solutes mainly by the PHLOEM. Some vascular bundles are described as *open* since they contain dividing CAMBIUM tissue between the xylem and phloem.

vascular cylinder, see STELE.

vascular plant, *n.* any member of the division TRACHEOPHYTA, possessing a specialized transport system formed of VASCULAR BUNDLES.

vascular system, *n.* **1.** (in animals) the blood circulatory system, including the arteries, veins, capillaries and heart. **2.** the water vascular system of echinoderms which serves to manipulate the tube feet for locomotion. **3.** (in plants) the tissue which serves to conduct water throughout plants. The tissue is mainly XYLEM and PHLOEM, and forms a continuous system that conducts water, mineral salts and food nutrients, and gives mechanical support.

vas deferens, *n.* a duct conveying sperm from the TESTIS to the exterior in invertebrates or to the cloaca or URETHRA in vertebrates, and hence to the outside.

vasectomy, *n.* a minor human operation, involving the cutting and separation of the cut ends of the VAS DEFERENS so that they cannot rejoin. Effectively, the operation prevents the sperm from mingling with the secretions of the accessory glands and thus acts as a means of BIRTH CONTROL. The resulting semen, therefore, lacks sperm, and although a normal ejaculation can be produced, there is no risk of fertilization. Sperm is reabsorbed in the vas deferens and there is usually no effect on sexual behaviour.

vas efferens, *n.* the tubules leading from the TESTIS to the VAS DEFERENS.

vasoconstriction, *n.* a narrowing of the blood vessels, often in reponse to cold, which occurs through a contraction of INVOLUNTARY muscles in the walls of the vessels brought about by a stimulus from the sympathetic nervous system (see AUTONOMIC NERVOUS SYSTEM).

vasodilation, *n.* the expansion of blood vessels by relocation of muscles mainly controlled by the sympathetic nervous system (see AUTONOMIC NERVOUS SYSTEM).

vasomotor, *adj.* (of sympathetic nerves) associated with the constriction and dilation of blood vessels.

vasomotor centre, *n.* an area of the stem of the brain that causes changes in blood pressure on stimulation of different parts, the pressor area giving a rise in pressure and the depressor area a fall.

vasopressin, see ADH.

vector, *n.* any organism that transmits a parasite. For example, the *Anopheles* mosquito transmits the MALARIA PARASITE.

vegetal pole or **vegetable pole,** *n.* the yolky end of the animal egg away from the cell nucleus.

vegetative, *adj.* of or relating to the nonsexual organs of a plant such as root, stem and leaves. See VEGETATIVE PROPAGATION.

vegetative cell, *n.* one of two primary cells formed within the pollen grains of GYMNOSPERMS, which elongates to form a slender POLLEN TUBE.

vegetative propagation or **vegetative reproduction,** *n.* a form of ASEXUAL REPRODUCTION in plants in which vegetative organs are able to produce new individuals. Natural methods of vegetative propogation are by means of, for example, RHIZOMES (stems), TUBERS (stems or roots) and RUNNERS (stems). Artificial methods include GRAFTING and CUTTINGS.

vegetative reproduction, see VEGETATIVE PROPAGATION.

vehicle, *n.* an inanimate carrier of an infection from one host to another.

vein, *n.* **1.** (in higher animals) that part of the BLOOD CIRCULATORY SYSTEM carrying blood back to the heart from the tissues. Veins are thin-walled, but have the same basic structure as ARTERIES, although veins are usually larger than the corresponding artery. Veins, unlike arteries, will collapse when empty, and are provided with a series of one-way valves that aid in maintaining the flow of blood back to the heart, assisted by skeletal muscles when moving. **2.** (in insect wings) the thickened parts of the CUTICLE that resemble a pattern of veins and enclose tubular airsacs (tracheae) and blood sinuses. Pumping blood into these veins early on in the adult stage causes the wings to take on their final form. **3.** (in vascular plants) any of the vessels within the blade of each leaf (made up chiefly of vascular tissue) that are continuous with the VASCULAR BUNDLES of the stem. The patterns of such veins are distinctive in each plant and are often used in classification.

velamen, *n.* a layer of dead cells, acting as water-absorbing spongy material, found on the outside of aerial roots of epiphytic plants.

veliger, *n.* a molluscan larva, similar to a TROCHOSPHERE, that develops a shell and other organs during later development. It is a ciliary feeder.

vena cava, *n.* (*pl.* venae cavae) one of the major veins of the BLOOD CIRCULATORY SYSTEM.

venation, *n.* **1.** (in insects) the arrangement of veins in the wing. **2.** (in plants) the arrangement of leaf veins.

venereal disease, *n.* any contagious disease transmitted usually during sexual intercourse, such as GONORRHOEA or SYPHILIS.

venter, *n.* the swollen base of the ARCHEGONIUM that contains the egg cell.

ventilation rate, *n.* the rate of respiration in terms of the volume of air breathed per minute.

ventral, *adj.* of or relating to the underside of an organism, or that side which is normally directed downwards in the usual stance or resting position. In bipedal primates such as humans, the ventral side is the front, but would obviously be the underside if a four-legged gait were assumed.

ventral gland, see PROTHACIC GLAND.

ventral root, *n.* any of the nerve roots issuing from the ventral side of the vertebrate brain or spinal cord, containing MOTOR nerves.

ventral tube, *n.* an appendage of the first abdominal segment of COLLEMBOLANS formed by the fusion of paired limbs. It may be associated with attachment or water uptake and the name *Collembola* (glue bar) is based on this structure.

ventricle, *n.* **1.** a chamber of the heart, having thick muscular walls, that receives blood from the ATRIUM, and pumps it to the arteries. **2.** four large spaces in the brain filled with CEREBROSPINAL FLUID. They occur in the CEREBRAL HEMISPHERES (lateral ventricles), FOREBRAIN (third ventricle) and MEDULLA OBLONGATA (fourth ventricle).

ventricular fibrillation, see FIBRILLATION.

venule, *n.* a small vein, differentiated from a capillary by possessing connective tissue (and sometimes smooth muscle) in the walls.

Venus flytrap, *n.*, *Dionaea muscipula*, a carnivorous plant in which a hinged leaf snaps shut on insects from which the plant obtains nitrogenous compounds.

Vermes, *n. obsolete collective term for* all worms of different phyla, ANNELIDS, PLATYHELMINTHS, NEMERTINES, etc.

vermiform appendix, see APPENDIX.

vernacular name, *n.* the common name for an organism, for example robin for *Erithacus rubecula*.

vernalin, *n.* a substance similar to a hormone produced in VERNALIZATION.

vernalization, *n.* a process in which young plants are treated with low temperatures (2–5°C) to induce a change to an older physiological state, thus shortening the interval between sowing and flowering. Vernalization can be reversed by high temperatures.

vernation, *n.* leaf arrangement in a bud.

versatile, *adj.* (of ANTHERS) having the filament attached near the centre so as to allow movement.

vertebra, *n.* (*pl.* vertebrae) one of the bony segments of the VERTEBRAL COLUMN.

vertebral column or **spinal column** or **backbone,** *n.* the series of vertebrae surrounding the SPINAL CORD. Each vertebra consists of a centrum which replaces the embryonic NOTOCHORD, a neural arch covering the spinal cord and often a haemal arch enclosing blood vessels. Transverse processes may also be present to which ribs articulate. Vertebrae are joined by the centra and often by projections of the neural arch. The column articulates with the skull by means of the atlas vertebra, with the ribs at the thoracic vertebrae, and with the pelvic girdle at the sacrum. See Fig. 305, LUMBAR VERTEBRA, ZYGAPOPHYSIS.

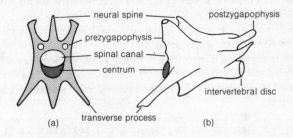

Fig. 305. **Vertebral column.** A lumbar vertebra of a mammal. (a) Front view. (b) Side view.

vertebrate, *n.* any member of the subphylum Vertebrata (= Craniata) in the phylum Chordata, including all those organisms that possess a backbone, such as fish, amphibia, reptiles, birds and mammals. In addition, they are characterized by having a skull which surrounds a well-developed brain and a bony or cartilaginous skeleton.

vertical classification, *n.* a form of CLASSIFICATION stressing common descent and placing organisms in a single higher taxon to demonstrate PHYLETIC relationships. Compare HORIZONTAL CLASSIFICATION.

vertical resistance, SEE RACE-SPECIFIC RESISTANCE.

vesicula seminalis, SEE SEMINAL VESSICLE.

vesicle, *n.* any small sac-like structure.

vessel, *n.* **1.** a tubular structure that transports blood. **2.** See XYLEM VESSEL.

vestibule, *n.* any passageway from one cavity to another, such as the depression leading to the mouth in *Paramecium* or from the VULVA to VAGINA in the female mammal.

vestibular apparatus, *n.* that part of the inner ear which together with the COCHLEA forms the membranous labyrinth (see EAR).

vestibular canal, *n.* the canal of the COCHLEA of the inner ear that connects with the oval window.

vestigial organ, *n.* any organ that during the course of evolution has become reduced in function and usually size. Examples include the pelvic girdle of the whale and wings in flightless birds. Often such organs have lost their original use and are used for other purposes. For example, the penguin's forelimbs are adapted for swimming.

viability, *n.* a measure of the likelihood that a ZYGOTE will survive and develop into a adult organism. For example, seeds become less viable as they get older, and a diminishing percentage actually germinates.

viable cell, *n.* a cell able to live.

vibrissa or **whisker,** *n.* any of the stiff sensitive hairs found around the mouth of mammals.

vicariad, *n.* any of several vicarious species which are closely related but form an ALLOPATRIC POPULATION.

Victoria mazonica, *n.* the giant water lily whose structure shows the close relationship between structure and function. The leaf has its edge upturned to prevent inundation, and its underside greatly strengthened by ribbing.

villous, *adj.* (of plants) shaggy.

villus, *n.* (*pl.* villi) a finger-like outgrowth, as in the lining of the SMALL INTESTINE. Villi effectively increase the surface area of the gut wall and contain (a) blood vessels for absorption of carbohydrates and amino acids which pass into the hepatic portal SYSTEM (b) lacteals which absorb fats into the LYMPHATIC SYSTEM.

vinegar, *n.* a dilute, impure acetic acid, made from beer or wine.

virion, *n.* see VIRUS.

viroid, *n.* a virus-like pathogenic structure consisting of an RNA molecule, but with no protein coat.

virulence, *n.* the collective properties of an organism that render it pathogenic to another one, the host.

virulent phage, *n.* a BACTERIOPHAGE which, following phage replication in the cell, results in LYSIS of the host. It cannot be integrated into the chromosome of the host, unlike a TEMPERATE PHAGE.

virus, *n.* the smallest organism ranging in size from about 0.025 μm to 0.25 μm. The mature virus, a *virion*, consists of nucleic acid (RNA or DNA) surrounded by a protein or protein and lipid coat. Viruses infect cells of bacteria, plants and animals, and whilst they carry out no METABOLISM themselves, they are able to control the metabolism of the infected cell. In infection the virus first adheres to the cell wall and then penetrates the wall with the 'tail'. The nucleic acid strand is injected into the cell where it first replicates itself and then forms new viruses which are released on the rupture of the cell wall.

virus pneumonia, *n.* a form of pneumonia caused by an infection of a pneumovirus that is closely related to measles. Children are particularly vulnerable to attack by the virus, which can spread as an EPIDEMIC.

viscera, *n.* the internal organs of the body cavity.

visceral arches, *n.* the skeletal structures lying between adjacent gill slits in vertebrates, or the division between the gill slits.

visceral hump, *n.* the dorsal part of molluscs that accommodates the visceral organs. It is usually covered by the mantle and shell.

visceral muscle, see INVOLUNTARY MUSCLE.

viscid, *adj.* (of plant parts) sticky.

viscosity, *n.* **1.** the property of stickiness by which substances resist change or shape. **2.** a measure of the ease with which layers of fluid pass each other.

viscus, *n.* any organ lying in the VISCERA.

visible spectrum, see ELECTROMAGNETIC SPECTRUM.

visual cortex, *n.* the thin outer layer of grey matter in the occipital region of the CEREBRUM that is concerned with the interpretation of information from the eyes.

visual purple, see RHODOPSIN.

vital capacity, *n.* the total amount of air that can be expired after a maximum inspiration (deep breath). This is calculated as the sum of:

(a) the *tidal volume*, the amount of air taken in with a normal breath.

(b) the *inspiratory reserve volume*, the amount of air which can still be taken in after a normal breath.

(c) the *expiratory reserve volume*, the amount of air which can be expelled after breathing out normally.

Typical results from an adult man at rest are shown in Fig. 306. Thus the vital capacity of a normal man is between 3.5 and 4.5 dm³, but can reach 6.0 dm³ in a trained athlete.

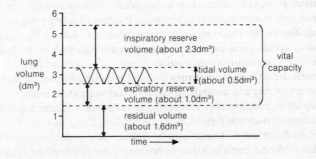

Fig. 306. **Vital capacity.** The vital capacity of an adult man at rest.

vitalism, *n.* a theory which postulates that biological phenomena cannot be expressed solely in physical and chemical terms.

vital staining, *n.* the staining of cells whilst alive, which has been used particularly for studying the movements of parts of embryos.

vitamin, *n.* an organic compound that is necessary in the diet for normal growth and health. Not all organisms require the same vitamins in their diet. For example, rats can synthesize vitamin C whereas humans cannot. Only small quantities are normally required in the diet, as vitamins usually act as COENZYMES or parts of coenzymes. They are divided into two types: fat-soluble (A, D, E and K) and water-soluble (B and C). See VITAMIN A, B–COMPLEX, ASCORBIC ACID, VITAMIN D, VITAMIN E, VITAMIN K.

vitamin A or **retinol,** *n.* a fat-soluble hydrocarbon, closely related to CAROTENOIDS, that occurs in liver, green vegetables and storage organs such as carrots. It is concerned with the normal functioning of the mucous membrane of the eye, respiratory and urinogenital tracts, and forms part of the photochemical reactions in the RODS of the eye. Deficiency of vitamin A leads firstly to *night blindness* then *xerophthalmia*, inflamed eyes and eyelids. The lining of the respiratory tract can become inflamed and sterility results. Infants obtain a large supply of the vitamin in the mother's first milk (COLOSTRUM).

vitamin B complex, see B–COMPLEX.

vitamin C, see ASCORBIC ACID.

vitamin D, *n.* a fat-soluble molecule found in fish liver oils, and also produced in the skin when subjected to ultraviolet rays from sunlight. The main function of the vitamin is to increase the utilization of calcium and phosphorus in bones and teeth. A deficiency results in RICKETS in children and *osteomalacia* (bone-softening) in adults, particularly women after several pregnancies.

vitamin E or **tocopherol,** *n.* a fat-soluble molecule found in many plants, such as wheatgerm oil, green leafy vegetables, egg yolk, milk and meat. The vitamin is known as an antioxidant, permitting the oxidation of, for example, unsaturated fatty acids and Vitamin A in the body. A deficiency can result in sterility.

vitamin K or **phylloquinone,** *n.* a fat-soluble molecule found in spinach, cabbage, kale and pig's liver. The vitamin is essential in the synthesis of prothrombin used in BLOOD CLOTTING. A deficiency causes an increase in clotting time.

vitamin M, see FOLIC ACID.

vitelline blood vessels, *n.* the blood vessels that convey nourishment to the embryos of vertebrates which possess yolk.

vitelline membrane or **primary egg membrane,** *n.* a FERTILIZATION

MEMBRANE surrounding and secreted by the ovum. Secondary membranes are secreted by other structures, such as the OVIDUCT or OVARY.

vitreous humour, *n*. the liquid jelly-like material occupying the space within the eyeball behind the lens of the eye.

viviparous, *adj*. **1.** (of animals) having the development of the young inside the body as in placental mammals. Compare OVIPAROUS, OVOVIVIPAROUS. **2.** (of plants) having seed germinating in the fruit. **3.** (of plants) having shoots that serve for vegetative reproduction rather than forming flowers.

vocal cord, *n*. the membranes that lie in the larynx of a mammal. Their vibration produces sounds which are altered by varying their position and tension.

voluntary muscle, SEE STRIATED MUSCLE.

voluntary response, *n*. a response under the control of the will of an organism. Compare REFLEX ARC.

volutin, *n*. granules of a highly refractile nature, made up of polyphosphates, which occur in many bacteria. Long chains may form storage material in some cells.

Volvocales, *n*. an order of colonial green algae made up of flagellated cells embedded in spheres of mucilage.

vomer, *n*. the thin flat bone forming part of the separation between the nasal passages in mammals.

von Baer's law, SEE HAEKEL'S LAW.

vulva, *n*. the external opening to the VAGINA in the female reproductive system of mammals.

W

waggle dance, *n*. the behaviour pattern by which bees convey information on the location of pollen supplies.

Wallace, Alfred Russel (1823–1913) British naturalist who was influenced by the ideas of MALTHUS and LYELL and corresponded with DARWIN about his ideas on natural selection. As a result he wrote a paper with Darwin which was read at the Linnaean Society in 1858, and founded modern evolutionary thought.

wall pressure, *n*. the pressure exerted by the cellulose cell wall of a plant, which is equal to the opposing PRESSURE POTENTIAL.

wandering cell, *n*. any cell such as an amoebocyte or macrophage, that actively moves about during the life of an organism.

Warburg manometer, *n*. an apparatus designed by the German biochemist Otto Warburg, that measures aerobic respiration rates in tissues and small organisms. Carbon dioxide released from respiring material inside a flask is absorbed by potassium-hydroxide, the resulting drop in pressure being measured by a manometer. The amount of CO_2 produced is equivalent to oxygen uptake by the tissue assuming a RESPIRATORY QUOTIENT of 1.0. The respirometer apparatus is very sensitive to temperature, which is controlled by immersing the flask in a water bath whose temperature is regulated by a very accurate thermostat. Variations in atmospheric pressure do not affect the results since the respirometer is a closed-circuit system.

warm-blooded animal, see HOMOIOTHERM.

warning coloration or **aposematic coloration,** *n.* any striking or conspicuous markings occurring on an animal and which indicate, presumably through learning, that the organism is distasteful to potential predators, i.e. the warning is to the predator. See MIMICRY.

water, *n.* a colourless, odourless liquid that is the most abundant component of any organism (over 60% by weight in man). Life almost certainly originated in water and it provides the medium for biological reactions to take place.

water culture, see HYDROPONICS.

water potential or (formerly) **diffusion pressure deficit (DPD)** or **suction pressure,** *n.* the tendency of a cell to draw in water from outside by OSMOSIS, the water moving from a higher to a lower water potential. Since pure water at one atmosphere has a water potential of zero, cells drawing in water have a water potential of less than zero. Thus water potential is measured as a negative value, which can be confusing. Readers should be careful when using the terms 'higher' or 'lower' water potential; 'more negative' or 'less negative' are more exact statements.

In mathematical terms water potential is the sum of the OSMOTIC POTENTIAL and the PRESSURE POTENTIAL of a cell:

$$\text{water potential} = \text{osmotic potential} + \text{pressure potential}$$

For example, when Osmotic Potential $= -9$ and Pressure Potential $= 4$, Water Potential $= -9 + 4 = -5$. See OSMOREGULATION.

waterproofing, *n.* the process of preventing the passage of water into and out of structures, usually brought about in a living organism by laying down LIPID layers, such as the insect cuticle.

water-soluble vitamin, *n.* any of a group of vitamins soluble in water. Most are components of coenzymes (e.g. those of vitamin B complex).

water stress, *n.* any WILTING exhibited by plants that are losing more water by TRANSPIRATION than they are taking up through the roots.

water uptake, *n.* the uptake of water into a plant through the roots.

Water only enters the plant if there is a deficit in the XYLEM caused by losses during TRANSPIRATION. The path of water from soil to leaf is as follows:

(a) movement of free soil water into the root hair by OSMOSIS.

(b) across the CORTEX of the root towards the STELE, by one of three methods: (i) via cell walls and intercellular spaces by hydration. (ii) via PLASMODESMATA by CYCLOSIS. (iii) via cell vacuoles by osmosis.

(c) through the endodermis (see CASPARIAN STRIP) across the PERICYCLE and into the XYLEM VESSELS. These latter contain ions which are actively pumped into the xylem to raise the concentration of solutes and thus the OSMOTIC PRESSURES.

(d) water moves up the plant in the xylem by a combination of: (i) TRANSPIRATION pull and cohesion-tension (see COHESION–TENSION HYPOTHESIS). (ii) capillarity. (iii) ROOT PRESSURE.

All water uptake is variable according to the plant, its age, environment and the time of year.

water vapour, *n.* water in the gas phase, especially when due to evaporation. Vapour is important in maintaining the atmospheric humidity essential to living organisms.

water vascular system, *n.* a system of canals found in ECHINODERMS, that contains fluid and is open to the sea or the body cavity. It supplies fluid to the tube feet for use in locomotion.

Watson, John (b.1928) American molecular biologist who, with Francis CRICK and Maurice WILKINS, put forward a model for the structure and functioning of DNA in 1953, work for which they shared a Nobel prize in 1962. Their double-helix model for DNA is now universally accepted, together with their general ideas for DNA replication.

wavelength, *n.* the distance between two successive points at which the wave has the same phase. For example, visible light has a wavelength of between 400 nm (violet) to 750 nm (red).

W-chromosome, *n.* the smaller of the two SEX CHROMOSOMES in female birds and LEPIDOPTERANS, being equivalent to the Y-chromosome in male mammals. See also Z–CHROMOSOME, SEX DETERMINATION.

Weberian ossicle, *n.* a small group of bones, probably derived from the vertebrae, that connects the air bladder of the OSTARIOPHYSI with the ear capsule. Named after the German physiologist Ernst Weber (1795–1878).

Weber–Fechner law or **Fechner's law,** *n.* a law stating that sensation increases arithmetically as the stimulus increases geometrically. Named after Ernst Weber (see WEBERIAN OSSICLE) and the German physicist Gustav Fechner (1801–87).

weed, *n.* any plant that competes for resources with a plant of importance to man. Weeds usually have a high VIABILITY and can use up disproportionate amounts of water, sunlight and nutrients. Where the weed is of a different type than the crop plant (e.g. the weed might be a DICOTYLEDON, such as ragwort, the crop a MONOCOTYLEDON such as wheat) selective herbicides are available which are effective against only broad-leaved plants.

weighting, *n.* **1.** a method of determining the importance of characters in CLASSIFICATION by attributing different values to them. **2.** The application of statistical loadings to data so as to enhance the value of some figures and reduce the value of others.

Weismannism, *n.* the theory, now considerably modified, which proposes that the germ cells are set apart at an early stage of the development and are uninfluenced by characteristics acquired during life. The known action of chemicals and physical factors on chromosomes has resulted in modification of the theory into the CENTRAL DOGMA. Named after the German biologist August Weismann (1834–1914).

Weinberg, W. see HARDY–WEINBERG LAW.

wen, *n.* a sebaceous cyst, especially of the human scalp.

Went, Fritz, Dutch biologist who identified the plant hormone AUXIN and devised a series of tests for the auxins. The BIOASSAY involves using a decapitated oat COLEOPTILE and an agar block containing the test substance to measure the possible effect of the test substance on growth.

wet rot, *n.* a brown rot of damp timber caused by the fungus *Coniophora cerebella*. The fungus can be controlled by application of a wood preservative such as creosote.

whale, see CETACEAN.

whalebone, see BALEEN.

wheel animalcule, see ROTIFER.

whisker, see VIBRISSA.

white blood cells, see LEUCOCYTE.

white fibrous cartilage, *n.* any CARTILAGE containing white nonelastic fibres of COLLAGEN.

white matter, *n.* the tissue of the CENTRAL NERVOUS SYSTEM, lying outside the GREY MATTER in the spinal cord but internal to grey matter in the brain of some vertebrates, and which contains the myelinated AXONS of nerves. The MYELIN SHEATHS give the tissue its white appearance.

whole mount, *n.* the preparation of an entire organism for microscopic examination.

whooping cough, *n.* an infectious disease, common in children, caused by the bacterium *Bordetella pertussis*, in which there is firstly a range of symptoms similar to the common cold; these then develop into severe coughing, usually ending in a characteristic 'whoop' as air is breathed in. There is frequent vomiting, and lung damage often occurs if ANTIBIOTICS such as erythromycin are not administered. In some cases, particularly in very young children, whooping cough can be fatal. Vaccination of infants is common although this procedure carries a small risk of brain damage.

whorl, *n.* a circular set (two or more) of leaves or SEPALS arising at the same level on the plant.

wild type, *n.* the 'normal' PHENOTYPE present in a natural or laboratory population, as distinct from a MUTANT(2) type which often can survive only under artificial conditions. Wild type alleles are usually given a 'plus' symbol. Thus the wild type allele of the vestigial wing mutation (vg) in *Drosophila* is vg^+.

Wilkins, Maurice (1916–) New Zealand biophysicist who carried out X-ray diffraction analysis of DNA, in collaboration with WATSON and CRICK, for which work they shared a Nobel Prize in 1962.

wilting, *n.* a plant state in which a loss of TURGOR by cells of the leaves and other soft tissues causes drooping. The condition is caused by a lack of water (see WATER STRESS), either through drought in the soil, or a disease (such as fungal wilt) which blocks the XYLEM VESSELS in the stem or leaves.

wing, *n.* **1.** either of the modified forelimbs of a bird that are covered with large feathers and specialized for flight in most species. **2.** one of the organs of flight of an insect, consisting of a membranous outgrowth from the thorax containing a network of veins. **3.** either of the organs of flight in certain other animals, especially the forelimb of a bat.

winkle, *n.* a gastropod MOLLUSC of the genus *Littorina*.

wisdom tooth, *n.* a 3rd molar in humans which appears at about the age of 20, and often has to be removed as it crowds or distorts other teeth.

wobble hypothesis, see CRICK.

Wolffian body, see MESONEPHROS.

Wolffian duct, *n.* the duct from the mesonephros in vertebrates. In fish and amphibia, it forms the urinary duct in females, and the urinogenital duct in males. In reptiles, birds and mammals it forms the VAS DEFERENS in males, and degenerates in females.

womb, see UTERUS(2).

woodlice, *n.* any terrestrial member of the crustacean order ISOPODA, being dorsoventrally flattened arthropods with limbs of similar structure.

Woodpecker finch, *n.* one of DARWIN'S FINCHES (*Geospinza*), which occurs in the Galapagos Islands and uses a cactus spine to remove insects from crevices. This is one of the very few examples of birds using a tool.

work, *n.* energy where mechanical effort is involved, measured in joules. A joule is defined in work terms as a force of 1 newton moving its point of application through 1 metre. See SI UNITS.

XYZ

xantho-, *prefix.* denoting yellow.

xanthophyll, see CAROTENOIDS.

X-chromosome, *n.* a type of SEX CHROMOSOME containing genes which, when found in a male (XY), always express. Examples of conditions controlled by X-linked genes are HAEMOPHILIA (2 loci), Duchenne muscular dystrophy, colour blindness (2 loci) and a form of diabetes. Compare Y-CHROMOSOME. See SEX LINKAGE.

xen or **xeno-,** *prefix.* denoting strange.

xenia, *n.* the changes due to foreign pollen in the appearance of ENDOSPERM. For example, in maize, when a form with white endosperm is pollinated by a form having dark yellow endosperm, pale yellow endosperm results.

xenograft, see HETEROGRAFT.

Xenopus, *n.* the African clawed toad that has been used to test for pregnancy in women. Injection of urine from a pregnant woman results in egg-laying in the toad.

xeroderma pigmentosum, *n.* a lethal genetical disorder of the skin in which affected individuals exhibit unusually heavy freckling and ulceration of the skin that has been exposed to sunlight. Such individuals lack an endonuclease enzyme that normally repairs thymine DIMERS produced in DNA by ultraviolet light. Affected individuals are thought to be homozygous for a pair of recessive alleles on an AUTOSOME, although there may be up to five different genes involved in the condition. Heterozygous individuals are often heavily freckled. There is no treatment and death occurs usually in childhood.

xeromorphy, *n.* the condition of having the appearance of a XEROPHYTE due to the possession of characteristics typical of such types.

xerophyte, *n.* a plant that is adapted to growing in areas with low or irregular supplies of water. Various modifications can be noted which reduce water loss by TRANSPIRATION: sunken STOMATA (e.g. *Pinus*); rolled leaves with the stomata on the inner surface only (e.g. MARRAM GRASS);

development of leaf spines (e.g. gorse); possession of small leaves (e.g. many heathers). Compare HYDROPHYTE, MESOPHYTE.

xerosere, *n.* a plant succession that begins on a dry site.

xiph-, *prefix.* denoting a sword-shaped.

Xiphosura, *n.* the group containing the king crabs, aquatic ARACHNIDS occurring as fossils as far back as the Palaeozoic era.

X-ray, *n.* an ionizing radiation that is a powerful MUTAGEN with wavelengths between 10^{-1} and 10^1 nm on the ELECTROMAGNETIC SPECTRUM. X-rays are produced by bombarding a metallic target with fast electrons in a vacuum, and are capable of penetrating various thicknesses of solids. Having passed through a solid they can act on a photographic plate producing a light/shade pattern indicative of the solid structure.

xylem, *n.* a woody plant tissue that is vascular in function, enabling the transport of water with dissolved minerals around the plant, usually in an upward direction. Xylem is characterized as having TRACHEIDS, xylem fibres for support, xylem PARENCHYMA and XYLEM VESSELS. The location of xylem is different in roots and stems, and the area of xylem is greatly increased by SECONDARY THICKENING.

xylem vessel or **vessel** or (rarely) **trachea,** *n.* an empty tube formed from the longitudinal fusion of several cells with strong walls reinforced with LIGNIN, whose function is the mass transport of water for TRANSPIRATION. Vessels are aggregated into XYLEM tissue within the VASCULAR BUNDLES of ANGIOSPERMS.

xylose, *n.* a PENTOSE SUGAR that is present in plant cell walls. Formula: $C_5H_{10}O_5$.

Y-chromosome, *n.* the smaller of the two SEX CHROMOSOMES, found in the male of mammals together with one X-CHROMOSOME. Although Y-chromosomes are responsible for SEX DETERMINATION in mammals, very few specific genes have been located. Such 'Y-linked' genes (a tentative example is a form of hairy ears in man) would only be inherited down the male line of a family.

yeast, *n.* a collective name for those unicellular ASCOMYCETE or BASIDIOMYCETE fungi of economic importance in the brewing and bread-making industries (see SACCHAROMYCES). Yeasts secrete ENZYMES that convert sugars into alcohol and carbon dioxide (see ALCOHOLIC FERMENTATION) and it is the CO_2 which causes bread to 'rise'.

yellow elastic cartilage, *n.* any CARTILAGE containing a network of yellow fibres composed of elastin.

yellow fever, *n.* an acute destructive disease of tropical and subtropical regions in which a virus causes cellular destruction in liver, spleen, kidneys, bone marrow and lymph nodes. The effects of the disease are serious, death occurring in about 10% of cases. The yellow fever virus is transferred from one human host to another via the mosquito *Aedes aegypti*. Control of the disease is principally by measures to reduce populations of the mosquito vector.

yellows, *n.* any of a number of plant diseases in which there is significant yellowing of the leaves. Yellows can be caused by a wide range of microorganisms, e.g. virus yellows of sugar beet.

yield, *n.* **1.** that part of the production of a crop which is removed by humans. **2.** the actual amount of substance obtained during the preparation of a substance, for example in fermentation or in a manufacturing process.

Y-linked, *adj.* (of a gene located on the Y-CHROMOSOME) one which can only be transmitted through the HEMIZYGOUS sex, i.e. males in mammals.

yoghurt, *n.* a product of fermented milk. Boiled milk is inoculated with *Lactobacillus bulgaricus* and *Streptococcus thermophilus* and incubated at 45°C until the lactic acid content is 0.85–0.9%.

yolk, *n.* the food store in the eggs of the majority of animals, made up mainly of fat and protein granules. Where yolk is present in the egg, as in chickens, there is meroblastic CLEAVAGE, but where it is absent, or nearly so, cleavage is holoblastic, as in *Amphioxus*.

yolk sac, *n.* the sac-like structure that contains YOLK and is in direct contact with the gut of embryos in fish, reptiles and birds. Though present in mammalian embryos, it does not contain yolk but absorbs uterine secretions until the PLACENTA(1) becomes functional.

Z-chromosome, *n.* the larger of the two SEX CHROMOSOMES in female birds and LEPIDOPTERA, being equivalent to the X-CHROMOSOME in mammals. See also W-CHROMOSOME, SEX DETERMINATION.

zeatin, *n.* a CYTOKININ type of plant hormone extracted from the ENDOSPERM of maize fruits.

zero order reaction, *n.* a chemical reaction in which the rate is independent of the concentration of the reactants.

zero population growth (ZPG), *n.* a state in a population in which births equal deaths, so that overall numbers remain in a steady state. ZPG (or even negative growth) has been achieved in many Western countries where the family size is low but has, so far, proved an impossible target for most Third World countries, the state being achieved only where there is widespread use of contraceptives (see BIRTH CONTROL, DEMOGRAPHIC TRANSITION).

Z-membrane, *n.* a transverse membrane found in the light banding of skeletal muscle, at each end of a SARCOMERE.

zoea, *n.* the larva of crabs.

zona pellucida, *n.* a mucoprotein membrane that surrounds the egg of mammals. It is secreted by the ovarian follicle cells.

zonation, *n.* the division of a BIOME into zones that experience particular physical conditions. This is particularly clear in the distribution of animals and plants on the rocky seashore, where, for example, the sublittoral zone (below low water mark) is followed by the lower shore, middle shore, upper shore and splash zone, each with their characteristic flora and fauna. See SEA ZONATION and Fig. 268.

zone, *n.* any division or specific area. See SEA ZONATION, ZONATION.

zoo-, *prefix.* denoting an animal.

zoochlorella, *n.* a unicellular green alga that lives in SYMBIOSIS with some sponges, coelenterates and annelids.

zooid, *n.* **1.** any individual polyp of a colony of invertebrate animals that are linked together. For example, COELENTERATES have feeding polyps (gasterozoids), and reproductive polyps (gonozoids). **2.** a motile cell or body, such as a gamete, produced by an organism.

zoology, *n.* the study of animals.

zoophyte, *n.* an animal having a plant-like appearance, such as a sea anemone.

zooplankton, *n.* that part of the plankton made up of animal life. Compare PHYTOPLANKTON.

zoosporangium, *n.* a SPORANGIUM in which ZOOSPORES are formed.

zoospore or **swarm spore,** *n.* any of the motile, flagellated SPORES (asexual cells) found in green and brown algae and in PHYCOMYCETE fungi. They are produced within a SPORANGIUM.

Z scheme in photosynthesis, the z-shaped passage of electrons in the LIGHT REACTIONS.

zwitterion, *n.* the chemical form of an AMINO ACID with a dipolar ion state. The pH at which the amino acid exists as a zwitterion is called its ISOELECTRIC POINT and at this pH it will not migrate in an electric field, produced, for example, by ELECTROPHORESIS. See Fig. 307.

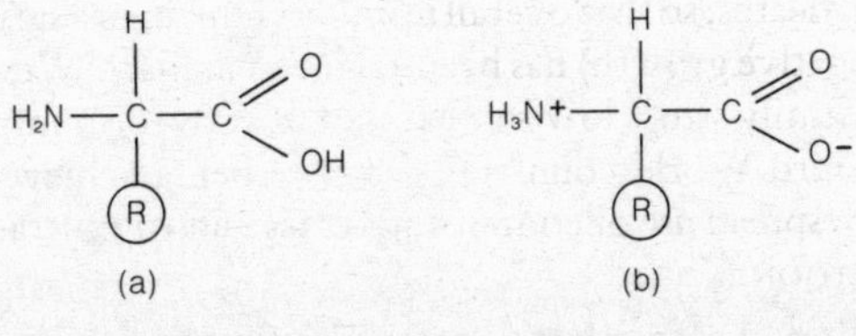

Fig. 307. **Zwitterion.** Molecular structure of (a) undissociated amino acid, (b) Zwitterion.

zyg- or **zygo-,** *prefix.* denoting joined.

zygadactylous, *adj.* (of birds) having two toes pointing forwards (second and third) and two backwards (first and fourth), for example the woodpecker.

zygapophysis, *n.* any facets of vertebrae that articulate with each other – usually two anterior zygapophyses which articulate with two posterior zygapophyses of the vertebra. See VERTEBRAL COLUMN and Fig. 305.

zygomorphic, *adj.* (of flowers) exhibiting BILATERAL SYMMETRY.

Zygoptera, see DAMSELFLY.

zygospore, *n.* a thick-walled resting SPORE that is formed from the union of similar gametes.

zygote, *n.* the DIPLOID(1) cell produced by the fusion of the nuclei of male and female gamete nuclei at FERTILIZATION.

zygotene, *n.* a stage of MEIOSIS near the beginning of Prophase 1, when the homologous chromosomes undergo SYNAPSIS so that identical genetical segments lie opposite to each other ready for CROSSING OVER. Although each chromosome is split into two CHROMATIDS, these are not usually visible.

zymase, *n.* the yeast enzyme that brings about fermentation by enabling the breakdown of hexose sugars to alcohol and carbon dioxide.

zymogen, *n.* an inactive precursor of an enzyme, particularly those concerned with protein digestion, for example, PEPSINOGEN and TRYPSINOGEN. Zymogens require ACTIVATION ENERGY to become functional.

zymogen granule, *n.* a cytoplasmic particle surrounded by a membrane formed by the GOLGI APPARATUS. It stores and secretes zymogen.

APPENDIX A: Geological Time

Phanerozoic Time

ERA	PERIOD	DURATION (millions of years BP)
Cenozoic 65 million years duration	Quaternary	2–present
	Tertiary	64–2
Mesozoic 175 million years duration	Cretaceous	134–65
	Jurassic	199–135
	Triassic	239–200
Palaezoic 350 million years duration	Permian	279–240
	Carboniferous	369–280
	Devonian	414–370
	Silurian	444–415
	Ordovician	514–445
	Cambrian	589–515

Cryptozoic Time (= Precambrian)

	Vendian	679–590
	Riphean	1699–680
	Aphebian	2599–1700
	Archaean	4600–2600

Many Periods are subdivided into epochs:

PERIOD	EPOCH		DURATION
Quaternary	Holocene		9999–present
	Pleistocene		2 million – 10,000
Tertiary	Pliocene	NEOGENE	6–2 million
	Miocene		25–7
	Oligocene	PALAEOGENE	37–26
	Eocene		53–38
	Palaeocene		64–54

APPENDIX B: SI Units

Measurement	Name of unit	Symbol
Length	metre	m
Mass	kilogram	kg
Time	second	s
Electric current	ampere	A
Thermodynamic temperature	kelvin	K
Luminous intensity	candela	cd
Amount of substance	mole	mol
Area	square metre	m^2
Volume	cubic metre	m^3
Velocity	metre/second	m/s^{-1}
Acceleration	metre/second^2	m/s^{-3}
Density	kilogram/metre^3	kg/m^{-3}
Mass rate of flow	kilogram/second	kg/s^{-1}
Volume rate of flow	cubic metre/second	m^3/s^{-1}
Momentum	kilogram metre/second	$kg\ m/s^{-1}$
Force	newton	N
Work (energy, heat)	joule	J
Kinetic energy	joule	J
Heat (enthalpy)	joule	J
Power	watt	W
Pressure (stress)	pascal	Pa
Surface tension	newton/metre	N/m^{-1}
Viscosity, dynamic	newton second/metre^2	$N\ s/m^{-2}$
Temperature	kelvin, degree Celsius	K, °C
Velocity of light	metre/second	m/s^{-1}
Electric charge	coulomb	C
Electric potential (potential difference)	volt	V
Electric resistance	ohm	Ω
Frequency	hertz	Hz

Appendix C: Relative Sizes of Structures, from Atoms to Eggs

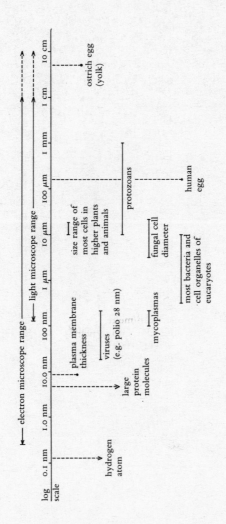